THERMODYNAMICS

Titles in the *Foundations of Engineering* Series

J.A. Cain and R. Hulse, *Structural Mechanics*
G.E. Drabble, *Dynamics*
R.G. Powell, *Electromagnetism*
P. Silvester, *Electric Circuits*
J. Simonson, *Thermodynamics*

Foundations of Engineering

Series Editor: G. E. Drabble

Thermodynamics

John Simonson

Formerly Senior Lecturer

Mechanical Engineering

City University

First published 1993 by
THE MACMILLAN PRESS LTD
Houndmills, Basingstoke, Hampshire RG21 2XS
and London
Companies and representatives
throughout the world

ISBN 0–333–55575–9

A catalogue record for this book is available
from the British Library

Printed in Hong Kong

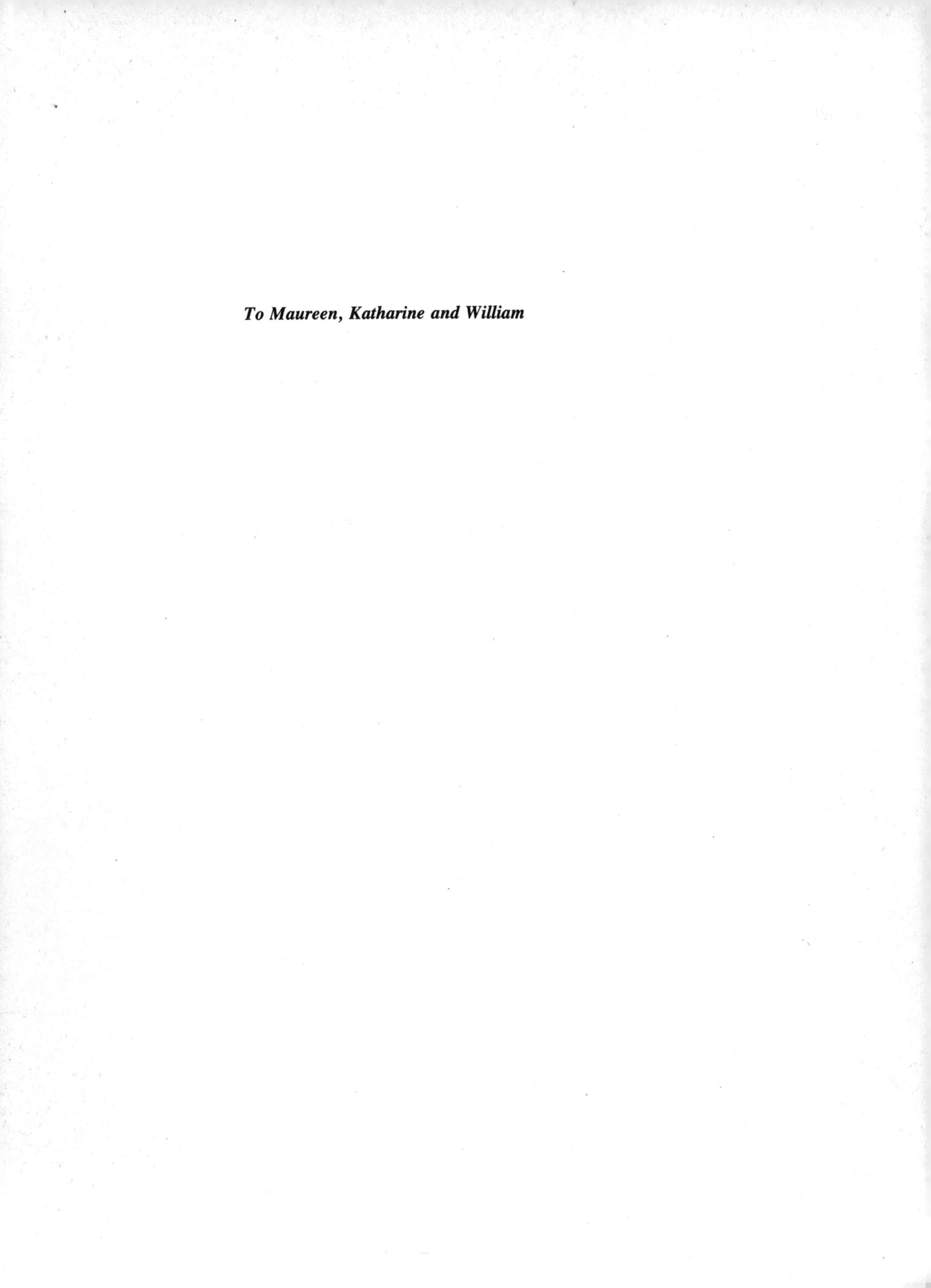

To Maureen, Katharine and William

CONTENTS

Series Editor's Foreword viii

Author's Preface ix

How to use this book xii

Thermodynamic Properties on Disc xiv

Programme

1 Introduction **1**

2 Work and Heat **37**

3 The First Law and Energy **113**

4 The Pure Substance **157**

5 The First Law: Vapours and Gases in Closed and Open Systems **215**

6 The Second Law: Thermodynamic Temperature and Entropy **289**

7 Vapour Power Cycles and Reversed Vapour Cycles **351**

8 Compressible Flow and Work Transfer in Turbine Blading **405**

9 Combustion **461**

10 Internal Combustion Engines, Gas Turbines and Fuel Cells **519**

Index 593

SERIES EDITOR'S FOREWORD

This series of programmed texts has been written specifically for students meeting a subject for the first time on their engineering degree courses. Each book covers one of the core subjects required by electrical, mechanical, civil or general engineering students, and the contents have been designed to match either the first, or in the case of this volume, the first and second year requirements of most universities, both old and new.

The layout of the texts is based on that of the well-known text, *Engineering Mathematics* by K. Stroud (first published by Macmillan in 1970, and now in its third edition). The remarkable success of this book owes much to the skill of its author, but it also shows that students greatly appreciate a book which aims primarily to help them to learn their chosen subjects at their own pace. The authors of this present series acknowledge their debt to Mr Stroud, and hope that by adapting his style and methods to their own subjects they have produced equally helpful and popular texts.

Before publication of each text the comments of a class of students, of some recent engineering graduates, and of some lecturers in the field have been obtained. These helped to identify any points which were particularly difficult or obscure to the average reader or which were technically inaccurate or misleading. Subsequent revisions have eliminated the difficulties which were highlighted at this stage, but it is likely that, despite these efforts, a few may have passed unnoticed. For this the authors and publishers apologise, and would welcome criticisms and suggestions from readers.

Readers should bear in mind that mastering any engineering subject requires considerable effort. The aim of these texts is to present the material as simply as possible and in a way which enables students to learn at their own pace, to gain confidence and to check their understanding. The responsibility for learning is, however, still very much their own.

G.E. DRABBLE

AUTHOR'S PREFACE

The broad aim of this book is to cover the first two years' requirement in Thermodynamics in engineering degree courses. In this series of books the material is provided in short sections which the average student can simulate at his own pace and in his own time. In endeavouring to meet this aim within the allotted space, it was found that some of the intended second year material had to be omitted. The content, therefore, will be found to be strongest in the fundamentals of Thermodynamics upon which the various applications are built.

Since in a two year syllabus the course content will vary between institutions, some readers may find some topics surplus to their needs, while others may be left wanting more. However, many students find the applications of Thermodynamics rather easier to grasp than some of the fundamentals, and it is hoped that where there is a deficiency in meeting a particular syllabus, the student will be able to cope.

This book appears at a time when traditional subjects, like Thermodynamics, are gradually being reduced in contact time in engineering courses to make way for the inclusion of modern technologies. Since Thermodynamics involves concepts that are new and difficult for the average student, it becomes increasingly difficult, both for the teacher to get across his subject, and for the student to absorb it, in the limited time available. Consequently, it is hoped that this text will fill a need by providing the means whereby, in private study, students can grasp the fundamentals of this subject in their own individual way.

This book is not intended to replace a course of lectures, in which students should gain an overall view of the subject, probably illustrated with coloured slides of relevant hardware. Such pictures do not find a logical place in a much more detailed development of ideas, as in this book, and are not included. In any case, the keen student of engineering will read technical magazines, and will become a student member of his professional institution, and go on visits to places of engineering interest to see for himself such things as large gas turbine engines and steam power plant.

On the whole, the general approach to the subject follows traditional lines established over the last few decades. In general, writers of educational textbooks are indebted to those who wrote books before them, and my own teaching for 28 years was largely based on the earlier editions of Spalding and Cole [1] and Rogers and Mayhew [2]. However, readers of this book should be aware that if they turn to the 1992 edition of Rogers and Mayhew, they will find that, compared with their earlier editions and the previously accepted practice which is followed here, and for reasons stated by the authors, the sign convention for the work interaction at a system boundary has been reversed. This means that the equation of the First Law in Frame 11 of Programme 3 would become:

$$Q + W = E_2 - E_1$$

The effects of this change then follow through the book.

In this book, following a suggestion by my colleague Dr Ian K. Smith, to whom I am very grateful, the importance of work ratio in engines and power plant is stressed. This is a parameter which will decide whether an engine will work or not, and in a historical perspective it is not surprising that early engines failed because they absorbed more work than they produced. Also the attainable thermal efficiency of any modern machine will be shown to be closely linked to its work ratio. After an initial study of Thermodynamics it is very instructive to read about the development of the subject, and about the evolution of power-producing machines: see for example, references [3] and [4]. The topic of work ratio is more fully developed in a recent paper by Dr Smith [5].

I am also grateful to WS Atkins Energy [6], for permission to use projected figures of future efficiencies of electrical power generation by gas turbine plant, combined cycle plant and composite plant involving fuel cells, given in Programme 10. The gradual, but continual, increase in thermal efficiency of power plant from the very low values in the early days of the Industrial Revolution, is a tribute to man's ingenuity within the limitations of the Second Law of Thermodynamics.

Over the years many discussions on thermodynamic matters with colleagues past and present, and with many students in tutorial classes, have contributed in unseen ways to the development of this book, and I wish to thank them all. The text has gone through various revisions based on the editor's comments, on comments received from students testing the book, and on comments from a thermodynamics reviewer. I am grateful for all this help in compiling the material into its present form and in eliminating errors. No doubt some points may still not be absolutely clear and some possible errors may still remain, so any further comments from readers would be most welcome.

Finally, I am very pleased to thank William Simonson for his drawing in Frame 5 of Programme 1, and last but not least, I would like to thank my wife for her support and forbearance during the time it has taken to produce this book.

REFERENCES

[1] D.B. Spalding and E.H. Cole, *Engineering Thermodynamics*, Edward Arnold, 3rd edition, 1973.
[2] G.F.C. Rogers and Y.R. Mayhew, *Engineering Thermodynamics, Work and Heat Transfer*, 2nd edition, 1967 (4th edition, 1992).
[3] D.S.L. Cardwell, *From Watt to Clausius. The Rise of Thermodynamics in the Early Industrial Age*, Heinemann Educational Books, 1971.

[4] A.F. Burstall, *A History of Mechanical Engineering*, Faber & Faber, 1963.
[5] I.K. Smith, Matching and Work Ratio in Elementary Thermal Power Plant Theory, *Proc. Part A, Journal of Power and Energy*, MEP, 1993.
[6] M.R. Fry, *Fuel Cells – R&D Review and Assessment of Composite Generation Options*, WS Atkins Energy, Epsom, Surrey, KT18 58W, June 1990.

HOW TO USE THIS BOOK

Like other books in this series, the material, having been divided into programmes on each major topic, is presented in short sections called frames. Each frame comprises either the development of a new idea, or the working of a calculation. Sometimes either of these may be too long for a single frame and the work is split into two or more parts. Where appropriate (when sufficient material has been given), a frame concludes with a question, or the requirement to do a calculation. The answer to the question, or the result of the calculation and the working involved follow at the beginning of the next frame. Ideally this should remain hidden until the task set has been attempted, and only when all is fully understood should you move on to following frames.

Programmes 1 to 6, on fundamentals, up to and including the Second Law, follow in sequence and therefore should be read in that order. Programmes 7 and 8 deal with vapour power cycles, and reversed heat pump cycles, together with compressible flow and work transfer in turbine wheels. Programme 9 on combustion can be read after Programme 6, and most of Programme 10 on engines and gas turbines can follow Programme 9, without first reading Programmes 7 and 8. The format of frames means that in general the text lacks subheadings, but use of the index at the back of the book will prevent one feeling lost in looking for any particular topic.

In order to use this book, students will need *Steam Tables*, together with the enthalpy–entropy diagram for steam and the pressure–enthalpy diagram for the refrigerant R-12. It is logical that the tables recommended by your course tutor be used. There are, however, a number of different publications available, and in this book guidance is given on the use of the following:

Tables

J.R. Cooper and E.J. Le Fevre, *Thermophysical Properties of Water Substance, Students' Tables in S.I. Units*, Edward Arnold, 1969. (Steam Tables only.)

R.W. Haywood, *Thermodynamic Tables in S.I. Units*, Cambridge University Press, 3rd edition, 1990. (Properties of steam, and of refrigerants R-12, methyl chloride, ammonia and carbon dioxide. This publication also includes the enthalpy–entropy chart for steam, and the pressure–enthalpy chart for R-12.)

G.F.C. Rogers and Y.R. Mayhew, *Thermodynamic and Transport Properties of Fluids, S.I. Units*, Basil Blackwell, Oxford, 4th edition, 1988. (Properties of steam and of refrigerants R-12, and ammonia.)

Diagrams or Charts

D.C. Hickson and F.R. Taylor, *Enthalpy–entropy Diagram for Steam, S.I. Units*,

Basil Blackwell, Oxford, 2nd edition, 1980.
D.C. Hickson and F.R. Taylor, *Pressure–enthalpy Diagram for R-12, S.I. Units*, Basil Blackwell, Oxford, 1977.
D.C. Hickson and F.R. Taylor, *Pressure–enthalpy Diagram for Ammonia, S.I. Units*, Basil Blackwell, Oxford, 1978.

THERMODYNAMIC PROPERTIES ON DISC

As a complement to this book, thermodynamic properties of steam and a number of other substances, including the common refrigerants, are available on disc for use with personal computers. This disc will be of value to students and practising engineers undertaking project work on vapour power cycles and heat pump cycles, exploring the potentialities of the range of fluids represented. After selecting a particular fluid, any two properties at a given state may be typed in to obtain all the remaining properties of the fluid at that state. A particular advantage of the disc is that from an initial state, properties at a final state after an isentropic, isenthalpic, isobaric or isochoric change may be found. The processes involved can require considerable work, and once the principles have been mastered and understood, the use of a disc is justified in project work, and is recommended.

For further details, please contact Dr I.K. Smith, Department of Mechanical Engineering & Aeronautics, City University, Northampton Square, London, EC1V 0HB.

Programme 1

INTRODUCTION

1

From your previous knowledge of science, the words 'heat', 'work' and 'energy' should be familiar to you. These words are in quotes because although they are familiar everyday words, they have special meanings and significance in the study we are about to undertake.

Energy can exist in different forms, and for example, you have probably heard of potential energy, kinetic energy, internal energy and chemical energy. Man-made machines use energy to do work, and work is done, for example, in raising weights, driving other machines and propelling vehicles.

Two of the kinds of energy mentioned are particularly useful because they can be used straight away to do work, without having to be converted first into another form of energy. Before looking at Frame 2, can you name them?

2

Potential energy Kinetic energy

The potential energy of an object or substance depends on its height above a given point, such as the level of the sea. If the object is moving, the kinetic energy is a function of the square of the velocity. Strain energy in a spring, which depends on the amount of stretching or compression, is another form of potential energy. Energy in these forms can be harnessed directly to do work.

Internal energy, which is the energy associated with temperature, and chemical energy, which is the energy 'contained' in fuel, cannot be used to do work directly in the same way. Thus it would be difficult and complicated to obtain work by any practical means from a bucket of hot water. Water with the same energy as potential energy can be made to do work very simply. Potential and kinetic energy occur naturally in the world around us. Before the Industrial Revolution, man invented the windmill to use the kinetic energy of the atmosphere, or the wind, to grind wheat and corn, raise weights and drive machinery.

Because of rain, water with useful energy is available to us in mountain streams and rivers. Think of two devices which use the potential and kinetic energy of water, and then look at Frame 3.

3

A water turbine A waterwheel

You also may have thought of wave power devices. These certainly use the energy of water and are of recent origin.

A water turbine uses the potential energy of water that has fallen as rain on hill sides and mountains. This energy is available to do work as the water flows back to sea level. As it does so, its potential energy decreases while its kinetic energy increases and the resulting momentum of the water turns the turbine wheel, doing work.

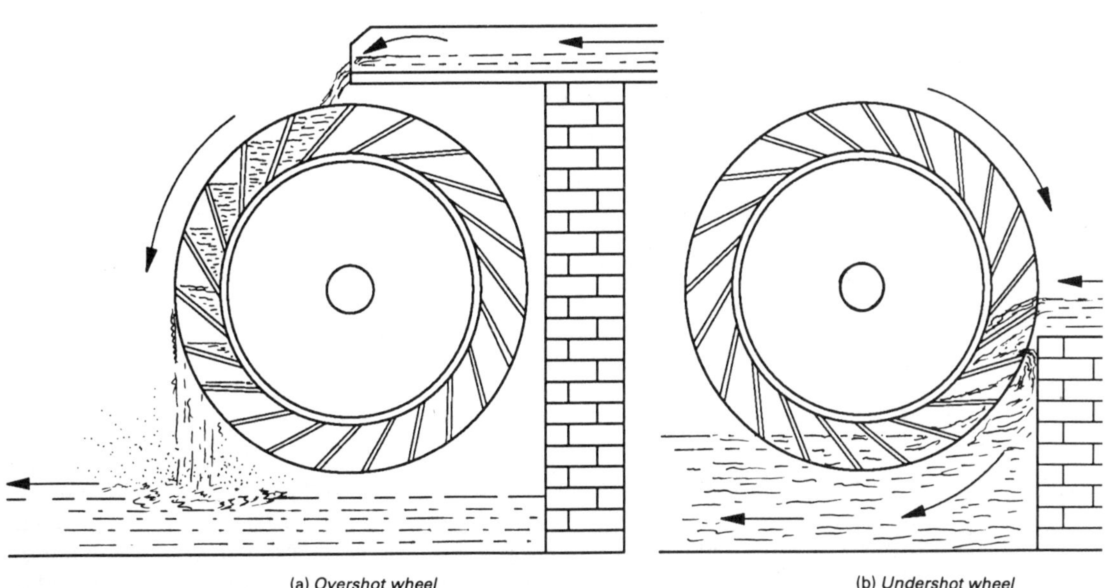

(a) *Overshot wheel* (b) *Undershot wheel*

The waterwheel is historically much older, and the two types known as (a) overshot and (b) undershot are shown in the diagram. Which form of energy does each type of wheel use?

4

(a) Mostly potential energy
(b) Mostly kinetic energy

The overshot wheel uses the potential energy of river water as it falls through a height of a little less than the diameter of the wheel. There may be a small extra contribution from the kinetic energy of the river flow as it enters the wheel. The undershot wheel uses mostly the kinetic energy of the flowing river, but also there may be a fall of water level considerably less than the diameter of the wheel.

Long before man learned to harness energy in these ways he learned to tame and control animals to do work for him. But unlike windmills and waterwheels, his domestic animals required to be fed.

The energy used by man from wind and water, and the energy obtained from digesting food by man and animal, all come from the same source – the sun. Do you know how the energy of the sun becomes available to us in (a) wind, (b) flowing water and (c) plant growth?

5

(a) Heat of the sun causes air to move, thus creating wind.
(b) Heat of the sun evaporates the sea, clouds are formed, the vapour condenses to form rain.
(c) Solar radiation falling on plant leaves causes photosynthesis, which leads to plant growth which may be eaten as food.

The processes are very much more involved and complex than are summarised here. The process of plant growth involves the plant absorbing water and minerals through its roots and carbon dioxide gas through its leaves. Through the action of solar radiation, photosynthesis produces carbohydrate material. Thus energy from the sun is stored in the plant material as chemical energy which can be released either by digestion by a living creature, or by burning of the plant material as fuel.

The combustion of fuel produces high temperature gases which give off heat. Heat and internal energy are not the same thing; we shall have to distinguish carefully between them later. But now we can define our subject. **Thermodynamics is the science associated with the production of work from heat**. One of the earliest examples of a device used to obtain work from heat in this way was the steam engine. Write down a brief summary of the principles of operation of the steam engine if you can.

6

> Water is boiled by heat from combustion of coal or other fuel. The pressure of the resulting steam is used to push a piston in a cylinder. The piston drives a crank which rotates a shaft.

These principles are illustrated in a diagram in the next frame.

The coal used came from plant life of the carboniferous age, some 300 million years ago, so work obtained from a steam engine is ultimately no different in origin from that obtained from a windmill or waterwheel.

7

Coal is burnt in the firebox and steam is produced as the hot gases pass through the boiler fire-tubes. The steam is admitted to alternate sides of the piston by the valve gear and the piston motion causes the crankshaft to rotate.

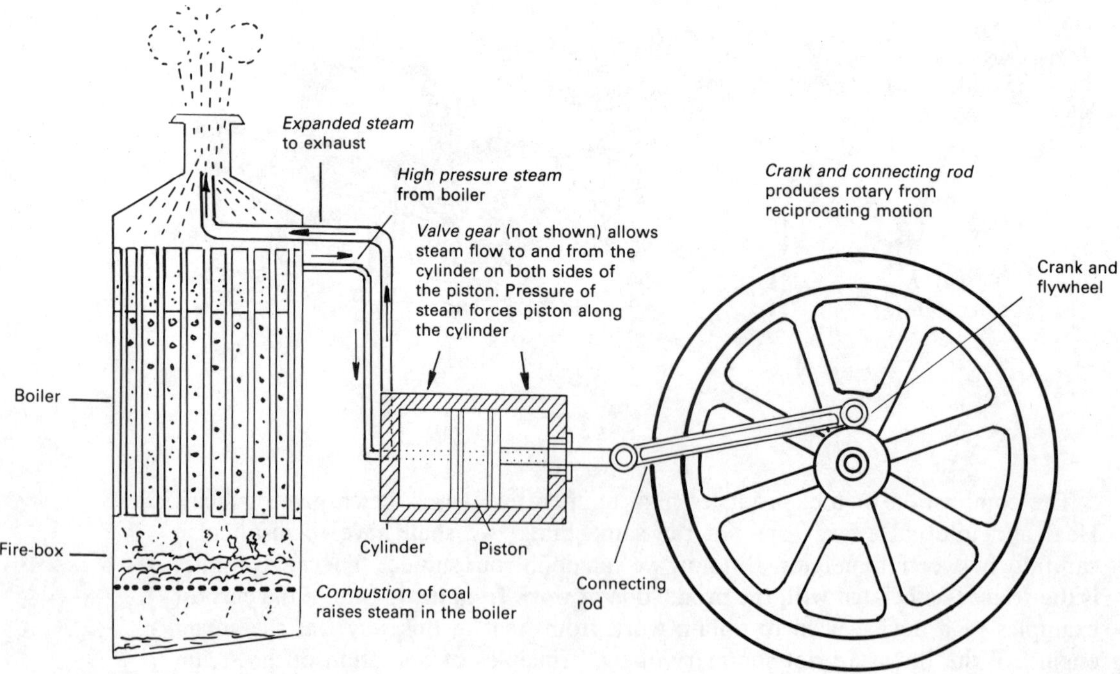

The development of the science of Thermodynamics was partly a consequence of the inability of the 19th century scientists and engineers to understand the principles of the steam engine. The steam engine is now largely historical, it has been replaced by more efficient inventions. Make a list of as many of these as you can.

8

A steam turbine power plant
The internal combustion engine
The gas turbine engine

You may have included electric motors, thinking perhaps of electric traction on railways. These use electricity generated most probably by either steam or gas turbine power plant.

Reciprocating engines, both petrol and diesel, and gas turbine engines may be referred to collectively as internal combustion engines since the fuel is burnt within the machine itself. A steam turbine can use steam generated from the heat of combustion of a fuel or from the heat of nuclear fission.

From the box of Frame 6 and by thinking about the box in this frame we can appreciate that the flow of substances through machinery, whether it be oil, petrol, powdered coal, air, water or steam, is involved with the production of work from heat.

Water and steam are two phases of the same substance. **A phase is a homogeneous entity or aspect of a substance**. There are three phases called solid, liquid and vapour. A gas is a special sort of vapour, to be considered later. Phase changes are often involved in producing work from heat.

Write down the names of the transitions between phases, and then look at Frame 9.

9

Phase transitions are called:
Solid to liquid – melting
Liquid to vapour – evaporation
Vapour to liquid – condensation
Liquid to solid – freezing
Solid to vapour – sublimation

You may not have heard of sublimation before. Snow can turn directly to vapour without melting first. The reverse process, also called sublimation, occurs, for example, when hoar frost appears.

When a substance changes phase, differences occur in the molecular bonding of the substance. Engineers studying thermodynamics generally take what is called a **macroscopic**, or 'large scale' view. They examine overall effects in changes taking place. In contrast, physicists and chemists tend to look at the microscopic aspects: they examine behaviour at the molecular level, often using statistical techniques to predict or analyse.

The chemical energy of fuel is in a sense a form of potential energy. When this energy is released by combustion, which forms of energy are formed?

10

Heat, internal energy, kinetic energy

In Frame 6 we said that water is boiled by heat from the combustion of coal. As we shall see, heat is energy in transfer, passing from one place to another. Thus when coal is burnt in a boiler, chemical energy is released and the coals become red or white hot, and a transfer of heat takes place to the water increasing its internal energy so that it boils. The pressure in the boiler rises, and the expanding steam can push a piston; in a turbine wheel, kinetic energy is absorbed from the expanded steam. Thus, indirectly, work is obtained from the chemical energy in the fuel.

Unfortunately, the process of obtaining mechanical work by burning fuel is not efficient. This efficiency may be defined as:

$$\text{Thermal efficiency} = \frac{\text{Energy equivalent of work output}}{\text{Energy equivalent of fuel input}}$$

Very early steam engines had efficiencies of no more than one or two per cent. What do you think happened to the remainder of the energy available in the fuel?

11

Lost as waste heat

Only one or two per cent of the chemical energy in the coal became useful work. The hot boilers lost heat to the surroundings, the expanded steam escaped into the atmosphere, and in using the steam leaving the cylinders to draw air through the boiler hot coals were sometimes blown up the chimney! Even today, modern power station steam plant has overall efficiencies of no more than 40–45 per cent, and the thermal efficiencies of internal combustion engines and gas turbines are in the general range of 25–45 per cent. So why is this?

It was not until well after the steam engine was invented that it was realised that there is an important limit law (the Second Law of Thermodynamics) on the conversion of heat into work. Thus, although steam pumping engines were first built around 1700, and the first railway, the Stockton and Darlington, was opened in 1825, the Second Law was not stated until 1850.

Over the years, a steady improvement in the thermal efficiency of all thermodynamic plant has been achieved by a steady increase in operating temperature and pressure. As we shall see later, this is an important link. The Second Law is so fundamental to the nature of energy that to help understand it we shall need to take a brief look at the microscopic view point.

These first eleven frames form an introduction to our subject of Thermodynamics, which really begins in the next frame. Just quickly go through them again and make sure you understand everything before reading on.

12

The whole of matter within the universe is subject to the laws of Thermodynamics, so as a starting point we have to learn to identify a particular parcel of matter that we wish to study. For example, this could be the gas contained in a cylinder by a piston. We shall call this gas an example of a **closed system**. It is shown in the figure below.

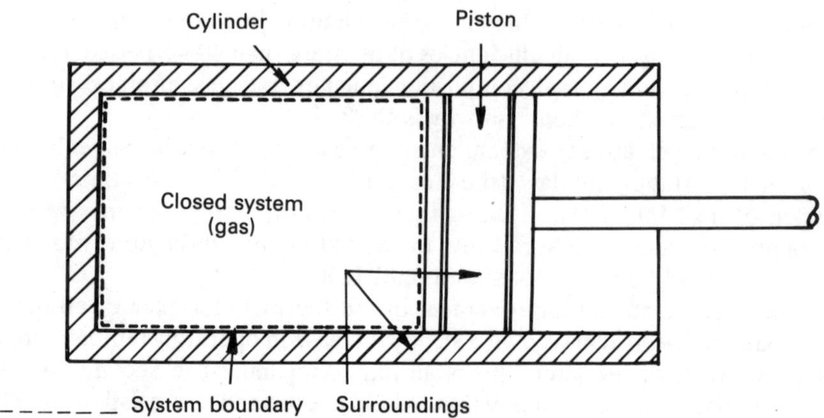

The **immediate** surroundings to the gas are the cylinder and piston, these together are called the **surroundings**. Beyond the surroundings is the remainder of the universe which we may assume to be unaffected by anything happening to the closed system or its surroundings.

A closed system is contained by a **boundary** which is imagined by the observer. In our example this is the contact surface between the gas and its enclosure. The closed system and boundary may change shape and move about, but the mass is fixed by definition, so no matter may cross the boundary.

A closed system has mass, volume, shape and position. Which of these features can change?

13

Volume, shape and position

The volume of the gas in the cylinder will change as the piston moves. The shape would change in the sense that the length/diameter ratio will change. The centre of mass moves as well, but this is only important in the vertical cylinder as the potential energy of the gas is affected. This is shown opposite.

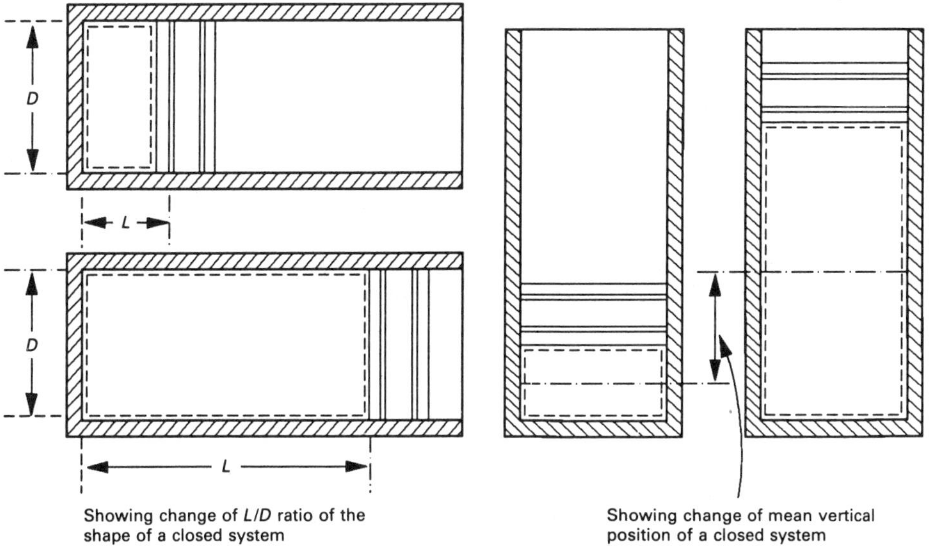

Showing change of L/D ratio of the shape of a closed system

Showing change of mean vertical position of a closed system

As the piston moves, an interaction takes place at the boundary between the closed system and its surroundings. Such interactions are of two types called **work** and **heat**, which are considered in the next programme. A heat interaction can also occur without any movement of the boundary.

The matter within a closed system may be of any phase, and a number of different objects may collectively be considered: for example, the gas and the piston together could be a closed system.

Sometimes it may be necessary to think of the atmosphere, or a lake or the sea, as the surroundings, but in such cases a distinction between the immediate surroundings and the rest of the universe is not important.

A rubber balloon is filled with helium and tethered in the atmosphere. How many closed systems can you identify? With each system, state the corresponding surroundings.

14

Closed system	*Surroundings*
1 The helium gas	The balloon and atmosphere
2 The balloon	The helium gas and the atmosphere
3 The helium gas and balloon	The atmosphere

15

In addition each of these may be reversed, so that the closed systems on the left become the surroundings for new closed systems on the right. Part of a closed system in one study (e.g. the balloon in the system in case 3) may form part of the surroundings in another (the balloon in the surroundings in case 1).

The need possibly to consider the atmosphere as a system or part of a system may not be apparent now, but this should become clear in later examples.

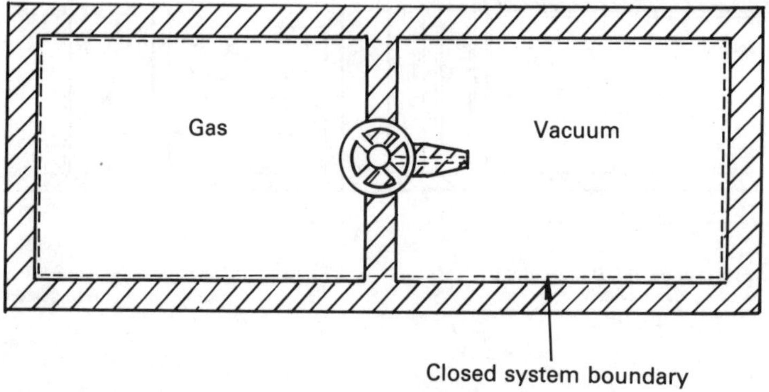

A vacuum could form part of a closed system as shown in the figure above. Industrial techniques involving a vacuum are quite frequently used, and need to be analysed. As a closed system must have mass by definition, a vacuum cannot form a closed system by itself.

Now let's think again about the closed system of gas or air contained by a piston in a cylinder, Frame 12. Where did that air come from, and how did it get there? It most likely came from the atmosphere and it would have entered the cylinder by an open valve that subsequently closed. Later it could be passed out to the atmosphere again through another valve. In a reciprocating engine running at speed, the flow of air entering the engine from the atmosphere is continuous. This is an example of a situation that requires another type of system.

16

In a petrol engine, air is sucked in via the carburettor, where petrol is introduced, to enter the cylinder through an open inlet valve. With inlet and exhaust valves closed the mixture is compressed, burnt and expanded to do work, before the exhaust valve opens to expel the exhaust gases back to the atmosphere. In an engine with, say, 4 or 6 cylinders running at speed, this gives a continuous flow of air and fuel to the engine and exhaust gas from the engine. The contents of any one cylinder may be considered as a **closed system** only during the piston movement when both valves are closed. When one of the valves opens, the cylinder contents either escape or are added to, and so it is difficult to follow what happens to the original closed system.

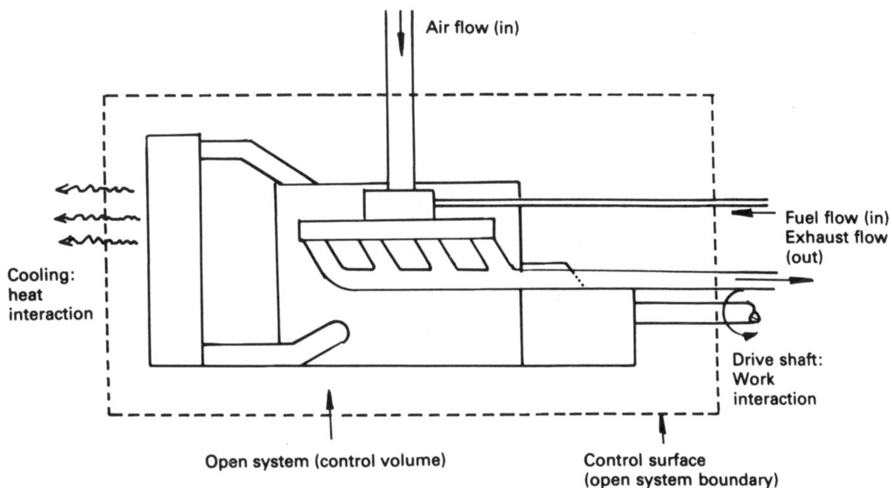

A second approach is required and this is to consider an **open system**. This is a defined space of fixed shape and position relative to the observer. It is limited by a **control surface**. The open system is also known as a **control volume**. Matter may enter and leave an open system across the control surface. Heat and work interactions may also take place at the fixed control surface.

In our engine example, an open system would contain the whole engine, and induction and exhaust pipes would cross the control surface. Work would be done where the drive-shaft crosses the control surface.

Now list the important differences between closed and open systems.

17

Closed system	*Open system*
May change shape and position	Of fixed shape and position
No matter may enter or leave	Matter may enter and leave

Now let's think more about a closed system. Consider some air in a cylinder, contained by a piston as shown. We wish to gather as much information about that quantity of air as we can. We may need to know its mass, volume, pressure and temperature. These terms are called **properties**, and the total of all the properties will define the **state** of the air.

18

We also use specific properties which are the properties of unit mass of the system, in this case the air.

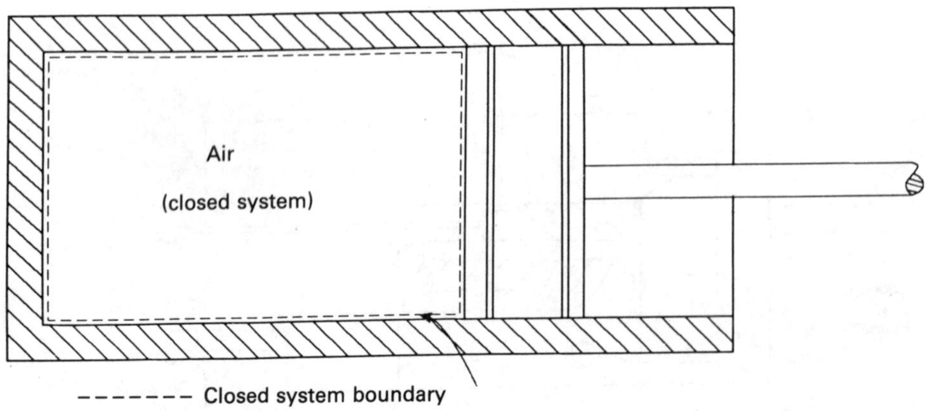

Since the volume of a closed system can be subdivided each with its own mass, we may say:

$$\text{specific volume} = \frac{\text{volume}}{\text{mass}}$$

The reciprocal of specific volume is density:

$$\text{density} = \frac{\text{mass}}{\text{volume}}$$

We shall use the following symbols for these properties:

Property	*Symbol*
Mass	m
Volume	V
Pressure	p
Temperature	T
Specific volume	v
Density	ρ

Do you think that 'specific pressure' and 'specific temperature' would be useful properties?

19

No

If we subdivide our original closed system into two or more parts, the pressure and temperature of each part will be the same as that of the original closed system. Pressure and temperature are known as **intensive properties**, and as such have no useful meaning in terms of unit mass.

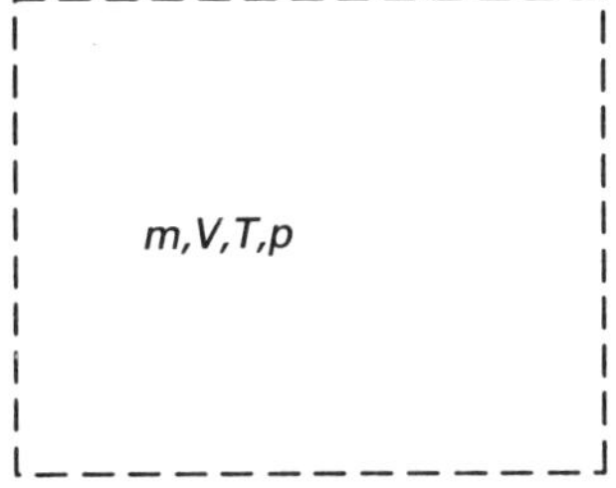

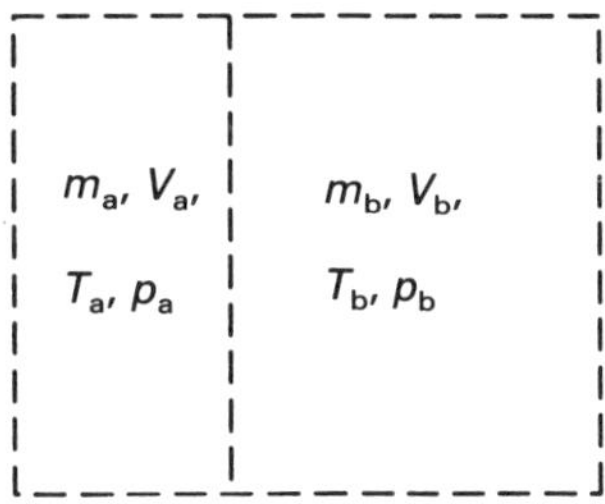

$$m = m_a + m_b \text{ (extensive)}$$
$$V = V_a + V_b \text{ (extensive)}$$
$$T = T_a = T_b \text{ (intensive)}$$
$$p = p_a = p_b \text{ (intensive)}$$

On the other hand, mass and volume, which can be subdivided and then added again to make the original closed system, are called **extensive properties**.

Are specific volume and density extensive or intensive properties?

20

Intensive

Thus, although volume is an extensive property, specific volume and density become intensive properties. Later we shall meet other extensive properties, and it is only extensive properties that can be expressed specifically (per unit mass). In the next frame we shall think about **units**.

21

A number of systems of units have been used in the past in different realms of science and engineering, however there is now a firm commitment to the International System of Units, S.I., first introduced by international agreement in 1977.

The units of mass, length and time are:

Physical quantity	*S.I. unit*	*S.I. abbreviation*
Mass	kilogram	kg
Length	metre	m
Time	second	s

There is also a set of prefixes to be used if these units are either too small or too large. A partial list of multipliers from 10^9 to 10^{-6} is given:

Multiplier	*Prefix*	*Symbol*
10^9	giga	G
10^6	mega	M
10^3	kilo	k
10^2	hecto	h
10^1	deca	da
10^{-1}	deci	d
10^{-2}	centi	c
10^{-3}	milli	m
10^{-6}	micro	μ

A km is therefore 10^3 m and a mm is 10^{-3} m. The units of specific volume are m^3/kg, and the units of density are kg/m^3.

22

Any equation involving physical quantities must have consistent units to be correct. It is always a good idea to check the units in an equation to be sure you are using the correct numbers. Where appropriate we will include units in equations and answers in the text in square brackets, for example:

$$\begin{aligned}\text{Volume } [m^3] &= \text{Length } [m] \times \text{Breadth } [m] \times \text{Depth } [m]\\ &= \text{Length } [mm] \times 10^{-3} \text{ } [m/mm] \times \text{Breadth } [m] \times \text{Depth } [m]\end{aligned}$$

If we had been careless with units, in the second case we would have got an answer 10^3 too big.

If a fluid has a specific volume of 1 m^3/kg, what is the density of the fluid in g/cm^3?

23

10^{-3} g/cm^3

Here is the working:

Units of specific volume are: 1 [m^3/kg]

Hence units of density are: 1 [kg/m^3]

$= 10^3$ [g/m^3]

$= 10^3/10^6$ [g/cm^3]

$= 10^{-3}$ [g/cm^3]

Before we can define pressure, we must first define force. From Newton's Second Law of Motion you should know that force (F) is proportional to the product of mass (m) and acceleration (a). If a suitable unit is adopted for force, then:

$$F = m \times a$$

The unit of force is chosen as that force to accelerate a mass of 1 kg at an acceleration of 1 m/s^2, so the units of F are kg m/s^2. This combination of units is called the newton, N. Thus:

$$1\ [\text{N}] = 1\ [\text{kg}] \times 1\ [\text{m/s}^2]$$

Now try these simple calculations:

(1) Calculate the force in newtons when a body of mass 10 kg has an acceleration of 2 m/s^2.

(2) A force of 4500 N acts on a body of mass 500 kg. Calculate the acceleration.

24

(1) 20 N and (2) 9 m/s^2

In case you did not get these answers, the working is:

(1) Force = 10 [kg] × 2 [m/s^2] = 20 [kg m/s^2] = 20 [N].

(2) $a = F/m$ = 4500 [N]/500 [kg] = 9 [kg m/s^2]/[kg] = 9 [m/s^2].

25

From force we can move on to consider weight. In a given gravitational field, the weight of a body is proportional to its mass. The constant of proportionality is g and in a standard gravitational field g in S.I. units is 9.81 m/s^2.

$$W = m\,g$$

Hence the weight of a body having a mass of 1 kg

$$= 1\ [\text{kg}] \times 9.81\ [\text{m/s}^2]$$
$$= 9.81\ [\text{N}]$$

You can get a feeling for the magnitude of a force by 'weighing' an object in your hand. Thus a newton is roughly the weight of a medium size apple.

Mass, you remember, is a property of a system which does not alter. However, the value of g can alter, it decreases with altitude above sea level, and consequently the weight of a body would decrease correspondingly.

The value of g is only 1.6 m/s^2 on the moon. Neil Armstrong was the first person to set foot on the moon in July 1969. If we guess his mass to be 80 kg, his earth weight is:

$$80\ [\text{kg}] \times 9.81\ [\text{m/s}^2]$$
$$= 784.8\ [\text{N}]$$

His moon weight was: $80\ [\text{kg}] \times 1.6\ [\text{m/s}^2]$

$$= 128.0\ [\text{N}]$$

These examples should now be just as easy:

(1) What is the weight of a man of mass 70 kg in a locality where $g = 9.75$ m/s^2?

(2) An object experiences a gravitational force of 1000 N where $g = 9.81$ m/s^2. What is the mass of the object?

26

(1) 682.5 N, (2) 101.94 kg

For the first answer: Weight = $[70\ \text{kg}] \times 9.75\ [\text{m/s}^2] = 682.5\ [\text{N}]$
and for the second: $1000\ [\text{N}] = m\ [\text{kg}] \times 9.81\ [\text{m/s}^2]$

Hence:

$$m = \frac{1000\ [\mathrm{N}]}{9.81\ [\mathrm{m/s^2}]}$$

$$= 101.94\ [\mathrm{kg}]$$

Now we can think about pressure. Pressure, like specific volume and density, is an intensive property involving two physical quantities. Pressure is the normal force F_n acting on the surface of a containing vessel divided by the area of the surface, A. The normal force is the force due to the fluid acting at right angles to the surface. Force is a vector quantity that can have direction as well as magnitude:

$$p = \frac{F_n}{A}$$

The unit of pressure, using S.I. units, must be newtons per square metre, N/m^2. This is given the special name pascal, Pa.

$$1\ [\mathrm{Pa}] = 1\ [\mathrm{N/m^2}]$$

This is a very small pressure: you can imagine the weight of an apple uniformly distributed over an area of a square metre.

For engineering needs we use multiples of this unit, the kPa, and the MPa, where:

$$1\ [\mathrm{kPa}] = 1\ [\mathrm{kN/m^2}] = 10^3\ [\mathrm{N/m^2}]$$

and $$1\ [\mathrm{MPa}] = 1\ [\mathrm{MN/m^2}] = 10^6\ [\mathrm{N/m^2}]$$

You are likely to meet any of these units, i.e. kPa, MPa, kN/m^2 or MN/m^2.

The force on the surface due to the fluid arises from the bombardment of the surface by the vast number of molecules of the fluid. Hence pressure is a macroscopic concept arising from the total contribution of all the molecules.

A gas of pressure 1500 Pa is contained in a cylinder by a piston of diameter 80 mm. What is the force in N on the piston?

27

7.54 N

28

$$\text{The piston area is: } \frac{\pi \times 80^2}{4 \times 10^6} \frac{[\text{mm}^2]}{[\text{mm}^2/\text{m}^2]} = 0.005\,03\ [\text{m}^2]$$

$$\text{The force is: pressure} \times \text{area} = 1500\ [\text{N/m}^2] \times 0.005\,03\ [\text{m}^2] = 7.54\ [\text{N}]$$

The pressure in a system containing fluid may be greater or less than the pressure of the surrounding atmosphere. The air of the atmosphere has mass which is acted upon by the earth's gravitational field. This produces what is called atmospheric pressure. This is highest at sea level and gradually reduces with altitude. Weather conditions cause regional variations, but a reference, or standard value is taken at sea level as 101 325 Pa. This is usually expressed as 101.325 kPa, (kilopascals), or 1.013 25 bar, where 1 bar = 10^5 Pa. The bar, expressed as 1000 mbar, is used by meteorologists in weather charts. Atmospheric pressure is sometimes approximated to 100 kPa, or 1 bar.

In engineering plant where high pressures often occur, the bar is sometimes used. This gives a 'feel' for the magnitude of the pressure involved, since the pressure in [bar] is simply a multiple of atmospheric pressure. However, being a 10^5 multiple of the Pa, the bar is not a preferred S.I. unit.

If an atmospheric pressure of 992 mbar is recorded, what is the pressure in Pa?

29

99 200 Pa

Since 992 [mbar] = 0.992 [bar] and 1 [bar] = 10^5 [Pa], it follows that 992 [mbar] = 99 200 [Pa]. When we measure pressure, we generally obtain a reading relative to the local atmospheric pressure. An instrument designed to measure pressure is called a pressure gauge, and the reading it gives is called **gauge pressure**. Since the instrument is situated within the atmosphere to measure the pressure of a closed system, logically it can only measure a value relative to the pressure around it. This reading when added to the local atmospheric pressure is called **absolute pressure**:

$$p(\text{absolute}) = p(\text{gauge}) + p(\text{atmospheric})$$

A bourden tube is usually found inside a pressure gauge. Its main feature is a flat section hollow curved tube closed at one end. The other end is open to the pressure to be measured, and this pressure causes the curved tube to try and straighten itself. The movement, which depends on the pressure, is transmitted by a linkage to a pointer, and a reading of the gauge pressure is obtained. Calibration is achieved by applying known pressures, and the range of pressure that can be measured depends on the stiffness of the tube.

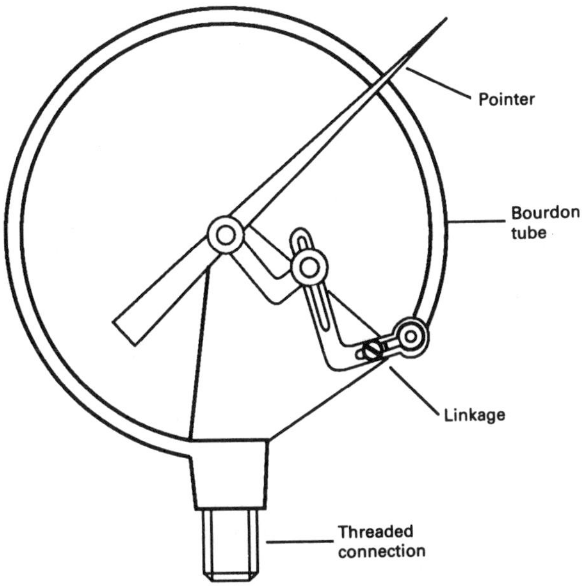

When we check the car tyres on a garage forecourt, the air line gives us a reading of gauge pressure. What is the pressure in a tyre when it is 'completely flat'?

30

Zero gauge pressure, or atmospheric pressure

A tyre that is flat has either had the valve stem removed or it has a hole somewhere, so it is unable to contain any pressure relative to the atmosphere. The pressure inside is therefore atmospheric.

A gauge pressure reading may be either positive, which is above the pressure of the atmosphere, or negative, which is below the pressure of the atmosphere. This is shown in the next frame.

31

The positive gauge pressure reading is shown in case (a), and the negative reading in case (b).

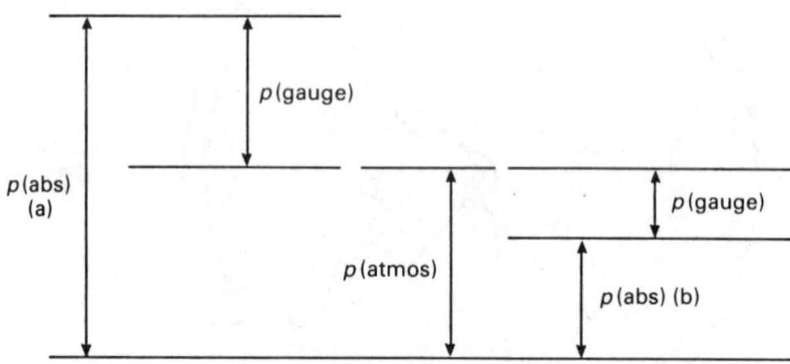

Bourdon tube pressure gauges are also used to measure absolute pressures less than atmospheric. The radius of the tube tends to decrease when such a pressure is applied. A gauge pressure that is negative is referred to as **vacuum**. Strictly, the word 'vacuum' by itself means zero absolute pressure, but in common usage we speak of 'so much' vacuum. Thus if the absolute pressure of a closed system is 70 kPa when the local atmospheric pressure is 101 kPa, there is said to be 31 kPa of vacuum. A real vacuum is a space in which there is no matter at all, and the pressure must be zero.

Now try these two simple questions:

(1) A steam condenser has 50 kPa of vacuum when the atmospheric pressure is 101 kPa. What is the absolute pressure in kPa?
(2) What is the absolute pressure in bar if the gauge pressure is 60 kPa when atmospheric pressure is 98 kPa?

32

(1) 51 kPa, (2) 1.58 bar

For (1): $p(\text{abs}) = -50\ [\text{kPa}] + 101\ [\text{kPa}] = 51\ [\text{kPa}]$

For (2): $p(\text{abs}) = 60\ [\text{kPa}] + 98\ [\text{kPa}] = 158\ [\text{kPa}] = 1.58\ [\text{bar}]$

Bourdon tube pressure gauges are robust instruments that may be used in everyday conditions. For measuring small gauge pressures (say up to about 1 bar) in laboratory type conditions, a manometer may be used. This works on the principle that the pressure at the base of a column of liquid of height L and density ρ is ρLg. This is simply the weight of the column per unit area.

In the figure, the absolute pressure p(abs) in the vessel is acting downwards at point 1 and atmospheric pressure is acting on the top of the column of liquid at point 3. The column of liquid of height l, between points 1 and 2, is common to

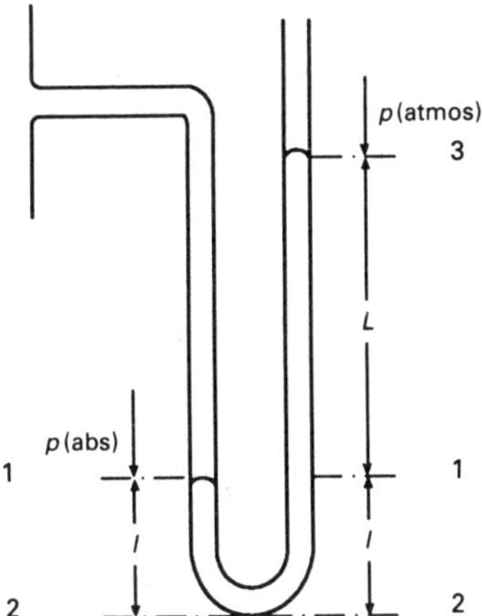

both sides of the manometer. A column of liquid of density ρ gives a pressure in force per unit area.

The units are:

$$[\mathrm{kg/m^3}] \times [\mathrm{m}] \times [\mathrm{m/s^2}] = [\mathrm{kg}] \times [\mathrm{m/s^2}]/[\mathrm{m^2}]$$
$$= [\mathrm{N/m^2}]$$
$$= [\mathrm{Pa}]$$

By balancing pressures:

$$p(\text{abs}) + \rho lg = \rho lg + \rho Lg + p(\text{atmos})$$

Hence: $$p(\text{abs}) = \rho Lg + p(\text{atmos})$$

So the height L gives a reading from which the gauge pressure can be calculated. In the diagram, if the pressure in the vessel were less than the local atmospheric pressure, the liquid column would be higher in the left-hand limb.

Water of density 1000 kg/m³ is used in a manometer. The height L noted in the formula above is 100 mm. What is the gauge pressure in Pa?

33

The answer is 981 Pa.

The gauge pressure is ρLg $= 1000\ [\mathrm{kg/m^3}] \times \dfrac{100}{1000}\ [\mathrm{m}] \times 9.81\ [\mathrm{m/s^2}]$

$$= 981\ [\mathrm{kg}]\ [\mathrm{m/s^2}]/[\mathrm{m^2}]$$
$$= 981\ [\mathrm{Pa}]$$

34

To record small pressures accurately, the column of liquid may be sloped as shown. It is usually possible to adjust the angle of tilt, α, allowing different ranges of pressure to be measured. Oil of density less than that of water may be used as the fluid.

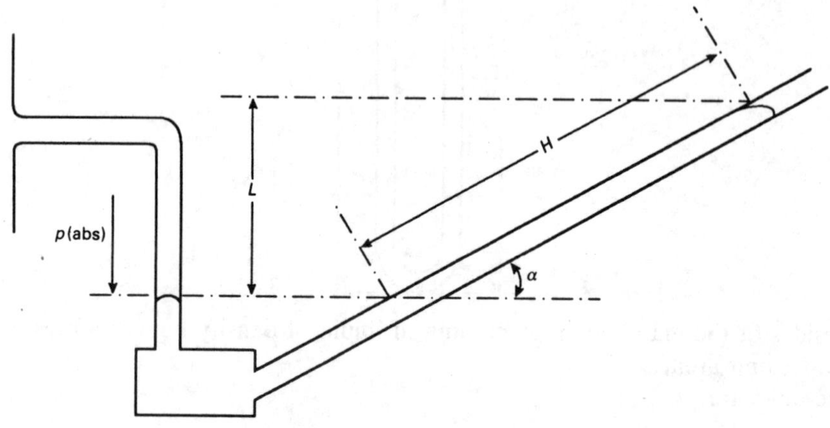

From the diagram it is seen that if a reading H is obtained, the vertical L is given by:

$$\frac{L}{H} = \sin \alpha$$

and so:

$$L = H \sin \alpha$$

The gauge pressure recorded is then $\rho L g$ [Pa], as before.

Oil of density 780 kg/m³ is used in a manometer in which the tube is set at 10° to the horizontal. A positive gauge reading of 350 mm is obtained. What is the gauge pressure in Pa?

35

465.05 Pa

The gauge pressure is $\rho(H\sin\alpha)g = 780\ [\text{kg/m}^3] \times \frac{350}{1000}\ [\text{m}] \times \sin 10°$

$$\times\ 9.81\ [\text{m/s}^2]$$

$$= 465.05\ [\text{Pa}]$$

When we want to measure the pressure of the atmosphere itself, we use a barometer. It consists essentially of a closed vertical tube inserted into a bowl of mercury so that the pressure of the atmosphere supports a column of mercury in the tube. A vacuum exists between the top of the mercury column and the closed end of the tube.

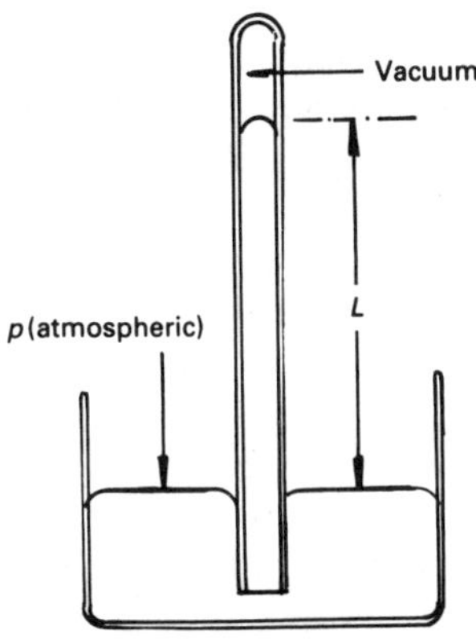

As atmospheric pressure changes, the height L changes correspondingly, either reducing or increasing the dead space of the vacuum. The atmospheric pressure will be given by:

$$p(\text{atmospheric}) = \rho L g$$

In a barometer, a pressure of one atmosphere, or 101.325 kPa, supports a column of mercury 760 mm high. What is the density of the mercury in kg/m^3?

36

13 590.46 kg/m^3

We have: $101\,325\ [\text{Pa}] = \rho\ [\text{kg/m}^3] \times \dfrac{760}{1000}\ [\text{m}] \times 9.81\ [\text{m/s}^2]$

Hence:

$$\rho = \frac{101\,325}{0.76 \times 9.81}\ \frac{[\text{kg m/s}^2]/[\text{m}^2]}{[\text{m}]\ [\text{m/s}^2]}$$

$$= 13\,590.46\ [\text{kg/m}^3]$$

In the next frame we shall think about the concept of temperature.

37

Temperature can be loosely defined as a 'measure of hotness'. Like pressure, temperature is a macroscopic concept related to the behaviour of the molecules of the substance. It is a function of the vibrational energy levels of the molecules. Because our bodies have temperature, we are sensitive to changes around us; by the sense of touch we can tell whether one object is hotter than another. Although temperature and heat are closely associated, temperature as a property is the more fundamental concept and must be considered first.

When you touch an object you experience heat at your finger tips, and your brain tells you something about the temperature of the object. If it is hotter than body temperature, heat flows into your fingers; if the object is colder than body temperature, then heat flows out from your fingers, and the nerve endings can detect the difference. Also we are familiar with the way 'hot' objects 'cool down' and 'cold' objects 'warm up' to the temperature of their surroundings. In either case, by the process of a transfer of heat, the object is said to have reached the same temperature as its surroundings.

Mass, length and time are measured by comparison with specified standards. Measurements of pressure and volume are dependent on these quantities. We have to devise similar methods to measure temperature: we have to specify temperature change by recording observable changes in other properties, such as volume and pressure, which can be measured.

38

When you take your own temperature using a mercury in glass clinical thermometer, you observe a change in the volume of the mercury. Initially the mercury column is barely visible since the thermometer is at room temperature. After half a minute or so you find the column is at or above the level regarded as normal. Thus the operation of a mercury in glass thermometer relies on the expansion of the mercury with temperature rise. The volume of the capillary stem of the thermometer is very small, but with magnification the visible change of volume of the mercury appears to be quite large. Also the response is rapid which makes it convenient to use. Can you think of another property of a substance that is frequently used as the basis of a thermometer?

39

Electrical resistance of a platinum wire

The electrical resistance of a platinum wire changes in a regular way with temperature.

In order to define a basis for temperature measurement we need to think about both temperature equality and temperature inequality.

You find in a room a thermometer hanging on the wall, and you see a certain level of mercury in that thermometer. It is not moving because the thermometer is in equilibrium with its surroundings, the room. **Temperature equality is said to exist between two systems when no change in any observable property occurs when they are brought into contact**. Because the volume of mercury in the thermometer is not changing, the thermometer and the room (the two systems) have the same temperature: by reading the thermometer we know the temperature of the room.

If the thermometer is now taken outside the house, three possible things can happen to the mercury level (or volume). What are they, and what do they mean?

40

(1) Remains the same: room and outside temperature the same
(2) Rises: outside temperature greater than room temperature
(3) Falls: outside temperature less than room temperature

Temperature inequality exists when there is an observable change in a property when two systems are brought into contact. Thus we are able to move from one point to another on a thermometer; this is followed by temperature equality at this new position or temperature on the thermometer.

Suppose we have a thermometer and four beakers of water, A, B, C and D. On placing the thermometer in each beaker and allowing equilibrium to be reached, we find the mercury levels are as shown. At equilibrium, the thermometer and beaker of water have the same temperature, and we see that B is hotter than A, C is cooler than A, and A and D both show the same reading.

This is shown in the figure in the next frame.

41

If the fluids A and D are equal in temperature to the thermometer, they must be equal in temperature to each other.

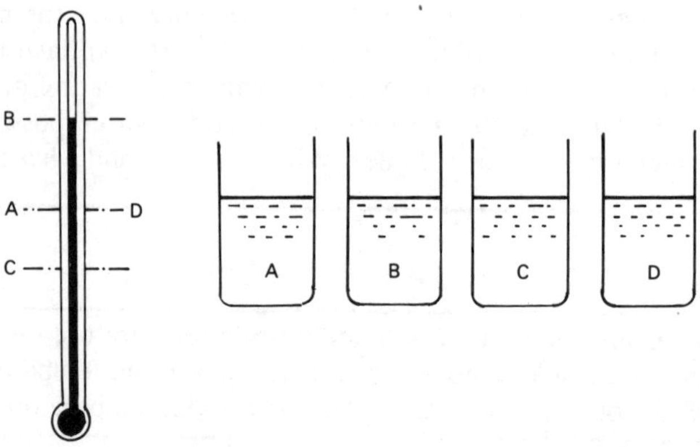

The observation that two beakers of water have produced the same mercury level on a thermometer leads to an extremely important result. It is the basis of the first fundamental law of Thermodynamics that we need to consider. This law states that **when two bodies have the same temperature as a third body, then they have the same temperature as each other**. Thus the third body, which in our experiment was the thermometer, may be used to relate the temperatures of other bodies, and we found two of the four beakers had the same temperature. This law gives us a basis for temperature measurement.

In the development of the science of Thermodynamics, the First and Second Laws on energy conservation and on the restriction on the conversion of heat into work, mentioned in Frame 11, date from the work of Joule (1843), Helmholtz (1847), Clausius (1850–1854) and Lord Kelvin (1851). It was the later work of Maxwell and Boltzmann (1860–1861) that lead to a fuller understanding of the importance of temperature equality, and the law given above is consequently generally known as the **Zeroth Law of Thermodynamics**. This is a law which, in terms of subsequent understanding, has to precede the First Law, so it is called the 0th Law, or the one before the first.

Now we shall put some numbers on our thermometer. This is done from the realisation that phase changes of pure substances under closely controlled conditions must take place at particular, and repeatable, temperatures.

Can you name two of the phase changes that are used in giving two particular points on a thermometer?

42

The melting of ice The boiling of water

Temperature scales are put on thermometers by establishing certain fixed points related to specified phase transitions, and then by dividing the distance between the fixed points into a number of divisions. The two most commonly known fixed points are those for melting ice and boiling water taking place under controlled and specified conditions.

The Celsius or Centigrade scale is used in the S.I. unit system, and the ice point is given the number 0 and the boiling point 100. The scale in between is divided into 100 divisions. On this scale the temperature of the human body is around 36.9°C. In 1954 the scale was redefined in terms of a single fixed point, the triple point of water, and the magnitude of a division or degree based on the Ideal Gas Temperature Scale. At the triple point of water, a state where ice, water and water vapour all exist in equilibrium, the assigned temperature is 0.01°C, and from the Ideal Gas scale the boiling point of water was found to be 100.00°C. Discussion of the triple point of water and the Ideal Gas Temperature Scale will follow later in the text, so you need not worry about these details at the moment.

43

The Fahrenheit scale preceded the Celsius scale historically, and is based on fixed points of 32°F for the ice point and 212°F for the boiling point of water. The story of how these numbers came about is given as an appendix to this programme.

It is possible to convert from one scale to the other:

$$\text{Temperature in } [^\circ\text{C}] = ([^\circ\text{F}] - 32) \times \frac{5}{9}$$

$$\text{Temperature in } [^\circ\text{F}] = [^\circ\text{C}] \times \frac{9}{5} + 32$$

Now do these conversions:

(1) What is 200°C in °F?
(2) What is 28°F in °C?
(3) At what temperature do the Fahrenheit and Celsius scales have the same numerical value?

44

(1) 392°F, (2) −2.22°C, (3) −40°C = −40°F

For (1): $[°F] = 200\ [°C] \times \frac{9}{5} + 32 = 392\ [°F]$

(2): $[°C] = (28\ [°F] - 32) \times \frac{5}{9} = -2.22\ [°C]$

(3): For equal numerical values: $([°C] - 32) \times \frac{5}{9} = [°C] \times \frac{9}{5} + 32$

Hence: $[°C]\left(\frac{9}{5} - \frac{5}{9}\right) = -32 \times \left(1 + \frac{5}{9}\right)$

and so: $[°C] = -40$ (= [°F] also)

This cross-over point of the two scales gives an alternative method for converting from one scale to the other:

$$[°F] = ([°C] + 40) \times 1.8 - 40$$
$$[°C] = ([°F] + 40)/1.8 - 40$$

Obviously, temperatures above 100°C and below 0°C need to be measured, and additional fixed points are required. The fixed points of the **International Temperature Scale** are based on the following phase changes taking place under carefully specified conditions:

Oxygen point: liquid–vapour	−182.97°C
Ice point: solid–liquid	0.0°C
Steam point: liquid–vapour	100.0°C
Sulphur point: liquid–vapour	444.6°C
Silver point: solid–liquid	960.8°C
Gold point: solid–liquid	1063.0°C

Just as there is international agreement about the kilogram, the metre and the second, there has to be international agreement about fixed points of temperature and how to interpolate between them. This is how we can be sure all thermometers will give the same reading at the same temperature.

What are these fixed points in °F?

45

$-297.35°F$, 32.0°F, 212°F, 832.28°F, 1761.44°F, 1945.4°F

Now we shall think about how we interpolate between the fixed points, as international agreement on this is just as important as it is on the fixed points themselves.

Variation in the properties of substances with temperature is very rarely linear. Alcohol-in-glass thermometers are often used for domestic temperature measurement. If we were to construct both an alcohol and a mercury thermometer by obtaining the ice and steam fixed points and then dividing the distance between the fixed points **equally** into 100 divisions, we would find that the thermometers would not agree precisely at intermediate temperatures, and both thermometers would be incorrect since the interpolation had not been carried out **according to agreed procedures**. The figure shows how the same object would give a reading of 60° on an alcohol thermometer and 61° on a mercury thermometer, but neither would be a correct reading in °C.

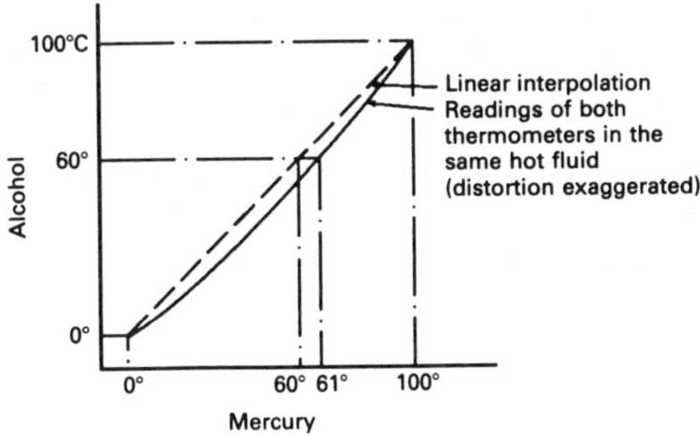

The International Temperature Scale requires interpolation between 0°C and 100°C to be carried out using a **platinum resistance thermometer**, so if both thermometers had been calibrated between 0° and 100° in this way, they would then agree at intermediate temperatures. This is also a consequence of the Zeroth Law.

46

All physical thermometric scales depend on various properties of substances, so they all suffer from the problem of non-linearity, including a scale based on the resistance of a platinum wire.

However, later we shall examine an absolute **thermodynamic temperature scale** that is independent of the physical properties of substances. It is called the **Kelvin** scale of temperature, and temperatures are measured in units of Kelvin, K, rather than in degrees. This temperature scale has an absolute zero.

47

As we have seen, temperature is a macroscopic concept related to the level of molecular activity of the substance. In the thermodynamic temperature scale there is an absolute zero of temperature at which all molecular activity ceases. The closest approach to this scale by any physical thermometer is achieved by the constant volume ideal gas thermometer, which will be considered later.

As we shall need to use absolute temperatures in K before we get to the point where we see how the concept came about, you will need to know the relation between temperatures in degrees Celsius and absolute temperatures in Kelvin:

Temperature in Kelvin = Temperature in °C + 273.15

Thus, the ice point is 273.15 K and the steam point is 373.15 K, when converting from °C.

What are the oxygen, sulphur, silver and gold points from the International Temperature Scale (Frame 44), expressed in Kelvin?

48

90.18 K, 717.75 K, 1233.95 K, and 1336.15 K

To get these results we simply take the temperatures in °C from Frame 44 and add 273.15 to each value.

We have now reached the end of this programme. Make sure you can do all these examples before moving on. In the following frame you will find a summary of important points to remember.

PROBLEMS

1. A quantity of fluid has a mass of 1099.8 kg and a volume of 1.3 m^3. What is (a) its specific volume and (b) its density? [(a) 0.00118 m^3/kg, (b) 846 kg/m^3]
2. A fluid of mass 9 kg occupies a certain volume which is divided into two portions having a volume ratio of 2:1. The density of the fluid is 20 kg/m^3. What is the volume of each portion? [0.3 and 0.15 m^3]
3. A vehicle of mass 1000 kg is brought to rest in a gravel-bed with a deceleration of 50 m/s^2. What is the retarding force? [50 000 N]
4. The piston of a reciprocating engine is momentarily at rest at the extremes of its stroke. As the piston starts to move, its acceleration is greatest. If the maximum acceleration is 7000 m/s^2, and the mass of the piston is 0.6 kg, what is the maximum force on the piston? [4200 N]
5. A spring balance for weighing sacks of potatoes is calibrated at sea level for which g = 9.81 m/s^2. It is subsequently used to check loads across a mountain

pass where $g = 9.7\ m/s^2$. If 4000 kg of potatoes are moved from a base at sea level, will there be an apparent surplus or shortage of potatoes at the check point, and by how much? [Apparent shortage of 44.85 kg]

6. A gas is contained by a piston in a cylinder and the piston is restrained by a spring and by atmosphere pressure acting on the outside of the piston. Atmospheric pressure is 101.325 kPa. The load in the spring is 10^6N and the piston diameter is 0.2 m. Calculate the gas pressure in bar. [319.32 bar]

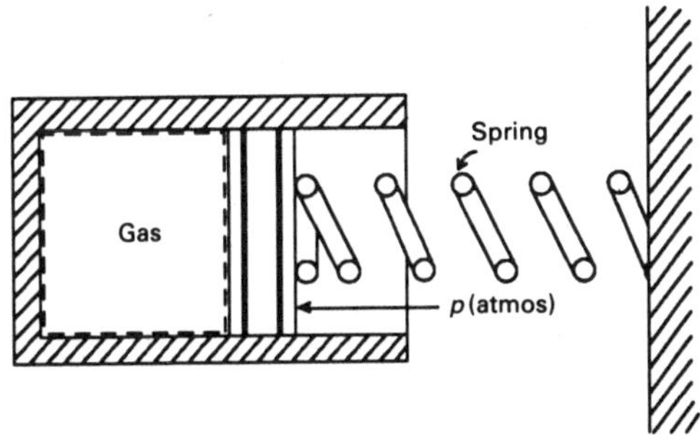

7. The pressure gauge on a gas cylinder reads 150 kPa. The gas cylinder is inside a pressure chamber and the gauge pressure of the pressure chamber is 2.5 bar. Atmospheric pressure is 1.01 bar. What is the absolute pressure in the gas cylinder? (*Hint*: find the absolute pressure in the gas chamber first and treat this as the atmospheric pressure surrounding the gas cylinder.) [5.01 bar]
8. A manometer, which uses oil of density 780 kg/m^3, is used to record the pressure drop across an orifice. It reads 300 mm, recording a pressure below atmospheric, which is 100 kPa. What is the absolute pressure recorded? [97.705 kPa]
9. Two thermometers, one alcohol-in-glass, the other mercury-in-glass, are calibrated together for fixed points in ice and steam, with the two resulting scales each divided equally into 100 divisions. The two thermometers then give readings of 60° (alcohol) and 61° (mercury), when placed in a pan of hot liquid. Discuss the question of which is correct. [Neither is correct; each should be calibrated against a platinum resistance thermometer]
10. The temperature in bright sunshine in a desert at mid-day is said to be 80°F in winter and 80°C in summer. What are these values in °C and °F respectively, and what is 80°C in K? [26.67°C, 176°F, 353.15 K]

49

Points to remember

We end with a summary of this programme, but remember that we have not yet defined work or heat. We shall do so in Programme 2.

Thermodynamics is the science associated with the production of work from heat. This takes place in purpose-built machines through which substances flow, often involving phase changes.

Thermodynamics is studied from a 'large scale' or macroscopic viewpoint.

Initially we consider closed systems, which are prescribed parcels of matter which may change shape and volume and move about.

An open system is a fixed and prescribed space through which matter may flow.

Heat and work are interactions which take place at the boundaries of both closed and open systems.

The properties of a closed system, its mass, volume, pressure and temperature, together define its state. The S.I. units of kilogram mass, metre length, and time in seconds are used.

Force = mass × acceleration; the unit of force is the newton, N, where

$$1 \text{ kg} \times 1 \text{ m/s}^2 = 1 \text{ N}$$

Weight = mass × gravitational acceleration, g, where the standard $g = 9.81$ m/s^2.

Pressure = normal force/area; the unit of pressure is the pascal, Pa, where 1 N/m^2 = 1 Pa. Multiple units: 1 kN/m^2 = 10^3 Pa, 1 MN/m^2 = 10^6 Pa.

$$\begin{aligned}\text{Standard atmospheric pressure} &= 101\,325 \text{ Pa}\\ &= 101.325 \text{ kPa}\\ &= 1.01325 \text{ bar}\end{aligned}$$

where 1 bar = 10^5 Pa. The bar, being 10^5 Pa, is not a preferred S. I. unit.

Pressure gauges, such as bourdon gauges and manometers, measure gauge pressure. Absolute pressure = atmospheric pressure + gauge pressure. Zero absolute pressure is a vacuum. Negative gauge pressure is stated as a measure of vacuum.

The pressure due to a column of liquid is $\rho L g$ N/m^2, where ρ is the density of the liquid in kg/m^3, with L in m.

The Zeroth Law of Thermodynamics states that when two bodies have the same temperature as a third body, they have the same temperature as each other.

Temperature scales involve recognised phase changes for fixed points and agreed interpolation procedures between fixed points. Two fixed points are melting ice and boiling water, giving 0° and 100° on the Celsius scale. These fixed points are 273.15 K and 373.15 K on the Kelvin scale of Absolute Thermodynamic Temperature.

APPENDIX

The origins of the fixed points on the Fahrenheit scale deserve an explanation and the following is based on an account given by James Cork [1]

An earlier scale was proposed by Professor Ole Römer (1644–1710) of Copenhagen. This scale had the ice point at 7.5° and the boiling point of water at 60°, and 0° was used for 'an artificial freezing mixture' of salt, ice and water.

Fahrenheit based his scale originally on two of these fixed points, at 0° and 7.5°, and in addition he used the temperature of the human body as a third fixed point, which he believed to be 22.5°. However, Fahrenheit wrote to his friend Professor Hermann Böerhaave of the University of Leyden and complained of the inelegance of the scale of Römer and its fractions. He proposed to multiply the figures by four to give 30° for the normal ice point and 90° for body temperature. He then thought that 96° would be a more suitable body temperature, simply on the grounds that it was divisible by 12.

This scale was published in 1724 [2]:

'. . . on placing the thermometer in a mixture of sal ammoniac or sea salt, ice and water, a point on the scale will be found which is denoted as zero. A second point is obtained if the same mixture is used without the salt. Denote this as 30°. A third point, designated as 96° is obtained if the thermometer is placed in the mouth so as to acquire the heat of a healthy man.'

Soon after this, however, Fahrenheit developed a thermometer to measure boiling points at atmospheric pressure, and he found the boiling point of water to be 212°. He therefore modified his scale [3] to include the boiling point of water as the upper fixed point at 212°, and to give a more rational 180° interval between the two fixed points, he made the ice point 32°. Body temperature is around 98.4° on this final version of the Fahrenheit scale.

[1] James Cork, *Heat*, Wiley, 2nd edition, 1942.
[2] D.G. Fahrenheit, *Phil. Trans. (London)*, Vol 33, 78, (1724).
[3] D.G. Fahrenheit, *Phil. Trans. (London)*, Vol 33, 179, (1724).

Programme 2

WORK AND HEAT

1

In Frame 17 of the previous programme you met the idea of *state*, and for a closed system, of, say, air contained in a cylinder, we saw that the total of all the properties defines the state of that air. We shall begin this programme by going a little deeper into this. The properties of the air as a closed system which are of interest, you will remember, are mass, volume, pressure and temperature.

Can you recall which of these properties are called intensive properties?

2

Pressure, Temperature

These properties are called intensive, because if we divide up the original system, the pressure and temperature, respectively, will be the same in each part. However, in reality this is often far from true.

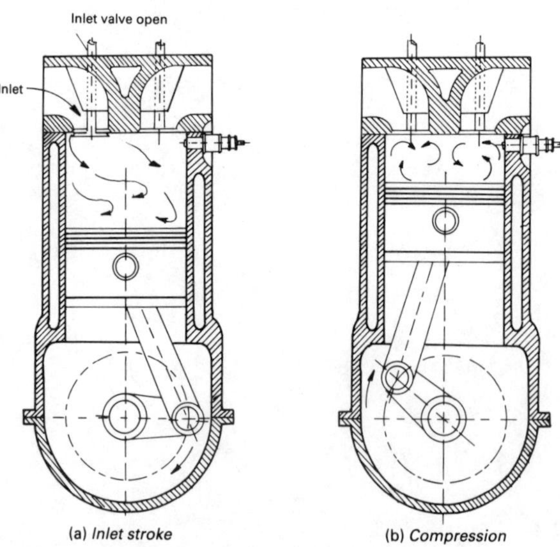

(a) *Inlet stroke* (b) *Compression*

A petrol engine works on a four stroke cycle. This means the piston moves up and down the cylinder twice, with two revolutions of the crankshaft, to make a complete cycle. In the diagram, (a) is the suction stroke. The inlet valve is open, the exhaust valve is closed, and fresh air–petrol mixture enters the cylinder. (b) is

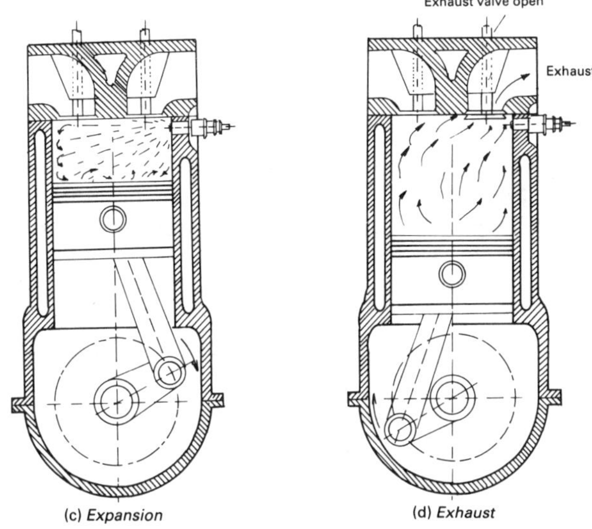

(c) *Expansion*

(d) *Exhaust*

Inlet valve closes at the end of (a) and exhaust valve opens at the end of (c)

the compression stroke with both valves closed, and ignition occurs at the end of (b) so that the power or expansion stroke (c) follows, again with both valves closed. The exhaust valve opens at the end of (c) and the cycle is completed by (d), the exhaust stroke. Finally the exhaust valve closes at the end of (d), and a new cycle begins. (Some engines work on a two stroke cycle, in which these events occur in only one revolution of the crankshaft.)

Try and picture the contents of a cylinder between the inlet valve closing and the exhaust valve opening. We have a closed system of fixed mass, (air and petrol which become exhaust gases), which for an engine running at 3000 rev/min will exist for approximately 0.02 seconds. Imagine a 'snap shot' of that contents just after the inlet valve has closed. There will be some swirl remaining in the cool air and petrol vapour that has just entered the hot cylinder, and motion in the fluid implies pressure differences. In addition there will be temperature gradients due to the hot cylinder walls.

At that instant in time, the changing pressure and temperature across the cylinder contents could be represented by average values from which we could define a **non-equilibrium state**.

3

During the 0.02 second life time of the closed system we have in mind, the air–petrol mixture will be rapidly compressed, burnt and expanded before the exhaust valve opens. The system will pass in rapid succession through a wide range of non-equilibrium states. At no time will any approach to an equilibrium state be possible.

An equilibrium state is defined as one in which the intensive properties are uniform throughout the system. So, at any instant, the pressure and temperature in the cylinder would have to be uniform throughout for the system to be in an equilibrium state.

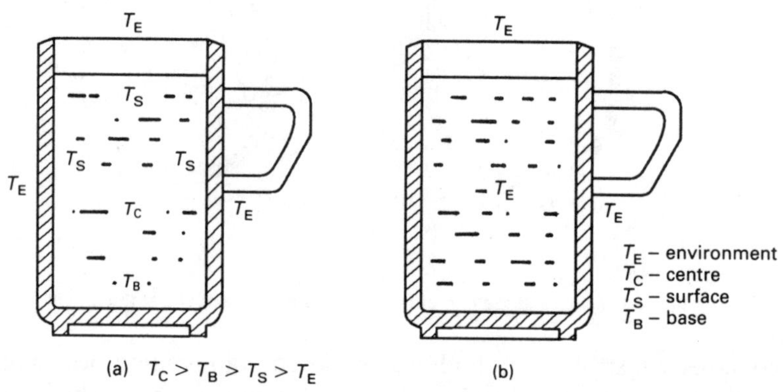

(a) $T_C > T_B > T_S > T_E$ (b)

Imagine two mugs of tea on the table. One, (a), has just been made, it is hot; the other, (b), has been there for some time and is cold. Both (a) and (b) are two closed systems, are they in equilibrium or non-equilibrium states?

4

(a) Non-equilibrium, (b) Equilibrium

As soon as the tea (a) has been made it will start to cool to the temperature of the surroundings; it will be cooler near the surface than at the centre and so it is in a non-equilibrium state. The cold mug of tea (b) has cooled to the temperature of the surroundings and has a uniform temperature; it is therefore in an equilibrium state. The difference between (a) and (b) has been caused by the passage of time, and an interaction between the closed system and its surroundings. There are two possible interactions between a system and its surroundings, given in Frame 13 of the previous programme. What are they, and which one has occurred here?

5

Work and Heat, Heat

Heat is one of the two possible types of boundary interaction, to which we shall return later. It causes the temperature of closed systems to change.

In our hot tea example in (a), not only was the intensive property of temperature not uniform throughout, but it was also somewhat higher than the temperature of the surroundings. This difference, as we shall see, is 'the driving force' for the heat interaction.

6

The temperature difference, or driving force, for heat in the fresh tea example will have a maximum of no more than 70–80°C. In the engine cylinder of Frame 2, the average temperature difference during sustained running between the cylinder contents and the cooling water in the cylinder jackets will be some hundreds of degrees. This will be the average driving force for heat transfer to the cooling water of the engine. Without this cooling the engine itself would rapidly overheat and become damaged.

The maximum temperature and pressure occur during combustion when the piston is close to its highest point in the cylinder. As the piston is pushed down the cylinder, the system boundary of the enclosed gases moves, the pressure falls and work is done.

What is the 'driving force' for this work?

7

Pressure difference

Since the pressure inside the cylinder is very much greater than atmospheric pressure outside, the pressure difference produces a considerable force acting on the piston face. As the pressure difference varies greatly with piston position, an average or mean pressure difference is often used in calculating work in engine cylinders. We shall return to this in a later frame.

So, conveniently, in engine cylinders we use an average pressure difference to calculate work, and an average temperature difference to calculate the heat to the cooling water. Thus we see that it is a difference in the intensive properties of pressure and temperature that causes the boundary interactions of work and heat. Would a pressure difference in itself cause work if the system boundary did not move?

8

No

You will already know that work is the product of force × distance, so if a pressure difference exists across a system boundary that does not move, no work is done. (We shall soon be considering work in much more detail.)

Consider the piston–cylinder arrangement in (a) below. The piston is prevented from moving, and no work is done even though the pressure in the cylinder is greater than it is outside. In (b), when the restraint is removed, the piston moves until the pressures are equal, and work is done.

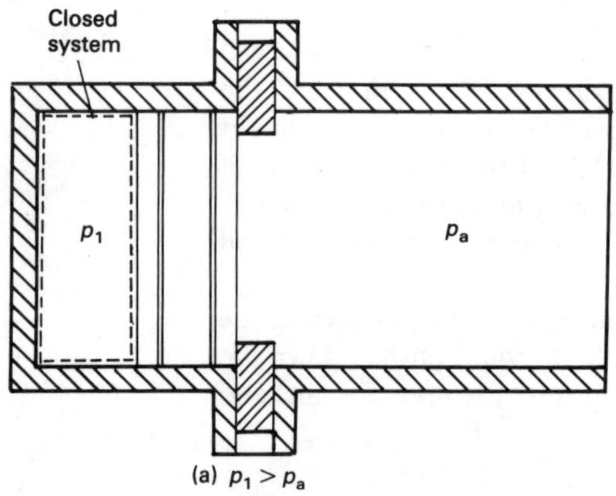

(a) $p_1 > p_a$

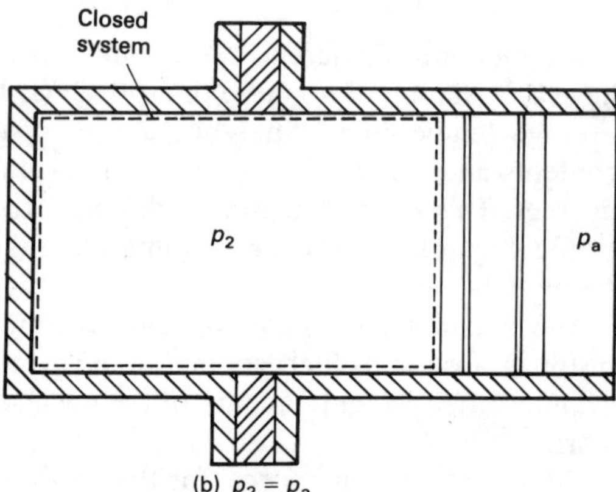

(b) $p_2 = p_a$

It would be nice to be able to impose a restraint to stop a heat interaction. Is there any way we can do that?

9

We can add insulation, but this will only **reduce the heat interaction**.

Unfortunately, heat always occurs when a temperature difference exists. We can only reduce it by insulating, we cannot stop it altogether.

To summarise so far, we have seen that it is possible for closed systems to exist in either equilibrium states, in which all intensive properties are uniform throughout, or in non-equilibrium states when this is not so. In addition, a boundary interaction of heat occurs when the 'driving force' of a temperature difference exists across the boundary, and work occurs when a pressure difference exists across a non-rigid boundary. Also, as we shall see, work can arise from a difference across the boundary in intensive properties other than pressure.

The interactions of heat and work may cause closed systems to pass between equilibrium states, or between a non-equilibrium to an equilibrium state, or between an equilibrium to a non-equilibrium state, depending on the conditions. In our study of Thermodynamics we are looking chiefly at interactions between equilibrium states.

10

A closed system, therefore, can exist in either equilibrium or non-equilibrium states and may experience work and heat at the boundary. In the example in Frame 2, conditions are changing very rapidly and the system is passing through a series of non-equilibrium states.

Imagine some air in a closed cylinder that expands against a piston that is moving very very slowly. The closed system (the air) will not be subject to any pressure waves or turbulence passing through it, so the pressure at any instant will be uniform. Any temperature change in the air may be allowed to equalise throughout, so at any instant the temperature will also be uniform. In this way, in imagination, the system will pass through a series of equilibrium states, from an initial to a final equilibrium state. This **process** may be shown on a graph of any two properties called a **property diagram**. The line joining the initial and final states is called a **state path**. Thus we may plot p against V and T against V for the process.

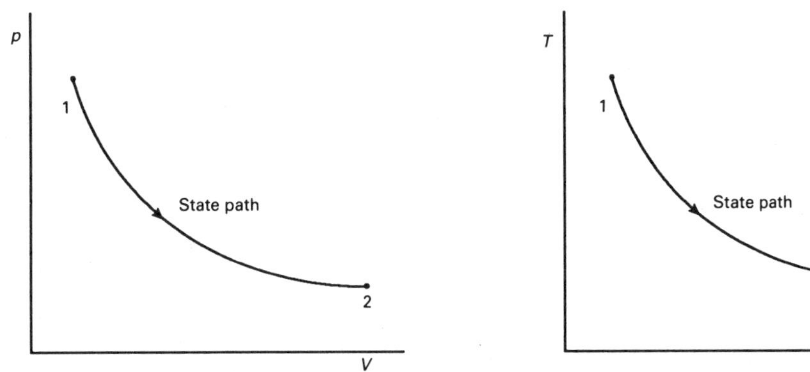

See if you can sketch the state path on a pressure–temperature graph.

11

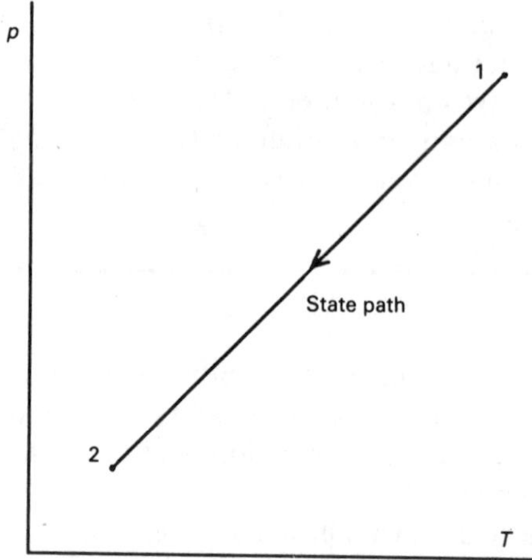

You can see that both the pressure and temperature are high when the volume is small, and both fall as the volume increases. We need not worry at this stage about the precise function between pressure and temperature.

If a closed system, in passing through a series of equilibrium states by gently applied work and heat at the boundary, could be made to retrace exactly the same state path by the exact opposite of the boundary interactions, then that closed system would have undergone a **reversible process**. In a return to the original state of the system, the state path followed must be exactly the same, and also the state of the surroundings finally must be exactly the same as it was originally. Thus we may imagine the gas in the cylinder returning to its initial state as a result of the exact opposite of the original interactions.

A reversible process is an important idea, but you must realise that no real process in nature or through man's inventiveness is ever reversible. **All real processes are irreversible**. In Thermodynamics we study imaginary reversible processes as being the ideal of actual real processes which take place.

12

You can probably appreciate that there can be varying degrees of irreversibility in a process. Thus in expanding a gas in a cylinder there could be more turbulence and fluid friction in one process, (a), than in another, (b), so that a state path using average non-equilibrium properties would be further removed for (a) than (b), from the state path of equilibrium properties for the reversible process.

Fluid turbulence and friction are natural phenomena which cannot be avoided, and as engineers we have to learn to use nature's laws to achieve our intended aims. As all real processes are irreversible, we try to engineer the expansion of gases or

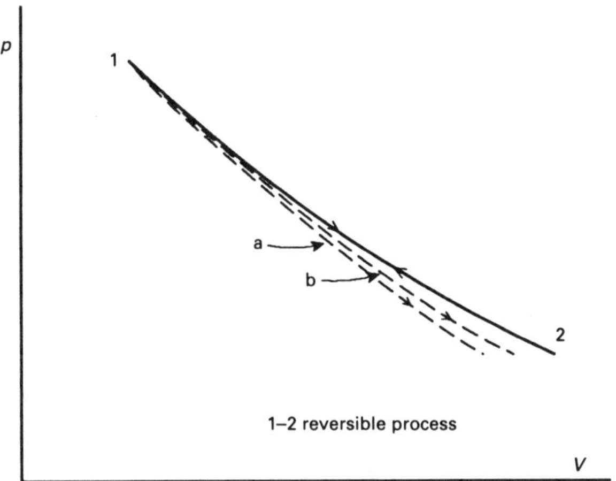

1–2 reversible process

vapours in cylinders, for example, to be as near as possible to being reversible.

Many familiar events, such as spilling coffee, breaking eggs and tearing your clothes are clearly irreversible, and other familiar, and clearly irreversible processes, are put to good use by engineers to generate work from heat. Can you think of any? (Think about the steam engine discussed in Programme 1.)

13

Combustion of fuel, boiling of water, expansion of steam

In Frame 11 we saw that it was easy to imagine a reversible expansion in a cylinder, and that perhaps one could get quite close to achieving it. In contrast, one cannot imagine all the processes involved in combustion going into reverse to recreate the original fuel. But we can grow new fuel by planting trees, which can use carbon dioxide released into the atmosphere by previous combustion.

During our discussion of work and heat, which now follows, you will learn how in an idealised way it is possible to achieve a reversible process in a compression or expansion, for example, and how we represent real irreversible processes on state point diagrams.

14

In Frame 11 of Programme 1 we found that when work is produced from heat, the efficiency of the conversion is relatively low, usually some 25–45 per cent, when carried out on a large scale. In contrast, the conversion of work into heat may be complete; thus the entire work output of an engine could be dissipated by friction to produce heat.

Since we can get a complete conversion of work into heat, but only a partial conversion of heat into work, and since much engineering practice is concerned with obtaining work from heat by burning fuels, we need to be very clear about what we mean by work. In the next frame we shall consider the mechanical definition of work. You ought to be able to define mechanical work. Do so.

15

Force × Distance

Mechanical work is defined as the product of force and the distance moved by the point of application of the force. Thus:

$$W = F\,x \text{ [N m]}$$

where F is the force in newtons and x is the distance moved in metres in the direction of the force. The product of units [N m] is the joule [J]. In raising a body of mass m, the force required is the weight of the body mg, and the height raised is h, so

$$W = mg\,h \text{ [J]}$$

This equation is used in this simple problem:

A man having a mass of 75 kg is raised through a height of 30 metres by a lift. Calculate the work done in kJ ($g = 9.81 \text{ m/s}^2$)

16

22.073 kJ

The working is: $W = 75.0 \times 9.81 \times 30.0$ [kg] [m/s^2] [m]
Notice how we have brought all the units together at the end of this equation.

So: $W = 22\ 072.5\ [J]$

$= 22.073\ [kJ]$

The man in this example is the system, and the immediate surroundings, the lift, has performed work on the man. To distinguish this from the lift lowering the man, a sign convention is required.

If a system does work on the surroundings, the work is positive for the system, and negative for the surroundings.

If a system has work done on it by the surroundings, the work is negative for the system and positive for the surroundings.

This may be shown pictorially:

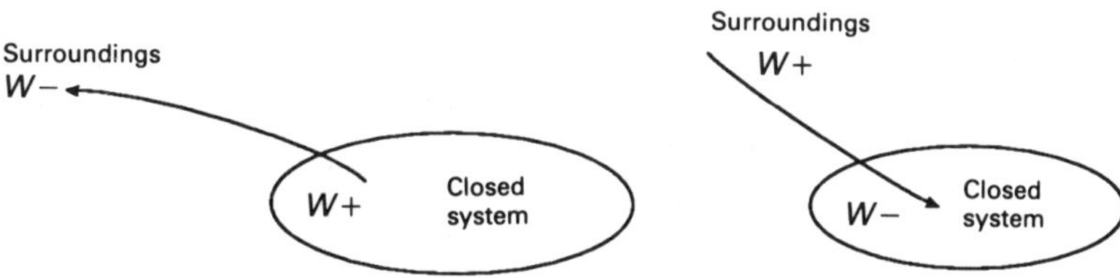

The man is the system, so what is the sign of the work interaction you have just calculated?

17

Negative

If the lift were the system, it would have performed +22.073 kJ on the man. If the man (the original system) were now lowered by the lift, the work interaction is +22.073 kJ for the man, and the lift has −22.073 kJ of work performed on it.

18

We have seen in Frames 16 and 17 that when a man is raised through a height by a lift there is negative work done on the man by the lift. Suppose he now runs up the stairs instead of using the lift. With the man as the chosen system, what are the surroundings, and is any work done on those surroundings? Remember that work is an interaction between a system and its surroundings.

19

There are no surroundings on which work is done. Work interaction is zero.

With the man as the chosen system there have been no forces acting through any distances as far as any surroundings are concerned. The man has not performed work on any external agent; he has raised himself through a height and in doing so has increased his own potential energy at the expense of other forms of energy stored in his body. Therefore, no work is done. This follows from the way we have defined work. On coming down the stairs again he would again use body energy in controlling his descent, but again there would be no work.

Suppose an object falls freely to the ground through the atmosphere. There is no drag force between the object and the atmosphere. Is work done?

20

No

Again, since there has been no interaction between the system (the falling object) and the surroundings (the atmosphere), no work is done. In reality, there would be a drag force on any object falling through the atmosphere, and work would be done in overcoming that force.

An object having a mass of 5 kg falls 10 m in a gravitational field of 9.81 m/s^2, against an average drag force of 8 N in the surrounding atmosphere. Calculate the work done by the object on the atmosphere. (The drag force and the distance fallen are the important pieces of information).

21

+80 J

The work done by the object is simply the product of drag force × distance fallen, that is, 8 N × 10 m = 80 J. The mass of the object and the value of g are irrelevant to the problem.

We now move on to think about the **rate** at which work is done.

The rate at which work is performed is called **power**. Thus:

$$\begin{aligned} \text{Power, } P &= 1\ [\text{N}] \times 1\ [\text{m}]/[\text{s}] \\ &= 1\ [\text{J/s}] \\ &= 1 \text{ watt } [\text{W}] \end{aligned}$$

If the lift in Frame 15 had raised the man 30 metres in 20 seconds, what would be the rate of work performed on the man? (Think of the total work done and the time taken.)

22

−1.104 kW

The amount of work is the same as before, 75.0 × 9.81 × 30 [J], and the sign is negative since work is performed on the man, so the rate of work output, or power is:

$$\begin{aligned} P &= -75.0 \times 9.81 \times 30.0/20.0\ [\text{J/s}] \\ &= -1103.63\ [\text{W}] \\ &= -1.104\ [\text{kW}] \end{aligned}$$

23

How fast can a man work? If a man lifts a mass of 4 kg through a height of 1.5 metres in 0.5 seconds, his power output is:

$$P = \frac{4[\text{kg}] \times 9.81[\text{m/s}] \times 1.5[\text{m}]}{0.5\ [\text{s}]}$$
$$= +117.72\ [\text{W}]$$

This is a fairly high rate of working for a man, particularly if continued for any length of time.

The total work done in a given time is the product of power and time:

$$W = P \times \text{time}$$

Thus work may be measured in units of watt hours [Wh], or kilowatt hours [kWh]. These are useful large units of work, and:

$$1\ [\text{Wh}] = 3600\ [\text{J}]$$
$$1\ [\text{kWh}] = 3600\ [\text{kJ}]$$

If the man in the previous example continuously lifted objects of mass 4 kg through 1.5 m and worked at a steady 25 W for 4 hours, how many objects did he lift? (Consider the total work done in 4 hours and the work to raise one object.)

24

6116

The total work is $25 \times 4 = 100$ [Wh] = 360 000 [J]
The work to raise one object is $4 \times 9.81 \times 1.5 = 58.86$ [J]

The number of objects is 360 000/58.86 = 6116. Working at 25 W for 4 hours does not sound very much, but lifting those 6116 objects through 1.5 m in that time would be pretty hard work.

So far we have only considered work which would be involved if a system were being moved about in a mechanical way, as in the lift raising the man. Work may occur between a system and its surroundings in a number of different ways, and it is not necessary for the system as a whole to move.

We shall think about this after the next frame. Do all the following problems before moving on.

25

PROBLEMS INVOLVING MECHANICAL WORK AND POWER

1. A crane raises an object of mass 200 kg to a height of 40 m and then deposits it on the roof of a building of height 30 m. Calculate the net work done by the crane. [+ 58.86 kJ]
2. A man pushes a lawn mower a total distance of 500 m in 4 min 20 secs. The mower exerts a resistive force of 100 N. Calculate the power developed by the man over that time. [+ 192.3 W]
3. (a) A bird flies to the top of a tree of height 20 m. The bird has a mass of 10 g and there is no resistive drag with the atmosphere. (b) The bird then glides back to the ground through a flight path of 80 m against a resistive drag of 0.5 N. In each case calculate the work done by the bird on the atmosphere. [(a) 0, (b) + 40 J]
4. A car travels at 60 km/hour on the level and it overcomes total resistive forces of 2100 N; calculate the power output of the car's engine. [35.0 kW]
5. A man walks downstairs through a height of 8 m carrying a case of mass 15 kg. What is the work done on the man by the case? [− 1177.2 J]

26

When work takes place between a system and its surroundings, the following may occur and more than one of these may be involved at any one time:
(1) Displacement work, in which parts of the boundary move.
(2) Shaft work, or the work of a rotating shaft, passing through the boundary.
(3) Fluid shear work, at the boundary.
(4) Solid friction, at the boundary.
(5) Electrical work, passing through the boundary.

This is not a complete list as there are other types of work, such as magnetic work, and work due to capillary attraction, which we shall not consider.

However, all will have to be covered by the thermodynamic definition of work, which is:

Thermodynamic work is that interaction at a system boundary which could be shown to be equivalent to an equal amount of mechanical work.

This definition rules out an interaction of which only part could be shown to be equivalent to mechanical work; this must be heat.

We will start by thinking about displacement work.

27

Displacement work occurs when the boundary of the closed system moves. Frequently only part of the boundary moves, when for example a gas expands against a piston in a cylinder. On the other hand, two or more parts of the boundary could move simultaneously (for example, when air is sucked into an engine cylinder from the atmosphere), or the whole boundary could move as in the expansion of a balloon.

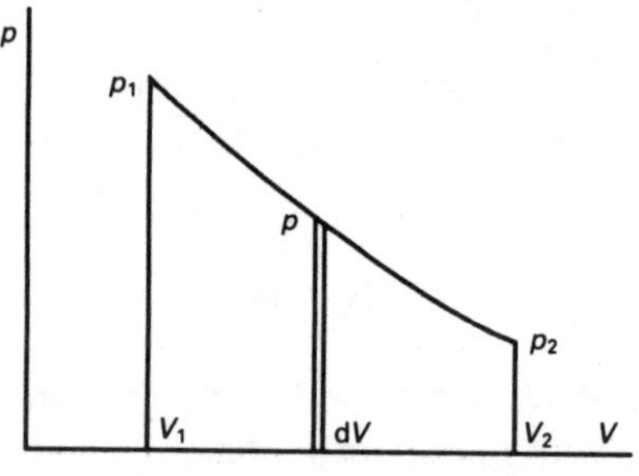

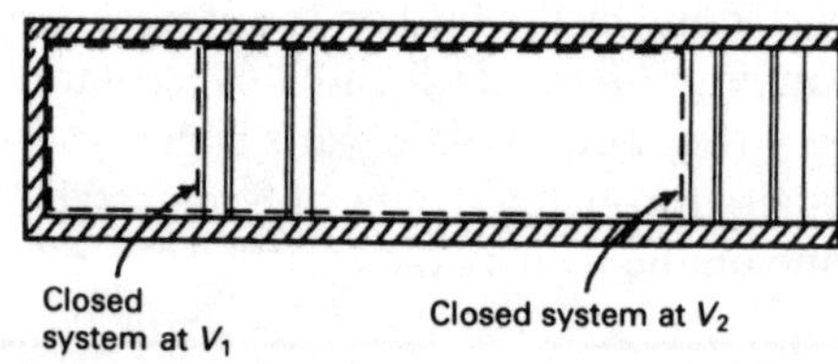

We shall assume that we have, from Frame 10, a pressure–volume plot of a reversible expansion between two equilibrium states 1 and 2. The piston positions at these states are shown in the diagram.

The displacement work is the product of force and distance moved at the point of application of the force. In this case only the piston moves, so:

$$\text{Force} = \text{Pressure} \times \text{Piston area}$$

The pressure varies with piston position, so for a very small piston movement:

$$\mathrm{d}W = p\,A\,\mathrm{d}x$$

Hence:
$$\mathrm{d}W = p\,\mathrm{d}V$$

since $\mathrm{d}V$, the volume change, is the piston area × movement of the piston, $A\,\mathrm{d}x$.

Hence:
$$W = \int_1^2 p\,\mathrm{d}V$$

This integral is the area under the curve of p against V.

A gas expands pushing a piston outwards. What is the sign of the displacement work for the gas?

28

Positive

The gas (the system) performs work on the piston, so it is positive work.

To calculate the displacement work, we need to know the relation between p and V. If this relation can be expressed as a mathematical function, then the integral $\int p\mathrm{d}V$ can be determined using calculus. There could be many different relationships each giving a different answer. In mathematical terms we say that $\mathrm{d}W$ is **inexact**, as it depends on the relation between p and V.

Since displacement work depends on the precise path between equilibrium states, we say that displacement work is a **path function**.

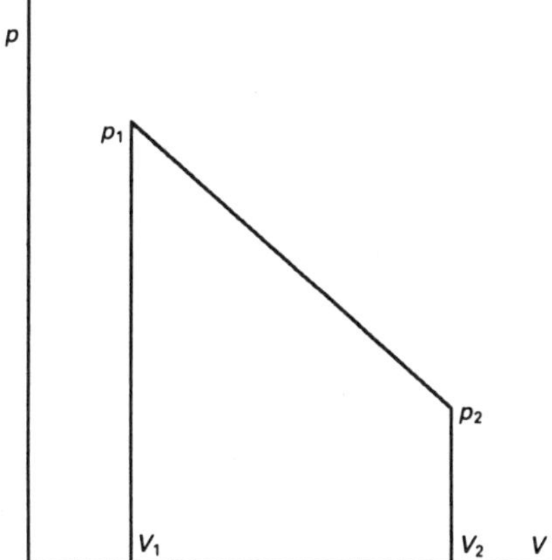

The simplest possible path between states 1 and 2 is a straight line. Write down an expression for the displacement work. (You won't need calculus for this.)

29

$$W = \frac{(P_1 + P_2)}{2} (V_2 - V_1)$$

The area under the straight line p–V relationship is the average pressure times the volume difference.

30

Now that we know how to calculate the displacement work of a reversible process, we shall have to consider all the likely pressure–volume relationships for reversible processes that may occur. Also we need to think about how to calculate the displacement work of real irreversible processes, and how close real processes may be to being reversible.

First we shall think about the differences between reversible and irreversible displacement work, and then we shall look at the sort of pressure–volume relationships that occur in real engines, and how close these are to being reversible. Then we shall consider the integral $\int p\mathrm{d}V$ for these relationships, taking them to be reversible.

In the next frame we shall try to construct in imagination a reversible expansion.

31

Suppose an expansion takes place between a state 1 to a state 2 in a vertical cylinder with a weighted piston. The piston, which is frictionless, has on it a large number of very small weights. We shall use thin sheets of steel, or laminates, one on top of another rather like a deck of playing cards.

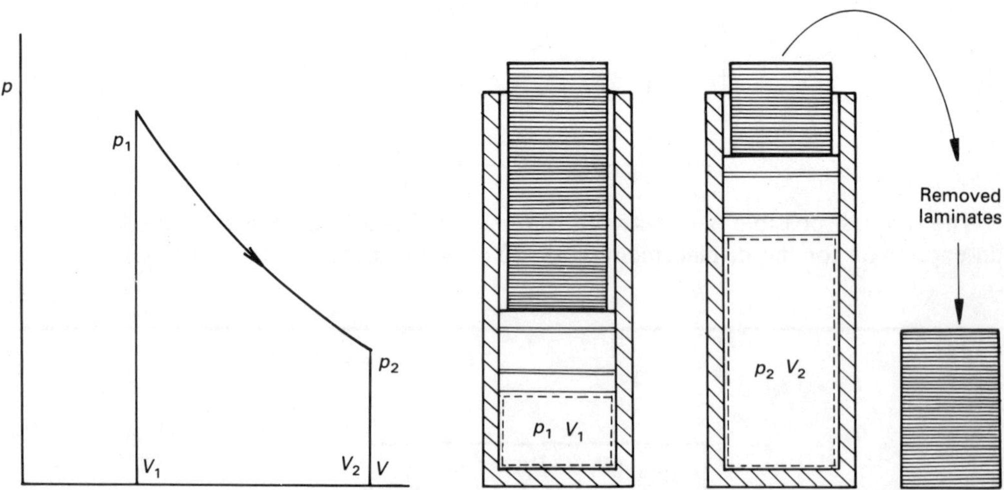

From state 1 the laminates are removed one by one so that the pressure of the gas (the system) slowly decreases by passing through a series of equilibrium states. The process is very slow so that the pressure of the gas is uniform at any intermediate point, and is exactly sufficient to support the weight of the piston. No turbulence or pressure disturbance is created in the gas, also any temperature change created by the expansion is allowed to equalise throughout the gas. At any point between state 1 and state 2 the process could be reversed by replacing the laminates one by one. The expansion from state 1 to state 2 and the corresponding compression from state 2 to state 1 would be examples of reversible displacement work, but clearly they are idealised processes that do not occur in practice. Reversible displacement work is said to be **fully resisted**.

Next we shall see what happens when the laminates are removed and added in finite steps of equal numbers.

32

Suppose now that the weights are removed in four equal bundles, instead of one at a time. After the first bundle is removed let the pressure exerted by the weight of the piston plus the remaining three bundles be p_A. The gas expands freely against this constant pressure and a final equilibrium position is obtained at a new volume V_A. Before this, however, the piston will oscillate about this new position as it rose quite rapidly from its original position at state 1. Point A becomes a new equilibrium position on the p–V diagram. The work done during this expansion is the area of the rectangle $p_A(V_A - V_1)$

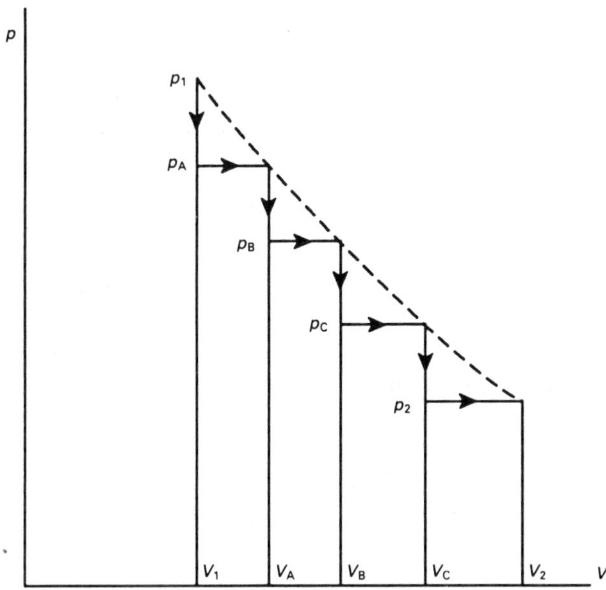

33

As the remaining bundles of laminates are removed, similar sequences of events occur leading to points B, C and 2 on the p–V diagram. In this way an irreversible expansion occurs from state 1 to state 2. The pressure at this state 2 will be the same as in a reversible expansion since p_2 depends only on the remaining weight of the piston. Without knowing anything about any heat interaction we cannot say whether V_2 will be more or less than the previous V_2 for the reversible expansion. The irreversible expansion may be represented by a broken line through the points 1–A–B–C–2.

The work for the other three expansions will be $p_B(V_B - V_A)$, $p_C(V_C - V_B)$, and $p_2(V_2 - V_C)$, so the total work is the sum of the four rectangular areas.

Is the work of an irreversible expansion equal to or less than $\int p dV$ for the reversible p–V curve passing through the same points?

34

W is less than $\int p dV$

The work of expansion in this example is the sum of the four rectangular areas and this is clearly less than the area under the curve.

You can see by studying the figure in the previous frame that if the weights were to be removed in smaller and smaller bundles, the magnitude of the displacement work would approach closer to that of the reversible work. This shows how there can be different degrees of irreversibility.

Can you sketch out the p–V diagram for an irreversible compression by adding the laminates in four equal bundles? (Refer back to Frame 32.)

35

As a bundle of laminates is added to the piston, the pressure due to the piston is suddenly increased, and compression takes place over a volume change with the pressure constant at this new value. In a reverse sequence, point 1 is the initial equilibrium state, points A, B and C are intermediate states, point 2 is the final equilibrium state and p_2 is equal to p_1 in the expansion considered in Frame 32. Again, a broken line may be drawn through points 1–A–B–C–2 to represent a p–V curve and the irreversible work of compression evaluated as:

$$p_A(V_A - V_1) + p_B(V_B - V_A) + p_C(V_C - V_B) + p_2(V_2 - V_C)$$

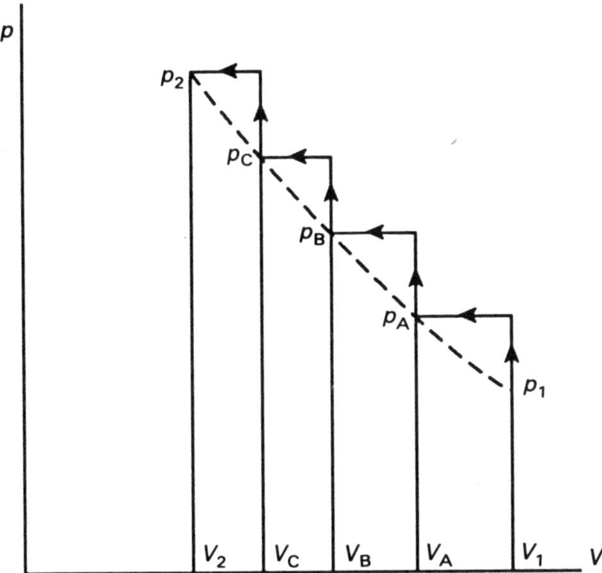

This will produce a negative result, as it should for a compression. Also it will be greater than $\int p\mathrm{d}V$ for the curve through points 1 to 2. We have these two important results:
The irreversible work of expansion **is less than** $\int p\mathrm{d}V$ for the p–V curve.
The irreversible work of compression **is greater than** $\int p\mathrm{d}V$ for the p–V curve.

36

If we now go back to the expansion again, what happens if all the weights are removed at once? The pressure in the cylinder would fall from p_1 to p_2 in one step, and the expansion would be at constant pressure at this lower value. There would be no meaningful p–V relationship since there are no intermediate points between the initial and final states. Sketch the p–V diagram of this sudden expansion, and can you write down an expression for the displacement work in this case?

37

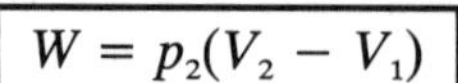

$$W = p_2(V_2 - V_1)$$

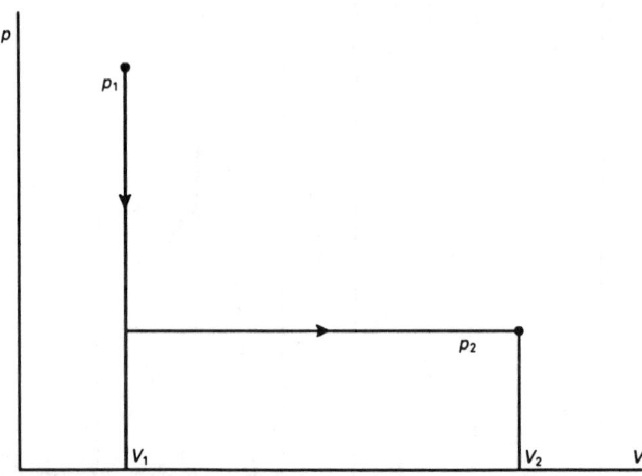

The work is the pressure due only to the piston weight times the total volume change.

You will recall that in the previous programme we showed how a vacuum could form part of a system. Now let's see what happens when a gas expands into a vacuum. The figure below shows a gas of volume V_1 at pressure p_1 restrained by a thin diaphragm beyond which is a vacuum. The diaphragm ruptures to allow the gas to occupy the whole volume V_2 when the pressure of the gas will be p_2.

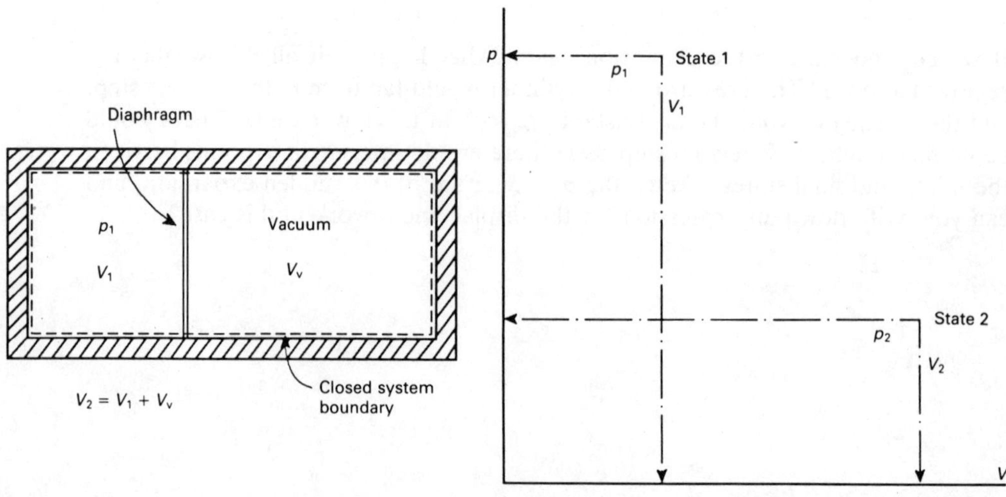

It will take some time for the gas to reach equilibrium at this new state 2, since as the diaphragm ruptures the gas will rush to fill the vacuum and the turbulence created will have to die away. There will be no intermediate equilibrium states at any volume between V_1 and V_2. The two end states are shown on the p–V diagram.

As there are no intermediate states, no meaningful line can be drawn between points 1 and 2. From what we learned in the previous frame we might be tempted to think that we could draw a horizontal line through p_2 and evaluate the work as $p_2(V_2 - V_1)$. Remembering that as the gas expands from state 1 to state 2 it is expanding against a vacuum which has no pressure, do you think this evaluation would be correct?

38

No

The system boundary which contained the vacuum did not move, so there was no work involved. We could redraw the system boundary to coincide with the diaphragm, and then the boundary moves to fill the vacuum. Since the movement of the boundary is not against any resistance (since there is no pressure), again no work is involved. This is the exact opposite of the fully resisted expansion discussed in Frame 31.

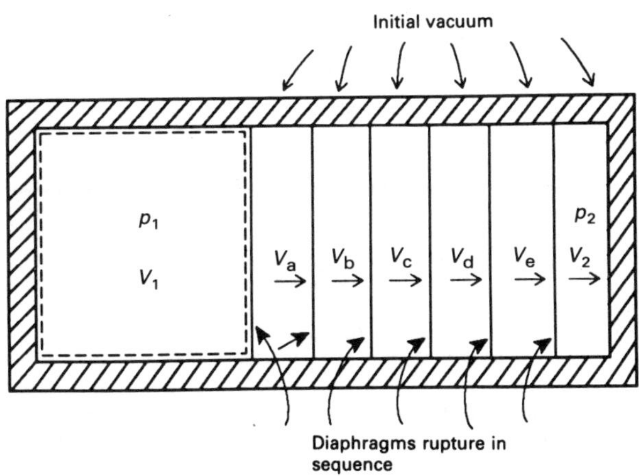

Next we shall think about what happens if there is a series of diaphragms within the vacuum, as shown, which are punctured one after another.

39

The diaphragms are punctured one after another, starting from the left, but each time the gas is allowed to reach an equilibrium state in its new increased volume. Consequently a series of equilibrium state points can be put on a pressure–volume diagram, and a broken line drawn between them, and it would be very similar in appearence to the expansion in Frame 32.

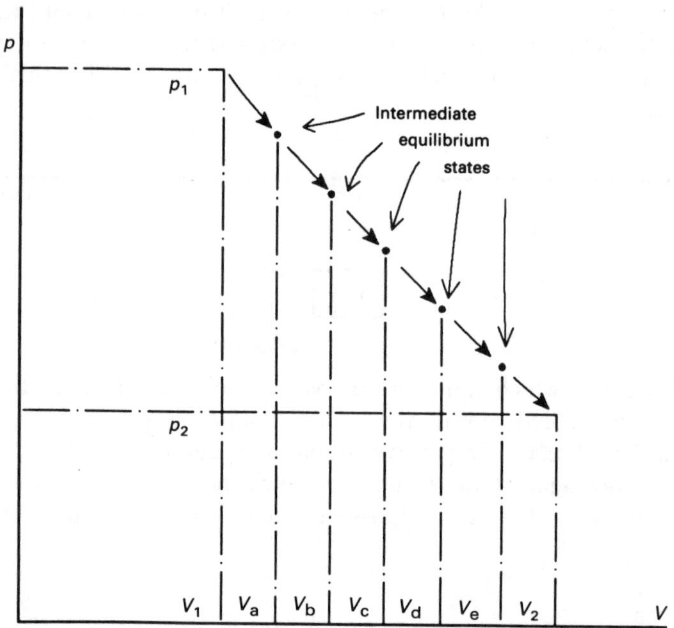

It would be possible to evaluate an integral $\int p\mathrm{d}V$ for the area under this curve. Would this integral be a measure of work done by the gas in expanding to fill the vacuum?

40

No

Again, since each expansion of the gas into a vacuum is unresisted, no work is done. If the closed system boundary is placed round the gas, the boundary moves against no resistive force, so no work is done. If the boundary is placed round the gas and vacuum, it does not move, so no displacement work is done.

Before moving on to evaluate the integral $\int p\mathrm{d}V$ for reversible processes, let us go on to think about real compressions and expansions in engine cylinders and how close they are to being reversible. We shall then have a better idea of the sort of pressure–volume functions that are important.

41

In a reciprocating engine, the average velocity of the piston, as it travels from one end of the cylinder to the other, is surprisingly low. Thus for a piston stroke of 120 mm for a speed of 3000 rev/min, the average velocity is 12 m/s. Any pressure wave in the gaseous charge would travel at the velocity of sound, some hundreds of metres per second. Hence the piston velocity itself is unlikely to produce very significant pressure changes across the closed system. However, temperature variations across the closed system will be fairly large because of cooling at the cylinder walls. This in turn would lead to additional pressure variations.

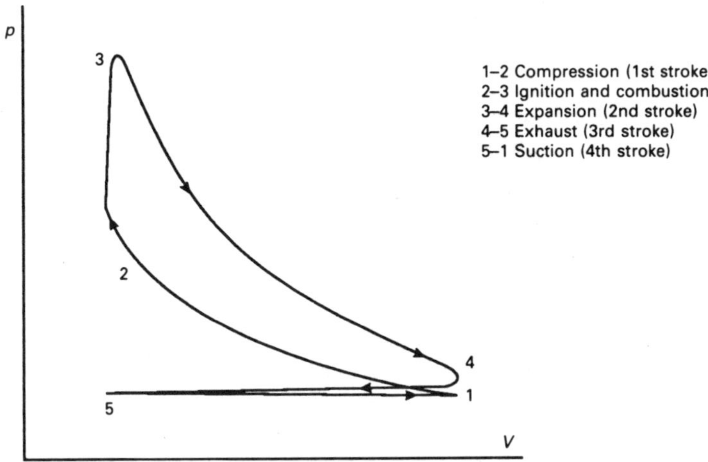

The pressure–volume plots of the four strokes of a petrol engine cycle are shown in the figure. The four stroke cycle was discussed briefly in Frame 2. The compression stroke 1–2 is not too far removed from being a reversible process, but the expansion stroke 3–4, with much higher temperatures and turbulence arising from the combustion, will have greater irreversibilities. The exhaust stroke 4–5 and suction stroke 5–1 are not processes of a closed system, since in each case the mass within the cylinder is changing as the piston moves.

Engine cylinder pressure volume diagrams of this type are called **indicator diagrams** since historically they were obtained by a device called an **engine indicator**. Such indicators were developed for the engines of the day, which were of very low speed by today's standards.

42

In Frame 7 it was said that the average or mean pressure difference in an engine cylinder is used to calculate the work done on the piston face. This mean pressure difference was originally measured using an engine indicator, and is called the **indicated mean effective pressure**. From this, the **indicated power** of the engine is found.

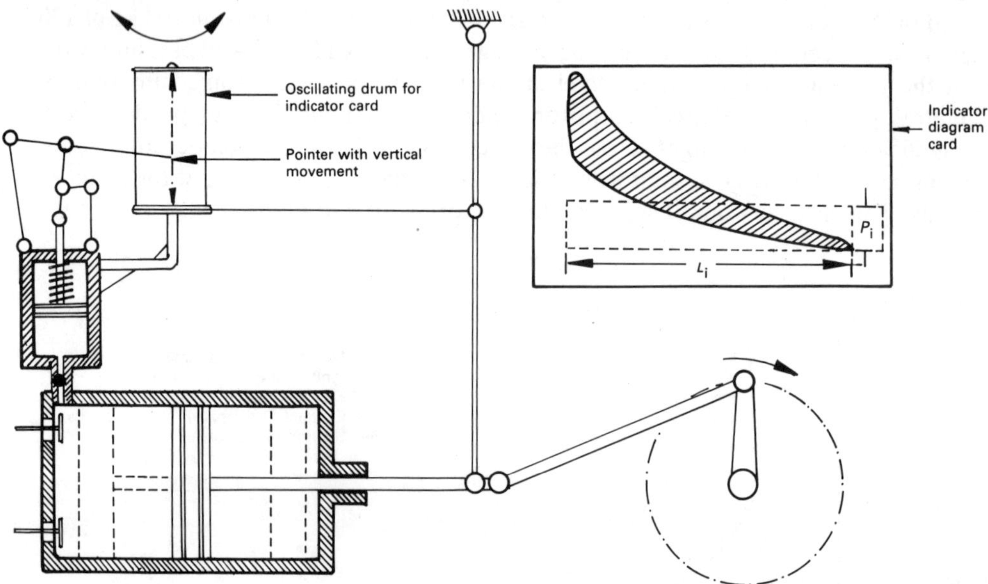

The figure above shows an indicator diagram, and in diagrammatic form, the indicator from which it is obtained. The actual pressure–volume changes in the cylinder are traced as a p–V diagram on a card using a special pointer. The card is wrapped round a drum that rotates 'to and fro', linked to the piston position. Thus the horizontal axis on the card represents cylinder volume. The pointer has only vertical movement which is proportional to the cylinder gauge pressure, set by a spring of suitable stiffness. The two movements combined produce the p–V diagram.

Let A be the measured area of the enclosed p–V diagram, in mm^2, and let L_i be the base line length of the diagram in mm. Then P_i is the mean height of the diagram, given by:

$$P_i = A/L_i$$

In this way the usual shape of a p–V diagram is converted to a horizontal rectangle, the height of which is proportional to the indicated mean effective pressure.

43

Let S_P be the cylinder pressure spring rate in kPa/mm, which means that every mm of vertical height on the diagram represents S_P [kPa] pressure. Then:

$$\text{IMEP} = S_P\ [\text{kPa/mm}] \times P_i\ [\text{mm}]$$
$$= S_P P_i\ [\text{kPa}]$$

This is the mean height of the diagram converted to a pressure, which will act on the piston for the whole of the expansion stroke.

Suppose a diagram is obtained, A is measured as 3040 mm^2, and L_i is found to be 120 mm. The spring rate is 15 kPa/mm, so we can calculate the IMEP

We have $P_i = A/L_i = 3040/120 = 25.33$ [mm]
Then, IMEP $= S_P P_i = 15 \times 25.33$
$= 380$ [kPa]

What is the IMEP obtained from a diagram of area 28 cm^2 with a base line of 9.5 cm, for a spring stiffness of 18 kPa/mm?

44

530.5 kPa

You will have noticed the mixture of units. The mean height of the diagram is 28/9.5 = 2.947 cm = 29.47 mm.

So: IMEP $= 29.47$ [mm] $\times$ 18 [kPa/mm]
$= 530.5$ [kPa]

Having calculated the IMEP we can now find the indicated work done on the piston face in one four-stroke cycle. Let the length of piston movement, or piston stroke, be L_P, and the piston area be A_P, then the mean force on the piston over the whole of its stroke is $A_P \times$ IMEP [kN]. Remember that [kPa] = [kN/m^2], so the piston area must be in [m^2].

Hence: $W = A_P \times \text{IMEP} \times L_P$ [kJ]. Next we can find the indicated power.

45

We have just found the net work done on the piston in every four stroke cycle, which takes two revolutions of the crankshaft.

Engine speed is usually measured in rev/min, N, so the number of crankshaft revolutions per second is $N/60$. Then the number of four stroke cycles per second is $N/(60 \times 2)$. (The number of two stroke cycles in a second would be $N/60$, so it is important to know what sort of engine you have.)

The power produced is then:

$$p = \frac{A_P \times \text{IMEP} \times L_P \times N}{60 \times 2} \text{ [kW]}$$

This is called **indicated power** because it is power calculated from the engine indicator. It represents power developed at the piston face, and this is greater than the power available from the engine output shaft which is called **brake power**. Power is lost in engine friction and in driving auxiliaries such as the cooling water pump. Since ideally we would like to use all the indicated power, the ratio of brake power to indicated power is called mechanical efficiency:

$$\text{Mechanical efficiency} = \frac{\text{Brake power}}{\text{Indicated power}}$$

For an IMEP of 380 kPa, we may calculate the power for a piston stroke of 0.3 m and a piston diameter of 140 mm, for a single-cylinder four stroke engine running at 500 rev/min.

The piston area is $\pi \times 0.14^2/4 = 0.0154 \text{ m}^2$

So:
$$p = \frac{0.0154 \times 380 \times 0.3 \times 500}{60 \times 2}$$
$$= 7.315 \text{ kW}$$

Calculate the indicated power output of a six-cylinder four stroke engine having an IMEP of 740 kPa, assumed equal in all cylinders. The piston diameter is 160 mm and the stroke is 240 mm. The engine speed is 600 rev/min. Also find the brake power, assuming a mechanical efficiency of 80 per cent.

46

107.13 kW, 85.70 kW

We calculate the indicated power in one cylinder and multiply by 6. All dimensions must be in metres, so the piston area is $\pi \times 0.16^2/4 = 0.0201 \text{ m}^2$.

The power in one cylinder is:

$$p = \frac{0.0201 \times 740 \times 0.24 \times 600}{60 \times 2}$$

$$= 17.854 \text{ [kW] per cylinder.}$$

Hence the total output is 17.854 [kW] × 6
= 107.13 [kW]
The brake power is then 0.8 × 107.13 = 85.70 [kW]

Engine indicators such as those shown in Frame 42 became obsolete as engines became smaller and faster.

47

In modern high speed engines it is usual to make use of separate sensors to record cylinder pressure and crank angle from which piston position may be determined later. These two signals are combined to give a diagram of pressure against crank angle on a computer screen. This will show clearly the pressure rise during combustion, and useful data, such as the rate of pressure rise, can be obtained. The diagram can be converted subsequently into the usual p–V form, from which relationships between pressure and volume could be found.

When the piston is at the closed end of the cylinder it is at top dead centre, and at the other extreme it is at outer dead centre.

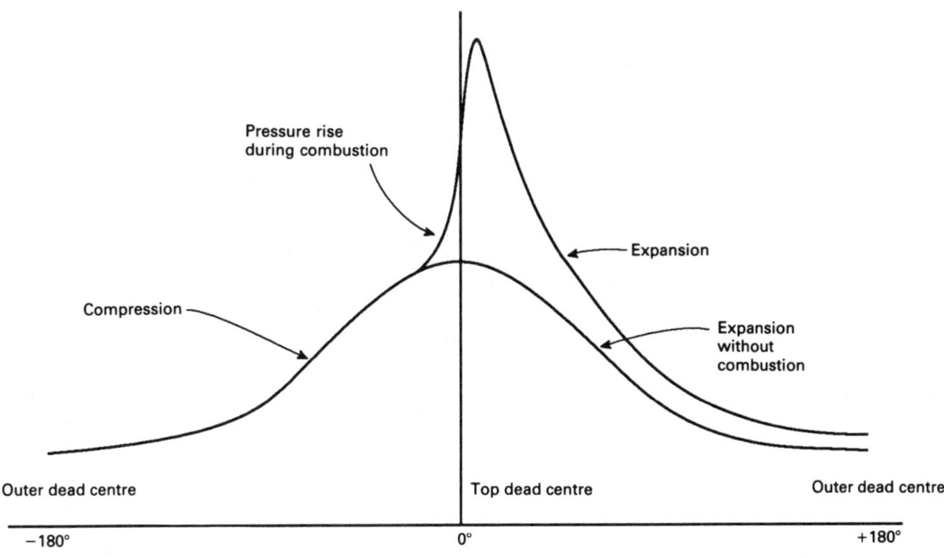

48

You have just seen how the pressure in an engine cylinder varies with volume during the different piston strokes of the cycle. The real curves shown could be approximated to constant volume during combustion, and to two different functions of pressure with volume for compression and expansion. The exhaust and suction strokes are more or less at constant pressure, but remember that in these strokes there is not a closed system within the cylinder. A constant volume process is shown in the figure.

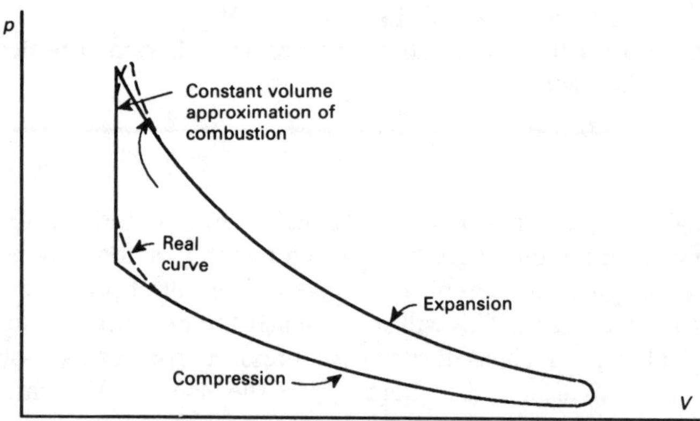

Is any work done during a constant volume process? Refer back to Frame 27 if you are in doubt.

49

No

In a constant volume process, there is just a pressure change in the fluid while the volume remains the same, hence $dV = 0$, and so:

$$W = \int p dV = 0$$

Thus we see that in the case of a **constant volume process**, the displacement work is zero. A constant volume process may seem unimportant if no work is done, but you will find later that we often use it in showing how engine cycles can be described in an ideal way.

50

The simplest straight line relationship on a p–V diagram is a volume change at constant pressure. Then:

$$W = \int_1^2 p\mathrm{d}V = p(V_2 - V_1)$$

If V_2 is the final volume and V_1 the initial volume, the result will have the correct sign for work, positive for an expansion (a) and negative for a compression (b). With volume in m^3 and pressure in Pa or N/m^2, the units of work will be J.

As we saw in Programme 1, pressures can be measured in bars. What is the work of an expansion when steam at 5 bar (absolute) expands from 0.1 m^3 to 0.15 m^3 at constant pressure? Give your answer in kJ.

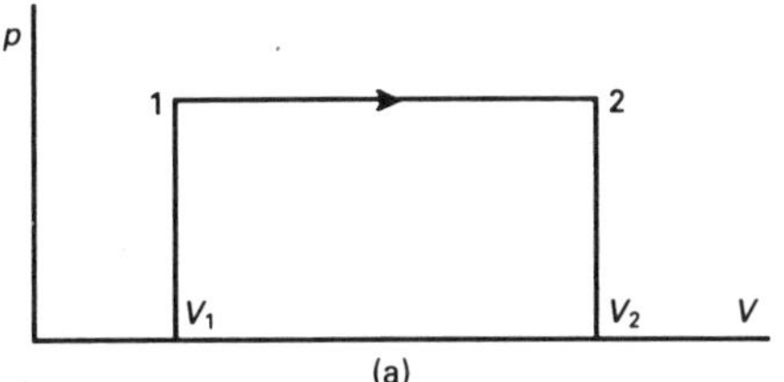

(a)

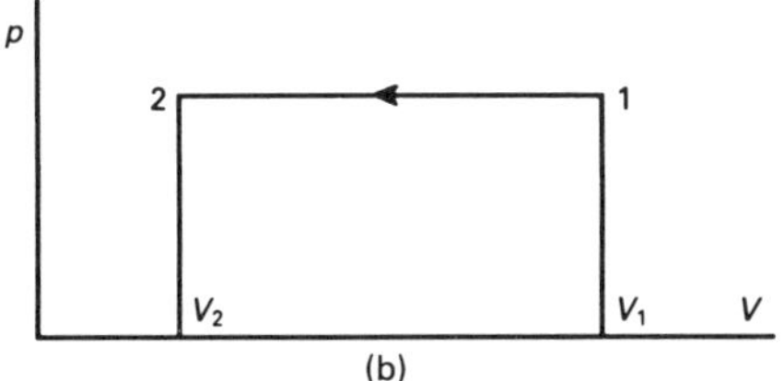

(b)

51

+25.0 kJ

The boundary of the closed system increases in volume by (0.15 − 0.10) m^3, at the constant pressure of 5 bar. This pressure is equal to 5×10^5 Pa. So the work in J will be:

$$W = 5 \times 10^5 \times (0.15 - 0.10)\ [\mathrm{J}]$$

$$= 25\,000\ [\mathrm{J}]$$

$$= 25.0\ [\mathrm{kJ}]$$

This is positive work in an expansion.

52

In the problem in Frame 50 the pressure was given as 5 bar absolute. If the atmospheric pressure was 1 bar, the gauge pressure would be 4 bar which would be the pressure difference acting across a piston containing the steam.

During the expansion, the piston would displace the atmosphere acting on its outer surface, so that the piston as the system would have work done on it by the steam, and in addition would do work on the atmosphere, since the atmosphere is displaced at the constant pressure of 1 bar.

Taking the piston as the system, what is the work done on the atmosphere, and what is the net work of the piston? (The previous answer was the work done by the steam on the piston, so you now need to find the work done by the piston on the atmosphere.)

53

−5.0 kJ, −20.0 kJ

The pressure–volume diagram for the work of the piston is shown below. The work done by the piston on the atmosphere is:

$$W = \frac{1\ [\text{bar}] \times 10^5\ [\text{Pa/bar}] \times (0.15 - 0.10)\ [\text{m}^3]}{1000\ [\text{J/kJ}]}$$

$$= +5.0\ [\text{kJ}]$$

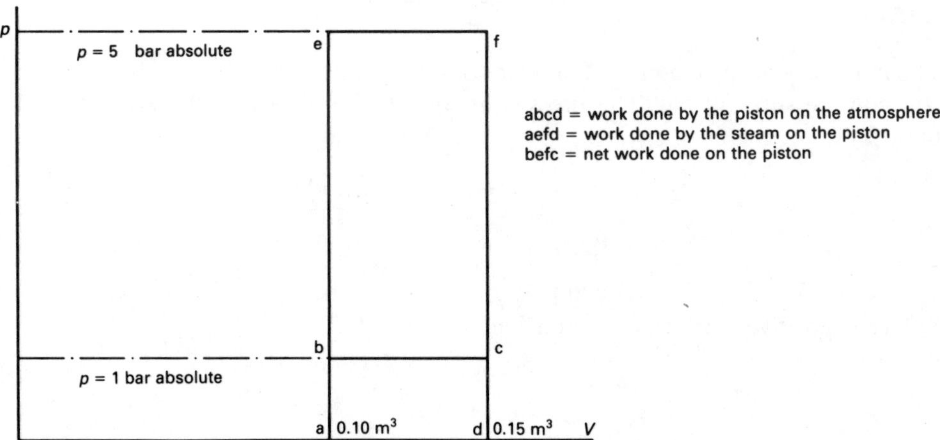

Hence the work done on the atmosphere is -5.0 kJ. The piston has -25.0 kJ of work done on it by the steam, so the net work of the piston is $-25.0 + 5.0$ kJ, or -20.0 kJ. We assume the piston does $+20.0$ kJ of useful work on the crankshaft of the engine, so in the absence of friction the net work of the piston is finally zero.

54

We often find it useful to calculate work per unit mass of the closed system. Since volume = mass × specific volume, we can say:

$$W = p \times m \times (v_2 - v_1)$$

where v_2 and v_1 are the specific volumes at the two end states.

Then:
$$\frac{W}{m} = p(v_2 - v_1) \text{ [J/kg]}$$

Work per unit mass, W/m, is called **specific work**, but remember work is not a property like other specific terms you have met.

Specific work is a useful idea; it enables us to compare different processes, or even different engines. A work-producing process having a higher specific work is the better process. Looking back at the problem in Frame 50, if the final specific volume is 0.3748 m^3/kg, find the mass of steam and the specific work. (Compare final v with final V to find the mass of steam.)

55

0.4002 kg, + 62.47 kJ/kg

The actual final volume is 0.15 m^3, and the final specific volume is 0.3748 m^3/kg so the mass of steam is 0.15/0.3748 = 0.4002 kg. We have just calculated the work as 25 kJ for 0.4002 kg of steam, so the specific work is 25.0/0.4002 kJ/kg, or 62.47 kJ/kg. Now repeat the calculation by finding first the initial specific volume of the steam.

56

The initial specific volume is 0.1/0.4002 = 0.2499 [m³/kg].

Hence: $\frac{W}{m} = 5 \times 10^5 \times (0.3748 - 0.2499)$ [J/kg]
$= 62\ 470$ [J/kg]
$= 62.47$ [kJ/kg]

In these problems we have considered a mass of steam expanding at constant pressure without any reference to changes in other properties that may take place. In fact, we shall see later that a constant pressure expansion of steam must involve a heat interaction as well. We will return to this when we are in a position to see the whole picture.

57

Now we shall look at what happens when the inlet valve of a petrol engine cylinder opens, with the piston about to go down the cylinder. A charge of air will be drawn into the cylinder, passing through the carburettor in which petrol will be introduced. If we only consider the air entering the cylinder, this will have been displaced at constant pressure by the atmosphere. The pressure in the cylinder will be less than atmospheric owing to a pressure drop through the inlet manifold and valve. The air in the cylinder, at this lower and more or less constant pressure, will also do work on the piston. The process is shown below.

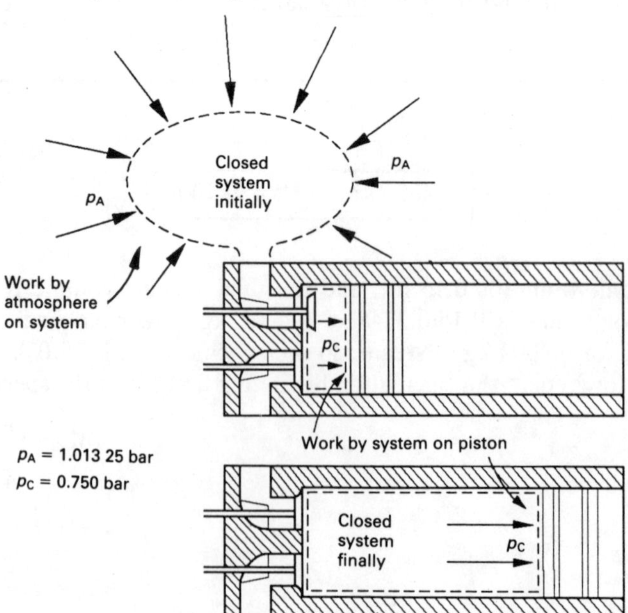

We wish to calculate the net displacement work of the air. Let m be the mass of air involved; it has pressure p_a and specific volume v_a in the atmosphere, and values of p_c and v_c in the cylinder. The volumes displaced in the atmosphere and in the cylinder are:

$$V_a = m\ v_a$$

and $$V_c = m\ v_c$$

so that the displacement work done *on* the air *by* the atmosphere is:

$$W_a = -p_a m\ v_a$$

and the displacement work done *by* the air *on* the piston is:

$$W_c = +p_c m\ v_c$$

The net work of the air that enters the cylinder is thus:

$$W = -p_a m\ v_a + p_c m\ v_c$$

What would be the net work of air of mass 0.0034 kg that enters a cylinder from the atmosphere at 1.013 25 bar, when the pressure in the cylinder is constant at 0.750 bar, and the specific volumes of the air are 0.850 m^3/kg and 1.066 m^3/kg in the atmosphere and cylinder respectively?

58

−20.97 J

Volume displacement in the atmosphere is $0.850 \times 0.0034 = 0.002\ 89\ m^3$.
In the cylinder it is $1.066 \times 0.0034 = 0.003\ 62\ m^3$.
The work done on the air by the atmosphere is $-1.013\ 25 \times 10^5 \times 0.002\ 89$
$= -292.8$ J
The work done by the air on the piston is $+0.750 \times 10^5 \times 0.003\ 62$
$= +271.8$ J
So the net work of the air $= +271.8 - 292.8$ J
$= -20.97$ J

This means net work is done on the air, as the work of the atmosphere on the air is greater than that done by the air on the piston.

59

In the exhaust stroke, the pressure in the engine cylinder is still quite high when the exhaust valve opens, close to the end of the expansion stroke. The pressure falls rapidly as the gases expand into the exhaust system, and much of the gas leaves the cylinder during this part of the exhaust stroke, the remainder being expelled by the piston as it moves up the cylinder.

A typical pressure–cylinder volume plot of an exhaust stroke is shown below but remember this does not represent a process for a closed system, as exhaust gases are leaving the cylinder during the exhaust stroke.

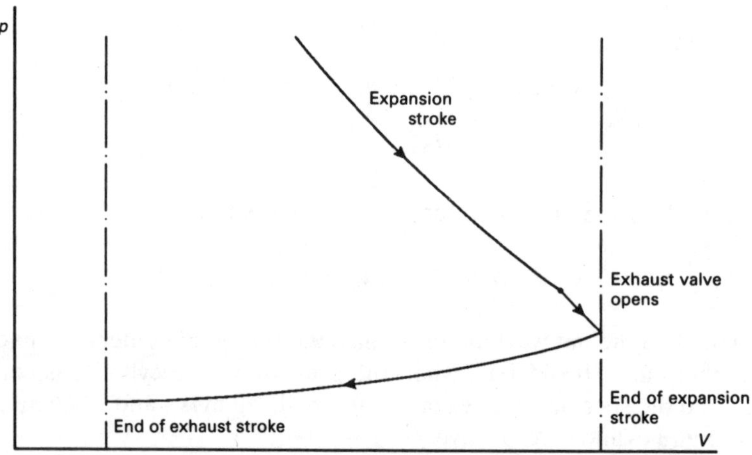

60

In Frame 41 we saw that in the compression and expansion strokes of reciprocating engines, the pressure and volume have inverse relationships, so that as the volume reduces in compression the pressure rises, and in expansion the pressure falls as the volume increases. The pressure–volume curves obtained in real engines can be analysed accurately, and it is found that actual curves (of irreversible compressions and expansions) fit, with reasonable accuracy, curves of the form:

$$pV^n = k$$

where k is a constant and n is chosen to allow this equation to fit an actual curve. **You must always use absolute pressures in this relationship.**

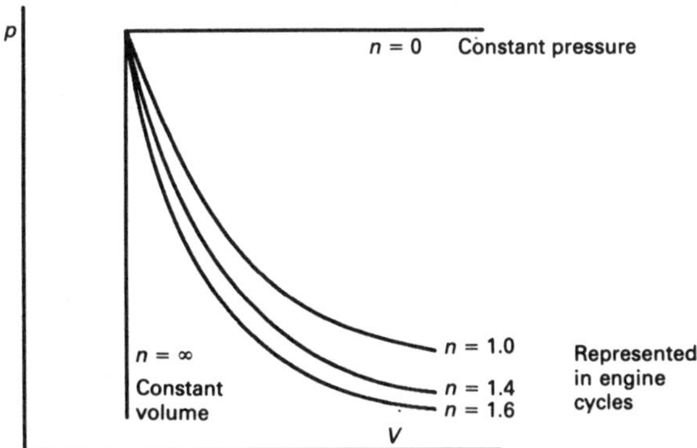

The index n can have a considerable range of values leading to a 'family' of processes, which are called **polytropic**, meaning many paths. These are shown in the figure.

The integration of $\int p\mathrm{d}V$ for the function $pV^n = k$, between an initial state 1 and a final state 2 is shown in the next frame. See if you can get the answer given before looking.

61

$$W = \frac{P_2V_2 - P_1V_1}{1 - n}$$

This is how we can find the work of a polytropic process:

$$W = \int_1^2 p\mathrm{d}V = k\int_1^2 \frac{\mathrm{d}V}{V^n} \text{ since } p = k/V^n$$

$$= k\left(\frac{V^{-n+1}}{-n+1}\right)_1^2$$

$$= \frac{k}{1-n}\left(V_2^{1-n} - V_1^{1-n}\right), \text{ and } k = pV^n \text{ at both states}$$

$$W = \frac{p_2V_2^n\,V_2^{1-n} - p_1V_1^n\,V_1^{1-n}}{1-n}$$

$$= \frac{p_2V_2 - p_1V_1}{1-n}$$

62

The denominator $(1 - n)$ is often negative since n is usually greater than 1, but this does not matter. The value of the integration will always have the correct sign for the work of the system, between an initial state 1 and a final state 2. Work calculated by this expression will be in J with V in m^3, and with p in Pa, or in kJ with p in kPa.

The values of n are usually in the range of 1.2–1.6, the lower value being found in a compression stroke in an engine, and the higher value in the expansion stroke. The value of $n = 1$ has special significance and will be considered later. Low values of n will also be found in the expansion of steam. Remember that these curves are close approximations to actual expansions or compressions which are irreversible, so the actual curves do not represent paths of equilibrium states.

If p_1 is 10 bar and expansion takes place over a volume ratio of 4, $(V_2/V_1 = 4)$, what are the values p_2 for $n = 1.2$ and 1.4 in the process $pV^n = k$?

63

1.8947 bar and 1.4359 bar

We have that $p_2V_2^n = p_1V_1^n$ for $V_2/V_1 = 4$

So: $p_2 = p_1 (V_1/V_2)^n = p_1 (1/4)^n$, for both values of n.

With $n = 1.2$: $p_2 = 10/4^{1.2} = 1.8947$ bar, and

for $n = 1.4$: $p_2 = 10/4^{1.4} = 1.4359$ bar

Now you should be able to calculate the work in a polytropic expansion of gas, using the formula from Frame 61, for $p_1 = 10$ bar, for $V_2/V_1 = 4$, and for $n = 1.4$. The initial volume is that of a cylinder of 150 mm diameter, with the piston 80 mm from the closed end. Your answer is to be in J, so watch your units.

64

1504.3 J

In any process, state 1 is always the initial state, so V_1 is the initial volume of the cylinder:

$$V_1 = \frac{\pi \times 0.150^2}{4} \times 0.080$$

$$= 1.4137 \times 10^{-3}\ [\text{m}^3]$$

Here we have used the diameter and cylinder length in metres. The pressures are given in bar, so we need to multiply by 10^5 to give our final answer in joules (1 bar = 10^5 [Pa], and 1 J = 1 [P]a × 1 [m^3])

$$\therefore \quad p_1V_1 = 10 \times 10^5 \times 1.4137 \times 10^{-3}$$

$$= 1413.7\ [\text{J}]$$

and

$$p_2V_2 = 1.4359 \times 10^5 \times 1.4137 \times 10^{-3} \times 4$$

$$= 811.98\ [\text{J}]$$

$$\therefore \quad W = \frac{811.98 - 1413.7}{1 - n}$$

$$= -\frac{601.72}{1 - 1.4}$$

$$= +\ 1504.3\ \text{J}$$

The positive answer is correct for an expansion of gas, the system.

The expression for polytropic work, from Frame 61, is the polytropic work of a reversible process. Therefore if we now use the same expression to calculate the work of compression from the final state back to the original state, we must get the same numerical answer, but with a negative sign for work done on the gas. Check that this is so. You only need to reverse the state point numbers.

65

$$\boxed{-1504.3 \text{ J}}$$

p_1V_1 is now 811.98, and p_2V_2 is 1413.7, so

$$W = \frac{1413.7 - 811.98}{1 - 1.4}$$

$$= \frac{+601.72}{-0.4}$$

$$= -1504.3 \text{ [J]}$$

Steam will also follow a polytropic process during its expansion in a cylinder, and the precise path will depend to some extent on its temperature. Thus n is low, about 1.1, if the steam temperature is at its boiling point for the initial pressure, and is higher at about 1.3, if the steam temperature is greater than boiling point. The steam is then said to be in a **superheated** condition. All these matters, concerning the properties of working fluids, will follow in a later programme. Historically, steam was the first working fluid to be used in an engine. Today steam is not so important, but *other* vapours, which boil at much lower temperatures, are being used increasingly to obtain work from low temperature heat sources, generally known as waste heat.

66

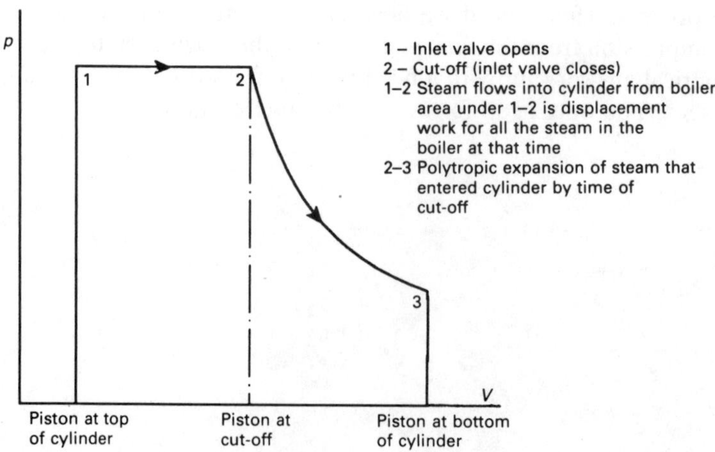

The polytropic expansion of steam (or a vapour) in an engine cylinder is shown in the figure on the previous page. Steam flows into the cylinder from the boiler, through a control valve. This valve controls the pressure of the steam while it is entering the cylinder; it is called a pressure reducing valve.

Steam enters the cylinder only for a certain fraction of the piston stroke. The inflow of steam ceases at this point of cut-off, and a polytropic expansion follows for the remainder of the stroke. We need to think of the whole process in two parts; up to the point of cut-off, and after the cut-off.

In order to calculate displacement work done on the piston, can you identify a closed system, (a) before cut-off and (b) after cut-off?

67

(a) Total steam content of the boiler, before the reducing valve is opened.
(b) Steam that has entered the cylinder.

When the reducing valve is opened, some of the steam in the boiler enters the cylinder at a constant pressure while the piston moves down to the position of cut-off. With the total steam as the system, the piston is the only part of the system boundary that moves. The work during this process is the product of the constant pressure and the volume displaced by the piston. After cut-off (b) the steam that has entered becomes a closed system of its own. Notice the two constant pressure lines at 1.013 bar and 0.5 bar; we will think about these later.

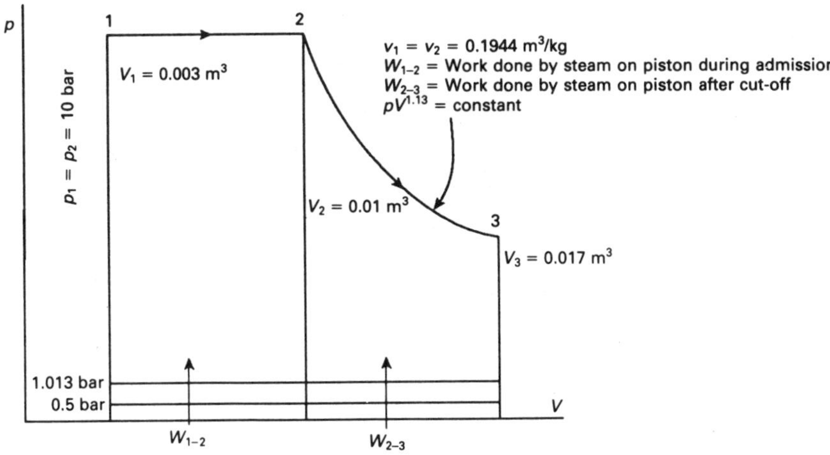

68

Let $p_2 = p_1 = 10$ bar (absolute pressure) and specific volume $v_2 = v_1 = 0.1944$ m^3/kg, with $V_1 = 0.003\ m^3$, $V_2 = 0.01\ m^3$ and $V_3 = 0.017\ m^3$.

What is (a) the displacement work up to the point of cut-off and (b) the mass of the steam that has entered the cylinder?

69

(a) 7000 J, (b) 0.0514 kg

During the period of steam admission, $V_1 = 0.003$ and $V_2 = 0.010\ m^3$. The specific volume is constant at $0.1944\ m^3/kg$. The pressure is 10 bar = 10×10^5 Pa, so the constant pressure work is:

$$\begin{aligned} {}_1W_2 &= 10 \times 10^5 \times (0.010 - 0.003) \\ &= 7000\ [\text{J}] \end{aligned}$$

The mass of steam is $V_2/v_1 = 0.01/0.1944$
$= 0.0514$ [kg]

Now calculate the polytropic work in the remainder of the piston stroke, using the equation from Frame 61, but with an initial state 2 and a final state 3, and the total work for the whole stroke.

70

5130.77 J, 12.131 kJ

After cut-off we have $p_2V_2^{1.13} = p_3V_3^{1.13}$, and $V_3 = 0.017$ m.

$$\begin{aligned} \therefore \quad p_3 &= p_2\left(\frac{V_2}{V_3}\right)^{1.13} \\ &= 10 \times 10^5 \times (0.010/0.017)^{1.13} \\ &= 5.49 \times 10^5\ \text{Pa} \end{aligned}$$

After cut-off, the work is given by:

$$
\begin{aligned}
{}_2W_3 &= \frac{p_3V_3 - p_2V_2}{1 - n} \\
&= \frac{5.49 \times 10^5 \times 0.017 - 10 \times 10^5 \times 0.01}{1 - 1.13} \\
&= \frac{9333.0 - 10\,000}{-0.13} \\
&= 5130.77 \text{ [J]}
\end{aligned}
$$

The total work is the sum of these two amounts:

$$
\begin{aligned}
W = {}_1W_2 + {}_2W_3 &= 7000.0 + 5130.77 \\
&= 12\,130.77 \text{ [J]} \\
&= 12.131 \text{ [kJ]}
\end{aligned}
$$

This represents the work done by the steam on the piston on its outward stroke. As we have to use absolute pressures in $pV^n = k$, the net work done on the piston will be reduced, as the piston displaces the atmosphere acting on its outer surface. Work by the piston is the volume displacement $\times$ atmospheric pressure. If atmospheric pressure is 1.013 bar, what is the net work done on the piston?

71

-10.713 kJ

The work done on the piston by the steam is -12.131 kJ, and the piston does constant pressure positive work in displacing the atmosphere of:

$$1.013 \times 10^5 \times (0.017 - 0.003) = 1418.2 \text{ J} = 1.418 \text{ [kJ]}$$

Hence the net work of the piston is $-12.131 + 1.418 = -10.713$ [kJ]

72

However, the piston will do positive work on the crankshaft of the engine, so the positive work transmission of the piston will be +10.713 kJ, and with no piston friction the net work of the piston is zero. The piston will now return up the cylinder to expel the expanded steam to the condenser; we may assume the pressure is constant at 0.5 bar (see the diagram in Frame 67), so for the same volume displacement the piston has a pressure of 0.5 bar on the steam side and 1.013 bar on the air side. So some of the work done on the atmosphere in the expansion stroke is recovered in the exhaust stroke. This will amount to constant pressure work between the 1.013 and 0.5 bar pressure lines.

What is the value of this work?

73

718.2 J

The area of the rectangle between the two pressure lines is:

$$(1.013-0.5) \times 10^5 \times (0.017-0.003) = 0.513 \times 0.014 = 718.2 \text{ [J]} = 0.718 \text{ [kJ]}$$

The cycle net work is 10.713 + 0.718 [kJ] = 11.431 [kJ].

Reciprocating steam engines (or engines using other vapours) operate on a two stroke cycle: a power or expansion stroke, and an exhaust stroke. So the cycle net work is repeated with every revolution.

The engine is running at 180 rev/min. What is the power produced?

74

34.293 kW

The power is the work of one cycle × cycles/second. At 180 rev/min, there are 3 cycles per second. So the power = 11.431 kJ × 3 cycles/sec = 34.293 kW.

Now let's look at the expression in Frame 60 again: $pV^n = k$. What happens when $n = 1$? It becomes a special case, which you must have seen before. Bear in mind that we are not talking about the properties of actual substances yet (this follows in Programme 4), but applied to gases this is a well known law. Whose is it?

75

Boyle's Law, $pV = k$

As we shall see in Programme 4, when gases expand or contract at constant temperature, we find that $pV = k$. This curve is called a rectangular hyperbola. If you refer back to the integration in Frame 61, you will find that we get $W = 0/0$, so we need to do the integral $\int p\mathrm{d}V$ again. This is how it is done.

$$\text{We have: } W = \int_1^2 p\mathrm{d}V = k\int_1^2 \frac{\mathrm{d}V}{V} = k\,[\ln V_2 - \ln V_1]$$

$$= k \ln \frac{V_2}{V_1} \quad \text{and then:}$$

since: $p_1V_1 = p_2V_2$

$$W = p_1V_1 \ln \frac{V_2}{V_1} = p_2V_2 \ln \frac{V_2}{V_1}$$

$$= p_1V_1 \ln \frac{p_1}{p_2} = p_2V_2 \ln \frac{p_1}{p_2}$$

Can you verify that the units of this work term are correct?

76

The natural logarithm of a ratio is a dimensionless number, so the product pV provides the units for work.

In the next frame we shall see from a figure that the work of an expansion for n greater than 1 is less than it is when n is equal to 1. So what controls the value of n and the amount of work we shall get in a particular expansion? As we shall see later, n is dependent on a second interaction of heat taking place simultaneously with the work interaction during an expansion or compression.

77

The figure below shows how the final pressure in an expansion decreases as the value of n increases. Consequently the area under the curve, the work of the expansion, decreases with n.

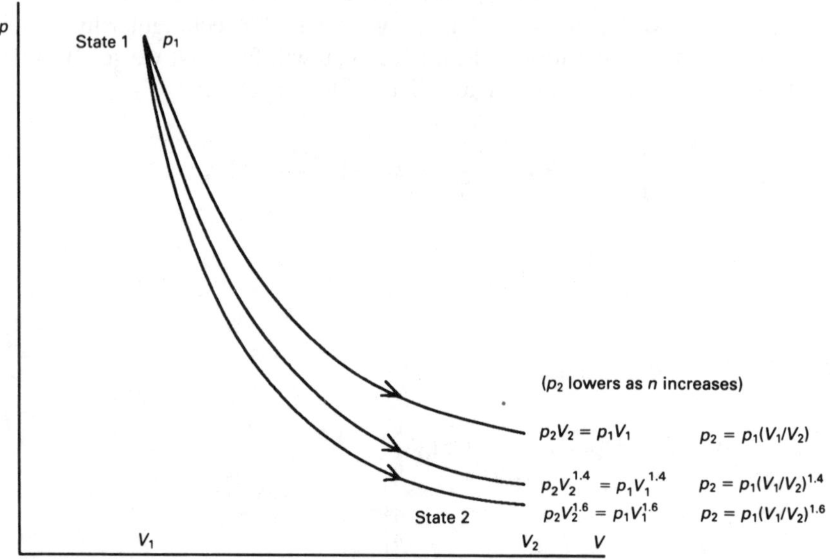

A gas expands from a volume of 0.2 m^3 and a pressure of 300 kPa, to a pressure of 100 kPa, according to $pV = k$. Calculate (a) the final volume and (b) the work done by the gas.

78

(a) 0.6 m^3, (b) 65.917 kJ

Using $p_2V_2 = p_1V_1$ we get $V_2 = 300 \times 0.2/100 = 0.6$ [m^3]

Then using $W = p_1V_1 \ln V_2/V_1$ we get: $W = 300 \times 0.2 \ln(0.6/0.2)$

$= 300 \times 0.2 \times 1.0986$

$= 65.917$ [kJ]

This is as far as we shall go with displacement work. Do these problems before moving on.

DISPLACEMENT WORK PROBLEMS

1. Gas at a pressure of 200 kPa is contained in a cylinder of diameter 150 mm. (a) Calculate the force on the piston. The piston moves 200 mm while the gas pressure remains constant. (b) Calculate the work done by the gas. [(a) 3.534 kN, (b) 706.8 J]
2. Air of volume 10 m^3 escapes from a cylinder into the atmosphere at 1.013 bar. Calculate the work done on the atmosphere. [1 013.0 kJ]
3. A steam engine runs at 500 rev/min. The cylinder diameter is 180 mm and the piston stroke is 300 mm. There is a power stroke in each crankshaft revolution. An indicator diagram is obtained, and the diagram area is 3800 mm^2. The spring rate of the indicator is 18 kPa/mm, and the base length of the diagram is 100 mm. Calculate (a) the IMEP in kPa and (b) the power developed at the piston face. [(a) 684 kPa, (b) 43.51 kW]
4. Gas at a pressure of 150 kPa, having a volume of 0.01 m^3, expands through a diaphragm into a vacuum having a volume of 0.015 m^3. Is work done by the gas? If not, why not? [No, see Frame 37]
5. Gas in a vertical cylinder has a pressure of 150 kPa and a volume of 0.01 m^3. The pressure is due to a weighted piston. The weight is removed and the pressure due to the piston only is 110 kPa. The gas expands to a total volume of 0.025 m^3. Calculate the work done by the gas. [+1.65 kJ]
6. Steam of specific volume 0.06 m^3/kg at a pressure of 1.6 bar expands at constant pressure to 0.12 m^3/kg. Calculate the specific work of the steam. [+9.6 kJ]
7. In the family of polytropic processes ($pV^n = k$), what is n for (a) a constant pressure process, (b) a constant volume process and (c) a process in which the pressure is inversely proportional to the volume? [(a) 0, (b) ∞, (c) 1]
8. Calculate (a) the final pressure, and (b) the work done on a gas when it is compressed according to $pV^{1.3} = k$, from an initial pressure of 120 kPa and volume of 0.08 m^3 to a final volume of 0.015 m^3. [(a) 1057.5 kPa, (b) −20.875 kJ]
9. In the previous problem, atmospheric pressure at 1.013 bar acts on the outside of the piston. Calculate the work done by the atmosphere on the piston, and the remaining work to be supplied to the piston by an external agent. Assume the piston is frictionless. [+6.584 kJ, +14.291 kJ]
10. In the induction stroke of an engine, 0.003 m^3 of air at 1.0133 bar is sucked from the atmosphere. In the engine cylinder, the pressure is constant at 0.85 bar and the piston displacement is 0.0035 m^3. Calculate the net work of the air that enters the cylinder. [−6.49 J]
11. Steam at 8 bar enters the cylinder of a steam engine. The cylinder diameter is 0.2 m and the piston stroke is 0.4 m. Cut off is at 40 per cent of the stroke. The specific volume of the entering steam is 0.24 m^3/kg. Expansion after cut off is polytropic, n = 1.13. Calculate the work done against the piston in a single

power stroke, and the rate of steam consumption in kg/h when running at 500 rev/min. [7.933 kJ, 628.5 kg/h]

12. In a petrol engine cylinder, the volume is 0.0036 m^3 at the outer end of the stroke and 0.000 36 m^3 at the top of the stroke. The cylinder pressure is 100 kPa at the beginning of compression. The compression and expansion strokes are polytropic, $n = 1.3$ in compression and $n = 1.5$ in expansion. At the end of compression, the pressure is increased by a factor of 1.6 at constant volume, and at the end of expansion, the pressure falls at constant volume. Calculate the net positive work done on the piston by the cylinder contents. [+0.7734 kJ]

80

We now move on to think about the second type of work listed in Frame 26, shaft work.

The work of reciprocating pistons causes the rotation of shafts, and where a rotating shaft cuts through a system boundary (whether it be an open or closed system), a work interaction with the surroundings takes place. This is shown in the figure.

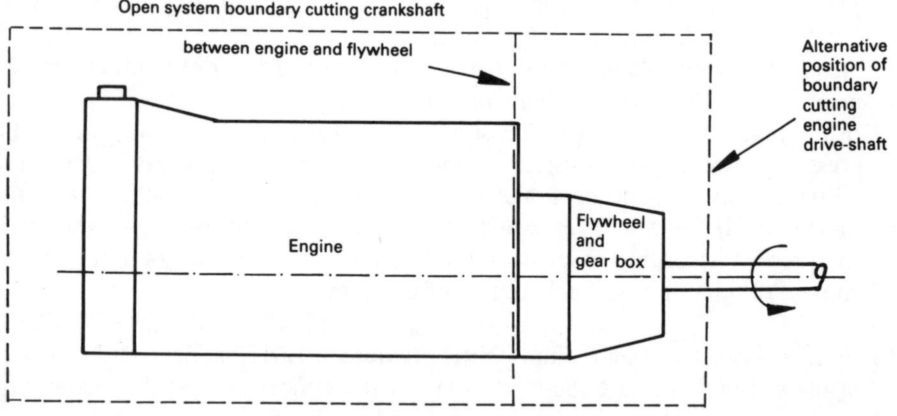

81

Like other forms of work, shaft work may be positive or negative with respect to the system. If we imagine the boundary of an open system to pass through the crankshaft of an engine between the crankcase and the flywheel, the engine which is the system, will, through the shaft, do positive work on the flywheel during expansion (power) strokes of the pistons, and the flywheel will do negative work on the system during the compression strokes.

You probably already know how to calculate torque, shaft work and shaft power, so this may be revision for you.

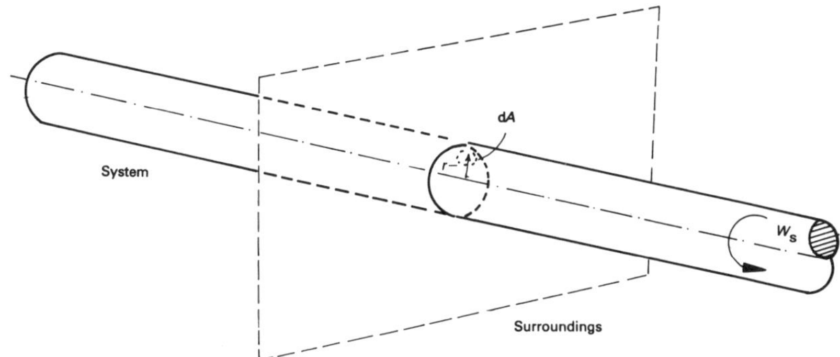

We can imagine the system boundary cutting the shaft and we have to see how work is transmitted from one side to the other across the imaginary cut. At an element of area dA at radius r of the shaft section, where the shaft cuts the boundary, there is a shear stress σ by which the shaft on one side of the boundary causes the shaft on the other side to rotate. When the element dA at radius r moves through a small angle dθ, shear work is done:

$$dW_s = (\sigma\, dA)(r\, d\theta)$$

Shaft work is the integration of this shear work over the area of cross-section of the shaft, and over the total angle of rotation, which must be in radians.

Hence: $$W_s = \int\int \sigma\, r\, dA d\theta$$

The shear stress σ acting on dA at radius r produces a force at radius r, or torque, dT_s:

$$dT_s = (\sigma\, dA)\, r$$

so that, on summing over the whole area:

$$T_s = \int \sigma\, r\, dA$$

and so: $$W_s = \int T_s d\theta$$

$$\therefore \quad W_s = T_s \times \theta$$

where the torque is assumed constant over θ radians of rotation.

Notice the formula is analogous to linear motion:
Linear: Work = force × linear displacement [N × m = J]
Circular: Work = torque × angular displacement, [Nm × radians = J]

Thus linear motion of a piston will lead to circular motion of a shaft through a crank, and the transmission of work by the shaft is due to shearing forces in the shaft.

A steady torque of 50 Nm is transmitted by a shaft that turns through 8 revolutions. What is the value of the work done?

82

2513.27 J

In 8 revolutions of the shaft, θ is $8 \times 2\pi = 50.27$ radians. So the work done is 50.27×50 Nm = 2513.27 J.

We usually measure the rate of rotation of shafts, rather than the number of revolutions, so we can calculate shaft power as:

$$P_s = T_s \frac{2\pi N}{60} \quad (\text{N m/s} = \text{W})$$

where N is the speed of rotation of the shaft, usually measured in rev/minute. The power output of an engine is determined from measurements of speed and torque. Torque is measured by a dynamometer.

A power transmission by a shaft rotating at 5000 rev/min is measured by a force of 100 N acting at a radius of 300 mm. Calculate the power transmitted and the work done in 30 seconds. (Calculate the torque first.)

83

15.708 kW, 471.239 kJ

The torque is force × radius: $T_s = 100 \times \dfrac{300}{1000} = 30$ [Nm]

so the power is torque × radians/sec: $P_s = 30\ [\text{Nm}] \times \dfrac{2\pi \times 5000}{60}\ [\text{sec}^{-1}]$

$= 15707.96$ [N m/s]

$= 15.708$ [kW]

The work lasts for 30 seconds, so $W_s = 15.708\ [\text{kJ/s}] \times 30\ [\text{s}]$

$= 471.238$ [kJ]

In the next frame we shall think about a negative form of shaft work called stirring work.

84

We have seen how shaft work is done at a system boundary where a rotating shaft cuts through that boundary. An example of positive shaft work is where an engine (an open system), is driving some machinery (the surroundings).

Negative shaft work would be involved when a motor (the surroundings) drives a compressor (the open system). Open systems, you will remember, are systems in which fluid flows through the control space. An electrical generator is an example of a closed system that has negative shaft work done on it.

Imagine we have a closed container of some fluid such as oil, and within the oil is a stirrer or paddle wheel driven by a shaft penetrating the wall of the container. The fluid creates a resistance to the rotation of the paddle, so negative shaft work would be done on the fluid. The figure shows the fluid as the system, with the system boundary cutting the shaft at the wall of the container. Negative stirring work can be calculated from the torque at the boundary, and the rotation of the shaft.

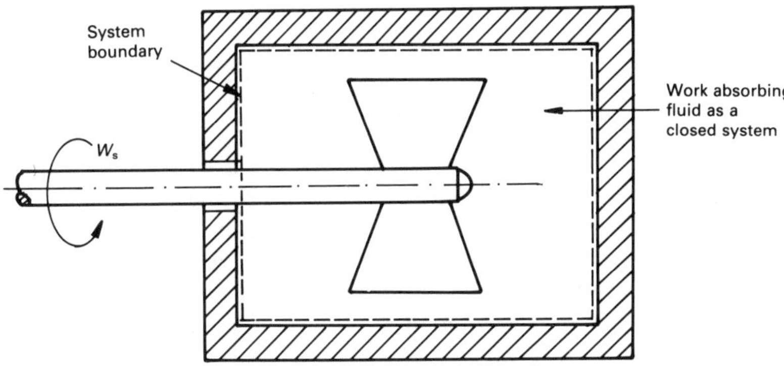

A torque of 5 Nm is transmitted through a shaft rotating at 2000 rev/min for 5 minutes to do stirring work on a container of fluid. What is the sign and magnitude of this work?

85

-314.16 kJ

The total number of revolutions is:

$$2000\ [\text{rev/min}] \times 5\ [\text{min}] = 10\,000\ [\text{revs}]$$

The total angular rotation is $2\pi \times 10\,000$ [radians].
So the work is: $2\pi \times 10\,000$ [radians] $\times 5$ [Nm] $= 314\,159$ [J] $= 314.16$ [kJ].
The sign is negative since it is work done on the system.

We have just seen an example of work done on a system in such a way that the system is not able to 'give back' any of that work. (A precise thermodynamic analysis would show that a tiny fraction of that work could be 'given back'.) From everyday experience we would know that the fluid in the vessel had become slightly warm as a result of the work done on it. Loosely we could say that 'the work had produced heat'. In the next frame we shall think about work in fluid shear. This is the third form of work given in Frame 26.

86

When a piston travels down a cylinder of an engine, the work done by the piston on the crankshaft is less than the work done on the piston by the expanding gases. This is because work is done in overcoming what is called 'piston friction'. There is a film of oil between the piston and the cylinder, and shear work is done on the oil. This causes the oil to get hot, and in high performance engines an oil cooler has to be used. If there was no oil film, we would have solid friction, and this would cause the piston to seize in a very short time.

The figure shows how, for a shaft turning in a bearing, work due to fluid shear will arise.

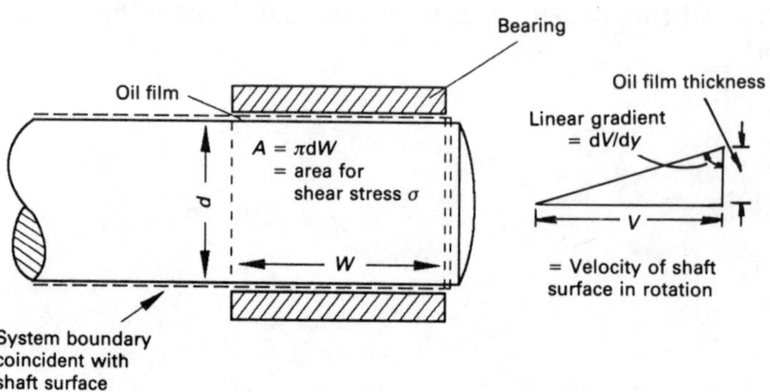

A shear stress will occur where there is a velocity gradient in the fluid:

$$\text{shear stress: } \sigma = \mu \frac{dV}{dy}$$

where μ is the coefficient of molecular viscosity of the fluid. It is another intensive property, and has units of N s/m^2 [= Pa s]. It is an easily recognised property, treacle is much more viscous than water.

The shear force on area A is σA, and at the system boundary in the fluid where the velocity is V, the movement is $V\,d\tau$ in time $d\tau$, so

$$dL = V\,d\tau$$

and the work done in fluid shear is shear force × distance moved:

$$dW_{fs} = \sigma A\, V\, d\tau$$

Hence: $$W_{fs} = \int \sigma A\, V\, d\tau$$

The rate of work or power is: $$\frac{dW_{fs}}{d\tau} = \sigma\, A\, V$$

So: $$P_{fs} = \mu \frac{dV}{dy} A V = \sigma\, A\, V$$

Using the formula for power absorbed in fluid shear, just given, calculate the power absorbed in a plain bearing 50 mm diameter by 100 mm in width, in which a shaft is rotating at 400 rev/min. There is an oil film of thickness 0.25 mm, with oil of viscosity 0.2104 Pa s. (Since the shaft is rotating in a fixed bearing, dV/dy = linear velocity of shaft surface/oil film thickness, and A is the cylindrical area of the bearing.)

87

14.49 W

The area of the bearing surface, A, is: $\pi \times 50 \times 100 \times 10^{-6} = 0.0157$ [m^2] The velocity of the shaft surface, V, is: $\pi \times 50 \times 10^{-3} \times 400/60 = 1.0472$ [m/s]

Hence: $$P_{fs} = 0.2104 \times \frac{1.0472}{0.00025} \times 0.0157 \times 1.0472$$

$$= 14.49 \text{ [W]}$$

88

This loss of shaft power causes the temperature of the oil to rise. This will lead to a heat interaction as the oil cools. Heat will be considered soon.

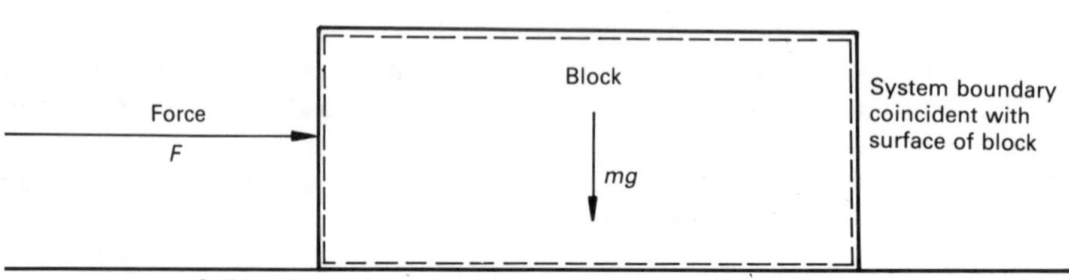

A heat interaction is also produced by sliding friction, and in this case work is done against a frictional force between a moving and a fixed object. The frictional force, F, for a body of mass m sliding on a horizontal surface with a coefficient of friction μ is:

$$F = \mu\, mg \text{ [N]}$$

Then work of solid, or sliding friction is done when this force acts through a distance:

$$W_{sf} = \mu\, mg\, L \text{ [J]}$$

89

We shall now think about electrical work at a system boundary. We may take an electric battery as our system and connect it to a motor, as the surroundings. This, on rotation, will do work, such as raising a weight.

The wires from the battery cross the system boundary at two points; if the potential between these points is V volts and if Q coulombs of electricity cross the system boundary, then, as you should know, the positive work done is QV [J]. One coulomb per second is a current of one ampere, so for a steady current of I [amps], the rate of work is VI [J/s] or VI [W]. Voltage is an intensive property, and we see that it is a difference in voltage which is the driving force for electrical work.

A battery of 12 volts passes a current of 2 amps for 20 seconds to a motor. Calculate the work done (voltage × current × time), and the height to which this would raise a mass of 20 kg (g = 9.81 m/s^2).

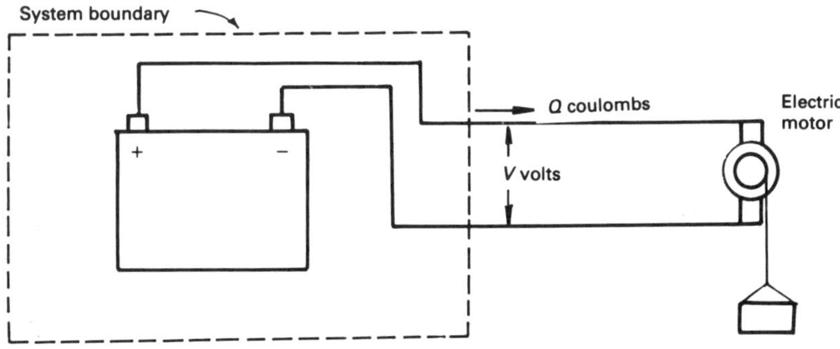

90

480 J, 2.446 m

The work from the battery $= 12$ [volts] $\times$ 2 [amps] $\times$ 20 [seconds]
$= 480$ [J]

The work of raising an object of mass 20 kg:
$= 20 \times 9.81 \times h$ [J]

where h is the unknown height.

$$\therefore \quad h = \frac{480}{20 \times 9.81} = 2.446 \text{ [m]}$$

We may now think of the motor as our system, for which there will be two work interactions. The motor has negative electrical work done on it, and it also does positive work in raising an object. In calculating a height of 2.446 m in our previous example, we assumed the motor would convert all the electrical work done on it into raising a weight. Thus the motor was assumed to act in an ideal way. However as you know, electric motors become warm, if not hot, in operation. This is because they experience electrical losses and the work output is always less than the electrical work input:

$$W_o = \eta_e W_{ei}$$

where W_o = work output of the motor, η_e = electrical efficiency and
W_{ei} = electrical work input.

This is another example in which negative work done on a system has produced a heating effect.

91

See if you can solve this problem on motor efficiency. (You need to calculate the electrical work done on the motor, and the mechanical work done by the motor.)

An electric motor raises an object of mass 10 kg through a height of 4 m in 15 seconds. It receives a current of 1.9 amps at 15 volts during this time. What is the electrical efficiency of the motor?

92

0.9179 or 91.79%

The negative electrical work done on the motor is:

$$15 \text{ [volts]} \times 1.9 \text{ [amps]} \times 15 \text{ [secs]} = 427.5 \text{ [J]}$$

The positive work done by the motor is:

$$10 \text{ [kg]} \times 9.81 \text{ [m/s]} \times 4 \text{ [m]} = 392.4 \text{ [J]}$$

The motor efficiency is:

$$\eta_e = \frac{392.4}{427.5}$$

$$= 0.9179 \text{ or } 91.79\%$$

The work interaction on the motor is seen to be irreversible because losses have occurred. We can see that 427.5 − 392.4 J = 35.1 J of work done on the motor have not been used in raising a weight. From our understanding of energy conservation, this 35.1 J of work must be accounted for, and it has been dissipated by the motor as electrical losses which cause a property of the motor, its temperature, to rise.

93

We shall return a little later on to examine this more closely. If the motor had been acting in a completely reversible way, the weight would have risen a height of h:

$$W = 427.5 \text{ [J]} = 10 \text{ [kg]} \times 9.81 \text{ [m/s}^2\text{]} \times \text{h [m]}$$

hence: $$\text{h} = 4.36 \text{ [m]}$$

Also, if the motor was able to operate as a dynamo, it would have produced a current of 1.9 amps at 15 volts for 15 secs as the weight fell 4.36 m.

An electric motor receives a current of 25 amps at 15 volts and its electrical efficiency is 90%. It runs at 195.33 rev/min. Calculate the torque at the output shaft.

94

16.5 Nm

The electrical power available as output power is:

$$25\ [\text{amps}] \times 15\ [\text{volts}] \times 0.9 = 337.5\ [\text{W}]$$

The output power is: $$\frac{T_s \times 2\pi N}{60} = 337.5\ [\text{W}]$$

$$\therefore \quad T_s = \frac{337.5 \times 60}{195.33 \times 2\pi} = 16.5\ [\text{Nm}]$$

We have now come across two cases, in stirring work, and in electrical work, where negative work done on a system has resulted in a temperature rise. This has caused a heat interaction with the surroundings. After the next frame we shall think about heat, and in the next programme we will meet the First Law of Thermodynamics and the relation between work, heat and energy.

95

PROBLEMS IN SHEAR WORK, FRICTIONAL WORK AND ELECTRICAL WORK

1. A torque of 80 Nm is transmitted by a shaft rotating of 600 rev/min. How much work is transmitted in 90 seconds? [452.39 kJ]
2. A torque is determined from a mass of 10 kg suspended at a radius of 200 mm. This torque is transmitted by a shaft rotating at 3000 rev/min. What is the power transmission? ($g = 9.81$ m/s^2) [6.164 kW]
3. Stirring work by a rotating shaft is done on a fluid in a container. A torque of 4 Nm is transmitted for 200 revolutions. What is the work transfer? [−5.027 kJ]
4. In a fluid there is a velocity gradient of 100 s^{-1} and for the fluid the coefficient of molecular viscosity is 0.79 N s/m^2. Calculate the shear stress. [79 N/m^2]
5. The shear stress in problem **4** acts over an area of 2000 mm^2 and the resultant shear force acts through a distance of 5 m. Calculate the work done in fluid shear. [0.79 J]
6. A plain bearing 40 mm diameter by 80 mm in width supports a shaft rotating at 300 rev/min. The oil film thickness is 0.2 mm. The coefficient of viscosity of the oil is 0.3 N s/m^2. How much work is absorbed in 10 minutes? [3571.9 J]
7. A force to overcome solid friction arises from a shear stress of 6000 N/m^2 acting over an area of 0.035 m^2. The force acts over a distance of 5 m. Find the work done in solid friction. [1050 J]
8. A battery of 12 volts passes a current of 5 amps to an electric motor for 12 minutes. The motor has an internal efficiency of 92%. Calculate the work output of the motor. [39.744 kJ]
9. What is the power output of the motor in problem **8**? [55.2 W]
10. An electric motor raises an object of mass 12 kg through a height of 5 m in 20 secs. The motor receives a current of 1.86 amps at 18 volts during this period. Calculate the electrical efficiency of the motor. [87.9%]
11. An electric motor is required to produce a torque of 20 Nm at 200 rev/min. It has an internal efficiency of 90% Calculate the current required at 12 volts. [38.785 amps]

96

In thermodynamics we define heat as follows:

Heat is defined as that interaction at a system boundary which occurs solely as a consequence of a temperature difference across the boundary.

Heat is an energy transfer at the boundary, solely due to a temperature difference; within the system, the word 'heat' has no meaning. Just as rain is a form of water in transfer between a cloud and the sea, and the word 'rain' has no meaning when it has become part of the sea, so does the word 'heat' have no meaning when the energy in transfer has become part of the energy of the system. A system does not 'contain' heat.

We associate the words 'transfer from' and 'transfer to' when talking of heat. This dates from the caloric theory of heat in which heat was thought of as a transferable substance, like water from a sponge.

We use Q to denote heat, and as for work, a sign convention is needed. The sign convention for heat is the opposite of what we use for work, and the conventions for both are given in the figure:

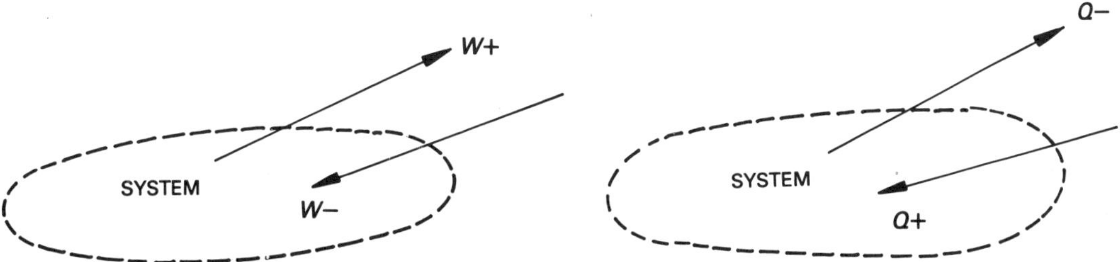

Heat to a system is positive
Heat from a system is negative
Work done on a system is negative
Work done by a system is positive

If (a) a block of ice from the refrigerator and (b) a block of steel from a furnace are placed on a steel bench at room temperature, what is the sign of the heat interaction in (a) and (b)?

97

(a) + (b) −

Heat transfer to the ice causes it to melt, and heat transfer from the block of steel causes it to cool. The melting ice example shows that a heat interaction does not necessarily mean a temperature change in a system, but there must always be a temperature difference with the surroundings. In this case the temperature will start to rise only after melting is complete.

98

Apart from electrical and frictional losses, which occur in work interactions, some other physical processes can cause temperature changes to occur which do not involve a heat interaction at the boundary. If our system is a dish of sulphuric acid and a dish of water, a temperature rise will occur when the acid is mixed with the water.

Having studied work in some detail you should be familiar with using the sign convention for work; in the same way you must get used to using the sign convention for heat. Positive heat transfer means the surroundings are hotter; negative heat transfer means the surroundings are cooler. First, do you think the following statements use the word 'heat' correctly?

(1) 'The heat of the electric fire warms the room.'
(2) 'The heat in the hot water pipes is obtained from burning gas in the boiler.'

99

(1) Yes, (2) No

In (1), because the element is very hot, there is a transfer of heat to the room. This has a warming effect in the room. The word 'heat' is used correctly. In (2) the word 'heat' is not correct; hot fluid in a pipe has a property called 'enthalpy', which will be defined later.

Historically, heat was measured in calorimetric experiments, in terms of the effect it had on raising the temperature of water. Quantities of heat so measured and defined depended on the units used. Thus a British Thermal Unit [BTU] was defined originally as the heat interaction to raise water of mass 1 lb through 1° on the Fahrenheit scale. Then, since scales are not linear, the current definition followed (the 60° BTU), the 1° rise being specified between 59.5° and 60.5° F. In a similar way a Centigrade heat unit [CHU], or pound calorie, is defined as the heat interaction to raise water of mass 1 lb from 14.5° C to 15.5° C. Another calorimetric unit of heat is the kilogramme calorie, kcal, which is the heat interaction to raise water of mass 1 kg from 14.5 to 15.5° C. This is more closely related to S.I. units than the CHU.

Given that the pound mass is 0.4536 kg mass, how many kcal is a heat interaction of 100 CHU?

45.36 kcal

100 CHU is the heat interaction to raise 100 lb mass of water through 1°C; which is equivalent to 45.36 kg mass of water through 1°C.

Hence 100 [CHU] = 45.36 [kcal].

The experiments of James Prescott Joule, carried out about 150 years ago, showed that identical effects could be produced by heat and work interactions. Thus a calorimetric experiment, in which a measured heat transfer took place, would produce a rise in water temperature which could be measured, and the same rise in temperature could be achieved by a paddle wheel driven by a shaft. By carefully measuring both the heat transfer and the work done by the rotating shaft, **a mechanical equivalent of heat** was found.

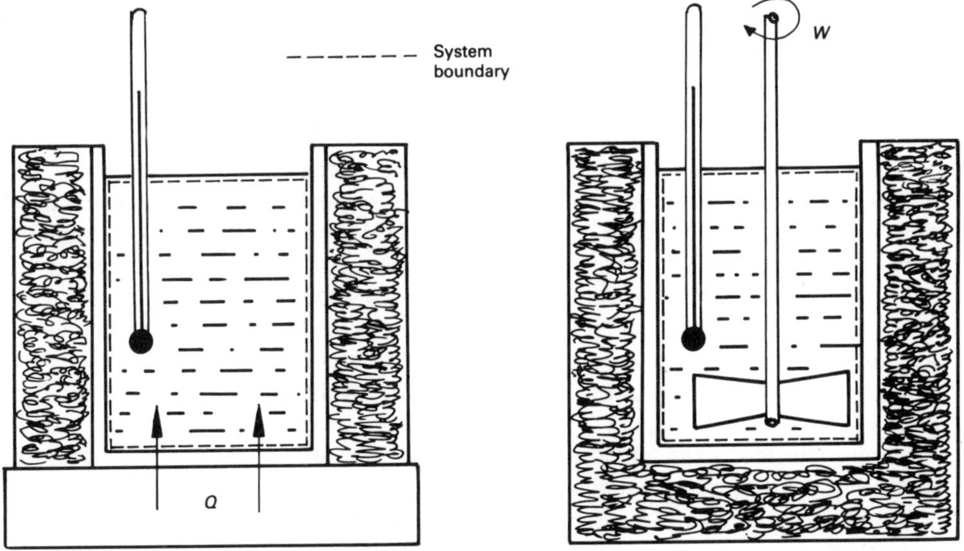

Identical effects of heat and work

Thus, in the British system of units, where the unit of work is the foot pound force [ft lbf], which is a force of 1 [lbf] acting through 1 foot, the mechanical equivalent of heat was found to be:

$$J = 778.17 \text{ [ft lbf/BTU]}$$

In the metric system of units:

$$J = 4186.8 \text{ [N m/kcal]}$$

101

For S.I. units where the unit of work is the joule [J], which has already been defined, the unit of heat is also taken as the joule. This means the unit of heat is related absolutely to the standards of mass, length and time, and is no longer related to temperature scales based on other physical properties. Thus we have identical units for heat and work, and $J = 1$. There are, however, no related units of mass and temperature rise that enable a joule to be measured in a calorimetric experiment.

What would be the mechanical equivalent of heat in units of [N m/CHU]?

102

1899.1 N m/CHU

$$J = 4186.8 \text{ [N m/kcal], and 1 [CHU]} = 0.4536 \text{ [kcal]}$$

So
$$J = 4186.8 \times 0.4536 \text{ [N m/kcal][kcal/CHU]}$$
$$= 1899.1 \text{ [N m/CHU]}$$

In engineering plant where heat transfers occur on a very large scale, as in the production of electricity in power stations, we often speak of a **heat source and heat sink**. Another name for either is a **heat reservoir**. It is a source or sink capable of either giving or receiving large heat transfers without any significant change in temperature. A jug of hot water is not a heat source since its temperature will fall as heat transfer occurs between it and the surroundings. If there is another interaction to maintain a limited source at a particular temperature for some purpose, then that could be a heat source or reservoir in the thermodynamic sense. An electric hot plate is an example.

On an industrial scale, the output of furnace gases at a controlled temperature would be regarded as a heat source.

In nature, examples of heat sinks are fairly widespread. Can you think of any examples?

103

A lake, a river, a sea, the atmosphere

Imagine a power-boat on a lake. Imagine that the engine draws in water from the lake for cooling; it passes through the engine once and is discharged back to the lake. The lake is a heat sink for waste heat from the engine, but the lake will not get warmer because of the boat. Normally an engine is cooled by a limited amount of water in the radiator, this will rapidly boil if its own circulation or cooling breaks down. In this case the atmosphere is the heat sink.

We must now think about how heat transfers occur. They take place naturally, and in the study of heat transfer we see how we can either promote the rate of transfer, as in boiler design, or reduce it, as in the insulation of buildings. The three natural processes of heat transfer are called conduction, convection and radiation, and they represent limits to what we can do.

104

In each of the three natural ways heat transfer occurs, heat always transfers from a high to a low temperature. The process of conduction in a solid is given by the experimental law of Fourier:

$$\dot{q} = -\mathrm{k}\left(\frac{T_2 - T_1}{\Delta X}\right) [\mathrm{W/m^2}]$$

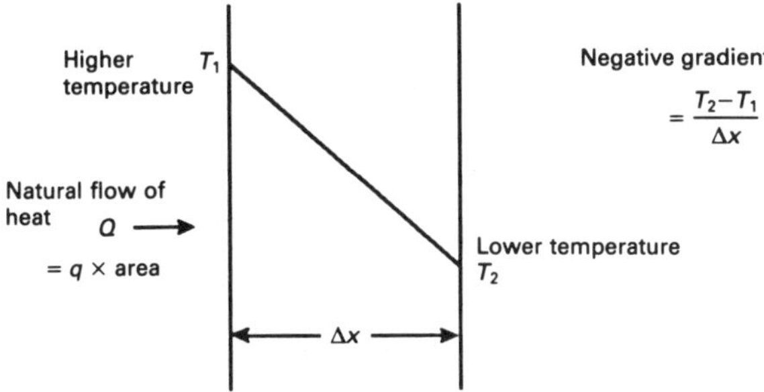

where $\dot{q}$ is the heat transfer rate per unit area, or heat flux, in [W/m^2], and ΔX is the thickness of the conducting material. The thermal conductivity of the material is k, having units of [W/m K]. The negative sign in the equation is to make the result positive even though the temperature gradient, as you can see, is negative. This law cannot be proved, it is simply the result of experimental observation.

105

Notice that we have used Q for a heat interaction, and now we are using $\dot{q}$ for a *heat flux*. We shall use the following nomenclature and units:

Q for *heat quantity*, in [J] or [kJ], and q for *heat quantity per unit area*, in [J/m^2] or [kJ/m^2]. Also:

$\dot{Q}$ for *heat transfer rate*, in [W] or [kW], and and $\dot{q}$ for *heat flux*, in [W/m^2] or [kW/m^2].

Thermal conductivity is another intensive property, which depends on the material. Metals, used in boilers and other heat transfer equipment, have high values, and non-metals, which are used for insulation, have low values. For example, k for glassfibre insulation is 0.037 [W/mK], and k for 1% chrome steel is 60.6 [W/mK], or 1638 times as much.

The thermal conductivity of brick is 0.7 [W/mK]. If the temperature of the surface of the wall inside a house is 22°C and outside it is 8°C, we may calculate the conduction rate through the wall, which is 260 mm thick.

In Fourier's equation, T_1 is 22°C and T_2 is 8°C; the wall thickness is 0.26 m, so the temperature gradient is:

$$(8 - 22)/0.26 = -53.856 \text{ [K/m]}$$

Hence:

$$\dot{q} = -0.7 \times (-53.856) \text{ [W/m}^2\text{]}$$
$$= 37.69 \text{ [W/m}^2\text{]}$$

If a temperature gradient of -100 [K/m] exists across a plate of steel for which the thermal conductivity is 48 [W/mK], what is the conduction rate for a plate area of 2 [m^2]?

106

9600 W

We already have the temperature gradient of -100 [K/m], so we can use Fourier's equation straight away, using the k of 48 [W/mK]:

$$\dot{q} = -48 \times (-100) \text{ W/m}^2$$

So for 2 m^2 the conduction rate is $2 \times (-48 \times (-100)) = 9600$ [W].

A heat transfer at the boundary of a system is a quantity of heat Q in [J] or [kJ], and as a conduction rate calculated by Fourier's law might vary over the time of the interaction, as temperatures change Q would be an integration over that time. For a constant heat transfer rate, Q is simply the product of the rate and time. We also need to add the correct sign to Q, thus although Q calculated by Fourier's law is always positive, this would then be made negative for heat transfer from a system and only positive for heat transfer to the system.

Thus in the example in the previous frame, we may assume a constant conduction rate from a building of wall area 42 m^2 for a period of 12 hours.

$$Q = (37.69 \times 42)\,[\mathrm{W}] \times (12 \times 3600)\,[\mathrm{secs}] = 1582.98\,[\mathrm{W}] \times 43\,200\,[\mathrm{s}].$$

Thus

$$Q = 68\,384\,736\,[\mathrm{J}]\ (\text{since } [\mathrm{W}] = [\mathrm{J/s}])$$

We can express this in [MJ] and add the correct sign for heat transfer from a system:

$$Q = -68.385\,[\mathrm{MJ}]$$

What is Q for conduction through the cylinder walls of an engine, assuming an average temperature gradient of -975 [K/m], an area for conduction of 0.2 [m^2], when conditions may be assumed steady for 45 minutes? Thermal conductivity of the cylinder walls is 73 [W/mK].

107

−38.435 MJ

The conduction rate is $-73 \times (-975) \times 0.2 = 14\,235$ [W]

This takes place for $45 \times 60 = 2700$ seconds. Since heat transfer from the system is negative, $Q = -14\,235\,[\mathrm{J/s}] \times 2700\,[\mathrm{s}] = -38\,434\,500\,[\mathrm{J}]$

So: $Q = -38.435\,[\mathrm{MJ}]$

It is beyond our immediate interest to look any deeper into conduction theory, books on heat transfer will take you further. The important point is to see how Q in thermodynamics is related to heat transfer rates found from the natural laws.

108

Convection is heat transfer between a fluid and a solid surface. The surface may be hotter or colder than the fluid. Conduction takes place in the fluid very close to the surface, and further away fluid motion carries energy to or from the surface region. The overall process is described by Newton's law, in which h is the convection coefficient with units [W/m² K]:

$$\dot{q} = h\,(T_1 - T_2)\ [\mathrm{W/m^2}]$$

The process of convection, from or to a surface, is shown below.

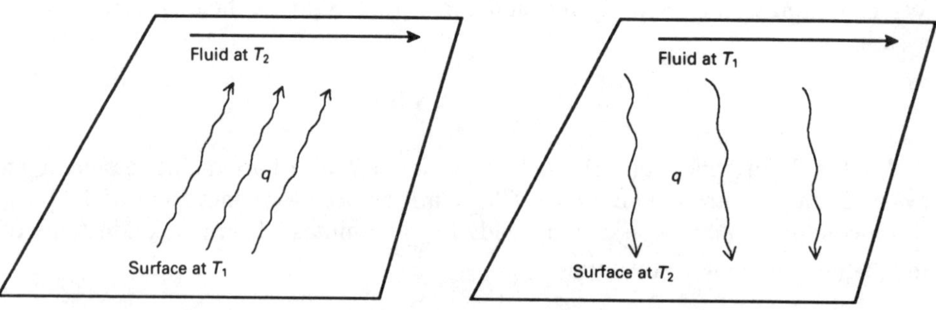

Newton's law, like Fourier's law, is based on experiment. Whereas in Fourier's law k is a property of the conducting material only, h in Newton's law depends not only on properties of the fluid (thermal conductivity, viscosity, density, specific heat), but also on the velocity of flow and the shape of the fluid/solid boundary. The convection coefficient h is found by computer analysis of the flow, or by measurement, and is then used in similar situations in design.

109

A wind sets up a convection coefficient of 8 [W/m² K] from the roof of a house. The roof temperature is 15°C and the air temperature is 5°C. For a roof area of 75 m², we may calculate the rate of heat loss.

The convection rate is the coefficient h times the temperature difference.

Thus:

$$\dot{q} = 8 \times (15 - 5) = 80\ [\mathrm{W/m^2}]$$

The area of the roof is 75 [m^2], so the rate of heat loss is:

$$\dot{Q} = 80 \times 75\ [\mathrm{W/m^2}]\ [\mathrm{m^2}]$$
$$= 6000\ [\mathrm{W}]$$
$$= 6\ [\mathrm{kW}]$$

In an experiment to find the convection coefficient from a 100 W light bulb, the surface temperature of the bulb is found to be 95°C in an air temperature of 15°C. Assume the convecting area is that of a 70 mm diameter sphere, and 84 W is lost by convection. What is h?

110

$$\boxed{h = 68.005\ \mathrm{W/m^2\ K}}$$

The surface area of the bulb is $4\pi r^2 = 4 \times \pi \times 0.035^2 = 0.0154$ [m^2]. Using Newton's law, we have $q = h \times (95 - 15) = h \times 80$ [W/m^2] and this takes place over the area of 0.0154 [m^2].

Hence: $84 = 0.0154 \times h \times 80$ [W]

So: $h = 84/(0.0154 \times 80)$

$= 68.005$ [W/m^2 K]

As well as depending on the fluid properties and the shape of the boundary, h also depends on whether the flow is **natural** or **forced**. Forced convection occurs where the fluid motion is produced by an external agent. In the example in Frame 109, although wind occurs naturally, the cooling would have been by forced convection. You hear people say 'heat rises' which means, for example, that air in a room which has been warmed by a convector will rise since its density is lowered. This is true natural convection, between the convector and the air. The velocity of flow is also important, giving rise to laminar (smooth) or turbulent (disturbed) flow. Turbulent forced convection gives the greatest values of h. Q for a heat interaction is found from $\dot{q}$, as we did for conduction.

A cup of tea is cooled by natural convection. The rate of heat loss is proportional to the temperature difference, so the temperature of the tea initially falls quicker. If you have to catch a train, and the tea is too hot, should you add cold milk straight away, or wait as long as possible, to get the greater cooling?

111

Wait as long as possible

As the tea cools faster when it is hot, greater overall cooling will be achieved by allowing it to cool as long as possible, and then add cold milk. If the milk had been added first, the subsequent cooling would have been slower.

The energy received on earth from the sun is transmitted by radiation. All radiation occurs best in a vacuum, and is partially absorbed in gases, particularly carbon dioxide and methane which have become known as 'the greenhouse gases'. When radiation is absorbed by a gas, the gas is warmed, so when solar radiation is absorbed by gases in the atmosphere, the atmospheric temperature rises, causing 'global warming'.

Radiation also depends on the nature of the surface. Thus a matt black surface radiates well, and a polished shiny surface like aluminium foil is a poor radiator.

Why is a shiny foil 'blanket' used to protect athletes after a marathon?

112

To reduce loss of body energy by radiation

If an athlete is exhausted, he or she cannot afford to lose energy, and this is reduced by limiting the radiation from the body.

Surfaces which radiate well also absorb radiant energy more easily, and white and shiny surfaces reflect radiant energy. Houses in hot countries are often painted white.

The idea of a perfect radiator arises. The emission of energy by radiation is linked to the kinetic energy of the molecules, and at a particular kinetic energy level, or temperature, there is a maximum radiation emission. A surface radiating in this way is said to be 'black', and it has an emissivity of 1. Most surfaces radiate less than this, and are said to be 'grey'. Black body radiation was studied by Stefan and Boltzmann, and the law of black body radiation is named after them. This is given in the next frame.

113

The Stefan–Boltzmann law is:

$$\dot{q} = \sigma \, T^4_{\text{abs}} \ [\text{W/m}^2]$$

where σ is the Stefan–Boltzmann constant, and has the value 5.67×10^{-8} [W/m^2 K^4] and T_{abs} is the absolute temperature in Kelvin, K.

The Stefan–Boltzmann law was established in two ways. Stefan was the first to show experimentally that black body emission depended on the fourth power of the absolute temperature. Max Planck determined the way in which black body radiation was distributed over the range of wave lengths making up the radiation. This gave a complicated function which Boltzmann was able to integrate to confirm the experimental law of Stefan.

The surface of the sun has a temperature of roughly 6000 K (it is very much hotter inside), and it radiates closely to the level of a black body. We may calculate the emission as:

$$\dot{q} = 5.67 \times 10^{-8} \times (6000)^4 \text{ [W/m}^2\text{]}$$

It is easier if we divide the temperature by 100 first:

$$\dot{q} = 5.67 \times (60)^4 \text{ [W/m}^2\text{]} = 5.67 \times (60)^4 \times 10^{-3} \text{ [kW/m}^2\text{]}$$

$$\therefore \quad \dot{q} = 73\,483 \text{ [kW/m}^2\text{]}$$

This is per m^2 of the sun's surface. The energy received on the surface of the earth has a maximum of about 1 kW per m^2 of the earth's surface. The much reduced level is due to the distance from the sun.

114

For any surface which is not black (which means most surfaces), the amount emitted depends on the emissivity of the surface, ε. This has a value less than 1, and is highest for rough surfaces like brick, and lowest for shiny surfaces like highly polished metal. Thus ε is about 0.8 for brick and 0.05 for polished aluminium.

The Stefan–Boltzmann Law is then modified:

$$\dot{q} = \varepsilon\, \sigma\, T^4_{abs} \text{ [W/m}^2\text{]}$$

At a given temperature the rate of radiation will be directly proportional to the emissivity, and as it depends on the fourth power of the absolute temperature, the effect of temperature level on radiation at higher temperatures is very striking. Compare the rates of radiation in [W/m^2] from a grey surface at 0°C, 300°C and 600°C for an emissivity of 0.75.

115

$236.73\ W/m^2$, $4588.99\ W/m^2$, $24\ 717.23\ W/m^2$

At 0°C, which is 273.15 K, $\dot{q} = 0.75 \times 5.67 \times 2.7315^4 = 236.7$ [W/m²]

At 300°C, which is 573.15 K, $\dot{q} = 0.75 \times 5.67 \times 5.7315^4 = 4589$ [W/m²], and at

600°C, which is 873.15 K, $\dot{q} = 0.75 \times 5.67 \times 8.7315^4 = 24\ 717$ [W/m²].

The three rates are in the ratio of 1, 19.4 and 104.4.

The equations of radiation used so far only tell us how much radiation is emitted by a surface. An equation to give the radiation passing between one surface and another is rather more complex. It is of the form:

$$\dot{Q}_{1-2} = \mathcal{F}_{1-2}\, A_1\, \sigma\, (T_1^4 - T_2^4)\ \text{[W]}$$

where $\mathcal{F}_{1-2}$ is a measure of the view surface 1 has of surface 2; it is also a function of the emissivities if the surfaces are grey. A_1 is the area of surface 1, and the equation involves the temperatures of both surfaces since surface 2 is radiating to surface 1 as well as receiving radiation from surface 1. This equation expresses a radiation rate between two temperatures, in the same way that Fourier's and Newton's laws do for conduction and convection respectively.

Consider a small laboratory oven. Inside the walls and top, together of area 0.74 m², at 300°C, radiate to metallurgical specimens on the floor which have a temperature of 150°C. The floor area and emissivities give an $\mathcal{F}_{1-2}$ of 0.156. We may calculate the radiation rate between the walls and the floor.

$T_1 = 573.15$ K, $T_2 = 423.15$ K and $A_1 = 0.74$ m², so using the equation just given:

$$\dot{Q}_{1-2} = 0.156 \times 0.74 \times 5.67 \times (5.7315^4 - 4.2315^4)\ \text{[W]}$$

Note how we always get rid of 10^{-8} in the Stefan–Boltzmann constant and divide the temperatures by 100. Then:

$$\dot{Q}_{1-2} = 0.6545 \times (1079.12 - 320.61)$$

$$\therefore \quad \dot{Q}_{1-2} = 496.44\ \text{[W]}$$

A black object of area 0.25 m² is at 400°C and is completely surrounded by a brick enclosure at 60°C. For this case $\mathcal{F}_{1-2} = 1$. What is the rate of radiation to the enclosure?

116

2735.9 W

We use $\dot{Q} = \mathcal{F}_{1-2} A_1 \sigma (T_1^4 - T_2^4)$ with $\mathcal{F}_{1-2} = 1$, $A_1 = 0.25$, $T_1 = 673.15$ K and $T_2 = 333.15$ K.

Hence:
$$\begin{aligned}\dot{Q} &= 1 \times 0.25 \times 5.67 \times (6.7315^4 - 3.3315^4)\\ &= 1.4175 \times (2053.276 - 123.185)\\ &= 1.4175 \times 1930.091\\ &= 2735.9\ [\text{W}]\end{aligned}$$

We have seen some simple examples of how heat interactions may be calculated. Heat flows naturally from a high to a low temperature, and a positive result is obtained. As we saw in Frame 106, to this we must add the correct sign for a heat interaction. Thus, if a system is hotter than its surroundings, the heat interaction is **negative**. If you want to cause a heat transfer from a cold body to a hot body, then you have to do work. A device which does this is called a heat pump. An example is a refrigerator.

Clearly the natural flow of heat is always irreversible, but in the next frame we shall try to imagine how a reversible heat transfer could occur.

117

Suppose a system were to have its temperature increased from T_A to T_B. Let the surroundings initially be at T_A, with the system in equilibrium. The temperature of the surroundings is controlled so that it is always an infinitesimal increment in temperature hotter than the system, to maintain a heat transfer to the system until T_B is reached. The heat transfer is infinitely slow so between these temperatures the system passes through a series of equilibrium states. The process could be reversed at any point by making the temperature of the surroundings an increment of temperature cooler. Thus a system could undergo a reversible heat transfer, but the process would be infinitely slow, and of no practical value.

118

In practice, a heat transfer in an industrial process may be made as efficient as possible by following this idea, but using a finite temperature increment. The figure shows how a fluid passing along a pipe has its temperature increased from T_3 to T_4 by another fluid flowing in the opposite direction, and as it does so falling in temperature from T_1 to T_2. The differences $T_1 - T_2$ and $T_4 - T_3$ are equal, with equal thermal capacities of both streams.

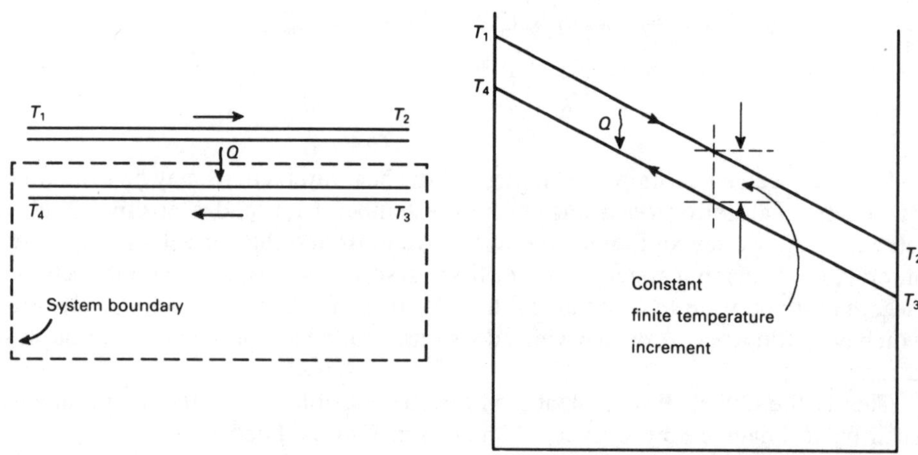

The fluid receiving a positive heat transfer is in a pipe forming part of a closed system within its system boundary. The efficiency of the actual heat transfer process is expressed as:

$$\eta = \frac{\text{temperature rise of heated stream}}{\text{maximum temperature difference}}$$

$$\eta = \frac{T_4 - T_3}{T_1 - T_3} = \frac{T_1 - T_2}{T_1 - T_3}$$

When the heat transfer is thought of as reversible, the temperature increment is infinitesimal and $T_4 = T_1$, and so the efficiency becomes:

$$\eta_R = \frac{T_1 - T_3}{T_1 - T_3} = 100\%$$

However, a 100% efficient heat transfer that is infinitely slow is of no practical use. As a finite temperature increment is increased, the efficiency drops, but the rate of heat transfer increases. The designer must try to achieve an optimum result from these opposing tendencies.

Using the above expression, what is the efficiency of a heat transfer for which $T_1 = 500°C$, $T_2 = 400°C$, $T_3 = 380°C$ and $T_4 = 480°C$?

119

83.33%

We have that: $\eta = \dfrac{480 - 380}{500 - 380} = \dfrac{100}{120} \times 100\% = 83.33\%$

You will recall how in Frame 28, we described work as a path function, so that:

$$W = \int_1^2 dW, \text{ where } dW \text{ is the area } pdV$$

In the same way, heat may be described as a path function, since the value of Q in any process depends on how the interaction occurs. We may write:

$$Q = \int_1^2 dQ$$

since dQ is inexact. Later we shall be able to show dQ graphically, by plotting one property against another, as we did for dW in plotting p against V.

120

Before leaving our discussion of work and heat, we return to the question of difficulty that may arise in seeing the difference between the two. In the example shown below, on moving the system boundary, a work interaction becomes a heat interaction. Can you see why this is?

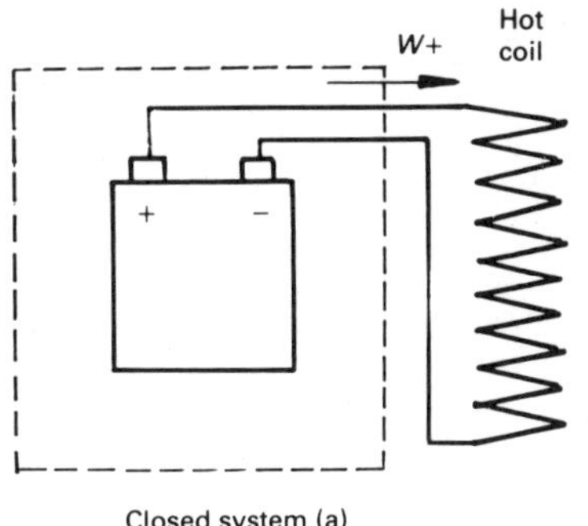

Closed system (a)

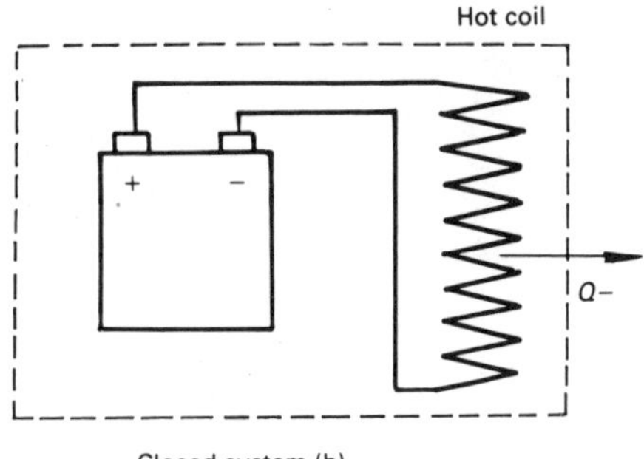

Closed system (b)

121

Look at figure (a). This shows an electric battery as a closed system connected to a coil which is red hot as a result of current passing through it. The battery is performing a positive work interaction on the surroundings which in this case is the coil, which gets hot. However, the same electrical work could be done on a motor working in an ideal way to give an equal amount of mechanical work.

In (b), the system boundary contains the coil, and there is no longer any electrical work involved at the boundary. In (a), electrical work is done by the system, even though in this case its effect on the surroundings is a heating effect by the coil. In (b), the boundary interaction is clearly heat, and only a fraction of this could be used to do mechanical work, owing to the limit law first mentioned in the previous programme.

See if you can show how, by sketching, that boundary interactions of fluid shear work and solid frictional work can be seen to become heat interactions if the boundary of the original system is altered.

122

The figures below show how heat flow can be shown to originate from work interactions of fluid shear and solid friction.

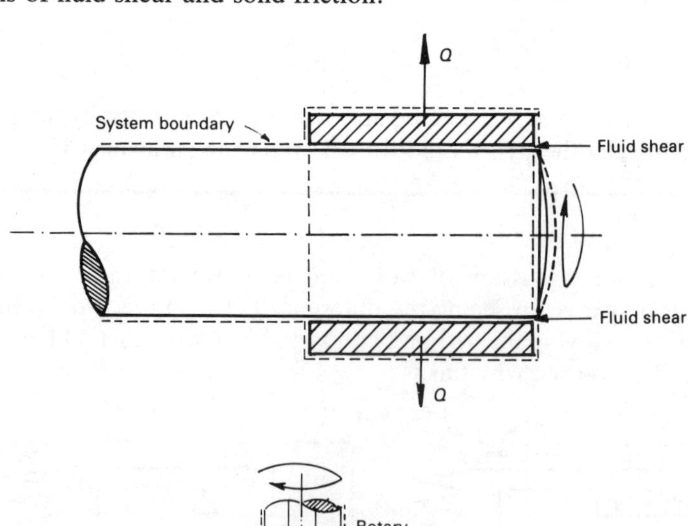

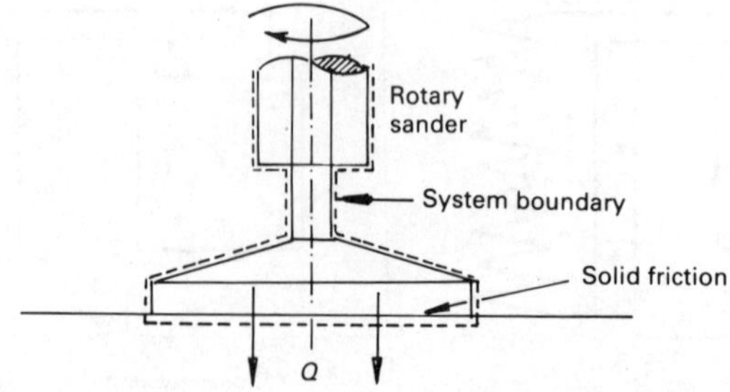

PROBLEMS INVOLVING HEAT

1. A brick wall is 300 mm thick and a temperature difference of 20 K exists across it. Thermal conductivity of brick is 0.85 W/m K. What is the rate of conduction through the wall, per m^2 of wall area? [56.67 W]
2. The wall in problem **1** is now insulated on the inside surface by adding a layer of insulation foam 30 mm thick, for which the thermal conductivity is 0.04 W/m K. Calculate the percentage reduction in the rate of heat flow through the wall. [68.0%]
3. Calculate the heat transfer in problem **2** over 24 hours for a wall area of 30 m^2, assuming the 20 K temperature difference is maintained. [47 000 kJ]
4. Heat exchanger tubes at a mean temperature of 200°C are cooled by water flowing across them at a mean temperature of 30°C. There are 100 tubes, 0.8 m long by 25 mm diameter. The convection coefficient is 460 W/m^2 K. Calculate the rate of heat transfer by convection. [491.35 kW]
5. A room is heated by natural convection from a wall panel at 60°C. The panel area is 0.8 m^2 and the room air temperature is 20°C. The convection coefficient is 30 W/m^2 K. Calculate the heat transfer to the room in 15 minutes. [864 kJ]
6. The area for cooling inside the water jackets of a petrol engine is 0.2 m^2. The convection coefficient to the cooling water is 800 W/m^2 K. The mean jacket surface temperature is 120°C and the mean water temperature is 60°C. Calculate the rate of heat loss to the cooling water. [9.6 kW]
7. A furnace wall is maintained at 500°C. The emissivity of the brick is 0.8. The total wall area is 12.8 m^2. Calculate the radiation from the wall. ($\sigma = 5.67 \times 10^{-8}$ W/m^2 K^4) [207.3 kW]
8. Calculate the temperature of the wall in problem **7** to reduce the radiation rate by 50%. [377°C]
9. An athlete is wrapped in a foil blanket of emissivity 0.05. Assuming the area of the blanket is 1.5 m^2 and that it takes up body temperature of 36.7°C, calculate the athlete's rate of heat loss by radiation. [39.12 W]
10. Consider the light bulb question in Frames 109 and 110. Assume the glass bulb at 95°C radiates as a black body, calculate the rate of radiation. [16.01 W]

Points to remember
A system is in an **equilibrium state** when all intensive properties are uniform. Temperature difference is the driving force for a heat interaction. Pressure difference is the driving force for displacement work.

A **reversible process** will follow a **state path** of equilibrium states on a property diagram. All **real** processes are **irreversible**.

A system does positive work on the surroundings; and has negative work done on it.

Displacement work = $\int p\mathrm{d}V$

The pressure-volume function of a polytropic process is pV^n = constant.

The work of a polytropic process is: $W = \dfrac{P_2V_2 - P_1V_1}{1 - n}$.

When $n = 1$, the function is a rectangular hyperbola pV = constant, which is Boyle's law for a gas. The work is: $W = p_1V_1\ln(V_2/V_1)$.

Heat transfer to a system is positive; from a system it is negative.

In S.I. units, the mechanical equivalent of heat is 1; heat, like work, is measured in joules.

Heat transfer by conduction is given by Fourier's law: $\dot{q} = -k\,(\mathrm{d}T/\mathrm{d}X)$ [W/m²] where $(\mathrm{d}T/\mathrm{d}X)$ is a negative temperature gradient and k is the thermal conductivity.

Heat transfer by convection is given by Newton's law: $\dot{q} = h(T_1 - T_2)$ [W/m²] where h is the convection coefficient.

Heat emission by radiation for a black body is given by Stefan-Boltzmann's law: $\dot{q} = \sigma T_{abs}^4$ [W/m²] where σ is the Stefan-Boltzmann constant = 5.67×10^{-8} [W/m² K⁴]. Heat transfer by radiation is given by: $\dot{Q} = \mathcal{F}_{1-2}A_1\sigma(T_1^4 - T_2^4)$ [W] for a body of area A_1 at T_1 radiating to a body at T_2, for which the view factor is $\mathcal{F}_{1-2}$, and T_1 and T_2 are absolute temperatures in Kelvin.

Programme 3

THE FIRST LAW AND ENERGY

1

In Frame 10 of the previous programme, we defined a process and state path, and you saw how a process between two equilibrium end states can be shown graphically. A process is caused by a heat or work interaction, or by both simultaneously, and a system could undergo a series of processes as a result of such interactions.

If after a series of processes, a system has a final end state identical to the initial state before the first process, then the system has executed a **cycle**. The minimum number of processes to complete a cycle is two, as shown in the figure.

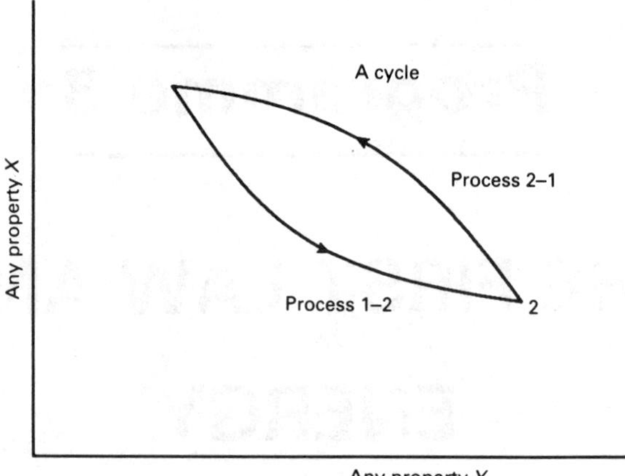

On a *p–V* diagram, a cycle can be built up from any of the pressure–volume relationships we have met. The simplest would be a rectangle, with lines of constant pressure and constant volume.

Sketch a cycle on *p–V* coordinates consisting of two constant pressure and two constant volume processes, involving a pressure ratio of 2 to 1 and a volume ratio of 3 to 1.

2

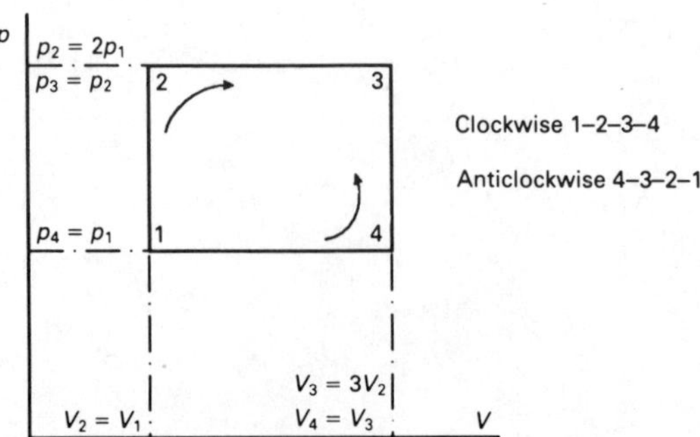

The cycle would look like this. It would be possible to go round this or any cycle in either a clockwise or anticlockwise direction. The direction is important, as we shall see later.

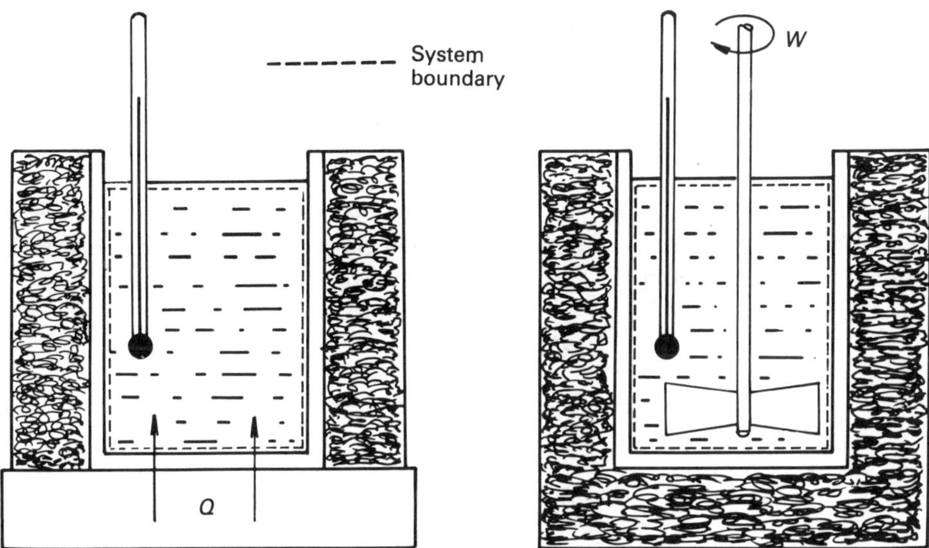

Identical effects of heat and work

In the Joule experiment, shown again above, the same temperature rise in the calorimeter is achieved by a heat transfer and by a stirring work interaction; if the work interaction is followed by a negative heat transfer, the water would return to its original state and a cycle would have been executed.

We have seen that the mechanical equivalent of heat is unity with heat and work having the same units, the joule. If in a Joule experiment the work is −4000 J (negative work done on the water system), to raise the temperature of the water slightly, what is the value of the heat transfer to restore the water to its original state?

3

−4000 J

To restore the water to its original temperature, the heat transfer must also be −4000 J, since it must be heat transfer to the surroundings.

4

This result is the basis of a statement of the **First Law of Thermodynamics**, which you must know already as the law of conservation of energy. The statement is:

When any system executes a cycle of processes, the net heat is equal to the net work.

Thus there is conservation of heat and work in a cycle. This is true no matter how many different heat and work interactions occur throughout the cycle. This statement may be expressed as:

$$\sum_{\text{cycle}} \delta Q = \sum_{\text{cycle}} \delta W$$

This is the first of two very important laws in Thermodynamics. Make sure you remember it. This expression may be written as:

$$\sum_{\text{cycle}} \delta Q - \sum_{\text{cycle}} \delta W = 0$$

Other ways of stating this law will follow, and they are all equivalent to the law of conservation of energy.

5

Thinking again about the Joule experiment, there is a restriction on the way the heat and work interactions occur, and a reverse of the Joule experiment is not possible. If we were to heat the water in the container, say by the same amount of 4000 J (the heat transfer is now +4000 J), we know that we could not obtain +4000 J of work while the temperature fell to its original value.

We may use the First Law to equate heat and work interactions in a cycle. Thus in a cycle of three processes, the first and third heat interactions are −700 J and 0 J, while the three work interactions are −700 J, 0 J and +1200 J. We may calculate the second heat interaction:

The sum of the heat interactions is:

$$\sum \delta Q = -700 + Q_2 + 0 = -700 + Q_2$$

The sum of the work interactions is:

$$\sum \delta W = -700 + 0 + 1200 = +500 \text{ [J]}$$

Hence: $-700 + Q_2 = +500$ [J]

So: $Q_2 = +1200$ [J]

Now, in a cycle of four processes, the heat interactions are 0, +10, 0 and −7.5 kJ; and the first, second and fourth work interactions are −6.0, 0 and 0 kJ. What is the value of the third work interaction?

+8.5 kJ

The sum of the heat interactions is:

$$\sum \delta Q = 0 + 10 + 0 - 7.5 = +2.5 \text{ [kJ]}$$

The sum of the work interactions is:

$$\sum \delta W = -6.0 + 0 + W_3 + 0 = -6.0 + W_3 \text{ [kJ]}$$

Equating the two: $-6.0 + W_3 = +2.5$ [kJ]

So: $W_3 = +8.5$ [kJ]

If we list out the interactions in a table, we can see that the work interactions were not accompanied by heat, and the heat interactions were not accompanied by work. Thus:

Process	Q	W
1–2	0	−6
2–3	+10	0
3–4	0	+8.5
4–1	− 7.5	0

The work interactions without heat are called **adiabatic** and the heat interactions without work would be at **constant volume**. On a p–V diagram the cycle would look something like this for a gas undergoing the cycle in a piston–cylinder arrangement.

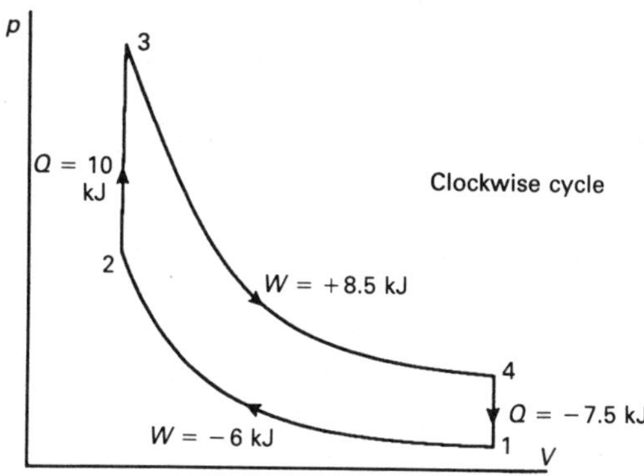

7

You can see there is net positive work from the p–V diagram, of +2.5 kJ, for a net positive heat of +2.5 kJ. This shows that the First Law is confirmed. The heat transfer of 10 kJ to the gas is at a high temperature, and is from the **source**, and the heat transfer of −7.5 kJ from the gas is at a low temperature and is to the **sink**. This shows the cycle proceeding in a clockwise direction. It would be possible for the cycle to go round the other way. The signs of all the interactions would be reversed. Now what are Q and W for the **same** (reversible) processes going in an anticlockwise direction?

8

Process	Q	W
2–1	0	+6
3–2	−10	0
4–3	0	−8.5
1–4	+ 7.5	0

The figure would be like this:

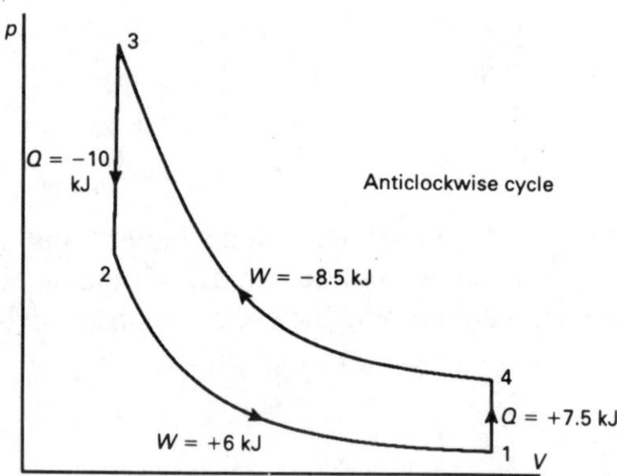

9

There is now a net negative work of −2.5 kJ and a net negative heat of −2.5 kJ, so we have in effect a repeat of the Joule experiment. You may also have spotted that the heat transfer of +7.5 kJ is now from the sink (Frame 7), so is at a low temperature, and the heat transfer of − 10 kJ is now to the source and is at a high temperature. The net negative work of 2.5 kJ causes 7.5 kJ of heat to leave the low temperature region, and 10 kJ of heat to enter the high temperature region. This is an example of a heat pump.

A cycle, as we have seen, is made up of processes, and so the next step is to see how the First Law can be formulated for a single process, or for a number of processes which do not constitute a cycle.

10

We already know that work is done by the expenditure of energy, and from Joule's experiment we also know that work and heat can produce the same effects on a system, and we may conclude that heat transfer will also occur by the expenditure of energy.

Thus heat and work interactions on a system will produce changes in the energy of a system. In a similar way that raising a body increases its potential energy, or compressing a spring increases its strain energy, so also heat and work interactions on some gas in a cylinder will change the energy of the gas. Remember (Frame 6) that an **adiabatic** process is one in which there is no heat transfer. Also remember that heat transfer depends on **time**, so if gas in a cylinder is compressed rapidly it will be nearly adiabatic because there is little time for any heat transfer. If we compress a gas adiabatically (in the ideal case because we cannot insulate perfectly), we will increase the energy of the gas by the amount of work done, and for a reversible compression, that gas could return to its original state and do an equal amount of work. The process would be like compressing a spring.

However when a gas is compressed, its temperature rises and heat transfer will naturally occur to try to restore the temperature to its original value. So during a non-adiabatic compression, heat transfer will occur simultaneously with the temperature rising and at the end of compression over the same reduction in volume the temperature and energy of the gas will be less than those for the adiabatic compression. If we then allow the gas to expand to its original state, to obtain the same work that was done in compression, the reverse of the heat transfer would also have to take place. We see then that in compressing gases, any heat transfer is important and has to be involved; in compressing springs, the strain energy involved is not affected by any heat transfer, so even if there is any heat transfer it is not included in energy calculations.

These events may be summed up by this statement:

When a system undergoes a change of state during a process the change in the energy of the system is numerically equal to the net heat minus the net work occurring at the system boundary during the process.

This is another statement of the law of conservation of energy, applied this time to a process, rather than to a cycle. We need to establish that energy is a property of a system.

11

The difference between an adiabatic and non-adiabatic compression of a gas is shown in the figure below.

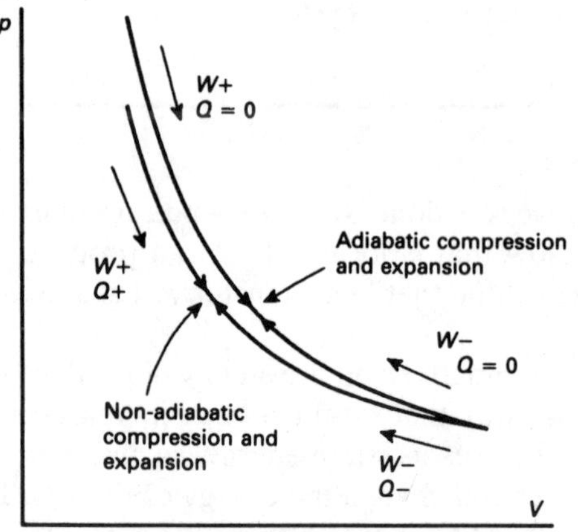

If we use E to represent energy, and we consider heat and work interactions to cause a system to change from state 1 to state 2, we may write:

$$Q - W = E_2 - E_1$$

where $Q = \sum_1^2 \delta Q$ and $W = \sum_1^2 \delta W$ for the process.

The units of this equation is the joule, the fundamental unit of work, heat and energy.

It is not necessary for both Q and W always to occur in a process, one or the other may be zero.

In the compression of a spring, where $Q = 0$, the equation is simply:

$$-W = E_2 - E_1$$

For the adiabatic compression of a gas, the same equation would apply; for the heating of water in a vessel, $W = 0$, and so:

$$Q = E_2 - E_1$$

Suppose a gas expands and 10.7 kJ of work is done by the gas. During the process there is a heat transfer of 4.2 kJ to the gas. We may find the change of energy. Both Q and W are positive, so:

$$4.2 - (+10.7) = E_2 - E_1$$

$$\therefore \quad E_2 - E_1 = 4.2 - 10.7$$

$$= -6.5 \text{ [kJ]}$$

If a gas is compressed and the work done on the gas is −8.5 kJ, and a heat transfer from the gas is −2.5 kJ, what is the change of energy of the gas?

12

+6.0 kJ

We use:

$$Q - W = E_2 - E_1$$

$\therefore$ $$-2.5 - (-8.5) = E_2 - E_1$$

hence: $$E_2 - E_1 = +6.0 \text{ [kJ]}$$

We now need to establish that **energy is a property**.
First we need to think about the characteristics of properties. A property of a system is recognised by the fact that its value depends only on the state of a system. Work and heat are path functions which have different values by different routes, but a property like temperature, pressure or energy, depends only on the state point and not on any route followed to reach that point.

The idea of a property is general, for example an aircraft can have latitude, longitude and altitude; it can also fly long or short distances between any two points. The latitude, longitude and altitude are properties of the aircraft indicating its state at a particular instant that the traffic controller is anxious to know, but the distance flown is a variable path function between the two points.

We will now look at the proof that **energy is a property**.

13

The figure shows two state points, 1 and 2, on a property diagram. The precise properties do not matter. Two cycles are shown: 1–A–2–C–1 and 1–B–2–C–1. Paths A and B are any two paths between points 1 and 2.

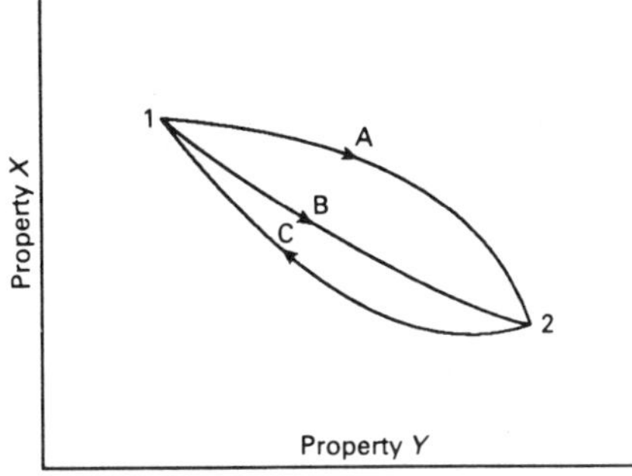

14

For these two cycles we may write:

$$\sum_{\text{cycle}} \delta Q - \sum_{\text{cycle}} \delta W = \sum_{1\,\text{A}}^{2} \delta Q - \sum_{1\,\text{A}}^{2} \delta W + \sum_{2\,\text{C}}^{1} \delta Q - \sum_{2\,\text{C}}^{1} \delta W = 0$$

for the cycle 1–A–2–C–1, and

$$\sum_{\text{cycle}} \delta Q - \sum_{\text{cycle}} \delta W = \sum_{1\,\text{B}}^{2} \delta Q - \sum_{1\,\text{B}}^{2} \delta W + \sum_{2\,\text{C}}^{1} \delta Q - \sum_{2\,\text{C}}^{1} \delta W = 0$$

for the cycle 1–B–2–C–1.

In these cycles path C is common, but paths A and B are different. By combining these two equations, we see that:

$$\sum_{1\,\text{A}}^{2} \delta Q - \sum_{1\,\text{A}}^{2} \delta W = \sum_{1\,\text{B}}^{2} \delta Q - \sum_{1\,\text{B}}^{2} \delta W$$

Paths A and B were any two different paths; this result must be the same for any other two paths between states 1 and 2.

Since $Q = \sum_1^2 \delta Q$ and $W = \sum_1^2 \delta W$ we have:

$$Q - W = \text{constant}$$

for any path between the same two end states. This constant is the change of energy between these two end states, which is independent of the path, and therefore energy is a property.

15

In Frames 11 and 12 we calculated changes of energy in a system resulting from given heat and work interactions for a particular process. Now that we know that the change of energy between two end states is the same for all processes between those end states, we can evaluate other possible heat or work interactions between those end states. Thus in Frame 11 we found:

$$E_2 - E_1 = -6.5 \text{ [kJ]}$$

as the result of a process. If Q is -7.8 kJ for a different path between the same two points, we may calculate W.

Thus:

$$Q - W = E_2 - E_1$$

$$-7.8 - W = E_2 - E_1 = -6.5$$

$$-W = -6.5 + 7.8$$

$$-W = +1.3$$

$$\therefore \quad W = -1.3 \text{ [kJ]}$$

If between two end states the change of energy is +5.0 [kJ], and for a process between these end states $W = -7.5$ [kJ], what is Q?

16

-2.5 kJ

We have, using:

$$Q - W = E_2 - E_1$$

$$Q - (-7.5) = +5.0$$

$$Q + 7.5 = +5.0$$

$$Q = -2.5 \text{ [kJ]}$$

Before moving on, do the following problems on the First Law applied to cycles and processes.

17

PROBLEMS ON THE FIRST LAW

1. In a cycle of two processes only, for the first process $W_1 = -8000$ kJ and $Q_1 = 0$; in the second process $W_2 = 0$. What is Q_2? [−8000kJ]
2. In a proposed cycle of two process only, $W_1 = 0$ and $Q_1 = +8000$ kJ. In the second process Q_2 is said to be 0 and $W_2 = +8000$ kJ. Explain why this is not possible.
3. In a cycle of three processes, $W_1 = -90$ kJ and $W_2 = 0$; $Q_1 = -140$ kJ, $Q_2 = +200$ kJ and $Q_3 = 0$. What is W_3 and what is the net positive work for this cycle? [+150 kJ, +60 kJ]
4. A cycle consists of four processes, and the heat interactions are 0, +12.0, 0 and −10.0 kJ respectively. The second, third and fourth work interactions are 0, +8.5 and 0 kJ. What is the first work interaction? [−6.5 kJ]
5. In a process $Q = +6$ kJ and $W = -4$ kJ. What is the change of energy of the system undergoing this process? [+10 kJ]

6. Between the same end states of problem **5**, in a different process $Q = +4$ kJ. What is W for this process? [−6 kJ]
7. If the cycle of problem **4** proceeds in the opposite direction, what are Q and W for the four processes?
[$Q = 0, -12.0, 0, +10.0$ kJ; $W = +6.5, 0, -8.5, 0$ kJ]
8. Some gas in a container undergoes a process which simultaneously involves stirring work, displacement work and heat. 8.0 kJ of stirring work is performed by an impeller inserted into the container, and the gas does 4.5 kJ of displacement work. The change of energy of the gas during the process is +5.5 kJ. What was the heat transfer? [+9.0 kJ]

18

Another consequence of the law of conservation of energy, when applied to systems, is that **the energy of a system remains unchanged if the system is isolated from its surroundings.**

The reason is that, since the system is isolated, both Q and W are zero, and so there can be no change in E. But the energy of the system, E, must include *all* forms of energy. So far we have spoken only of potential energy and kinetic energy, without actually defining them. What other forms of energy are there?

Imagine a closed system consisting of a ball bearing rolling in a curved dish, inside an insulated container. This will be an isolated system, with no heat or work interactions at the boundary. As the ball rolls back and forth in the dish there will be a continuous interchange between potential and kinetic energy, but E will remain constant. As the ball rolls there will be friction. What will be the effect of this?

19

The ball will come to rest at the bottom of the dish

Friction will occur between the ball and the dish and between the ball and the air, and gradually the ball will come to rest at the bottom of the dish. We may continue to assume the system is isolated so that both Q and W are zero, and yet we find that the ball once having potential energy or kinetic energy, now resting at the bottom of the dish, has neither. Yet we say that the energy of the system has not changed. In what form is this energy?

Careful observation would tell us that the temperature of the ball, the dish and the air in the system has risen very slightly. **The form of energy associated with**

temperature change in a system, and directly affected by a heat interaction, is called internal energy.

Internal energy is that mode of energy related to temperature, and on a molecular level, internal energy is associated with the kinetic energy of the molecules of the system. In any natural or man-made process, all forms of energy ultimately become degraded into internal energy. In a system in equilibrium, the kinetic energy levels, or the temperatures, are uniform throughout the system, and always any non-equilibrium state is followed by other states approaching equilibrium.

In this case there has been no heat interaction, but the kinetic and potential energy have been converted into internal energy. A heat interaction would only have made the ball and dish warmer, it would not have increased the potential or kinetic energies, so the ball would not roll faster or rise higher.

20

We hear people say 'friction produces heat', but this is not correct. You should say that friction produces that mode of energy which is directly affected by a heat interaction. In the previous programme we have seen how work may be 'lost' in a system and a heating effect produced; thus stirring work on a system increases the internal energy of a system, and electrical losses in a motor increase the internal energy of the motor, to raise its temperature.

We now have:

$$U = \text{internal energy} \quad [\text{J}]$$

and

$$E = U + \text{P.E.} + \text{K.E.} \quad \text{where:}$$

$$\text{Potential energy} = \text{P.E.} = m\,g\,h \quad [\text{J}]$$

$$\text{Kinetic energy} = \text{K.E.} = \tfrac{1}{2}\,m\,V^2 \quad [\text{J}]$$

Assuming no friction we can calculate the highest velocity achieved, if the ball falls from say, 100 mm. Thus:

$$\text{P.E.} = \text{K.E.}$$

So:

$$m \times 9.81 \times 100 \times 10^{-3} = \tfrac{1}{2} \times m \times V^2$$

$$\therefore V^2 = \frac{9.81 \times 100 \times 2}{1000}$$

$$\therefore V = 1.4007 \text{ m/s}$$

In this calculation we did not need to know the mass of the ball. What is the kinetic energy of the ball at this velocity, assuming the mass is 20 g?

21

0.019 62 J

The kinetic energy is $\frac{1}{2} m V^2 = \frac{1}{2} 20 \times 10^{-3} \times 1.4007^2$

$$= 0.019\ 62\ [\text{J}]$$

If we did not have the mass of the ball bearing, we could only have found the **specific kinetic energy**, or the kinetic energy per unit mass, $V^2/2$. We see then that energy is an **extensive property**, and specific energy is an **intensive property**. We have, then:

$$\frac{E}{m} = \frac{U}{m} + \frac{\text{P.E.}}{m} + \frac{\text{K.E.}}{m}\ [\text{J/kg}]$$

$$\therefore \qquad e = u + \text{p.e.} + \text{k.e.}\ [\text{J/kg}]$$

where e = specific energy

u = specific internal energy

p.e. = specific potential energy

k.e. = specific kinetic energy

This exercise illustrates the redistribution of energy in an isolated system:

The piston, crank, and flywheel of an engine has 40 kJ of kinetic energy. The assembly is brought to rest by compression of gas in the engine cylinder. Treating the gas and moving assembly as an isolated closed system (both Q and W are zero), what is the change of internal energy of this system?

22

+40 kJ

From Frame 11 we have: $\quad Q - W = E_2 - E_1$

and from Frame 20: $\quad E = U + \text{P.E.} + \text{K.E.}$

$$\therefore \qquad Q - W = (U + \text{P.E.} + \text{K.E.})_2 - (U + \text{P.E.} + \text{K.E.})_1$$

In this equation Q and W are zero, from the statement of the problem, and in the energy terms potential energy will be absent, or at least, the same in states 1 and 2. Hence the equation becomes:

$$0 = (U + \text{K.E.})_2 - (U + \text{K.E.})_1$$

or:

$$U_2 - U_1 = (\text{K.E.})_1 - (\text{K.E.})_2$$
$$= 40 - 0$$
$$= 40 \text{ [kJ]}$$

Thus, the kinetic energy of the moving piston, crank and flywheel assembly has been converted entirely into an increase of internal energy of the gas. If the closed system had been the gas only, we would have said that the gas had been compressed adiabatically and had work done on it by the piston, causing an increase in internal energy of 40 kJ.

23

This example has shown how our view of a problem changes if we move the system boundary. We can have a similar change of view with respect to heat and internal energy. At the beginning of Programme 2 we spoke of non-equilibrium states in which the intensive properties of temperature and pressure were not uniform throughout the closed system. If we consider a perfectly insulated closed system which has temperature variations through it, then in the course of time the temperature will become uniform at all points in the system. Initially the specific internal energy varied with position in the system; finally it became uniform. There has been, as in the previous example, a redistribution of energy in a closed system, except in this case only internal energy was involved. We may wish to look more closely and see how much internal energy is transferred from one part of the system to another. To do this we need to consider different parts of the original system as new closed systems, and apply the First Law to each.

24

Imagine two equal blocks of metal with different temperatures (a), in close metallic contact and perfectly insulated around the outside. Together they form an isolated system in a non-equilibrium state which if left to itself will in time achieve a uniform temperature throughout.

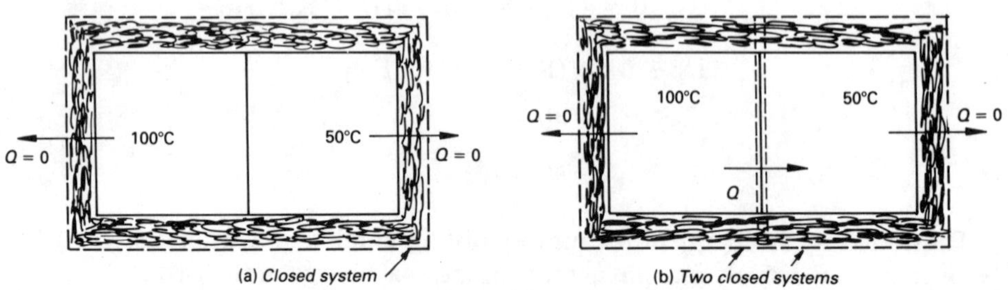

(a) *Closed system* (b) *Two closed systems*

We now wish to find the change of internal energy of each block as equilibrium is achieved. To carry out this work of accountancy we have to divide the original system into two, one new closed system for each block. There is now a common system boundary at the plane of contact of the two blocks (b), and we can apply the First Law:

$$Q = U_2 - U_1$$

in turn to each block, where Q is the heat interaction at the system boundary common to both blocks. Since Q is by definition an interaction at a system boundary, Q did not exist in the first instance because the system boundary did not exist.

Let one block in (a) have a temperature of 100°C and the other a temperature of 50°C. Let the internal energies of the two blocks be 19 000 J and 9500 J, so that the internal energy of the system is 28 500 J. There will be a redistribution of internal energy, until the temperature of each block is the same, which, since the blocks are equal in all respects, will be 75°C. The internal energy of each block will then be 28 500/2 J = 14 250 J.

In case (b) where we have two systems, what will be the magnitude of the heat transfer, Q, for (1) the left-hand block and (2), the right-hand block?

25

−4750 J, +4750 J

The two blocks will both achieve equilibrium at 75°C as before. We know from case (a) that U_1 for the left-hand block was 19 000 J, and U_2 was 14 250 J:

$\therefore \qquad Q - W = 14\,250 - 19\,000$

and $\qquad Q = -4750$ [J], since $W = 0$

For the right-hand block, U_1 was 9500 [J] and U_2 was 14 250 [J]:

$$Q - W = 14\,250 - 9500$$

$$Q = +4750 \text{ [J], since } W = 0$$

In Frame 20 we saw that the property energy E consists of internal energy, kinetic energy and potential energy. There is also chemical energy which manifests itself in combustion, to be considered in Programme 9. There are other forms of energy as well but in many thermodynamic problems only one or two energy forms are involved. Thus only internal energy, U, is involved in closed system problems with gases or vapours. In the next frame we shall think about a new property related to internal energy, called **specific heat**.

26

Measurements of heat interactions under closely controlled conditions have given much information on system properties during the development of thermodynamics. Such a measurement, related to a system of a pure substance of known mass, gives rise to a term called **specific heat**.

Suppose we have as a system a quantity of a pure gas such as nitrogen. This can undergo a process between equilibrium end states, and the First Law equation applies:

$$Q - W = E_2 - E_1$$

Since we are thinking of a gas of a certain mass at rest, the only relevant term in E is U.

$$\therefore \qquad Q - W = U_2 - U_1$$

or: $\dfrac{Q - W}{m} = u_2 - u_1$, and at constant volume with no work involved:

$$\frac{Q}{m} = u_2 - u_1$$

27

The heat interaction Q/m is measured for constant volume conditions over very small temperature ranges to give in the limit the differential change in u with T, or

$$c_v = \left(\frac{\partial u}{\partial T}\right)_v$$

where c_v is the **specific heat at constant volume** [J/kg K] or [kJ/kg K], which may be expressed as the rate of change of internal energy with temperature. The word **specific** is used since specific heat is the heat transfer necessary to change 1 kg of substance through 1 degree Kelvin of temperature, in this case under constant volume conditions so that work effects are excluded.

28

The measured heat interaction gives rise to a change in state of the system under constant volume conditions, and one of the properties that has changed is internal energy. In fact, both the temperature and pressure will have risen. As specific internal energy is an intensive property, the specific heat at constant volume must also be intensive. You can deduce this from its units. In the next programme we shall see that specific heat at constant volume is independent of pressure.

Since internal energy is a property of a system depending only on the end states and not on any path, the change of internal energy between two end states depends only on the temperature change and the specific heat at constant volume. Between the same two end states we might have a constant volume process (used to measure the specific heat), or we might have any number of processes involving both heat and work. The change of internal energy will be the same.

29

The use of c_v is illustrated in the following problem:

In a process, gas of mass 0.05 kg has heat and work interactions of -400 J and -2000 J. c_v for the gas is 718 J/kg K. We shall calculate the temperature change of the gas during the process.

We have:

$$Q - W = U_2 - U_1 = m\, c_v(T_2 - T_1)$$

$$-400 - (-2000) = 0.05 \times 718 \times (T_2 - T_1) \quad \text{[J]}$$

$$\therefore \quad +1600 = 35.9 \times (T_2 - T_1)$$

$$\therefore \quad T_2 - T_1 = 44.57 \text{ K}$$

Now try this example:

The process in the previous problem is followed by a heat interaction of +400 J under constant volume conditions. What is the change of temperature for this process?

30

11.142 K

For constant volume conditions:

$$Q = U_2 - U_1 = m\,c_V(T_2 - T_1)$$

So: $$+400 = 0.05 \times 718 \times (T_2 - T_1)$$

$\therefore$ $$(T_2 - T_1) = 400/(0.05 \times 718)$$
$$= 11.142 \text{ K}$$

Another special condition under which a heat interaction could be measured is at constant pressure. For our system of a mass of pure gas, a work term would be involved, and this would be $p(V_2 - V_1)$, since the pressure is constant. The First Law equation for the process becomes:

$$Q - W = U_2 - U_1$$
$$Q - p(V_2 - V_1) = U_2 - U_1$$

$\therefore$ $$Q = (U_2 + pV_2) - (U_1 + pV_1)$$

The combination of properties $(U + pV)$ is found to be very useful. This is called **enthalpy** and, being a combination of properties, it is also a property itself. The symbol for enthalpy is H, and the units are [J] or [kJ]. Enthalpy, like internal energy and volume, is an extensive property, leading to specific **enthalpy**, the intensive property, having units of [J/kg] or [kJ/kg]. Thus the First Law equation for the process may be written:

$$Q = H_2 - H_1 \text{ where } H = U + pV$$
$$= m(h_2 - h_1) \text{ where } h = u + pv$$

So a heat interaction at constant pressure will cause a change in enthalpy.

31

If in a constant pressure process, the heat interaction is carefully measured for a known mass of pure substance, a second specific heat term is obtained, and this is called the **specific heat at constant pressure**.
Thus:

$$c_P = \left(\frac{\partial h}{\partial T}\right)_P$$

The units are the same as before, [J/kg K] or [kJ/kg K]. For most real substances, specific heats are found to depend on temperature, and for c_p it depends to a small extent on the pressure as well. The special relation between c_p and c_v for gases will be considered in the next programme. Later in this programme we shall need to calculate enthalpy changes when we deal with open systems. The enthalpy change depends only on change of state, and not on any particular heat or work interaction leading to that change.

32

We sometimes need to calculate the property specific enthalpy from its constituent properties of pressure, specific volume and internal energy. In doing so you need to be careful about the units. Thus: a fluid has the following properties: u = 2506 kJ/kg, v = 1.694 m^3/kg and the pressure is 0.1 MPa. What is the specific enthalpy?

In consistent units, the pressure is 0.1 [MPa] $\times$ 10^3 [kPa]/[MPa] = 100.0 [kPa]

$$\therefore \quad h = 2506 + (100 \times 1.694)$$
$$= 2675.4 \text{ [kJ/kg]}$$

A gas has a specific heat at constant volume of 718 J/kg K, a pressure of 150 kPa, a specific volume of 0.574 m^3/kg and a temperature of 300 K. What is the specific enthalpy of the gas relative to a datum of 275 K? (Calculate the internal energy for a 25 K rise above the datum.)

33

104.05 kJ/kg

The specific heat at constant volume consistent with the units of pressure is 0.718 kJ/kg K.

So:

$$h = u + pv$$
$$= 0.718 \times 25 + 150 \times 0.574$$

$$= 17.95 + 86.10$$
$$= 104.05 \text{ kJ/kg}$$

In this result u is 17.95 kJ/kg relative to the datum of 275 K.

The following problem combines together some of the things you have learned in the last few frames:

In a constant pressure process, 0.01 kg of gas having a c_v of 718 J/kg K increases in temperature by 300 K as its specific volume increases from 0.086 to 0.172 m³/kg at a constant pressure of 1000 kPa. We shall calculate Q and c_P, and to do this we apply the First Law for a process: $Q - W = U_2 - U_1$.

We can calculate the work, and then the change of internal energy, and then find Q. From Q and the temperature change at constant pressure, we can find c_P. First, we find the work:

$$W = p(v_2 - v_1)m = 1000 \text{ [kPa]} \times 10^3 \text{ [Pa/kPa]} \times (0.172 - 0.086) \text{ [m}^3\text{/kg]} \times 0.01 \text{ [kg]}$$
$$= 860.0 \text{ [J]}$$

Then we find the change of internal energy:

$$U_2 - U_1 = m\, c_v(T_2 - T_1) = 0.01 \times 718 \times 300 = 2154.0 \text{ [J]}$$

Hence: $Q - 860.0 = 2154.0$

$\therefore$ $Q = +3014$ [J]

Also: $Q = 0.01 \times c_P \times 300 = 3014$ [J]

for the constant pressure process.

$\therefore$ $c_P = 3014/(0.01 \times 300)$

$= 1004.7$ [J/kg K]

Now try this:

A gas of mass 0.05 kg is compressed in such a way that the work of compression is 2.492 kJ and the temperature of the gas rises by 52.0 K. c_v for the gas is 718 J/kg K. Is there any heat transfer, and if so what is it?

34

$$\boxed{-625.2 \text{ J}}$$

As always in problems of this type, we need to use the First Law equation for a process:

$$Q - W = U_2 - U_1$$

The work is: -2492 [J], since it is done on the gas,

$$\text{and } U_2 - U_1 = 0.05 \times 718 \times 52$$

$$= 1866.8 \text{ [J]}$$

Hence: $Q - (-2492) = 1866.8$ [J]
Thus there is a heat transfer, and,

$$Q = 1866.8 - 2492$$

$$= -625.2 \text{ [J]}$$

To summarise this programme so far, from the ideas of work and heat developed in the previous programme, we have seen how the First Law of Thermodynamics may be applied to cycles and closed systems. This has led to our ideas about energy and, insofar as we can without looking in detail at the properties of pure substances, we have applied the First Law to processes in closed systems. Later we can build on this after we have considered the properties of pure substances, like gases and vapours.

You should attempt all the following problems before moving on.

35

PROBLEMS ON FIRST LAW APPLIED TO CLOSED SYSTEMS

1. A ball bearing rolls back and forth in a curved dish. The ball and dish are a closed isolated system. The ball is released 150 mm vertically above the lowest point and rolls to a height of 135 mm on the other side of the dish. The ball has a mass of 100 gms. What is (a) the increase of internal energy of the system and (b) the change of energy of the system? [(a) +0.0147 J, (b) 0]
2. A tin of processed food at 25°C is placed in hot water at 90°C in a vacuum jug. We may assume there is no heat loss from the jug and its contents. The masses of the tin of food and water are 150 gs and 1.4 kg respectively; the specific heats at constant volume are 3.9 and 4.185 J/kg K respectively. Calculate the equilibrium temperature of the jug contents. [84.1°C]
3. A bullet travels at 700 m/s at a height of 100 m above sea level. The mass of the bullet is 40 g; its specific heat is 380 J/kg K and its temperature is 70°C. (a) What forms of energy does the bullet have? (b) What is the total energy of the

bullet relative to the atmosphere at 20°C at sea level, and what percentage of this is kinetic energy?

[(a) Potential, kinetic and internal energy. (b) 10 599.24 J, 92.46%]

4. A gas is compressed adiabatically over a certain volume ratio and the work is −378 kJ. The gas has a mass of 1.15 kg and a specific heat at constant volume of 718 J/kg K. What is the temperature rise of the gas after compression? [457.8 K]
5. The same gas is now instead compressed polytropically over the same volume ratio and there is a heat transfer of −85.23 kJ. The temperature rise of the gas is now only 298.5 K. What is the work done in compression? [−331.7 kJ]
6. The same gas from problem **5** now expands following the same polytropic path of problem **5** to its original state and +331.7 kJ of work is done. What is the heat transfer? [+85.23 kJ]
7. A fluid has a specific internal energy of 2456 kJ/kg K at a pressure of 150 bar. The corresponding specific volume is 0.010 35 m^3/kg. What is the enthalpy of 4 kg of this fluid? [10 445 kJ]
8. A fluid expands from a specific volume of 0.001 55 m^3/kg to a specific volume of 0.013 49 m^3/kg at a constant pressure of 125 bar. The initial and final specific internal energies are 1492 and 2505 kJ/kg K. What are the work and heat transfers for fluid of mass 1 kg? [+149.25 kJ, +1162.25 kJ]
9. A gas having a specific heat at constant pressure of 1004 J/kg K is compressed polytropically and its temperature increases by 246 K. Calculate the increase in specific enthalpy of the gas. [246.98 kJ/kg]
10. Steam at 1 MPa pressure, having a specific volume of 0.1944 m^3/kg and specific internal energy of 2584 kJ/kg, expands polytropically with $n = 1.15$ to 0.2 MPa pressure. At this state the specific internal energy is 2306 kJ/kg. Calculate the work and heat transfers per unit mass of steam during the expansion. [+245.48 kJ/kg, −32.53 kJ/kg]
11. Gas of mass 0.0058 kg expands from a pressure of 3500 kPa, a volume of 0.001 m^3 and a temperature of 2100 K through a volume ratio of 12 according to a polytropic relationship with $n = 1.6$. During the expansion, the heat transfer is 2.1387 kJ from the gas. Calculate the work done by the gas and the final temperature. c_v for the gas is 0.718 kJ/kg K. [+4.52 kJ, 473 K]

36

We now wish to use the First Law to analyse machines and other devices such as heat transfer equipment through which matter is flowing continuously, so we need to use an **open system** (see Programme 1, Frame 16). But we have so far only used the First Law for a closed system, in which the quantity of matter is fixed. We get round this by extending our closed system boundary, at the point where matter enters, and at the point where it leaves. The diagram shows this; the closed system boundary moves over a short time, and as it does so m_1 enters the control surface and m_2 leaves. But within the system boundary, the matter is constant.

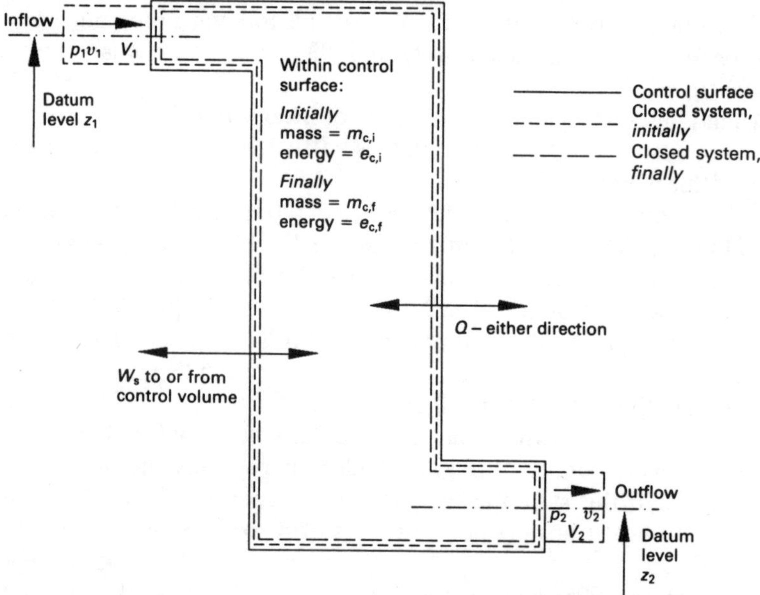

37

The figure represents a volume contained by a control surface with an inlet flow at position 1 and an outflow at position 2. These may be at different heights, and the difference could be significant. The ventilation system of an underground railway is a possible example. The chosen closed system is shown by the initial position of the system boundary, and after a time lapse, by the final position of the boundary. There is fluid of mass m_1 contained between the control surface and the initial position of the system boundary, and fluid of mass m_2 between the control surface and the final position of the boundary. The mass of fluid within the control surface initially and finally is $m_{c,i}$ and $m_{c,f}$ respectively. During the period of the time lapse, the mass m_1 enters the control surface, and mass m_2 leaves. Between the initial and final positions of the system boundary, the mass in the closed system must be constant, therefore:

$$m_1 + m_{C,i} = m_2 + m_{C,f}$$

If there is a *continuous steady flow* then m_2 will equal m_1 and $m_{C,f}$ will equal $m_{C,i}$
During the period of time between the initial and final states of the system there may have been heat and work interactions, and we may apply the First Law equation for the process:

$$Q - W = E_2 - E_1$$

Both Q and W may be of either sign, and both represent all the heat and work interactions across the **control surface** between the initial and final states of the system. In addition to W_s, which represents shaft work, there will be displacement work at the inflow and outflow points where the system boundary moves. The total work is therefore:

$$W = W_S - p_1 v_1 m_1 + p_2 v_2 m_2$$

In the next frame we shall think about the system energy terms E_1 and E_2.

38

The initial energy of the system is:

$$E_1 = e_1 m_1 + e_{C,i} m_{C,i}$$

and e_1 may be assumed to consist of specific internal energy, kinetic energy and potential energy, so:

$$E_1 = m_1\left(u_1 + \frac{V_1^2}{2} + gz_1\right) + e_{C,i} m_{C,i}$$

and, similarly: $$E_2 = m_2\left(u_2 + \frac{V_2^2}{2} + gz_2\right) + e_{C,f} m_{C,f}$$

where z_1 and z_2 are heights above a fixed datum at the entry and exit points. We will put all the terms together in the First Law equation in the next frame.

39

We have then:

$$Q - (W_S - p_1v_1m_1 + p_2v_2m_2)$$

$$= m_2\left(u_2 + \frac{V_2^2}{2} + gz_2\right) + e_{C,f}m_{C,f} - m_1\left(u_1 + \frac{V_1^2}{2} + gz_1\right) + e_{C,i}m_{C,i}$$

The displacement work terms for m_1 and m_2 may be combined with the internal energies u_1 and u_2 to give enthalpy, so:

$$Q - W_S = m_2\left(h_2 + \frac{V_2^2}{2} + gz_2\right) - m_1\left(h_1 + \frac{V_1^2}{2} + gz_1\right) + e_{C,f}m_{C,f} - e_{C,i}m_{C,i}$$

This is a complete form of the energy equation for a control volume, and it covers all possible situations. It has units of energy. The use of this equation will be explored in the following frames.

40

Notice that subscript 1 refers to the system boundary initially, and it also refers to the state of fluid entering at position 1. Subscript 2 refers to the system boundary finally and to the state of the fluid at position 2. If we assume conditions are steady, it must be that:

$$e_{C,i} = e_{C,f}, \quad m_{C,i} = m_{C,f}, \quad \text{and } m_1 = m_2$$

Steady conditions can readily be seen for flow through pipes, turbines, rotary compressors etc., but for reciprocating machinery, steady conditions would imply that the initial and final states of the system, as required in the formation of the equation, would have to occur at identical positions of the piston and crank in successive cycles. For steady state flow, m_1 and m_2 become a mass flow rate $\dot{m}$ in [kg/s], and $\dot{Q}$ and $\dot{W}_S$ in [W] or [kW] replace Q and W_S in [J] or [kJ].

For steady conditions the energy equation is given below.

$$\dot{Q} - \dot{W}_S = \dot{m}\,(h_2 - h_1) + \dot{m}\left(\frac{V_2^2}{2} - \frac{V_1^2}{2}\right) + \dot{m}(gz_2 - gz_1) \text{ [W] or [kW]}$$

41

We can illustrate the use of this equation with this problem:

The flow of steam is used in a district heating scheme. Measurements are recorded at two locations in the system, as follows:

	Station 1	Station 2
Enthalpy	2811 kJ/kg	2506 kJ/kg
Velocity	15 m/s	35 m/s
Height above datum	+3 m	−4 m

Between these two stations the total heating is 618 kW to the consumers. This is a negative heat interaction in our equation.

The mass flow of steam is 2 kg/s. Calculate the power used in pumping steam between these two stations.

$\dot{W}_S$ is the power input; it is the only unknown in the steady flow equation. Substituting into the equation we have:

$$-618 - \dot{W}_S = 2(2506 - 2811) + 2(35^2 - 15^2)/(2 \times 10^3) + 2 \times 9.81(-4 - 3)/10^3$$

The first term on the right is the enthalpy difference, the second term is the kinetic energy difference and the third term is the potential energy difference. The last two terms are divided by 10^3 to give consistent units of [kJ/s] or [kW]. Thus, we get:

$$-618 - \dot{W}_S = -610 + 1.0 - 0.137$$

$$\therefore \qquad \dot{W}_S = -8.863 \text{ kW}$$

This is the power to circulate the steam between the two stations; it is negative since it is rate of work done on the steam. Now try this exercise:

The ventilation system for an underground store draws in air at 30°C at ground level with a velocity of 30 m/s. The air is discharged at 150 m below ground level at 18°C and with a velocity of 5 m/s. The power input of the fan is 40 kW. The mass flow rate is 4 kg/s. c_P for air is 1.005 kJ/kg K.

Calculate the rate by which the air is cooled before being discharged.

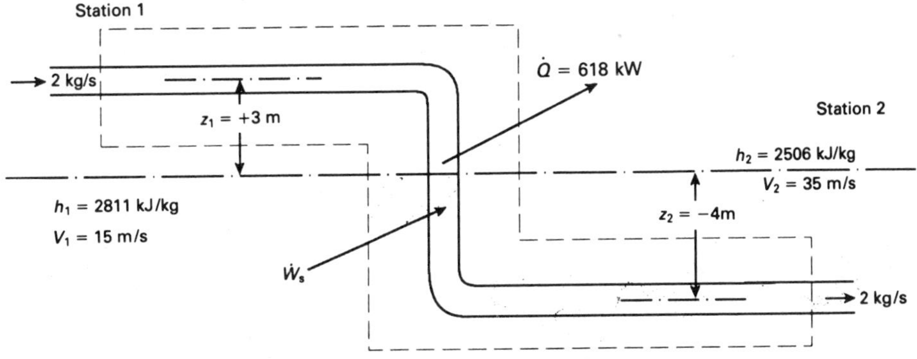

42

$$\boxed{-95.9 \text{ kW}}$$

Using the energy equation:

$$\dot{Q} - (-40.0) = 4.0 \times 1.005 \times (18-30) + \frac{4.0 \times (25 - 900)}{2 \times 10^3} + \frac{4.0 \times 9.81 \times (-150 - 0)}{10^3}$$

$$\dot{Q} + 40.0 = -48.24 - 1.75 - 5.89 \text{ [kW]}$$

$$\dot{Q} = -95.88 \text{ [kW]}$$

Clearly the air is being cooled by this ventilation system, so the heat transfer is from the control volume. It is interesting to see that the kinetic energy difference amounts to only 1.83% of the final answer, and the potential energy term is 6.14% of the answer. This is a case where the potential energy term, even for air, is significant.

The complete form of the energy equation given in Frame 39 can be used in transient situations. An overall process that takes some time can be considered, and in the next frame we shall see how we can solve a problem on filling an air receiver.

43

In the storage of compressed air, a receiver contains initially air of mass 2.323 kg at 27°C. It is filled to a higher pressure to contain air of mass 3.982 kg at 77°C. Air enters the receiver at a constant temperature of 140°C at a velocity of 20 m/s. Calculate the heat transfer from the receiver between these two conditions of the receiver. For air, $c_P = 1.005$ kJ/kg K and $c_v = 0.718$ kJ/kg K.

The problem is illustrated in the diagram on the next page. We have an air compressor which has a constant output of air at state 1 and it is used to fill a storage vessel with compressed air. There is no outflow, so terms for state 2 are not used.

The equation becomes:

$$Q - W_S = -m_1\left(h_1 + \frac{V_1^2}{2} + gz_1\right) + (m_1 + m_{C,i})e_{C,f} - m_{C,i}e_{C,i} \text{ [J] or [kJ]}$$

The final mass in the receiver, $m_{C,f}$, is the sum of $m_{C,i}$ and m_1, where m_1 is the mass that has entered.

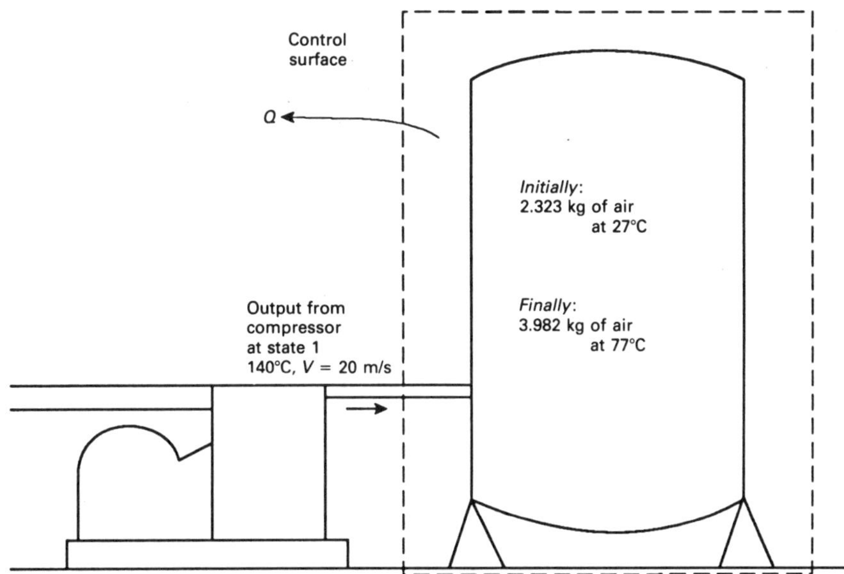

Since the control volume does not include the compressor, no work W_s is involved, and the term gz_1 may be neglected since it is small, so we get:

$$Q = -m_1\left(h_1 + \frac{V_1^2}{2}\right) + (m_1 + m_{C,i})e_{C,f} - m_{C,i}e_{C,i} \text{ [J] or [kJ]}$$

The terms $e_{C,f}$ and $e_{C,i}$ would consist of internal energy only, so you can see that the enthalpy and kinetic energy of the fluid entering at state 1 become internal energy of the fluid once it is inside the vessel.

This equation can now be used directly to solve the problem.

44

Using the energy equation developed in the previous frame:

$$\begin{aligned} Q &= -1.659 \times \left(1.005 \times 140 + \frac{20^2}{2 \times 10^3}\right) + 3.982 \times 0.718 \times 77 \\ &\quad - 2.323 \times 0.718 \times 27 \text{ [kJ]} \\ &= -1.659(140.7 + 0.2) + 220.149 - 45.034 \text{ [kJ]} \\ &= -58.638 \text{ [kJ]} \end{aligned}$$

Note the small size of the kinetic energy term.

45

We may now alter the limits of the control volume to include the compressor, so that the compressor work now enters the energy equation, and the state of the air entering the compressor then becomes state 1 in the equation of Frame 43:

$$Q - W_S = -m_1\left(h_1 + \frac{V_1^2}{2}\right) + (m_1 + m_{C,i})e_{C,f} - m_{C,i}e_{C,i} \text{ [J] or [kJ]}$$

The new diagram is given below.

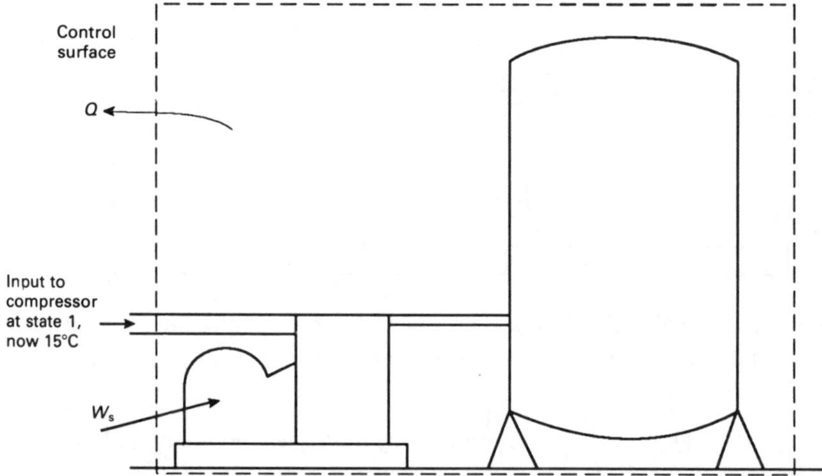

If we assume that air enters the compressor at 15°C with negligible velocity, and with all other details remaining the same as before, including the heat interaction, we can calculate the size of the work input, W_S.

The solution is:

$$Q - W_S = -1.659 \times 1.005 \times 15 + 220.149 - 45.034$$

$$-58.638 - W_S = -25.009 + 220.149 - 45.034$$

$$-W_S = 208.744 \text{ [kJ]}$$

or $$W_S = -208.744 \text{ [kJ]}$$

The negative answer is expected, since it is work of compression which is work done on the air. The work input of 208.744 kJ was used in the compression of 1.659 kg of air into the receiver. We could say, therefore, that the specific work of compression in this case was:

$$\text{Specific work} = \frac{W_S}{m_1} = \frac{208.744}{1.659} \text{ kJ/kg} = 125.83 \text{ kJ/kg}$$

This is a basis of comparison between air compressors. An air compressor with a smaller specific work is a better buy, since it will cost less to run.

46

When the valve on a compressed air cylinder is opened, the air will rush out. The general energy equation can explain what happens. We need only the terms of initial and final energy in the cylinder and state 2 terms of air leaving the cylinder, and we assume the process is adiabatic ($Q = 0$):

$$0 = m_2 \left(h_2 + \frac{V_2^2}{2} \right) + e_{C,f} m_{C,f} - e_{C,i} m_{C,i}$$

or:

$$m_2 \left(h_2 + \frac{V_2^2}{2} \right) = m_{C,i} e_{C,i} - m_{C,f} e_{C,f}$$

Thus, for the air that has left the cylinder, m_2, its original internal energy, has become enthalpy and kinetic energy. Since the velocity can be quite high, the enthalpy will be low; this means the temperature is low. If you let the air out of a bicycle tyre, you will find it is quite cold.

47

Consider a cylinder of compressed air at 10 bar pressure, and at a temperature of 300 K. The air escapes into the atmosphere at 1 bar, until the pressure in the cylinder is 9 bar, when the temperature is 291 K. The air masses originally and finally in the cylinder are 0.1161 and 0.1045 kg respectively. As the air escapes it has a velocity of 300 m/s. What will the temperature be? We may take c_P and c_V for air as 1005 and 718 J/kg K respectively.

We may use the last equation of Frame 46 directly. With temperatures in Kelvin we can use a datum of zero absolute temperature.

The mass of air that has escaped is $0.1161 - 0.1045 = 0.0116$ kg.

So:

$$0.0116 \times (c_P T_2 + 300^2/2) = 0.1161 \times 718 \times 300 - 0.1045 \times 718 \times 291$$

$$\therefore \quad 0.0116 \times 1005 \times T_2 + 0.0116 \times 300^2/2 = 25\,007.94 - 21\,834.02$$

$$\therefore \quad 11.658\, T_2 + 522 = 3173.92$$

$$\therefore \quad T_2 = 227.48 \text{ K}$$

$$= -45.67°\text{C}$$

48

Because air at low pressure would not rush out of the cylinder, we might feel that the energy of the air under pressure was greater. Air under pressure flows out at the expense of its internal energy, so that its final temperature must be low.

Now see if you can do the problem in the next frame. It involves air leaving a compressed gas cylinder, but an air turbine is included in the control volume, so that some work is extracted.

49

A cylinder of compressed air at a pressure of 10 bar and having a temperature of 400 K is used to drive an air turbine. The cylinder contains 3 kg of air initially, and 0.952 kg finally at a temperature of 252 K and a pressure of 2 bar. Air leaves the turbine at a constant 252 K and at a velocity of 40 m/s. Assume adiabatic conditions. For air, $c_p = 1.005$ and $c_v = 0.718$ kJ/kg K.

Calculate the total work output of the turbine. (In the energy terms, use absolute zero temperature as your datum.)

50

$$\boxed{W = +\,169.04 \text{ kJ}}$$

The equation you need is similar to the first one in Frame 46, except that the work term appears on the left:

$$-W_S = m_2\left(h_2 + \frac{V_2^2}{2}\right) + e_{C,f}m_{C,f} - e_{C,i}m_{C,i}$$

The air that has left the chamber is 3.0 − 0.952 = 2.048 kg. We have:

$$-W_S = 2.048(1.005 \times 252 + 40^2/(2 \times 10^3)) + 0.952 \times 0.718 \times 252 - 3 \times 0.718 \times 400$$

$$\therefore \quad -W_S = 520.31 + 172.25 - 861.6 = -169.04$$

or: $W_S = +169.04$ [kJ]

We now return to look again at the steady flow energy equation in Frame 40 and to think of its application in a number of different situations.

First we may look at rotary machinery, such as axial or radial flow turbines and compressors. The term axial means that flow through the machine is in an annulus of blades concentric with the axis of the machine, and the diagram shows for the turbine and compressor the annular passage in section, without the blades, within the control volume between entry at state 1 and exit at state 2. A section of the radial flow compressor is also given, with axial entry at state 1 and radial flow exit at state 2. Axial flow compressors and turbines are used in aircraft jet engines, and axial flow steam turbines are used in power stations. Radial flow is used in smaller machines, such as in motor vehicle superchargers, or in small gas turbines.

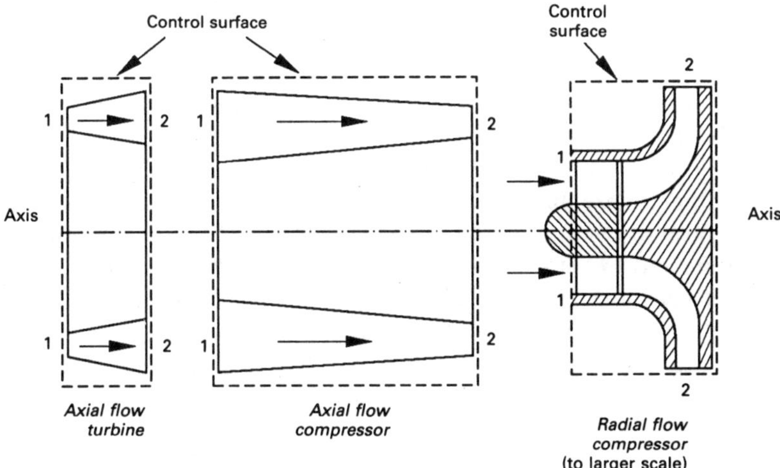

We usually assume that there is no significant heat loss, so $\dot{Q} = 0$, and the steady flow energy equation, becomes:

$$-\dot{W}_S = \dot{m}\,(h_2 - h_1) + \dot{m}\left(\frac{V_2^2 - V_1^2}{2}\right) + \dot{m}(gz_2 - gz_1)\ [\text{W}]\ \text{or}\ [\text{kW}]$$

where $\dot{W}_S$ is positive for a turbine (work producer) and negative for a compressor (work absorber). Although the potential energy term is included, it is usually very small. Compressors and turbines obviously get hot in operation, but generally adiabatic conditions are assumed as $\dot{Q}$ is small in relation to the work term.

52

We will now apply this equation to a turbine problem:

The enthalpy of steam entering a turbine is 2845 kJ/kg; the steam has negligible velocity, but it leaves the turbine at 200 m/s. It then has an enthalpy of 2584 kJ/kg. The steam exit is 4 m lower than the entry level. With no heat loss, see if you can show that the power output is 602.6 kW for a mass flow of 2.5 kg/s.

$$\text{We have that: } -\dot{W}_S = 2.5 \times (2584 - 2845) + \frac{2.5 \times 200^2}{10^3 \times 2}$$

$$+ 2.5 \times 9.81 \times (-4 - 0)/10^3$$

$$-\dot{W}_S = -652.5 + 50.0 - 0.0981$$

$$\dot{W}_S = +602.6 \text{ [kW]}$$

The kinetic energy term is quite large at about 8% of the turbine work term. The outlet velocity of 200 m/s is high by standards of good practice. We can see that we could have neglected the potential energy term.

Here is another rotary machine problem:

In an axial flow air compressor, air enters at 300 K and at atmospheric pressure, and it is compressed to a higher pressure to leave at a temperature of 520 K. The entry and exit velocities are 100 and 35 m/s respectively. For assumed zero heat transfer, calculate the specific work of compression.

53

−216.71 kJ/kg

Remembering that we can express enthalpy as $c_P \times$ T, we can put:

$$-\dot{W}_S/\dot{m} = c_P(T_2 - T_1) + \left(\frac{V_2^2 - V_2^2}{2}\right)$$

we have: $$-\dot{W}/\dot{m} = 1.005 \times (520 - 300) + (35^2 - 100^2) / (2 \times 10^3)$$

$$\therefore \quad -\dot{W}_S/\dot{m} = 221.1 + (1225 - 10^4) / (2 \times 10^3)$$

$$= 221.1 - 4.3875$$

$$= 216.7125$$

$$\therefore \quad \dot{W}_S/\dot{m} = -216.71 \text{ kJ/kg}$$

This is negative, as it should be, for work of compression.

In this problem we could have combined the enthalpy and kinetic energy terms together to give a term called **stagnation**, or **total head enthalpy**, h_T.

$$h_T = h + \frac{V^2}{2}$$

The units of this equation are [J/kg]. For $V^2/2$, the units are [m^2/s^2], and for enthalpy the units of [J/kg] are [N m/kg] = [kg m^2]/[s^2 kg] = [m^2/s^2], so the units are self-consistent.

For gases, where $h = c_P T$, this equation will give us:

$$T_T = T + \frac{V^2}{2c_P}$$

where T_T is the **total head** or **stagnation temperature**. T is then the **static temperature** and the group $V^2/2c_P$ is the **dynamic temperature**. The total head temperature is the temperature the fluid would have if it was brought to rest or diffused to zero velocity in adiabatic conditions.

We can see then that for adiabatic machines with no potential energy terms:

$$-\dot{W}_S = \dot{m}\,(h_{T_2} - h_{T_1})$$

The difference in specific total head enthalpy is therefore equal to the specific work term. The concept of total head temperature finds its main use in gas turbine stages and in rotary compressors, when there is then no need to refer specifically to either inlet or outlet velocity of the gas. This is also useful from a practical point of view, as total head temperature is easier to measure than static temperature.

Consider the turbine wheel of a gas turbine engine. The work output is high relative to any heat transfer from the wheel; the expansion of gases through the wheel is regarded as adiabatic. The inlet and outlet total head temperatures are 1100 K and 650 K, respectively. The mass flow through the wheel is 30 kg/s, and c_P for the gas flow is 1.15 kJ/kg K. What is the work output? The solution is given in the next frame.

55

15 525 kW

We have: $-\dot{W}_S = \dot{m}c_P\,(T_{T_2} - T_{T_1})$

$\therefore$ $-\dot{W}_S = 30 \times 1.15 \times (650 - 1100)$

$= -15\,525$ [kW]

$\therefore$ $\dot{W}_S = +15\,525$ [kW]

Now here is another example. Air enters an axial flow air compressor with a total head temperature of 305 K and leaves at a velocity of 60 m/s with a temperature of 580 K. The air mass flow is 10 kg/s. Calculate (a) the exit total head temperature and (b) the power input to the compressor.

56

(a) 581.79 K, (b) −2 781.75 kW

Here is the solution:

The total head temperature $= 580 + (60 \times 60)/(2 \times 10^3 \times 1.005)$

$= 580 + 1.791$

$= 581.79$ K

For the power: $-\dot{W}_S = 10 \times 1.005 \times (581.79 - 305)$

$= 2781.75$

$\therefore$ $\dot{W}_S = -2781.75$ [kW]

We shall make further use of total head temperature and enthalpy in rotating machinery in Programme 10.

Now we look at heat exchange equipment. Often it is necessary to use the flow of a hot fluid to raise the temperature of the flow of a cold fluid, or perhaps just as often the reverse is needed. In a car engine cooling system, the flow of air through the engine compartment is used to cool the water in the radiator, before it is returned to the engine. With the flow of water controlled by a thermostat, the engine maintains a comfortable working temperature. Normally for fluids flowing in pipes, differences in kinetic and potential energies may be neglected, so we have:

$$\dot{Q} = \dot{m}\,(h_2 - h_1)$$

or: $$\dot{Q} = \dot{m}\,c_P(T_2 - T_1)\ [\mathrm{W}] \text{ or } [\mathrm{kW}]$$

Water enters a pipe at 20°C and leaves at 80°C, and the heat transfer to the water is 703.25 kW. What is the mass flow of water? c_P for water is 4.186 kJ/kg K. Use $\dot{Q} = \dot{m}\,c_P(T_2 - T_1)$, with a positive heat transfer rate.

57

2.8 kg/s

The working is very simple:

$$703.25 = \dot{m} \times 4.186 \times (80 - 20)$$

$$\therefore \qquad \dot{m} = \frac{703.25}{4.186 \times 60}$$

$$\therefore \qquad \dot{m} = 2.8 \text{ kg/s}$$

In a heat exchanger, two fluid streams exchange enthalpy when the negative heat transfer of one stream is the positive heat transfer of the other. Two control volumes share a common surface across which the heat transfer occurs.

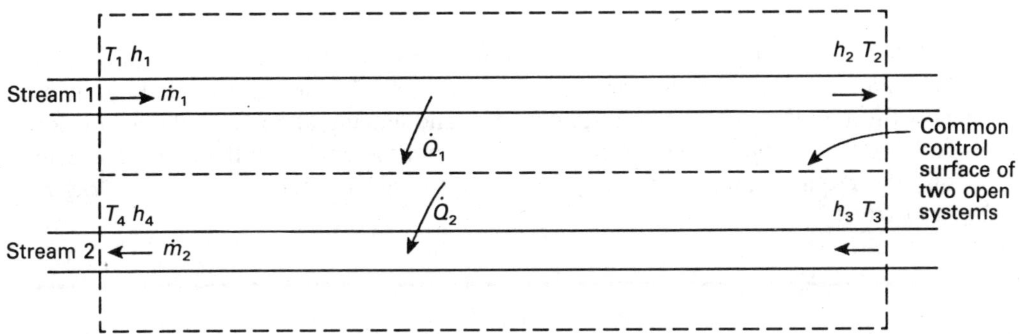

We need three equations which are given in the next frame.

58

For stream 1: $\dot{Q}_1 = \dot{m}_1 (h_2 - h_1)$

For stream 2: $\dot{Q}_2 = \dot{m}_2 (h_4 - h_3)$

and $-\dot{Q}_1 = \dot{Q}_2$

We may use these equations to solve this problem:

A stream of hot oil is cooled from 95°C to 20°C by a flow of water entering at 15°C and leaving at 60°C. If the oil flow is 5 kg/s, calculate the water flow required. c_P for oil is 1.98 kJ/kg K and for water is 4.186 kJ/kg K.

$$\text{For the oil stream } \dot{Q}_1 = 5.0 \times 1.98 \times (20 - 95)$$
$$= -742.5 \text{ [kW]}$$

Now what is the mass flow rate of water?

59

3.942 kg/s

$$\text{For the water: } \dot{Q}_2 = -\dot{Q}_1 = +742.5 \text{ [kW]} = \dot{m}_2 \times 4.186 \times (60 - 15)$$

$$\therefore \quad \dot{m}_2 = \frac{742.5}{4.186 \times 45} = 3.942 \text{ kg/s}$$

In a car radiator, heat transfer is from water to air. Water enters the radiator at 90°C and leaves at 45°C to flow back to the engine. The water is cooled by the air flow through the radiator, entering at 18°C and leaving at 55°C. If the water flow rate is 0.079 63 kg/s, calculate the heat loss from the engine to the cooling system, and the required air flow rate. c_P for water and air are 4.186 and 1.005 kJ/kg K, respectively.

60

15 kW and 0.4034 kg/s

$$\text{For the water flow: } \dot{Q} = 0.079\,63 \times 4.186 \times (45 - 90)$$
$$= -15.0 \text{ [kW]}$$

Heat transfer from the water is negative.

For the air: $Q = \dot{m} \times 1.005 \times (55 - 18)$

so: $15 = \dot{m} \times 37.185$

$\dot{m} = 0.4034$ [kg/s]

You may wonder why a car radiator is so called. Very early car radiators tended to rely more on radiation cooling, and designs changed when it was realised that more efficient cooling was achieved by convection. In today's cars, relatively very little cooling is by radiation, but the name has stuck.

61

Another application of the steady flow energy equation may be found in the adiabatic duct, nozzle or diffuser. There is no shaft work W_S and there is no significant change in z, so with $\dot{Q} = 0$, the equation becomes:

$$0 = \dot{m}\,(h_2 - h_1) + \dot{m}\left(\frac{V_2^2 - V_1^2}{2}\right)$$

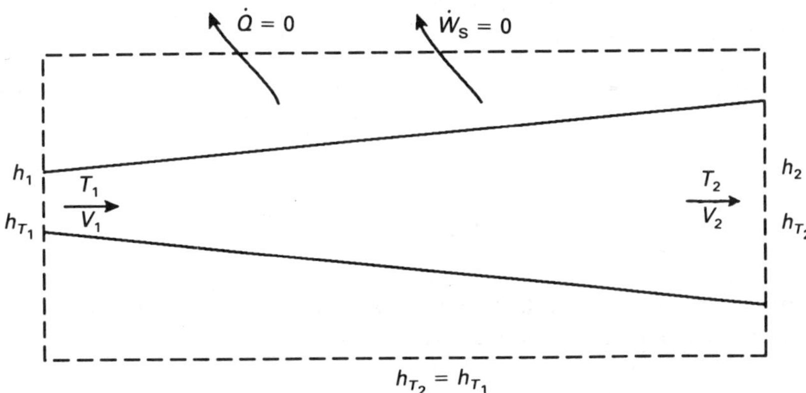

These terms can be regrouped to give constant stagnation enthalpy.

62

Thus we get: $$h_2 + \frac{V_2^2}{2} = h_1 + \frac{V_1^2}{2}$$

and the grouping of properties $h + \frac{V^2}{2}$ is, as we have seen, called **stagnation enthalpy**. For gases, where $h = c_P T$, this equation is then:

$$c_P T_T = c_P T + \frac{V^2}{2}$$

where T_T is the **total head** or **stagnation temperature**.

If a gas of temperature 150°C has a velocity of 100 m/sc, we may find its total head temperature, given that c_P = 1.005 kJ/kg K.

$$T_T = 150 + 100^2/(2 \times 1.005 \times 10^3)$$
$$= 150 + 4.975$$
$$= 154.975°C$$

Air in a machine room is at rest and has a temperature of 20°C. It is sucked into a compressor at 40 m/s. What is the temperature of the air as it enters the compressor? (Take c_P as 1.005 kJ/kg K.)

63

19.204°C

The temperature of the air in the room at rest is the stagnation temperature, so the temperature of the air that is moving is:

$$T = 20.0 - (40 \times 40)/(2 \times 10^3 \times 1.005)$$
$$= 20.0 - 0.796$$
$$= 19.204°C$$

Air enters a nozzle at 300°C with a velocity of 50 m/s. It leaves the nozzle at 200 m/s. What is the air temperature? c_P for air is 1005 J/kg K. Use the formula from Frame 61 to solve this problem.

64

281.34°C

We have:
$$1005 \times T_2 + \frac{200^2}{2} = 1005 \times 300 + \frac{50^2}{2}$$
$$1005\ T_2 + 20\ 000 = 301\ 500 + 1250$$
$$T_2 = 281.34°C$$

Last of all, in some cases of adiabatic flow, for example when liquids flow through a pressure drop, the flow is called **adiabatic throttling**. The velocities are usually small so the kinetic energy term is very small compared with the enthalpy term. So we get the very simple result:

$$h_2 = h_1$$

In a throttling process, $h_1 = 100$ kJ/kg and the velocity is 3 m/s. After the expansion the velocity is 10 m/s. It would be interesting to calculate the true h_2 and the percentage error in assuming it to be 100 kJ/kg.

65

We have:
$$100 + \frac{3^2}{2 \times 10^3} = h_2 + \frac{10^2}{2 \times 10^3}$$

$\therefore$
$$h_2 = 100 + \frac{9 - 100}{2 \times 10^3}$$
$$= 100 - 0.0455$$
$$= 99.9545\ [\text{kJ/kg}]$$

The error is 0.0455 on 100, which is 0.0455%. Thus the constant enthalpy assumption in such a process involves only a very small error.

A fluid having an enthalpy of 117 kJ/kg and a negligible velocity expands to a lower pressure when it has a velocity of 8 m/s. Calculate (a) the kinetic energy of the fluid after expansion and (b) the true enthalpy.

66

(a) 0.032 kJ/kg, (b) 116.968 kJ/kg

The kinetic energy is the $V^2/2$ term, which is $8 \times 8/(2 \times 10^3)$, hence:

$$\text{Kinetic energy} = 0.032 \text{ kJ/kg}$$

Taking into account this kinetic energy we have:

$$117 = h_2 + 0.032$$

hence: $$h_2 = 116.968 \text{ kJ/kg}$$

In Programmes 2 and 3 we have considered work, heat and energy, the First Law, and problems involving the First Law applied to closed and open systems. The exercises undertaken with gases and steam have involved changes of state, but so far we have not been concerned with finding any actual state of these fluids. The next programme will consider pure substances and their states, so that we shall be able to undertake, in programme 5, more advanced problems of the type we have met so far.

After that, in programme 6, you will meet the very important property called **entropy**, and the **Second Law of Thermodynamics**.

Before that, however, you should make sure you can do all the problems that have been set in this programme.

67

PROBLEMS ON OPEN SYSTEMS

1. An air conditioning system for a computer room in a tower block draws in air on the roof at a height of 100 m with a velocity of 25 m/s. The air is at 28°C. The air is discharged at a height of 10 m with a velocity of 2 m/s at 14°C. The mass flow rate is 2 kg/s, and a heat transfer of −40.73 kW cools the air before it is discharged. Calculate the rate of work for the air passing through the system. c_P for air is 1005 J/kg K. [−10.203 kW]
2. 5 kg of steam having an enthalpy of 2530 kJ/kg enters a closed vessel which already contains 2 kg of steam having an internal energy of 2456 kJ/kg. The steam enters at a velocity of 50 m/s. No heat or work transfer is involved. Calculate the final specific internal energy of the steam in the vessel. [2509.75 kJ/kg]
3. Combustion gases flow through a turbine at the rate of 10 kg/s. They enter the turbine at 450°C and at a velocity of 50 m/s, and leave at 220°C at 240 m/s. There

is a small rate of heat loss from the turbine of 15 kW. c_P for the gases is 1.15 kJ/kg K. Calculate the power output of the turbine. [2354.5 kW]

4. Feed water to a boiler leaves the condenser at a height of 5 m below floor level with an enthalpy of 192 kJ/kg. It passes through a feed pump and leaves at 12 m above floor level with an enthalpy of 197.6 kJ/kg. The mass flow of water is 8.4 kg/s. Calculate the rate of work output of the motor driving the feed pump, assuming 100% efficient power transmission. There is no heat transfer and velocities are negligible. [+48.4 kW]
5. A steam turbine has a power output of 33 MW. Steam enters the turbine with an enthalpy of 3433 kJ/kg and at a velocity of 125 m/s. It leaves the turbine with an enthalpy of 2148 kJ/kg at a velocity of 280 m/s. Calculate the mass flow rate of steam, assuming no heat transfer. [26.32 kg/s]
6. Transformer oil at 80°C has to be cooled to 25°C by a flow of water entering at 12°C. The oil flow rate is 20 000 kg/h, and the water flow is 5 kg/s. c_P for oil and water are 2.06 and 4.186 kJ/kg K respectively. Calculate (a) the water temperature at exit and (b) the heat transfer rate between the oil and the water. [(a) 42.07°C (b) 629.38 kW]
7. Steam enters a condenser with a specific enthalpy of 2584 kJ/kg. It is condensed to leave as saturated water with an enthalpy of 192 kJ/kg. The steam flow is 6 kg/s, and it is condensed on a surface cooled by the water. The temperature rise of the water is 20 K. Calculate (a) the heat transfer rate in the condenser and (b) the mass flow rate of the water in kg/h. c_P for water is 4.186 kJ/kg K. [(a) 14 352 kW (b) 617 142.9 kg/h]
8. In a steam nozzle, steam enters with negligible velocity with a specific enthalpy of 2840 kJ/kg and leaves with a specific enthalpy of 2780 kJ/kg. Calculate the velocity of steam leaving the nozzle. [346.4 m/s]
9. Steam has a specific enthalpy of 2800 kJ/kg and a velocity of 120 m/s. What is its total head specific enthalpy? [2807.2 kJ/kg]

68

Points to remember

A series of processes that start and finish at the same state is called a cycle. In a cycle:

$$\sum_{\text{cycle}} \delta Q - \sum_{\text{cycle}} \delta W = 0$$

and this is an expression of the First Law of Thermodynamics; when any system executes a cycle of processes, the net work is equal to the net heat.

For a process, there is a property called energy which undergoes a change of state during a process which is numerically equal to the net heat less the net work at the system boundary during the process.

$$Q - W = E_2 - E_1$$

The form of energy associated with temperature change and affected by a heat interaction is called internal energy. Energy may consist of internal energy, kinetic energy and potential energy:

$$E = U + \text{K.E.} + \text{P.E.}$$

Energy is an extensive property, so that $u = U/m$ = specific internal energy.

Specific heat at constant volume, $c_V = \left(\frac{\partial u}{\partial T}\right)_V$.

Enthalpy is an extensive property, $H = U + pV$, so $h = H/m = u + pv$.

Specific heat at constant pressure, $c_P = \left(\frac{\partial h}{\partial T}\right)_P$.

The energy equation for a control volume is:

$$Q - W_S = m_2\left(h_2 + \frac{V_2^2}{2} + gz_2\right) - m_1\left(h_1 + \frac{V_1^2}{2} + gz_1\right) + e_{C,f}\, m_{C,f} - e_{C,i}\, m_{C,i}$$

The steady state equation, or the steady flow energy equation is:

$$\dot{Q} - \dot{W}_S = \dot{m}\left(h_2 - h_1\right) + \dot{m}\left(\frac{V_2^2 - V_1^2}{2}\right) + \dot{m}\,(gz_2 - gz_1)$$

and total head or stagnation enthalpy and temperature are:

$$h_T = h + \frac{V^2}{2} \text{ and } T_T = T + \frac{V^2}{2c_P}$$

Programme 4

THE PURE SUBSTANCE

1

In the previous two programmes we have developed the ideas of work, heat and energy. We have also seen how the energy of a system changes with work and heat interactions at the boundary. So far in the problems you have met, the state of a system has been completely specified; this means all properties have been given. In this programme you will learn how to calculate, or look up in tables or charts, those properties which need to be found. This may be done from formulae or from tables of data acquired and refined over many years. It will enable you to tackle a much wider range of problems on systems and the First Law, which will follow in Programme 5.

2

In the previous programme we looked in some detail at processes in closed and open systems. In each case a **working fluid** was involved, either a gas or a vapour. So that we can establish all the properties of the fluid at any particular state, we need to know precisely what the fluid is. Is it a single substance, or a mixture of substances? If it is a mixture, does its composition vary within the system boundary?'Can a mixture be treated as a single substance? Will the composition of a mixture be the same at all possible states of the system? We need to be able to look at our system, and to know how to consider these questions. To begin, we shall think about the idea of a pure substance.

3

A pure substance is a system which at all times is homogeneous in respect of its composition and chemical combination. This means that the following statements are true:

(1) *Composition*. The proportions of chemical elements in each part of the system are the same.

(2) *Chemical combination*. The chemical elements are combined together in the same way in each part of the system.

(3) *Time*. During the time of any process undergone by the system, the **composition and chemical combination** do not change.

In the following frames we will look at examples, and learn how to identify pure substances.

4

Frequently we shall be dealing with steam in closed and open systems. Steam is the vapour phase of water, which can freeze and become ice. However, ice, water and steam all have the same composition and chemical combination, H_2O, and even though the matter can exist in three phases, it satisfies the definition of a pure substance. Later we shall see how the state of the ice–water–steam system can be represented on property diagrams in which one property is plotted against another, and we shall see that ice, water and steam can all exist together at a particular state called the **triple point**. We shall also learn how to use tables and charts to find the properties of steam, water and steam–water mixtures.

5

Vapours and gases are of the same phase and are generally invisible. We shall be dealing with gases later in the programme; essentially, they are vapours at temperatures much greater than their boiling points. However, we often see clouds in the sky, or experience a mist or fog. A cloud or a mist is a mass of tiny water droplets suspended in air, and may involve dust particles on which the droplets have formed. In a cloud, atmospheric conditions may increase condensation of water vapour in the air to an extent that the water droplets are no longer suspended, and they fall as rain.

A mist or cloud consisting of tiny droplets of a liquid such as water suspended in a single gas such as nitrogen would be a pure substance provided that: (a) the droplets are of uniform size, (b) the number and spacing of droplets per unit volume is constant throughout and (c) the conditions of (a) and (b) remain constant with time under all conditions.

6

We shall often be dealing with air as a system, but as you know, air is a mixture of gases. In a mixture of gases, the molecules of individual gases are uniformly mixed up together, and as we shall see, provided we can specify precisely the amounts of each gas present, we shall be able to calculate the properties of the mixture in terms of known properties of each gas. A mixture of gases will satisfy the first two conditions of our definition of a pure substance. Provided the states at which the mixture is considered are such that no chemical reaction takes place, then condition (3) is also satisfied. When this is so, the gas mixture is a pure substance, which is useful because a mist or cloud generally involves air rather than a single gas.

7

If we specify the composition of air in a system in terms of amounts of each gas present, then we can deduce properties of that air in terms of the properties of the individual gases, and we are dealing with a pure substance that may be treated as a single gas. For most practical purposes we deal with air having the approximate composition of 76.7% nitrogen and 23.3% oxygen by mass, but the exact composition of air in the atmosphere involves small amounts of other gases which may vary.

A system consists of a cloud of water droplets that satisfies the conditions listed in Frame 5 in a specified composition of air. Is it a pure substance?

8

Yes

Since the air of specified composition is a pure substance and is treated as a single gas, then that gas carrying water droplets of uniform size and distribution is also a pure substance.

We have seen that water and steam present together as a system is a pure substance, but suppose a closed system consisted of air and liquid air at very low temperature.

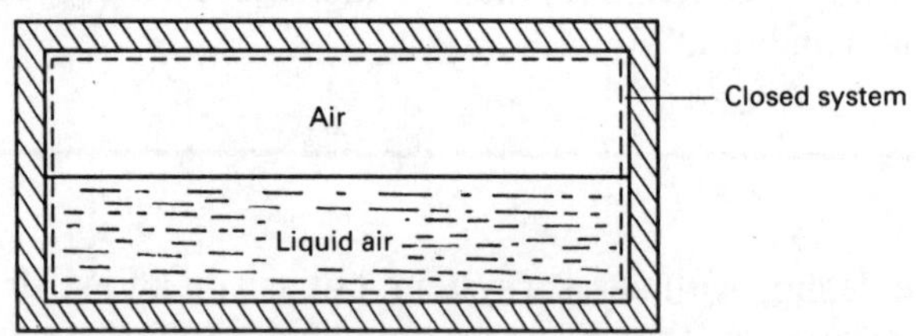

Oxygen boils at 90 K and nitrogen at 77 K, so the oxygen–nitrogen ratio in liquid air will vary with temperature in this range and with the pressure of the vessel containing the vapour phase in equilibrium with the liquid. In dealing with air–liquid air in equilibrium, we would only have a pure substance at temperatures below 77 K as above this temperature the proportion of oxygen present in the vapour and liquid will depend on the properties of temperature and pressure. A substance whose composition varies within the system and depends on its properties is not a pure substance.

9

If hydrogen and oxygen gases in a proportion of 2:1 are ignited, an explosion occurs and steam is produced. We have seen that steam in equilibrium with water is a pure substance, but in the case of the two gases over water we can see that condition (2) of Frame 3 is not satisfied. Although hydrogen and oxygen are present in the right quantities to satisfy condition (1), we do not have a pure substance for which we can define a state.

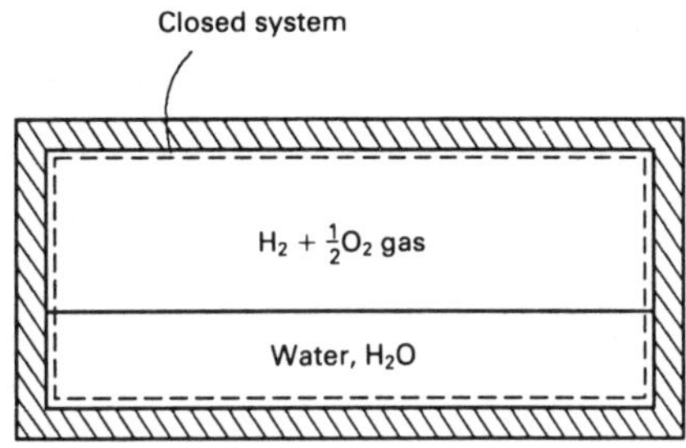

Bearing in mind that a vapour behaves like a gas, is a closed system of a uniform mixture of petrol vapour and air a pure substance?

10

Yes

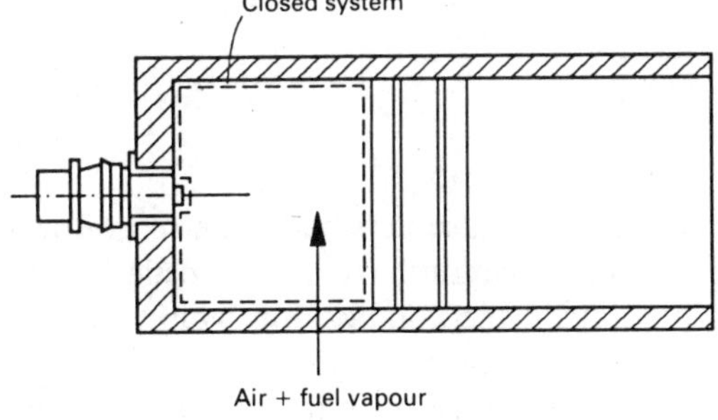

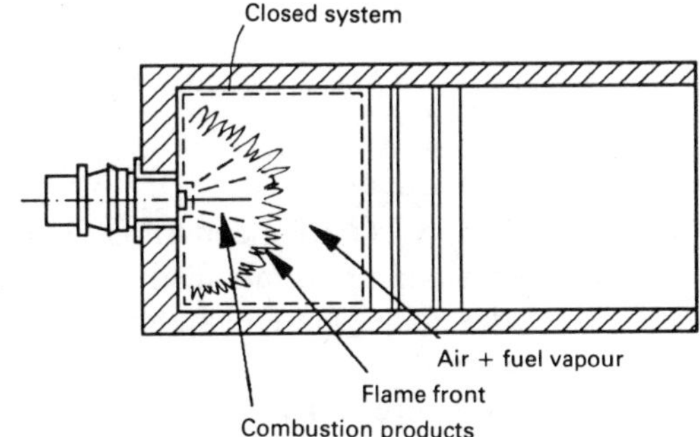

In the next frame we think about the system shown above after combustion.

11

For practical purposes, petrol vapour in air may be regarded as a gas mixture, and may be treated as a pure substance. If droplets of petrol are also present and spread randomly throughout the system, then the system is not a pure substance. In an engine cylinder this mixture is present only up to the time at which combustion starts; while the flame front is travelling across the mixture, condition (3) is not satisfied and we no longer have a pure substance. After combustion we have unused air and combustion products which form a different pure substance having different properties.

12

The picture below shows a cylinder into which diesel fuel is being injected. The spray of fuel is initially a cloud of droplets; these are not of equal size and are not uniformly distributed throughout the volume of the cylinder, and very quickly the droplets evaporate and a non-uniform gas mixture is produced. Do we have a pure substance?

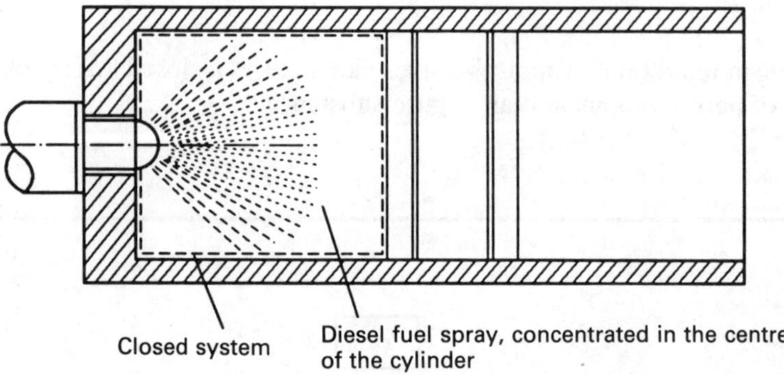

13

No

In this example, condition (1) is not satisfied. If the droplets had been of uniform size and uniformly distributed, then condition (1) would have been satisfied up to the point where evaporation started. Then, again the composition would start to vary.

Before moving on, consider the following examples on pure substances. The answers are in the next frame.

State whether each of the following is a pure substance, and if not on what grounds, either failing condition (1), (2) or (3) in Frame 3.

(a) A mixture of air and methane gas.
(b) A mixture of air and propane vapour, with liquid propane in the bottom of the container.
(c) Steam with water in the bottom of the container.
(d) Air with liquid air in the bottom of the container, in equilibrium at 80 K.
(e) A mixture of air and petrol vapour in the cylinder of an engine, with oil droplets on the cylinder walls.
(f) Air and fuel vapour through which a flame front is passing.
(g) Gaseous combustion products including water vapour.
(h) Gaseous combustion products from which water has condensed in the bottom of the container.

14

Here are the answers:
(a) Yes, (b) No – 1, (c) Yes, (d) No – 1, see Frame 8, (e) No – 1, (f) No – 2 and 3, (g) Yes, (h) No – 1

We will now look in detail at the properties of the ice/water/steam system as a pure substance.

Suppose we have some water in a closed cylinder with a weighted piston as shown. The weight is constant so the pressure in the cylinder is constant. To start with the water is at room temperature, and then we heat it until it boils and becomes steam. We read the temperature at intervals and note the movement of the piston. From this we find the volume and specific volume of the cylinder contents. The readings enable us to draw a graph of temperature against specific volume, for that one particular constant pressure. The heating is carried out slowly so that the graph shows a path between equilibrium states on a state point diagram.

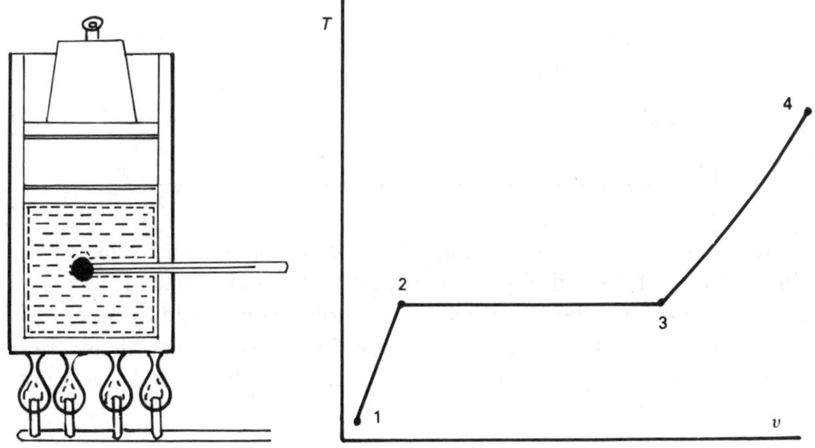

What do you think is happening between points 2 and 3 on the diagram?

15

The water is boiling

When the water boils at constant pressure, the volume increases very substantially from its original volume as water, to the volume of the same mass of steam. When this happens at the standard atmospheric pressure of 1.01325 bar, what is the temperature in degrees Celsius?

16

100.0°C

As you should know, water boils at 100°C at atmospheric pressure. The figure of Frame 14 is repeated below, showing more detail.

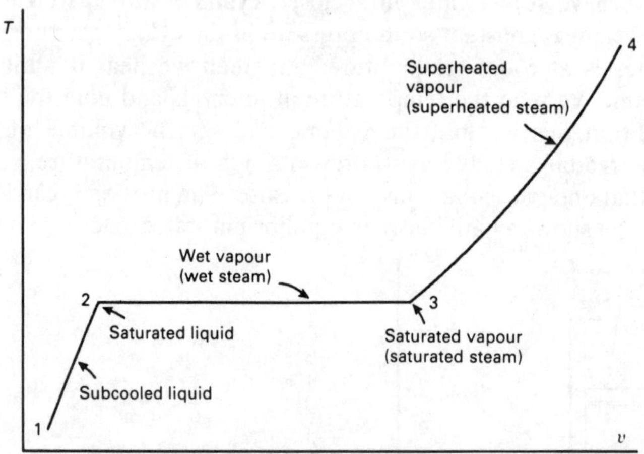

The water starts to boil at point 2; this point is called **saturated liquid**. At point 3, the water is completely boiled; this point is called **saturated vapour**. Between points 1 and 2, before the water boils it is called **subcooled liquid**, and between points 3 and 4 we are heating steam, which is called **superheated vapour**, or **steam**. Between points 2 and 3 there is a mixture of water and steam which is called **wet vapour**, or **steam**.

17

Suppose we now repeat this experiment of heating water, at say two values of higher pressure. We will get two more sets of data so that we can draw two more lines of temperature against specific volume. The three lines are shown below.

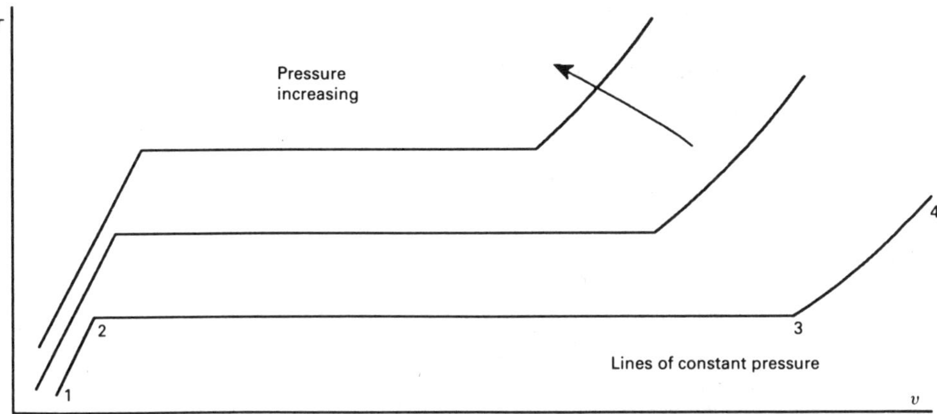

From this diagram what can you say about the way the temperature of the boiling point of water changes with pressure?

18

Boiling point increases with pressure

You can see from the figure in the previous frame that as the pressure increases, the horizontal portions of the curves where boiling is taking place occur at higher temperatures. Also you can see that the specific volumes at which boiling commences and is completed are moving closer together with rising pressure. The temperature at which boiling occurs at a particular pressure is called the **saturation temperature**, the corresponding pressure is called **saturation pressure**. These two properties are not independent of each other. At a particular pressure there can only be one temperature at which boiling occurs.

What is the saturation pressure for water at 100.0°C?

19

1.013 25 bar

This question was answered for you in Frame 16. If we continue to raise the pressure and repeat the constant pressure heating experiments, we see the horizontal portion of the temperature–specific volume line getting shorter as the pressure increases until we reach a pressure when it disappears altogether.

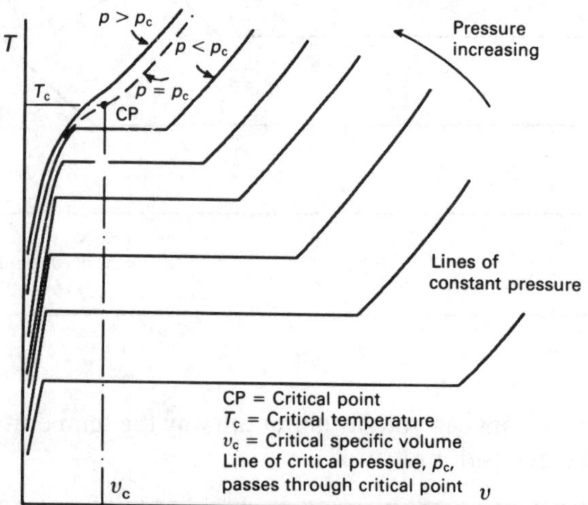

The pressure at which this happens is called the **critical pressure**. The corresponding temperature is called the **critical temperature**, and these are two properties of the **critical state**. The specific volume of the critical state is also shown on the figure. The critical state is usually known as the **critical point**.

20

You may have noticed that the points where boiling starts and finishes in the previous figure would trace out two curves which meet at the critical point. These two curves are shown in the figure as heavy solid lines, and each represents a locus of a point of transition.

The left-hand line is the locus of points where water becomes saturated liquid, and the right-hand line is the locus of points where the boiling water becomes saturated steam. The lines are known as the **saturated liquid line** and the **saturated vapour line**, respectively, and they meet at the critical point.

Lines of constant pressure are also shown. The saturated liquid line has not been drawn accurately in this figure; can you suggest what is wrong? (Remember, this shows temperature against specific volume for *water*.)

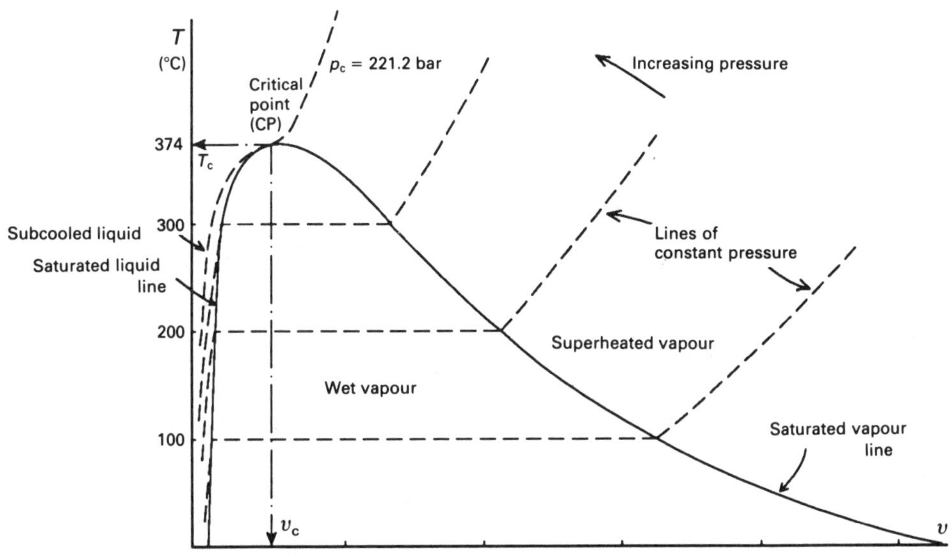

21

The slope is not steep enough

It is common practice to show this line as you see it here, but in fact it is nearly vertical. This means that the constant pressure lines in the subcooled liquid region are indistinguishable. The following will give you an impression of the different slopes of the two saturation lines:

		Specific Volume	
Temperature	*Pressure*	*Sat. liquid*	*Sat. vapour*
7.0°C	0.01 bar = 1.0 kPa	0.001 m³ /kg	129.2 m³ /kg
45.8°C	0.10 bar = 0.01 MPa	0.00101	14.67
100.0°C	1.013 bar = 0.1013 MPa	0.001044	1.673
179.9°C	10.0 bar = 1.0 MPa	0.001128	0.1944
269.9°C	55.0 bar = 5.5 MPa	0.001302	0.03563
374.15°C	221.2 bar = 22.12 MPa	0.00317	0.00317

The range of values of specific volume mean that in practice the commonly used sketches of diagrams showing the specific volumes of the saturation lines of water are, at the best, crude approximations.

At the last pressure, both specific volumes are the same. What point is this?

22

The critical point

The conditions of the critical point are:

Pressure: 221.2 bar = 22.12 MPa

Temperature: 374.15°C

Specific volume: 0.00317 m^3/kg

The pressures and temperatures used in steam plant have advanced with progress in materials technology. The upper limit of the pressure range is now between about 170 bar and 240 bar, or 17–24 MPa, so **supercritical** conditions are actually used.

To establish the temperature–specific volume diagram we have been studying, heating of water was carried out under constant pressure conditions.

What other specific property of a closed system changes when it is heated at constant pressure?

23

Enthalpy

We have seen from Frame 30 of Programme 3 that:

$$Q = H_2 - H_1$$
$$= m(h_2 - h_1)$$

for a constant pressure process, so these constant pressure heating experiments give us information about enthalpy changes, as well as the T–v relationships at different pressures. The experiments have given us, therefore, values of the following properties: pressure, specific volume, temperature and enthalpy, together with particular values at the transition points of saturated liquid and vapour, and the critical point.

From enthalpy, specific volume and pressure, another property may be found. What is it?

24

Internal energy

Also from Frame 30 of Programme 3 we have that:

$$H = U + pV$$

or:

$$h = u + pv$$

in terms of specific properties, so that if we know h, p and v, we have:

$$u = h - pv$$

All the important properties we need for the water/steam system may be obtained from carefully conducted constant pressure heating experiments.

25

Take another look at the T–v diagram in Frame 20. If you look carefully at states on the saturated liquid line, on the saturated vapour line, in the subcooled liquid region and in the superheated steam region, you will see that only two properties are needed to pin-point the state of the system. This gives rise to the **two property rule**, which states:

The state of a pure substance in equilibrium and of given mass can be established by specifying two properties only.

Thus, since we have lines of constant pressure on our T–v diagram, we need T and v, p and v, or T and p as our two properties to find the position of our state on the diagram, from which we could find the third property.

Now look at the diagram in Frame 20 in the wet steam region. If the two properties we know are temperature and pressure, can we find the volume?

26

Clearly the answer is no. This is because the temperature and pressure are not independent and so are only one property. The two property rule allows us one more property to find the state in the wet steam region. This will have to be a new property which we must now investigate.

27

Consider again the constant pressure heating of the water–steam system. The curve from the figure of Frame 16 is given again below.

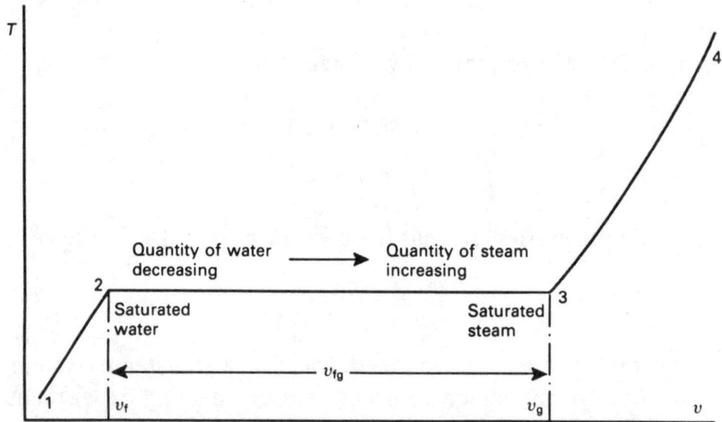

The specific volume of the boiling water is changing rapidly between points 2 and 3 as the water turns into steam. We only know that at point 2 we have saturated water and at point 3 we have saturated steam. In between, knowing only the specific volume at any point, we do not know how much water there is still to boil, so we need another property for the water/steam system in the wet steam region.

We may define:

$$v_f = \text{specific volume of saturated liquid}$$
$$v_g = \text{specific volume of saturated vapour}$$
$$\text{and} \quad v_{fg} = v_g - v_f$$

where v_{fg} is the change of specific volume between the two saturated states.

The actual specific volume depends on how much water or steam there is, so we introduce the term x, which is the **quality** or **dryness fraction** of the steam. **The dryness fraction is the ratio of the mass of steam to the total mass of substance**. It is an intensive property of wet steam.

What is the dryness fraction for (a) saturated water, and (b) saturated steam?

28

(a) 0, (b) 1

For saturated water there is no steam present, and for saturated steam there is no water present. The dryness fraction is expressed as:

$$x = \frac{\text{mass of vapour}}{\text{mass of vapour + mass of liquid}}$$

From this we can show that the ratio of the mass of vapour to mass of liquid will be:

$$\frac{\text{mass of vapour}}{\text{mass of liquid}} = \frac{x}{1 - x}$$

If a water–steam system consists of 2 kg of saturated water and 6 kg of saturated steam, what is the dryness fraction, x?

29

0.75

We have:

$$x = \frac{\text{mass of vapour}}{\text{mass of vapour} + \text{mass of liquid}}$$

$$= \frac{6}{6 + 2}$$

$$= 0.75$$

We can also see that the ratio of mass of vapour to mass of liquid is:

$$\frac{\text{mass of vapour}}{\text{mass of liquid}} = \frac{0.75}{1 - 0.75} = 3$$

Since for unit mass the mass of vapour is x and the mass of liquid is $1 - x$ we can find the specific volume of unit mass of wet steam of dryness fraction x using the definitions from Frame 27:

$$v = (1 - x)v_f + x\,v_g \qquad (1)$$

$$= v_f + (v_g - v_f)x$$

so

$$v = v_f + x\,v_{fg} \qquad (2)$$

Also:

$$v = v_g - (1 - x)v_g + (1 - x)v_f \quad \text{(from (1), by adding and subtracting } v_g\text{)}$$

$$= v_g - (1 - x)(v_g - v_f)$$

and

$$v = v_g - (1 - x)v_{fg} \qquad (3)$$

30

These three equations are important and are frequently used in steam or vapour calculations. It is useful to memorise them.

Here is an example: for wet steam at 200°C, $v_f = 0.001\ 57\ m^3/kg$ and $v_g = 0.1273\ m^3/kg$; the specific volume is $0.09\ m^3/kg$. What is the dryness fraction?

Since we know v_f, v_g and v, we have to use equation (1).

$$v = x\,v_g + (1 - x)v_f$$

$$0.09 = x\,0.1273 + (1 - x)0.001\ 57$$

so: $$x\,(0.1273 - 0.001\ 57) = 0.09 - 0.001\ 57$$

and: $$0.125\ 73\,x = 0.088\ 43$$

hence: $$x = 0.7033$$

Now see if you can do this example:

At a certain pressure, $v_g = 0.0356\ m^3/kg$ and $v_{fg} = 0.0343\ m^3/kg$. The dryness fraction is 0.4. What is v?

31

$$\boxed{v = 0.0150\ m^3/kg}$$

Since we know v_g and v_{fg}, we need to use equation (3).

Thus: $$v = v_g - (1 - x)v_{fg}$$

$$= 0.0356 - (1 - 0.4) \times 0.0343$$

$$= 0.0150\ m^3/kg$$

Values of v_f and v_g for all values of saturation pressure and temperature are given in steam tables, which we shall be meeting shortly. So it is equation (1) of Frame 29 which you will probably use most.

It may happen in a problem that you are given a value for v and a temperature or pressure which is not specified as being a saturation value. The figure shows part of the saturated vapour line between the pressures of 26 bar and 30 bar. A specific volume to the left of this line will be in the wet region; one to the right of the line will be in the superheated region.

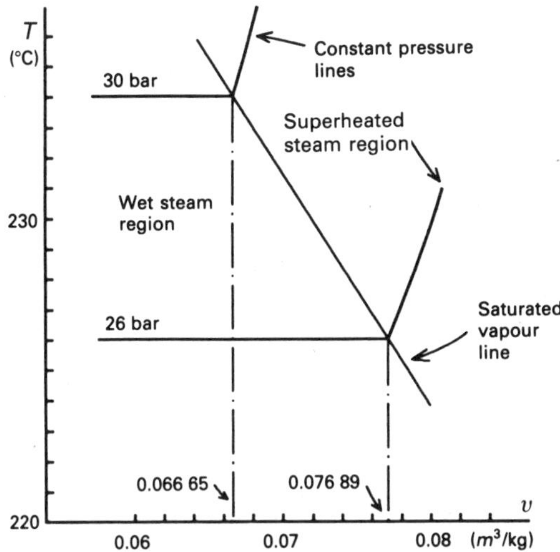

What is the condition (either wet or superheated) of steam of specific volume 0.071 42 m³/kg at (a) 30 bar and (b) 26 bar?

32

(a) Superheated, (b) Wet

Since 0.071 42 m³/kg at 30 bar is greater than 0.066 65 m³/kg (v_g), the steam must be superheated; and since 0.071 42 m³/kg at 26 bar is less than 0.076 89 m³/kg (v_g), the steam must be wet.

You may have been wondering where the names of 'saturated water' and 'saturated steam' have come from.

The use of the term 'saturated' in saturated liquid and saturated vapour can best be understood if we think of the reverse of the heating process. The term arises from the early days of steam engines when steam was condensed simply by spraying water into it. Thus superheated steam was cooled this way, and when the temperature stopped dropping it became saturated steam (since it could take up no more water). The water being used to condense steam would be subcooled water originally; its temperature would rise as it condensed the steam and when it could not condense any more steam it became saturated water.

33

We have already seen that the locus of points of saturated liquid and saturated vapour meet at the critical point. This means that at the critical point $v = v_f = v_g$ and x has no meaning. The locus of points of constant dryness fraction over the range of temperatures in the wet steam region also all meet at the critical point, as you can see from the diagram.

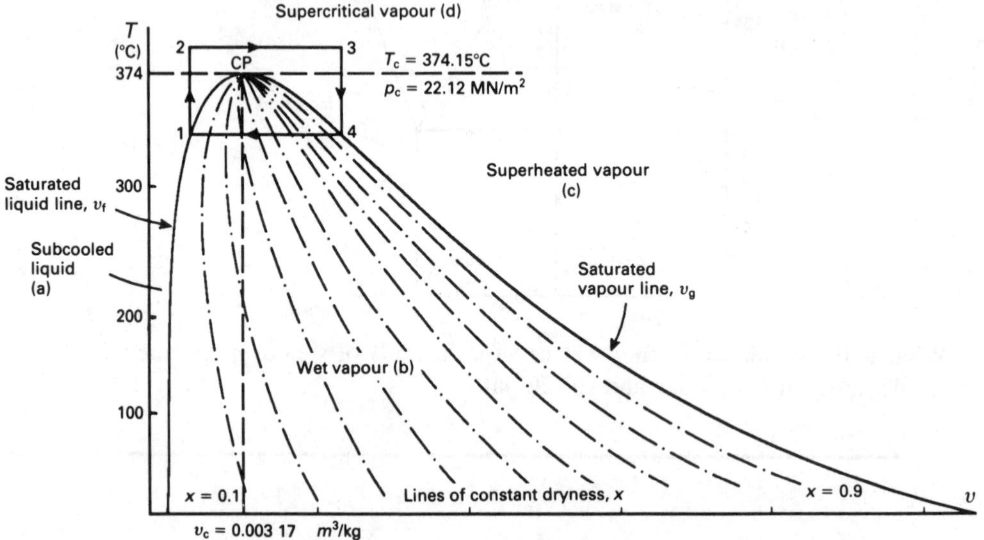

At any temperature higher than the critical temperature of 374.15°C the condition is called **supercritical**, and you can see that this region extends from the left at small volumes, where below the region is of subcooled liquid, through to the right, where below the region is of superheated vapour. In the supercritical region there are no phase changes. It is possible to imagine a cycle of two constant volume processes, 1–2 and 3–4, plus two constant temperature (isothermal) processes, 2–3, and 4–1. One may proceed round the cycle repetitively through a condensation process 4–1 without any corresponding boiling process.

Proceeding from 1 to 4 via 2 and 3, the liquid will gradually change to saturated vapour without any clear observable phase change.

34

It can be useful, having a set of properties of a particular state, to be able to say whether that state is, for example, saturated or superheated. We had an example of this in Frame 31. The properties of the critical point are given in the figure in the previous frame: $p_c = 22.12$ MN/m², $T_c = 374.15$°C and $v_c = 0.003\ 17$ m³/kg. What is the condition of steam for which $T = 374.15$°C, and $v = 0.024\ 46$ m³/kg? We see that the temperature is the same as the critical temperature, and the specific volume is greater than the critical point value. Therefore we can see the condition is on the line between superheated and supercritical vapour.

Secondly, two properties of a state are: p = 15 MN/m^2 and v = 0.0012 m^3/kg, at which pressure v_f is 0.001 664 m^3/kg. We can see that the specific volume is less than the saturated liquid value, so the condition is of subcooled liquid. Now try these:

Name the condition of the water–steam system, i.e. (a), (b), (c) or (d), from the diagram in Frame 33, bearing in mind the position of the critical point, in the following examples:

(1) T = 400°C, v = 0.003 17 m^3/kg

(2) p = 150 bar, v = 0.015 m^3/kg (v_g = 0.010 35 m^3/kg)

(3) T = 200°C, v = 0.003 17 m^3/kg.

35

(d), (c), (b)

In (1), the temperature is greater than the critical value, the volume is the same as the critical value, so the condition is supercritical (d).

In (2), the pressure is lower than the critical pressure, and the specific volume is greater than the specific volume of saturated steam, so the condition must be superheated (c).

In (3), the temperature is less than the critical value, the specific volume is the same as the critical value, so the state is immediately below the critical point, in the wet steam region, (b).

36

From all the information on the temperature–specific volume diagram, including the constant pressure lines shown in the figure in Frame 20, we could construct a pressure–specific volume diagram.

On this diagram, the isotherms, or lines of constant temperature, become the state paths of processes, showing particular p–v relationships. The saturated liquid and saturated vapour lines (heavy solid) and lines of constant dryness, (broken lines), are respectively the loci of those conditions. The p–v diagram for the water–steam system is given in the next frame.

37

When studying the p–v diagram below, bear in mind the range of values of specific volume given in Frame 21; the lines of saturated vapour and liquid below are only an approximation of their true shape.

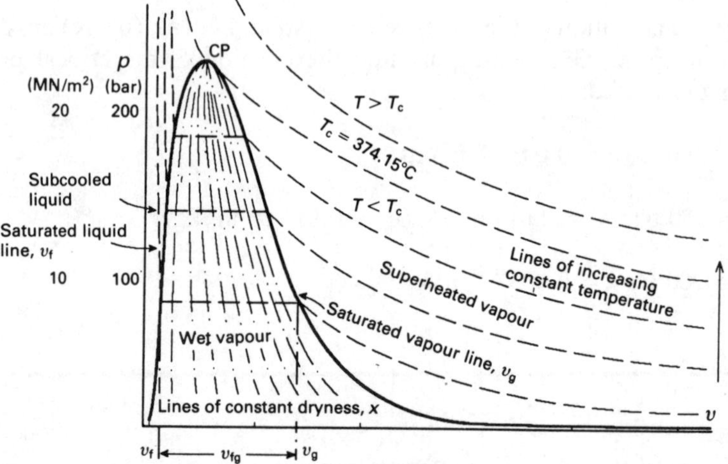

In the T–v and p–v diagrams, one of the properties, v, is the specific version of the extensive property V. If we plot the two intensive properties p and T we obtain a rather different form of diagram. Thus since p and T are not independent in the wet steam region, we obtain a single line from the critical point separating the liquid and vapour phases. The wet vapour states cannot be shown on p–T coordinates only.

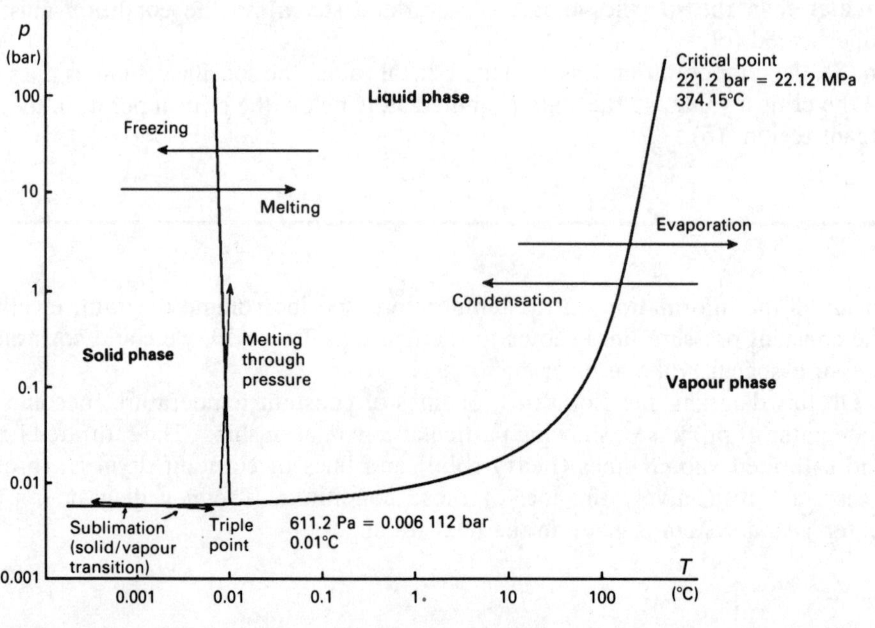

This diagram shows regions of lower temperature and lower pressure than were shown previously. The line between the liquid and vapour phases is extended to temperatures at which water freezes. At the higher pressures we cross another line between the liquid and solid phases; this line is closer to the vertical than is shown, it meets the line between the liquid and vapour phases at the triple point. Here all three phases, solid, liquid and vapour, exist in equilibrium. The triple point for ice/water/steam is at 611.2 Pa or 0.006 112 bar, and at 0.01°C.

38

At temperatures below the triple point temperature, we can see two interesting phenomena. The first is that, as you can see from the diagram, pressure causes melting. This happens in ice skating, and the water produced acts as a lubricant. The second is the process of sublimation as solid turns to vapour, or vapour to solid. Snow will disappear by sublimation on clear cold winter days, and hoar frost will appear at night.

In the *p–T* diagram in the previous frame, the line of liquid–vapour phase transition is of most importance to us, and this may be converted into a pressure–enthalpy, or *p–h* diagram. To do this we need to measure the amount of heat transfer to the system at constant pressure, relative to a datum, for the points at which water starts and finishes its boiling process.

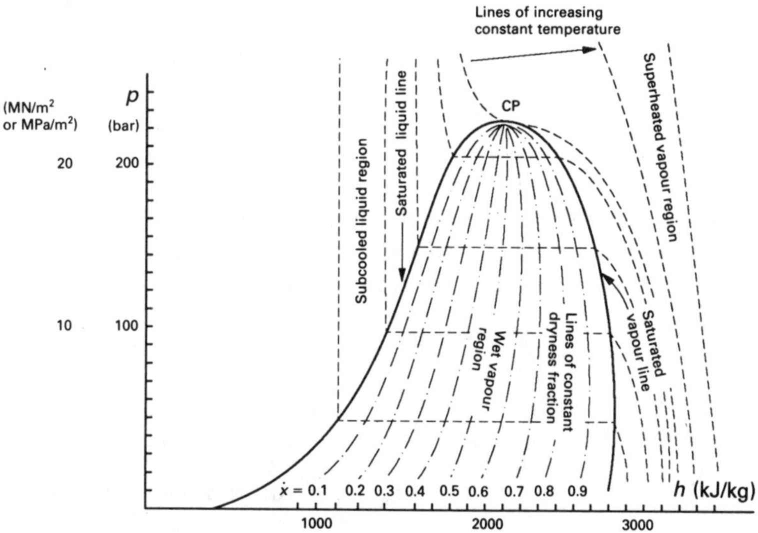

This diagram would be obtained. It shows the saturated liquid and vapour lines (heavy solid lines) lines of constant dryness fraction x (chain dotted lines), and lines of constant temperature (broken lines). As in the *T–v* and *p–v* diagrams, a region occurs for the wet vapour condition, with x increasing from left to right.

The *p–h* chart is of particular interest and value when dealing with calculations on refrigerators and heat pumps, as we shall see later.

39

In Frame 29 we expressed v in terms of v_f, x and v_g within the wet region. We can do the same for specific enthalpy, h. Thus:

$$h = (1 - x)h_f + x\,h_g \quad (1)$$
$$= h_f + x\,h_{fg} \quad (2)$$
$$= h_g - (1 - x)h_{fg} \quad (3)$$

where h_f = specific enthalpy of saturated liquid, h_g = specific enthalpy of saturated vapour and $h_{fg} = h_g - h_f$. These formulae follow a similar sequence to those for specific volume, which makes them easy to remember.

The saturation pressure for steam at 200°C is 1.555 MPa. If x for wet steam at this pressure is 0.7033, h_f = 852.67 kJ/kg and h_{fg} = 1940.42 kJ/kg, calculate h. [Use formula (2).]

40

2217.37 kJ/kg

Using formula (2) for h:

$$h = 852.67 + 0.7033 \times 1940.42$$
$$= 2217.37 \text{ [kJ/kg]}$$

Since we know p and v for wet steam conditions and we know the relationship between properties that gives h in terms of u, p and v, we can now find values for u, the specific internal energy in the wet steam region.

We have: $h = u + pv$

so $u = h - pv$

hence $u_f = h_f - pv_f$, for saturated liquid

$u_g = h_g - pv_g$ for saturated vapour

and $u_{fg} = u_g - u_f$ (by definition)

$= (h_g - h_f) - p(v_g - v_f)$

$= h_{fg} - pv_{fg}$

Specific internal energy within the wet steam region is given by three relationships similar to those for specific volume and specific enthalpy.

Thus:

$$u = (1 - x)u_f + x\,u_g \quad (1)$$

$$= u_f + x\,u_{fg} \quad (2)$$

$$= u_g - (1 - x)u_{fg} \quad (3)$$

41

We will calculate u in kJ/kg for steam from the previous example, with $h = 2217.37$ kJ/kg, for which $v = 0.09$ m³/kg and $p = 1.555$ MPa.

The units of pv must be in kJ/kg, so we have to multiply the pressure by 10^3 to give kPa. Then the product of [kPa] and [m³/kg] gives [kJ/kg].

Hence $$u = 2217.37 - 1.555 \times 10^3 \times 0.09$$

$$= 2217.37 - 139.95$$

$$= 2077.42 \text{ kJ/kg}$$

Now do the same example by calculating u_f and u_{fg} separately and using x. You are given that $h_f = 852.67$ kJ/kg, $h_{fg} = 1940.42$ kJ/kg, $v_f = 0.001\,57$ m³/kg and $v_g = 0.1273$ m³/kg. First calculate u_f and u_{fg}.

42

$$\boxed{u_f = 850.23 \text{ kJ/kg}; \; u_{fg} = 1744.91 \text{ kJ/kg}}$$

We have: $$u_f = h_f - pv_f = 852.67 - 1.555 \times 10^3 \times 0.001\,57$$

$$= 850.23 \text{ kJ/kg}$$

and: $$u_{fg} = h_{fg} - p(v_g - v_f)$$

$$= 1940.42 - 1.555 \times 10^3\,(0.1273 - 0.001\,57) = 1744.91 \text{ kJ/kg}$$

43

The calculation can then be concluded:

$$u = u_f + x\,u_{fg}$$
$$= 850.23 + 0.7033 \times 1744.91$$
$$= 2077.42 \text{ kJ/kg.}$$

We have now seen how it is possible for data for the water–steam system to be obtained by practical measurement, and how T–v, p–v, p–T and p–h diagrams may be prepared and how they will look. The other very important property of specific entropy, s, will be introduced in Programme 6. In engineering practice, tabulated values of properties are used for making calculations. In addition, charts of specific enthalpy–specific entropy, h–s, and pressure-specific enthalpy, p–h, are used for obtaining rapid but less accurate results. These will also be introduced later.

For steam, a number of publications are available suitable for student use, and any of the publications listed in the Preface are recommended. These will be referred to as the Rogers and Mayhew tables, the Haywood tables, and the Cooper and Le Fevre tables. However, the principal sources of data in these tables varies, and if a comparison is made it will be found that property values close to the critical point vary slightly with the data source.

44

In discussing the use of these tables, at this stage we shall consider only the properties p, T, v, h and u. For saturated and wet steam conditions, the saturation pressure and temperature appear in parallel columns as the independent variable, then other properties, the dependent variables, may be read off from the subsequent columns. You should carefully study the tables of your choice, or those recommended to you, since the layout varies, as also do the units used for p and T. Thus in the Rogers and Mayhew tables, p is in bar, and T in degrees Celsius. In the Haywood tables, p is in kN/m^2 up to atmospheric pressure, and is in MN/m^2 thereafter, while T is in degrees Celsius. In the Cooper and Le Fevre tables, p is in MPa, and temperature T is absolute in units of Kelvin. In these tables, θ is also used where $\theta = T - 273.15$ K. Thus θ may be regarded as temperature in degrees Celsius. In selecting the data required it is very important to distinguish between the p and T columns and to note the units used.

45

The property values of specific volume, specific enthalpy and specific internal energy are tabulated for values of p and T in discrete steps of one or the other. For example, in the Rogers and Mayhew tables, between the triple point and atmospheric boiling point, there are two tables, the first in discrete values of T, and the second in discrete values of p. Interpolation is needed to find data in between the values given.

In dealing with property values for v, h and u we know that v_{fg}, h_{fg} and u_{fg} are respectively the difference between the vapour and liquid saturation values (see Frames 29, 39 and 40); we generally find the more frequently used h_{fg} values are given, but not the other two. You will find the symbol e is used for specific internal energy in the Cooper and Le Fevre tables.

46

When both u and h values are given, it is possible to check the consistency of the data. Thus in the Rogers and Mayhew tables, page 3, we can calculate u_g from h_g and check with the value given.

Thus at 0.01 bar, using $u_g = h_g - pv_g$

we have:

$$u_g = 2514 - 0.01 \times 10^2 \times 129.2$$
$$= 2514 - 129.2$$
$$= 2384.8 \text{ [kJ/kg]}.$$

Remember that a bar is 10^5 Pa (or N/m^2), which is 10^2 kPa (or kN/m^2). The answer has been rounded off in the tables as 2385 kJ/kg. Now repeat the above calculation for the pressure of 0.015 bar.

47

$$\boxed{u_g = 2393.03 \text{ kJ/kg}}$$

As before, we have $u_g = h_g - pv_g = 2525 - 0.015 \times 10^2 \times 87.98$

$$= 2393.03 \text{ [kJ/kg], rounded off as } 2393 \text{ [kJ/kg]}.$$

48

In Frame 45, interpolation was mentioned. In fact we often have to do this. From the diagram in Frame 37 you can see that the specific volume of saturated steam, v_g, follows a curved relationship with pressure. But for a small pressure increment, say between 22 kN/m^2 and 24 kN/m^2 (0.22 bar and 0.24 bar), the curvature of the graph would be negligible, and it would be reasonable to assume a straight line relationship over this small step. We can use the principle of **linear interpolation** between any two successive points in the tables.

49

Suppose we wish to calculate v_g at 23.5 kN/m^2, when we have $v_g = 7.00$ m^3/kg, at 22 kN/m^2, and $v_g = 6.45$ m^3/kg at 24 kN/m^2. (Data from the Haywood tables, page 11.) If you are not used to interpolating, it helps to sketch a diagram; after some practice you can probably manage without. The following figure shows what we are going to do:

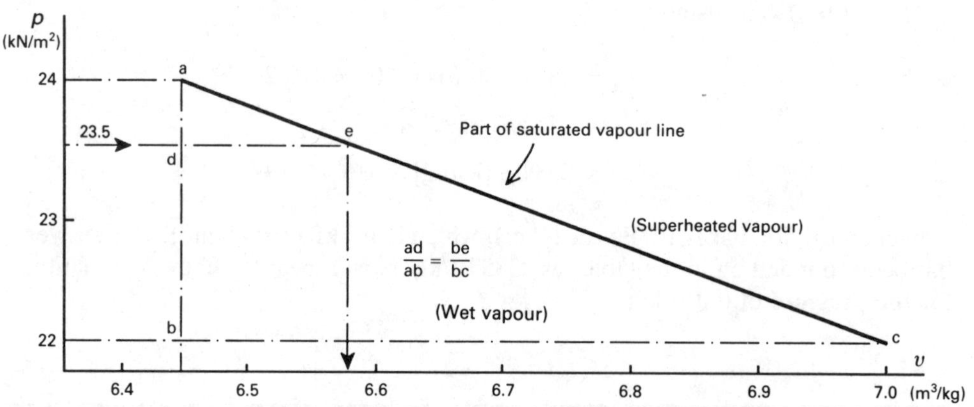

From the diagram we see that from similar triangles:

$$\frac{\text{ad}}{\text{ab}} = \frac{\text{de}}{\text{bc}}$$

$$\therefore \qquad \text{de} = \text{bc} \times \frac{\text{ad}}{\text{ab}}$$

and consequently: v_g at 23.5 kN/m^2 = 6.45 m^3/kg + de

$$= 6.45 + (7.00 - 6.45) \times \frac{(24 - 23.5)}{(24 - 22)}$$

$$= 6.45 + 0.55 \times \frac{0.5}{2.0}$$
$$= 6.59\ [\mathrm{m^3/kg}]$$

Using the Rogers and Mayhew tables on both pages 2 and 3 in turn, calculate h_g for a temperature of 82.5°C, given that from the first table h_g = 2643.2 kJ/kg at 80°C and 2651.5 kJ/kg at 85°C, and from the second table h_g = 2645 kJ/kg at 81.3°C and 2649 kJ/kg at 83.7°C. (You could devise a similar exercise for yourself from other tables.)

50

2647.35 and 2647 kJ/kg

From the first table (page 2), $h_g = 2643.2 + (2651.5 - 2643.2) \times \frac{2.5}{5}$

$$= 2647.35\ [\mathrm{kJ/kg}]$$

From the second table (page 3), $h_g = 2645 + (2649 - 2645) \times \frac{(82.5 - 81.3)}{(83.7 - 81.3)}$

$$= 2645 + 4.0 \times (1.2/2.4) = 2647\ [\mathrm{kJ/kg}]$$

The first answer is more accurate since the data in the first table are to a higher order of accuracy.

51

In the Rogers and Mayhew tables, values of v_f are given in a separate table on page 10. These are for discrete steps of temperature, so interpolation is needed to find a value at a particular pressure. As v_f is small it is tabulated as $v_f \times 10^2$ for convenience. Here is another exercise: calculate v_f at 20 bar given v_f = 0.001 173 m³/kg at 19.08 bar, and v_f = 0.001 190 m³/kg at 23.2 bar.

52

0.001 1768 m³/kg

The figure shows the interpolation, and we see that:

$$v_f = 0.001\,173 + (0.001\,190 - 0.001\,173) \times \frac{(20.00 - 19.08)}{(23.10 - 19.08)}$$

$$= 0.001\,173 + 0.000\,017 \times \frac{0.92}{4.02}$$

$$= 0.001\,1768\ [\mathrm{m^3/kg}]$$

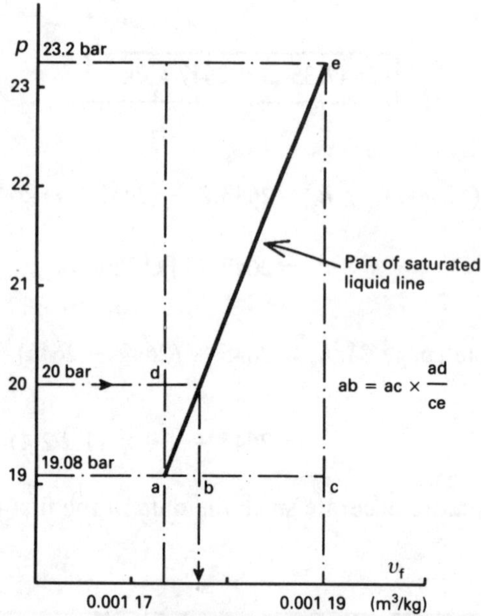

After the tabulations of properties for saturated water and steam, all tables follow with property values for superheated steam. Here the saturation temperature is the independent variable, together with the corresponding saturation pressure. Steam is superheated at higher temperatures at the same pressure, and values of specific enthalpy, h, internal energy u, volume v (and entropy s) may be found. The difference between the actual and saturation temperature is called the **degree of superheat**.

From the Rogers and Mayhew tables, page 6, at 1.5 bar, the saturation temperature is 111.4°C and v_g = 1.159 m³/kg. At 150°C, v = 1.286 m³/kg. (a) What is the degree of superheat at this condition? (b) What is v at 125°C?

The answers are (a) 38.6 K of superheat and (b) 1.204 m³/kg, and the working follows in the next frame.

53

(a) The degree of superheat is the actual temperature less the saturation temperature. So the degree of superheat is 150.0°C − 111.4°C = 38.6 K.

(b) $$v \text{ at } 125°\text{C} = v_g + (v \text{ at } 150°\text{C} - v_g) \times \frac{(125.0 - 111.4)}{(150.0 - 111.4)}$$

$$= 1.159 + (1.286 - 1.159) \times \frac{13.6}{38.6}$$

$$= 1.204 \text{ [m}^3\text{/kg]}$$

A plot of temperature against specific volume is given below.

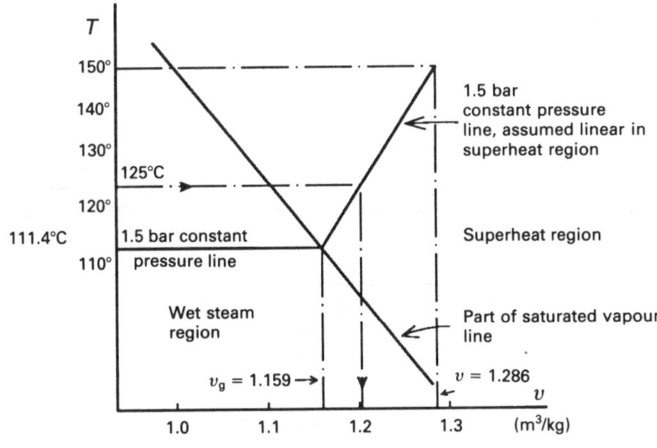

54

Both the Haywood tables and the Cooper and Le Fevre tables present data for the superheat states in tables for a single phase. Thus in the Haywood tables you will find separate tabulations for specific enthalpy, specific internal energy, density (rather than specific volume) and specific entropy. In each tabulation, columns are in steps of pressure in MN/m², from 0.01 MN/m² up to 100 MN/m², or 1000 bar, while rows are in steps of Celsius temperature, from 0°C up to 800°C. This means that property values for both the single phases of **water** and **superheated steam** phases are given in a single table. By referring back to the saturation tables, you can establish, if you are in doubt, the phase at a particular state. In any case the two phases are separated in the tables by a stepped line.

Thus, at a pressure of 1 MN/m², or 10 bar, with a saturation temperature of 179.9°C, at any temperature below this value we have the water phase, and above this temperature we have superheated steam. State the phase in each of the following cases: (a) 0.01 MN/m², 25°C; (b) 0.1 MN/m², 75°C and 475°C; (c) 10 MN/m², 300°C; (d) 20 MN/m², 350°C; (e) 22.12 MN/m², 375°C; (f) 100 MN/m², 800°C.

55

(a) water, (b) water and superheated steam, (c) water, (d) water, (e) superheated steam, (f) supercritical steam

In the previous example, both the temperature and pressure are greater than critical values, hence the condition is supercritical.

For accurate work in dealing with superheat states it is often necessary to interpolate between both pressure and temperature. Thus to find h at 0.25 bar and 125°C (Rogers and Mayhew Tables), we would carry out two interpolations to find h at 0.1 bar and 0.5 bar at 125°C, then a vertical interpolation to find h at 0.25 bar and 125°C. See if you can do that.

The values of h needed are: 0.1 bar, 2688 and 2783 kJ/kg at 100°C and 150°C, and at 0.5 bar, 2683 and 2780 kJ/kg at 100°C and 150°C respectively.

56

2734.0 kJ/kg

$$h \text{ at 0.1 bar and } 125°\text{C} = 2688 + (2783 - 2688) \times \frac{(125 - 100)}{(150 - 100)}$$

$$= 2735.5 \text{ [kJ/kg]}$$

$$h \text{ at 0.5 bar and } 125°\text{C} = 2683 + (2780 - 2683) \times \frac{(125 - 100)}{(150 - 100)}$$

$$= 2731.5 \text{ [kJ/kg]}$$

Notice that h at 0.5 bar and 125°C is less than that at 0.1 bar and 125°C. If you look at the p–h diagram in Frame 38 you can see that would be so. Lines of constant temperature in the superheat region show the enthalpy increasing as the pressure falls. We can now do the final interpolation:

$$h \text{ at 0.25 bar and } 125°\text{C} = 2735.5 - (2735.5 - 2731.5) \times \frac{(0.25 - 0.1)}{(0.50 - 0.1)}$$

$$= 2735.5 - 1.5$$

$$= 2734.0 \text{ [kJ/kg]}$$

We could have found this value by interpolating between pressures first. You should confirm, as another exercise, that the same answer must be obtained. First find h at 0.25 bar and 100°C (2686.125 kJ/kg), and h at 0.25 bar and 150°C (2781.875 kJ/kg), and then finally h at 0.25 bar and 125°C.

Using the Haywood tables, interpolating between 0.01 MN/m^2 (0.1 bar), at 125°C for which h = 2735, and 0.05 MN/m^2 (0.5 bar), at 125°C for which h = 2731 kJ/kg, the result is 2733.5 kJ/kg.

57

The water state which appears in the same tables as superheated steam in the Haywood tables and Cooper and Le Fevre tables, may be referred to as a **subcooled** or **compressed** liquid state. It is a state that occurs to the left of the saturated liquid line in the property diagrams we have used.

For example, water at 10 bar or 1 MN/m^2, and at 100°C, may be regarded as water subcooled from the saturation temperature of 179.9°C at 10 bar, or it may be regarded as saturated water at 100°C compressed from 1.013 25 bar to 10 bar. The properties of water in this state may be found directly or by interpolation in these tables. A separate table of properties of compressed water is given on page 11 of the Rogers and Mayhew tables.

If we regard water in this state as compressed water, we can calculate the approximate specific enthalpy, from the saturation value, h_f.

We use the relation between properties:

$$h = u + pv$$

so that in differential form we may write:

$$\mathrm{d}h = \mathrm{d}u + p\mathrm{d}v + v\mathrm{d}p$$

We then assume that the compression of the saturated liquid state to the compressed liquid state takes place at constant temperature, so in calculating the increase of enthalpy, du is zero; also we assume that the change of volume is very small indeed so that $\mathrm{d}v = 0$, and so $p\mathrm{d}v = 0$. This means that the increase of enthalpy $\mathrm{d}h$ is $v\mathrm{d}p$.

There is no special symbol for h in the compressed liquid state, so we write:

$$h(\text{compressed liquid}) = h_f + v_f\,(\Delta p)$$

You should bear in mind that this result is only approximate and is most accurate if Δp is small. However it does give us a simple formula which is convenient to use.

What is h at 10 bar, or 1 MN/m^2, and 100°C, from tables, and by calculation?

58

419.7 kJ/kg, 420.04 kJ/kg

By inspection, from the Haywood tables, page 16, h = 419.7 kJ/kg. From the Cooper and Le Fevre tables, page 6, it is 419.8 kJ/kg.

To interpolate from the compressed water table, page 11 of the Rogers and Mayhew tables, we see that $(h - h_f)$ between 1.013 25 bar and 100 bar at 100°C is +7 kJ/kg. Hence at 10 bar and 100°C, we have:

$$h = 419.1 + 7 \times \frac{(10 - 1.013\,25)}{(100 - 1.013\,25)}$$

$$= 419.74 \text{ [kJ/kg]}$$

Using: h(compressed liquid) = $h_f\ v_f\ (\Delta p)$, with v_f = 0.001 044 m^3/kg, we have

$$h = 419.1 + (10 - 1.013\,25) \times 10^2 \times 0.001\,044$$

$$= 420.04 \text{ [kJ/kg]}$$

The next frame is a set of problems on using your steam tables. These exercises will help you to find your way around, as it is important that you are fluent in using the tables before you start on problems involving steam. Depending on which tables you are using, interpolation may or may not be needed, and answers (which were obtained using the Rogers and Mayhew tables) may vary slightly from those given. This applies particularly to solutions using properties of states near the critical point.

59

PROBLEMS ON THE USE OF THE STEAM TABLES

1. Find specific enthalpy of steam at the following states:
   (a) dry saturated at 0.25 MN/m^2, or 2.5 bar
   (b) dry saturated at 75.9°C
   (c) dry saturated at 365.7°C
   (d) superheated at 5000 Pa and 50°C
   (e) superheated at 15 bar and 600°C
2. Find the specific volume of dry saturated steam at the following states:
   (a) saturation temperature of 38°C. What is the saturation pressure?
   (b) saturation pressure of 15 MN/m^2. What is the saturation temperature?
3. Find the specific internal energy of saturated water at
   (a) 200 bar and (b) at 373.7°C.

4. Find the specific volume and specific enthalpy of steam at 3 MN/m^2 and 400°C.
5. At what condition is h_{fg} equal to zero?
6. Calculate the specific enthalpy of steam at 100°C and 0.6 dryness fraction.
7. Calculate the internal energy of 3 kg of steam at 10 MN/m^2 and 0.85 dryness fraction.
8. Calculate the specific volume of saturated water at 0.6 MPa.
9. Calculate the specific volume of wet steam at 150 bar and 0.9 dryness fraction.
10. What is the specific enthalpy of dry saturated steam at 17.2 MN/m^2?
11. What is the specific volume of wet steam at 39 bar and 0.4 dryness fraction?
12. Calculate the internal energy of 2.5 kg of steam at 1 MPa and 320°C.
13. Calculate the specific volume of steam at 16 bar and 380°C.
14. Calculate the enthalpy of 10 kg of compressed water at 60.1°C and 130 bar.

60

ANSWERS TO PROBLEMS IN FRAME 59

1. (a) 2717, (b) 2636, (c) 2411, (d) 2594, (e) 3694 kJ/kg.
2. (a) 21.63 m^3/kg, 0.066 24 bar, (b) 0.010 35 m^3/kg, 342.1°C.
3. (a) 1786, (b) 1949 kJ/kg.
4. 0.0993 m^3/kg, 3231 kJ/kg.
5. The critical point.
6. 1773.12 kJ/kg.
7. 7116.6 kJ.
8. 0.001 100 6 m^3/kg.
9. 0.009 481 m^3/kg.
10. 2540.8 kJ/kg.
11. 0.021 195 m^3/kg.
12. 7066 kJ.
13. 0.1863 m^3/kg.
14. 2642.01 kJ.

61

In Frame 37 we considered the p–T diagram of the water/steam system, and we saw that the properties were extended into the solid phase, which involved the triple point. We must now see how the solid phase region appears on a p–v diagram.

62

From our knowledge already, we see that volumes reduce with falling temperature, and when this is extended into freezing we would expect contraction to take place. A pressure–specific volume diagram showing this is given below.

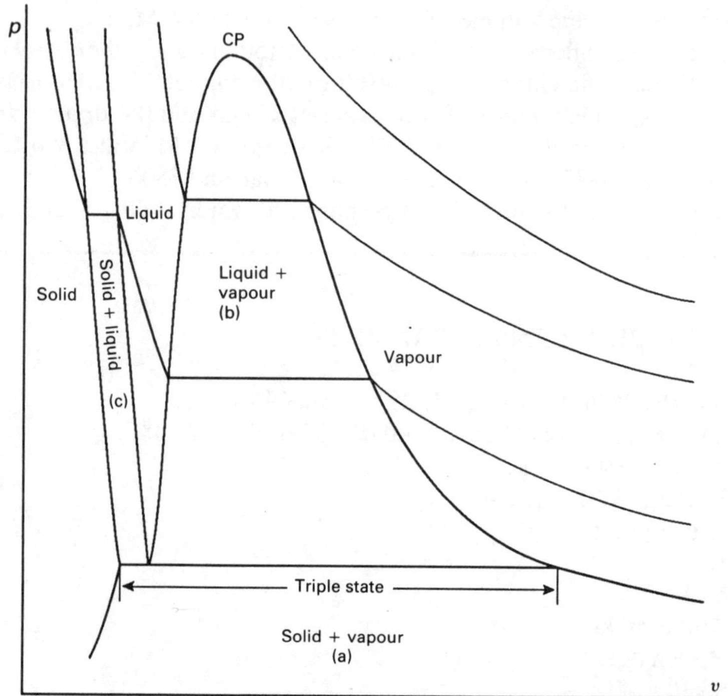

Compare this carefully with the p–T diagram in Frame 37. Consider any point on a line separating two phases on the p–T diagram. At this saturation pressure and temperature, a volume change occurs with the change of phase; this volume change becomes a certain increment on the volume axis at the same value of pressure. Thus the three lines making up the p–T diagram become three bands of volume change, and the triple point becomes a horizontal line for the state of the triple point. Region (a) corresponds to the solid/vapour boundary on the p–T diagram; region (b) corresponds to the liquid/vapour boundary; and region (c) corresponds to the solid/liquid boundary.

63

The p–T diagram of Frame 37 and the complete p–v diagram from the previous frame may be combined to give a p–v–T diagram. This is shown on the next page.

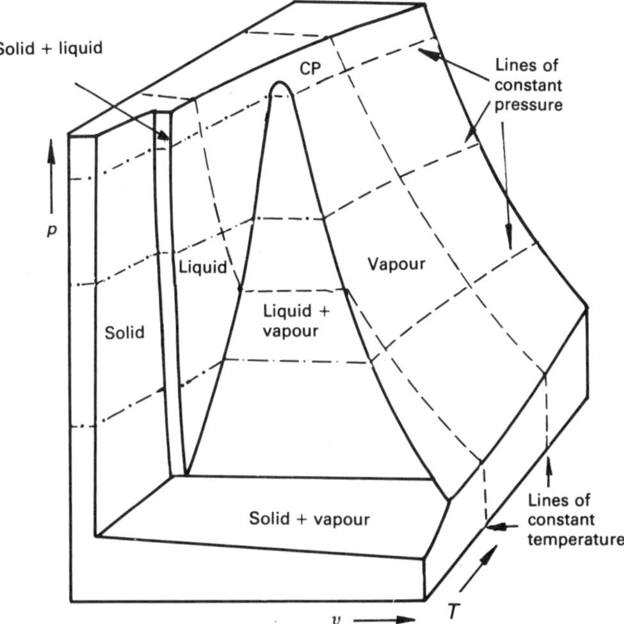

It is essentially a perspective drawing of a three dimensional shape, and it is known as a *p–v–T* surface. If you look at the *p–v* face you will see the diagram from the previous frame; and if you look at the *p–T* face you will see the diagram of Frame 37. Study this carefully, you should be able to see how the characteristics of the two separate diagrams appear on a single surface.

This is a *p–v–T* surface for a substance that contracts on freezing. Bearing in mind the fact that ice floats on water, is this *p–v–T* diagram correct for the ice/water/steam system?

64

No

Since ice floats on water, ice at 0°C and at atmospheric pressure is less dense than water at the same conditions. This means that a given mass of ice has a larger volume than the same mass of water in an equilibrium state, and the water has consequently expanded on freezing. The expansion is about 9% in terms of volume increase, so that roughly 90% of an iceberg is below water level. Because of this characteristic of water, we need a different *p–v–T* diagram which is shown in the next frame.

65

The fact that ice expands on freezing is an important quirk of nature. Otherwise ponds would freeze from the bottom upwards, trapping marine life. It is also responsible for burst water pipes, though the damage is usually not apparent until after temperatures rise.

Since the *p–v–T* diagram of Frame 63 is not correct for the ice/water/steam system; an alternative diagram must be used which shows expansion on freezing. This is difficult to see on a *p–v* diagram but may be shown more clearly on the *p–v–T* surface which is given below.

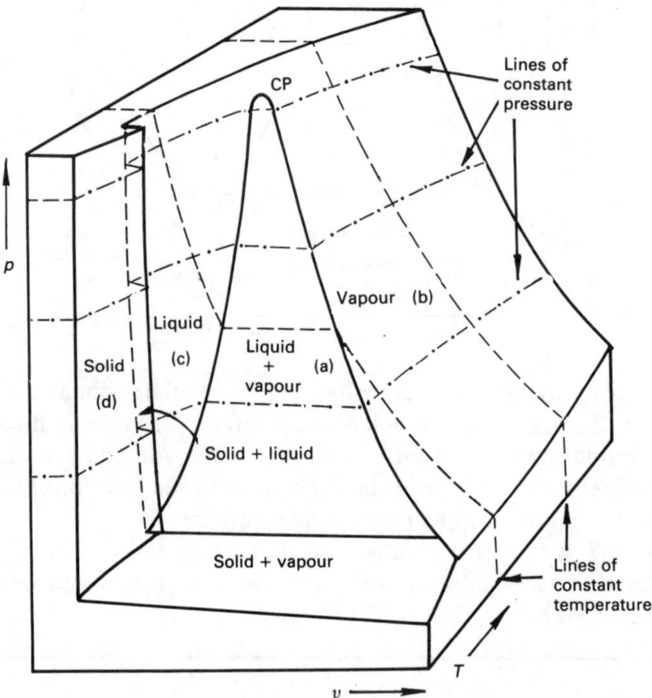

Apart from the volume increase on freezing, all other features are similar to the *p–v–T* diagram in Frame 63. The liquid + vapour surface (a) is roughly at right angles to the liquid (c) and vapour (b) surfaces which merge above the critical point. As the water is cooled the specific volume of the liquid decreases so that surface (c) disappears behind the solid surface (d). The surface becomes hidden from view because of the expansion on freezing.

In the diagrams we have considered, the vapour and liquid phases of a pure substance are of most importance. What about gases? In the following frames we shall see where gases occur on the phase diagrams, and learn about their properties.

66

In Thermodynamics we think about real gases and ideal gases. An ideal gas is a concept to which real gases more or less closely conform. We shall meet the ideal gas rule which is a simple relationship between the properties of an ideal gas in an equilibrium state. This rule is an approximation to the actual behaviour of real gases. Much more accurate and complicated laws exist between the properties of real gases, but we shall leave these aside.

The gas region is shown shaded on the *p–v* diagram. It occurs to the right of the saturated vapour line up to the critical point, and to the right of the critical temperature isotherm (constant temperature line) at pressures greater than the critical pressure.

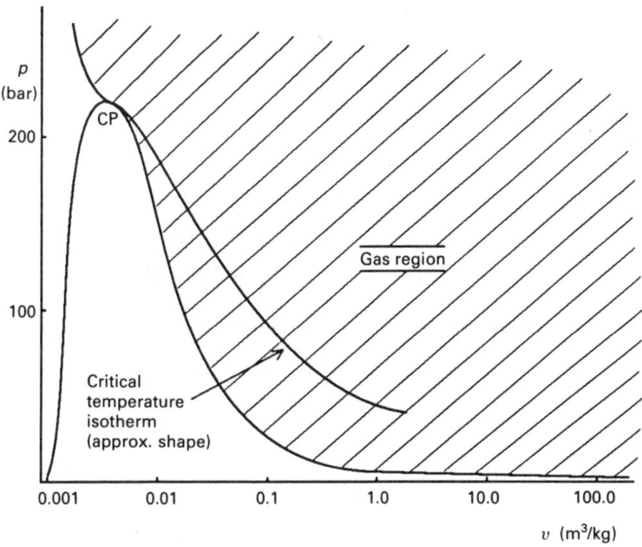

The gas region covers two areas identified previously. You should be able to name them.

67

Superheated and supercritical

The superheated region is below the critical temperature and the supercritical region is above it. As the gas region overlaps the superheated vapour region, is there any difference between the two? A gas is a vapour in a state sufficiently far removed from the critical point and saturated vapour boundary that its properties can be related by a simple algebraic equation. The ideal gas rule is such an equation.

68

The critical temperature and pressure of four common gases are given in the table below.

	Critical temperature	*Critical pressure*
Carbon dioxide, CO_2	31.1°C	73.91 bar = 7.391 MN/m^2
Oxygen, O_2	−118.6°C	50.40 bar = 5.040 MN/m^2
Nitrogen, N_2	−146.8°C	33.92 bar = 3.392 MN/m^2
Hydrogen, H_2	−239.7°C	12.98 bar = 1.298 MN/m^2

Thus, at atmospheric conditions the last three gases exist at well below the critical pressure and well above the critical temperature, in the superheated vapour region. The same state point for carbon dioxide would be found between the saturated vapour line and the critical temperature isotherm in the *p–v* diagram in Frame 66.

At room temperature, carbon dioxide is below its critical temperature. By looking at the *p–T* diagram of Frame 37, can you tell whether it is possible to liquefy carbon dioxide by pressure alone at room temperature?

69

Yes

If you refer back to the figure in Frame 37, you can see that provided the carbon dioxide is below 31.1°C it may be liquefied by increasing the pressure. This is because on the *p–T* diagram we are to the left of the critical temperature, and on moving upwards we pass from the vapour to the liquid regions. By increasing the pressure to the right of the critical point we do not meet the liquid region. The other gases, well above their critical temperatures, were once called 'permanent' gases since they could not be liquefied by pressure alone.

Water vapour in the atmosphere is really superheated steam at very low pressure. The water content of the atmosphere is called **humidity** and humid air is a pure substance which may be treated as a gas.

If the atmosphere at 15°C contains water vapour having a saturation pressure of 0.01 bar, what is the degree of superheat of this water vapour?

70

8 K

From your steam tables you should be able to deduce that the saturation temperature is 7°C at 0.01 bar pressure. The degree of superheat is therefore 15°C – 7°C = 8 K.

Suppose now we take a particular gas, say nitrogen, and measure its specific volume over a range of pressures for a particular absolute temperature, we find for each temperature that:

$$pv = \text{constant}$$

This, as you know, is Boyle's Law. Then, if we repeat the measurements for a range of constant absolute temperatures, and then plot the group pv/T against p for each temperature, we shall get a series of curves which converge to a single pv/T value as the pressure falls to zero. This is shown in the figure.

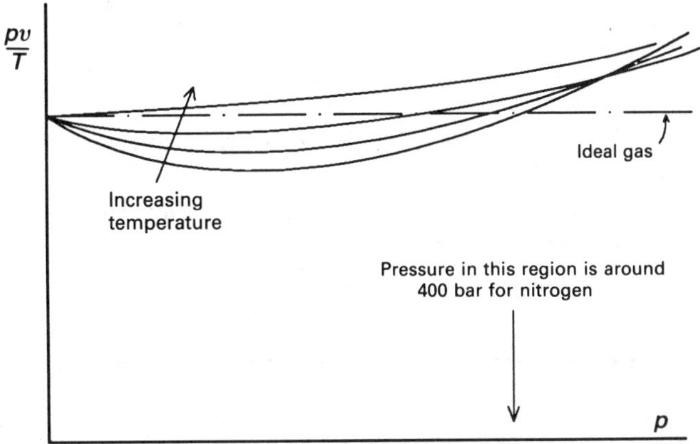

71

This single value of pv/T at zero pressure applies only to this gas. If we repeated the experiment for another gas we would get another value of pv/T at zero pressure. This single value for a given gas is called the gas constant for that gas.

For each different gas: $\dfrac{pv}{T} = R =$ a constant for that gas. So each different gas has its own value of R, the gas constant for that gas.

72

Although we see that pv/T does vary with pressure, in fact this variation is small at low pressures. The behaviour of a real gas conforms fairly accurately to this formula. If it conformed *exactly* we would call it an **ideal gas**, something that does not actually exist.

The ideal gas rule is expressed as:

$$pv = RT$$

where R = the gas constant. Each gas has its own value of gas constant.

If the pressure remains constant we see that specific volume varies with temperature, i.e. v/T is constant. This is often referred to as Charles' law, but when Charles' experiments took place in 1787 the concept of absolute temperature had not been defined. His experiments were actually based on temperatures measured using an equally divided mercury in glass thermometer.

Typical values of the gas constant R are given below:

Gas	R	
Carbon dioxide	188.9 J/kg K	= 0.1889 kJ/kg K
Oxygen	259.8	= 0.2598
Nitrogen	296.8	= 0.2968
Hydrogen	4124.0	= 4.1240
Air	287.0	= 0.2870

Later we shall see how R for air, which is a mixture of oxygen and nitrogen, may be calculated from the individual values of R.

The units of R follow from the pressure in Pa, or N/m^2, the specific volume in m^3/kg and the absolute temperature, K.

Check the consistency of the units in the ideal gas rule, $pv = RT$.

73

J/kg = J/kg

The left-hand side units are: $\dfrac{[\text{Pa m}^3]}{[\text{kg}]} = \dfrac{[\text{N m}^3]}{[\text{m}^2\text{kg}]} = \dfrac{[\text{J}]}{[\text{kg}]}$, since [J] = [Nm]

The right-hand units are: $\dfrac{[\text{J K}]}{[\text{kg K}]} = \dfrac{[\text{J}]}{[\text{kg}]}$

Obviously, with pressure in kPa, we would get kJ/kg units in the equation.

74

Whereas with steam the two property rule enables us to look up the third property in tables (with possible interpolation), the ideal gas rule enables us to calculate the third property.

For example, taking air as a pure substance, for which $R = 287$ J/kg K, $v = 0.8163$ m³/kg and $p = 101.325$ kPa, we may calculate the temperature to give the third property to define the equilibrium state completely.

As the pressure is in kPa, we will need R in kJ/kg K, which will be 0.287 kJ/kg K. Using $pv = RT$, we get:

$$101.325 \times 0.8163 = 0.287 \times \mathrm{T}$$

$$\therefore \quad T = \frac{101.325 \times 0.8163}{0.287}$$

$$= 288.19 \text{ [K]}$$

What is the mass of 4 m³ of carbon dioxide at 350 K and at a pressure of 150 kPa? R for carbon dioxide is 0.1889 kJ/kg K

75

9.075 kg

Since $v = V/m$, we can express the ideal gas rule as:

$$p(V/m) = RT$$

or:

$$pV = mR\mathrm{T}$$

$$\therefore \quad m = \frac{pV}{RT} = \frac{150 \times 4}{0.1889 \times 350} = 9.075 \text{ [kg]}$$

You may have calculated the specific volume first using $pv = RT$, and got an answer of 0.440 77 m³/kg, and then divided this into the volume of 4 m³.

Since we have: $\dfrac{pV}{T} = R$ = a constant

we can say that if a gas follows a process from a state 1 to another equilibrium state 2, then we know that p_2v_2/T_2 is the same as p_1v_1/T_1, so if we know all the properties at state 1 and two of the properties at state 2 we can calculate the third property at state 2.

76

Air is in the equilibrium state given by $p = 101.325$ kN/m^2, $v = 0.8163$ m^3/kg and $T = 288.19$ K.

If it is compressed isothermally to twice the pressure, what is the final specific volume? You should have remembered that isothermally means at constant temperature, so this means that:

$$pv = \text{a constant, so } p_2v_2 = p_1v_1$$

We know that p_2 is twice p_1 so $p_2 = 202.65$ kPa

hence: $$202.65 \times v_2 = 101.325 \times 0.8163$$

$$\therefore \quad v_2 = 0.408\ 15\ [\text{m}^3/\text{kg}]$$

Now let the air in the same original state be expanded at constant pressure to twice the specific volume, what is the final temperature?

77

576.38 K

Since v/T = a constant (the process is at constant pressure), with $v_2/v_1 = 2$, by cross-multiplying we get:

$$\frac{v_2}{v_1} = \frac{T_2}{T_1} = 2$$

$$\therefore \quad T_2 = 2 \times T_1 = 576.38 \text{ K}$$

So far, then, we have the ideal gas rule which involves a constant R which has a different value for each gas. We have seen in Frame 75 that we can rewrite the law for a gas of mass m, and we get:

$$pV = mRT$$

where V is the volume of gas, m^3, which has a mass of m kg. We may then write the law for M kg of the gas, where M is the molecular mass.

This will give us:

$$pV_M = MRT$$

where V_M is the volume of the molecular mass at pressure p and temperature T. Molecular mass is defined in the next frame.

78

The molecular mass, M, is defined as:

$$M = \frac{\text{(mass of one molecule of substance)}}{1/12 \times \text{(mass of one atom of carbon-12)}}$$

so M is just a ratio of two masses. The molecular mass of carbon-12, which is an isotope or particular form of carbon, is exactly 12. The mass of one molecule or atom is extremely small, but the numbers we use for molecular mass are all in proportion to a substance's actual molecular mass. You may recall from your knowledge of chemistry that a mass equal to the molecular mass in g is known as a mole.

A mole is defined as the amount of substance which contains as many molecules as there are atoms in 12 g of carbon-12.

Carbon-12 is monatomic which means one molecule consists of one atom. As we are dealing in kg we need the kg mole written as kmol which is then 10^3 times the mole. We have then that V_M is the volume of a kmol at pressure p and temperature T.

We may calculate V_M at 101.325 kPa and 288.19 K for a kmol of nitrogen (28.016 kg) having R of 296.8 J/kg K. Thus:

$$V_M = \frac{28.016 \times 296.8 \times 288.19}{101.325 \times 1000}$$

$$= 23.65 \text{ [m}^3\text{]}$$

Calculate V_M for a kmol (44.011 kg) of carbon dioxide at the same state.

79

23.65 m³

As before:

$$V_M = \frac{44.011 \times 188.9 \times 288.19}{101.325 \times 1000}$$

$$= 23.65 \text{ [m}^3\text{]}$$

Were you surprised to get the same answer?

80

For *any* gas we would find that V_M is the same.

Let X be the mass of one atom of carbon-12 in kg. Consider a substance A which has a molecular mass of M_A. From the definition of molecular mass, the actual mass of one molecule of substance A must be $M_A \times 1/12 \times X$ kg. So if we take a kmol of substance A, which is M_A kg of substance A, the number of molecules, N_A, must be:

$$N_A = \frac{M_A}{M_A \times 1/12 \times X}$$
$$= 12/X$$

Similarly we can show that for any substance B of molecular mass M_B the number of molecules in M_B kg of the substance must be:

$$N_B = \frac{M_B}{M_B \times 1/12 \times X}$$
$$= 12/X$$

Thus a kmol of any substance contains the same number of molecules. Thus we have proved the statement defining the mol in Frame 78, since this must be true for the mol as well.

81

We have already seen that the volume of a kmol of any substance is the same under the same conditions of temperature and pressure, so these equal volumes each have the same number of molecules of gas.

This is expressed by **Avogadro's Hypothesis** (1811) which states that **equal volumes of all gases under the same conditions of temperature and pressure contain the same number of molecules.**

We expressed the number of molecules for a kmol as $12/X$. **Avogadro's number**, which is 6.023×10^{23}, is the number of molecules in one g mol (10^{-3} of kmol). We need not concern ourselves how this number was obtained.

Now recalculate V_M at 0°C (273.15 K) and 101.325 kPa.

82

22.41 m^3

Since this is a volume change at constant pressure, we may write:

$$V_M = 23.65 \times \frac{273.15}{288.19}$$

$$= 22.41 \ [m^3]$$

For a g mole this is 22.41 litres, which is known as **Avogadro's volume**.

Since V_M is constant at a given p and T for M kg of gas it follows that:

$$pV_M = R_M T$$

where R_M = universal gas constant = MR = 8314.5 [J/kmol K]

As V_M is the volume of a kmol of a gas, we may introduce the number of kmol, n, and write:

$$pV = nR_M T$$

This is the ideal gas rule in terms of the universal gas constant and the number of moles. Any quantity of gas can be expressed in terms of kmol, thus 2 kg of carbon dioxide is 2/44.011 = 0.045 44 kmol, where 44.011 is the molecular mass of carbon dioxide.

Calculate the temperature of 16 kg of oxygen (molecular mass = 32.000), having a volume of 7 m^3 and a pressure of 200.0 kPa. (R_M = 8.3145 kJ/kmol K)

83

336.76 K

The number of kmol, n, is 0.5, so we have: $200.0 \times 7.0 = 0.5 \times 8.3145 \times T$

$$\therefore \quad T = \frac{200.0 \times 7.0}{0.5 \times 8.3145} = 336.76 \ [K]$$

84

In order to use the ideal gas rule with the universal gas constant, we need to know the molecular masses of common gases. The definition of molecular mass was given in Frame 78; this was agreed internationally and, as we have seen, was based on the carbon-12 isotope with a molecular mass of 12. This means that other molecular masses are then very close to whole numbers.

Some common molecular masses are:

Oxygen	O_2	31.999	(32)
Nitrogen	N_2	28.013	(28)
Hydrogen	H_2	2.016	(2)
Carbon dioxide	CO_2	44.010	(44)
Air		28.970	(equivalent molecular mass)

The reasons for the precise values of molecular mass need not worry you. It is quite in order to work from the whole numbers given.

85

In Frame 72 we met the ideal gas rule, which involved absolute temperature, T. You will recall that at the end of Programme 1 we first came across the idea of an absolute thermodynamic temperature which is independent of the physical properties of real substances. We will now look a little closer at the behaviour of gases in relation to absolute temperature.

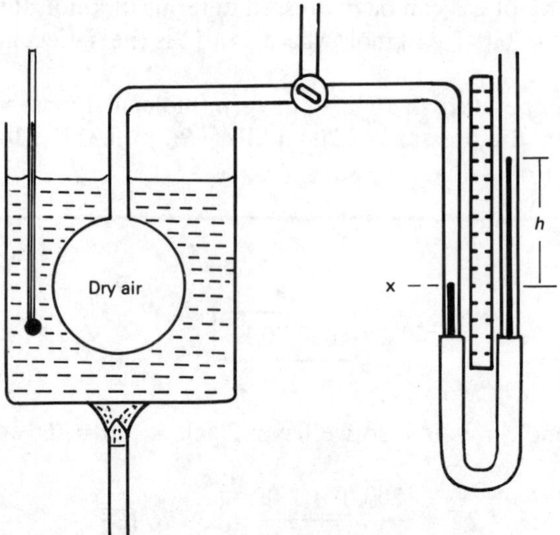

The apparatus shown is a simple constant volume gas thermometer. The bulb of dry air is enclosed in an ice–water mixture at atmospheric pressure and at 0°C. The mercury level is adjusted to position X and height h is measured as h_0 mm mercury.

The ice–water mixture is now brought to the boil by heating and h_{100} is measured after the mercury is again adjusted to position X. This gives a second pressure reading corresponding to 100°C. If we obtained intermediate values of h at temperatures measured on a thermometer calibrated according to agreed procedures, we would find that a plot of absolute gas pressure against temperature in °C would give a straight line.

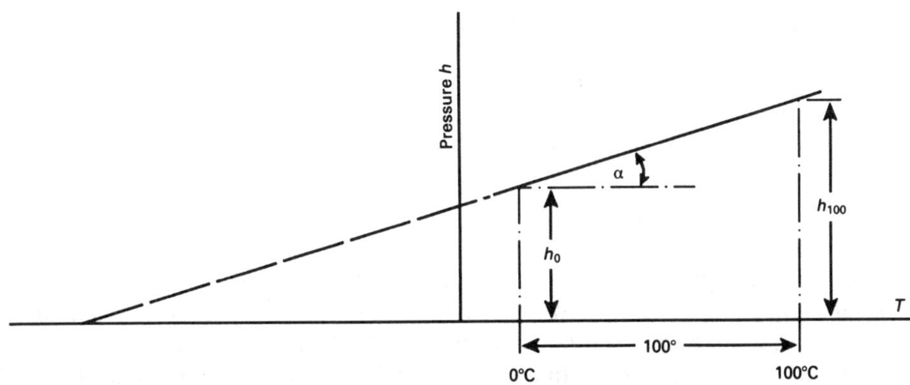

86

The slope of the line is α, where

$$\alpha = \frac{h_{100} - h_0}{100}$$

An intermediate h in terms of temperature T (between 0° and 100°C) is:

$$h = h_0 + \alpha T$$

so:

$$h = h_0[1 + (\alpha T/h_0)]$$
$$= h_0 [1 + \alpha T]$$

where:

$$\alpha = \frac{h_{100} - h_0}{100 h_0}$$

In the next frame we shall calculate α from some experimental measurements and see what we can do with the result.

87

If in a gas thermometer $h_0 = 770$ mm mercury at 0°C and $h_{100} = 1051.9$ mm at 100°C (absolute pressures), we can find the temperature at an intermediate h of 960 mm.

$$\text{We have } \alpha = \frac{1051.9 - 770}{100 \times 770} \quad \text{since } h_0 = 770 \text{ and } h_{100} = 1051.9 \text{ mm}$$

$$= 0.003\,661$$

$$\text{Then: } 960 = 770[1 + 0.003\,661 \times T]$$

$$\therefore \quad 1.246\,75 = 1 + 0.003\,661 \times T$$

$$\therefore \quad 0.246\,75 = 0.003\,661 \times T$$

$$\therefore \quad T = 67.4°\text{C}$$

We see from the pressure–temperature graph in the previous frame that we have a pressure reading at 0°C and a straight line of slope α which suggests that we could extrapolate the line to zero pressure to find an absolute zero temperature.

Rather than extrapolate straight away we may repeat the procedure described by using smaller masses, but equal volumes of gas (i.e. at lower pressures), so that a series of values of α is obtained. By plotting against pressure and now extrapolating to zero pressure, a value of α at zero pressure is found. Repeating with different gases we find the same value of α for all the 'permanent' gases.

The slope of our line α is found to be constant for all gases. Why is this? In Frame 82 we found the ideal gas rule could be expressed as:

$$pV_M = R_M T$$

where both V_M and R_M were seen to be constants for all gases. In that case the ratio p/T must be a constant for all gases, and p/T is the slope of our line α.

What is the reciprocal of the α that we have just calculated?

88

273.15

89

The reciprocal of α is 1/0.003 661 = 273.15. By looking at the equation for h in terms of α and T, you can see the reciprocal of α is the number of degrees of temperature to give an h value equal to h_0, and since h is measured from absolute zero pressure, $1/\alpha$ is the temperature at h_0 measured from absolute zero. Thus 0°C is 273.15° absolute, and 100°C is 373.15° absolute.

We now have a straight line temperature–pressure relationship which is known as the ideal gas constant volume temperature scale.

At the end of Programme 1 we stated that there is an absolute thermodynamic temperature scale in degrees Kelvin, K, in which 0°C and 100°C are 273.15 K and 373.15 K respectively. We still have to see whether the ideal gas temperature scale and the thermodynamic temperature scale are one and the same thing.

What are 246.94°C and −58.78°C in degrees K ?

90

520.09 K and 214.37 K

$$246.94 + 273.15 = 520.09 \text{ K and } -58.78 + 273.15 = 214.37 \text{ K}$$

We now turn to the question of specific heats of gases, both real and ideal. In 1845, Joule carried out a simple experiment which proved that internal energy was dependent on temperature only, and not on pressure or specific volume. The apparatus is shown below and described in the next frame.

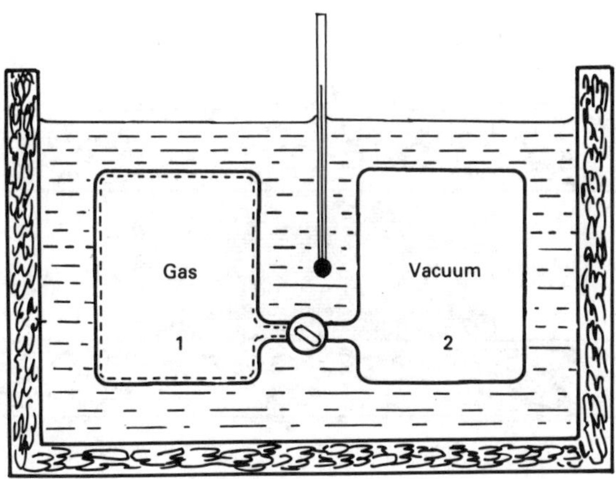

91

In the apparatus shown, gas is contained in vessel (1) under pressure while vessel (2) is evacuated. The two vessels are joined by a line with a closed valve and the whole is placed in water in an insulated container. The system is the gas in (1), finally filling (1) and (2) after the valve is opened. Joule found no change in the water temperature after the expansion, so $Q = 0$. As we know, $W = 0$ for this expansion, so:

$$Q - W = U_2 - U_1$$

$$0 = U_2 - U_1$$

and so:

$$0 = m\,c_v(T_2 - T_1)$$

where m is the mass of the gas in the closed system and c_v is the specific heat at constant volume. Hence U does not depend on p or v of the gas, but only on temperature, as we have seen in the previous programme. Consequently c_v is also seen to be independent of p or v, but may depend on temperature.

92

Later, in 1852, Joule and Thomson carried out a steady flow version of the same experiment. Gas was allowed to flow through a porous plug in an insulated tube, so that again both Q and W are zero. (Refer back to the previous programme if you cannot remember the terms in the steady flow energy equation.)

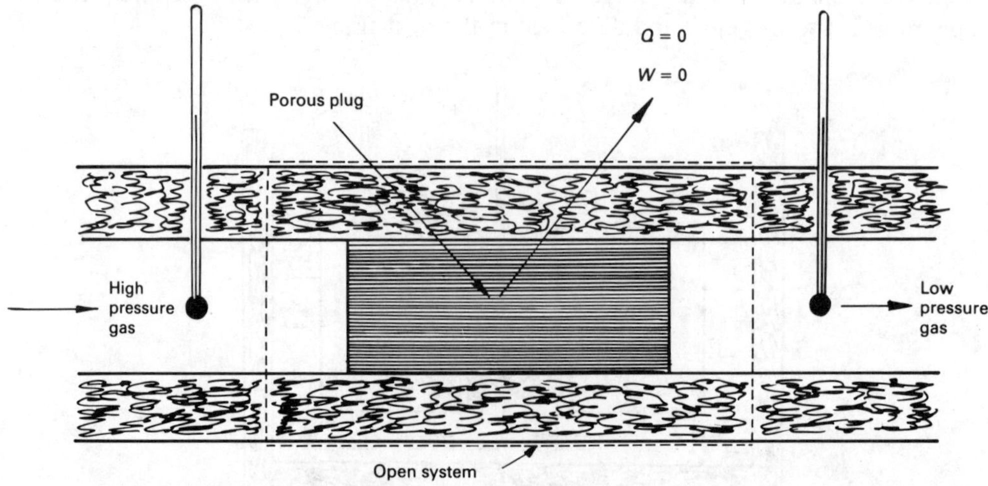

With negligible velocities, the kinetic energy terms were also zero, so the steady flow energy equation becomes:

$$0 = H_2 - H_1$$

where station 1 is for the high pressure gas and station 2 is for the low pressure gas. For gas flow the equation becomes:

$$0 = \dot{m}\, c_P(T_2 - T_1)$$

Accurate measurements of temperature did reveal very small changes, showing that even though the enthalpies were equal on either side of the expansion, the temperatures were not. Thus, enthalpy, and c_P for a gas do depend to a slight extent on pressure. However, for many practical purposes it is good enough to ignore these small dependencies, and to say that both c_P and c_V for a gas are independent of pressure.

93

The specific heats at constant volume and constant pressure were defined in the previous programme. We saw that:

$$\text{Specific heat at constant volume} = c_V = \left(\frac{\partial u}{\partial t}\right)_V$$

$$\text{Specific heat at constant pressure} = c_P = \left(\frac{\partial h}{\partial t}\right)_P$$

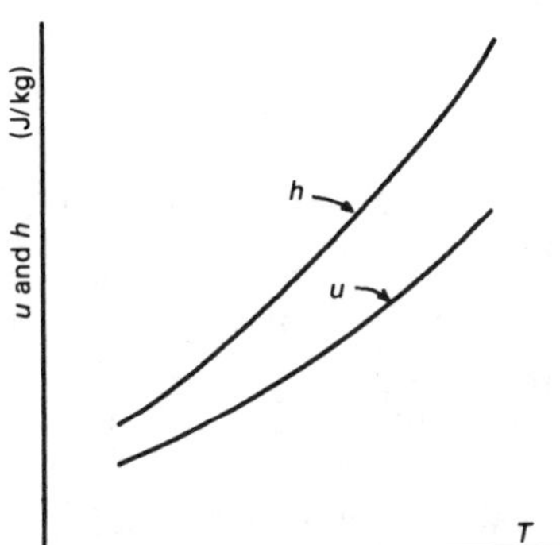

Both the internal energy–temperature and enthalpy–temperature diagrams for air at low pressure show slight increases in curvature with temperature, which means that both specific heats increase slightly with temperature. There are small changes for all gases, but the changes are not great, and are often ignored. This leads to two gas definitions under the general heading of the ideal gas.

94

A perfect gas is a gas which obeys the ideal gas rule, and has a constant specific heat at constant volume:

$$pv = RT$$

$$c_v = \text{constant}$$

A semi-perfect gas is a gas which obeys the ideal gas rule:

$$pv = RT$$

but with: c_v = a function of temperature

Thus, without any actual ruling on the specific heats of semi-perfect gases, these may be allowed to follow the variations found in real gases. Real gases at low pressures follow semi-perfect gas behaviour quite closely.

95

We have seen from the two-property rule, Frame 25, that the state of a pure substance can be defined by two independent properties. Then any third property must be a function of these two. The ideal gas rule is an example:

$$pv = RT$$

or: $T = pv/R$ (where R is the gas constant)

so that T is a function of p and v. In general terms, this may be expressed:

$$z = f(x,y) \tag{1}$$

If as the result of *any* infinitesimal process, x changes to $(x + dx)$, and y changes to $(y + dy)$, then the change in z is:

$$\mathrm{d}z = \left(\frac{\partial z}{\partial x}\right)_y \mathrm{d}x + \left(\frac{\partial z}{\partial y}\right)_x \mathrm{d}y \tag{2}$$

In this equation, dz is the total differential of z, and:

$\left(\frac{\partial z}{\partial x}\right)_y \mathrm{d}x$ is the partial differential of z with respect to x, and:

$\left(\frac{\partial z}{\partial y}\right)_x \mathrm{d}y$ is the partial differential of z with respect to y.

This is a basic mathematical principle which is illustrated in the figure below.

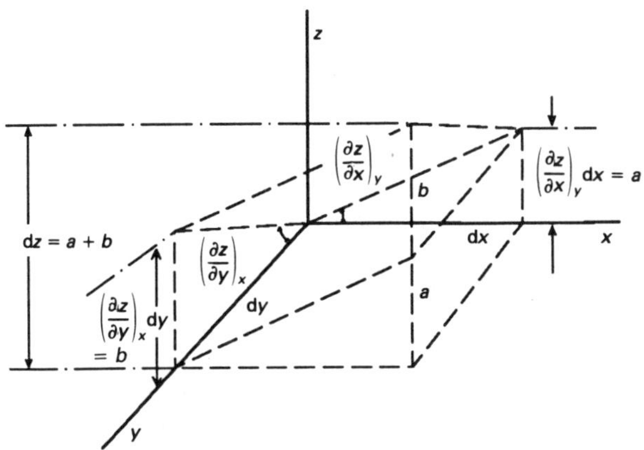

96

From the ideal gas rule, we can say:

$$dT = \frac{\partial T}{\partial p} dp + \frac{\partial T}{\partial v} dv$$

But, from $pv = RT$:

$$\frac{\partial T}{\partial p} = \frac{v}{R} \quad \text{and} \quad \frac{\partial T}{\partial v} = \frac{P}{R}$$

$$\therefore \quad dT = \frac{v}{R} dp + \frac{p}{R} dv$$

$$\therefore \quad R\, dT = v\, dp + p\, dv$$

or: $$p\, dv + v\, dp = R\, dT$$

In a similar way, if the property h is defined in terms of u and the product pv, such that:

$$h = u + pv$$

for any process causing small changes in u, p and v, the change in h is given by:

$$dh = du + p\, dv + v\, dp$$

97

The expression for dz in Frame 95, equation (2), may be expressed more generally as:

$$dz = M\,dx + N\,dy \tag{3}$$

When we come across equations like this in Thermodynamics, we may test to see whether an exact differential is involved, that is, whether:

$$z = f(x,y) \tag{1}$$

We will do this in the next frame.

98

In this case x, y and z are all thermodynamic properties, which depend only on end states and not on the path between states. If an exact differential is involved, then this must follow:

$$\left(\frac{\partial M}{\partial y}\right)_x = \left(\frac{\partial N}{\partial x}\right)_y \tag{4}$$

With the exact differential, from equation (2) in Frame 95:

$$M = \left(\frac{\partial z}{\partial x}\right)_y \text{ and } N = \left(\frac{\partial z}{\partial y}\right)_x \text{ dy, so it follows that:}$$

$$\left[\frac{\partial}{\partial y}\left(\frac{\partial z}{\partial x}\right)_y\right] = \left[\frac{\partial}{\partial x}\left(\frac{\partial z}{\partial y}\right)_x\right]$$

or:

$$\frac{\partial^2 z}{\partial x \partial y} = \frac{\partial^2 z}{\partial y \partial x}$$

This is true since the order of differentiation makes no difference to the result. Consider this example. From the ideal gas rule we have:

$$R\,dT = p\,dv + v\,dp \tag{5}$$

which we compare with equation (3):

$$dz = M\,dx + N\,dy \tag{3}$$

We see then that dT is dz, p/R is M, dv is dx, v/R is N, and dp is dy. Show that equation (4) is shown to be true for this case.

99

We have to show that:

$$\frac{\partial M}{\partial y} = \frac{\partial N}{\partial x}$$

Now $M = p/R$, so:
$$\frac{\partial M}{\partial y} = \frac{\partial M}{\partial p} = \frac{\partial (p/R)}{\partial p} = 1/R$$

and $N = v/R$, so:
$$\frac{\partial N}{\partial x} = \frac{\partial N}{\partial v} = \frac{\partial (v/R)}{\partial v} = 1/R$$

Therefore, as we know, our original equation is an expression of a relation between thermodynamic properties. In a more in-depth study of Thermodynamics, this fact of the equivalence between thermodynamic properties and partial derivatives is frequently used to obtain the more advanced properties (which we have not yet met), in terms of the derivatives of the measurable properties like pressure, volume and temperature. This leads to what are called **the thermodynamic relations**, the further study of which will appeal to the more academically minded.

100

Another example of equation (3) is:

$$dQ = p\,dv + du$$

This is the differential form of the equation of the First Law for a closed system, with the work term dW expressed as $p\,dv$. This means the work interaction is reversible and is an exact differential and may be expressed in terms of properties. However, the equation is not of the form of equation (2) in Frame 95 since dQ is not an exact differential and is not a property. The test we did in the previous frame would not work in this case. Later we shall see that if the heat transfer were reversible as well, then dQ could be expressed in terms of properties, and the equation then becomes an expression of a total derivative from which some thermodynamic relations may be obtained. Simple examples of this will be found in Frames 77 and 78 of Programme 6.

101

In the next programme we shall explore the use of perfect gases and vapours in processes in closed systems and in flow through open systems, governed by the First Law. To do this with gases we shall need to look at some more relationships between the properties, and we shall also see how a gas mixture can be treated as a single pure substance at normal temperatures.

Before moving on, make sure you can do all the following examples on the state of ideal gases.
R for air = 287 J/kg K = 0.287 kJ/kg K; R_M = 8314.3 J/kmol K; molecular mass for air = 28.97

1. An engine cylinder 120 mm diameter by 110 mm long contains air at 320 K and 200 kN/m^2. What is the mass of air? [2.709×10^{-3} kg]
2. What is the volume of 1 kg of air at atmospheric pressure, 1.01325 bar and at 288 K? [0.8157 m^3]
3. Carbon dioxide gas of mass 0.2 kg has a volume of 0.08 m^3 and a pressure of 150 kPa. R = 188.9 J/kg K. What is the temperature? [317.63 K]
4. What is the difference in mass between 10 m^3 of air and 10 m^3 of hydrogen at a pressure of 90 kN/m^2 and a temperature of 273 K? R for hydrogen is 4124 J/kg K. [10.687 kg]
5. A pressure vessel contains nitrogen at 200 bar and 300 K. R = 0.2968 kJ/kg K. What is the specific volume? [0.004 452 m^3/kg]
6. Air at 300 kPa and 400 K expands to 100 kPa with the volume doubling. What is the final temperature? [266.67 K]
7. The pressure of a gas doubles under isothermal conditions. What is the ratio of volume change? [0.5]
8. A gas has a volume of 0.5 m^3 and a temperature of 800 K. It changes its volume to 0.2 m^3 at constant pressure. What is the final temperature? [320 K]
9. The pressure of a mass of gas is 100 kPa when the temperature is 300 K. It increases its pressure at constant volume until the final temperature is 620 K What is the final pressure? [206.67 kPa]
10. The volume of 0.2 kmol of air is 4.45 m^3 at a temperature of 400 K. What is the pressure? [149.47 kPa]
11. A mass of air at 180 kPa and at a temperature of 300 K occupies a volume of 10 m^3. What is the kmol of air present? [0.7216]
12. A gas of mass 0.154 kg has a pressure of 120 kN/m^2, a volume of 0.2 m^3 and a temperature of 300 K. What is the molecular mass of the gas? [16.0]

102

Points to remember

A pure substance is a system which at all times is homogeneous in respect of composition and chemical combination.

A fluid at a particular pressure boils at its saturated temperature. For this temperature, the pressure is the saturated pressure.

Learn the shape of the T–v diagram (Frames 20 and 33), and learn to identify all areas, boundaries and lines. A fluid can exist as a subcooled liquid, saturated liquid, wet vapour, saturated vapour, superheated vapour and supercritical vapour.

The dryness fraction of a wet vapour is x, where $x = \dfrac{\text{mass of vapour}}{\text{mass of (liquid + vapour)}}$.

Learn the form of the p–v diagram (Frame 37), the p–T diagram (Frame 37) and the p–h diagram (Frame 38).

For saturated liquid, v_f = specific volume, u_f = specific internal energy and h_f = specific enthalpy.

For saturated vapour, v_g = specific volume, u_g = specific internal energy and h_g = specific enthalpy.

For wet vapour of dryness fraction x:

$$v = (1 - x)v_f + xv_g = v_f + xv_{fg} = v_g - (1 - x)v_{fg}, \text{ where } v_{fg} = v_g - v_f$$

$$u = (1 - x)u_f + xu_g = u_f + xu_{fg} = u_g - (1 - x)u_{fg}, \text{ where } u_{fg} = u_g - u_f$$

$$h = (1 - x)h_f + xh_g = h_f + xh_{fg} = h_g - (1 - x)h_{fg}, \text{ where } h_{fg} = h_g - h_f$$

The point where the saturated vapour and saturated liquid lines meet is the critical point. At the critical point v_{fg}, u_{fg} and h_{fg} are all zero.
In the compressed liquid state, approximately $h = h_f + v_f(\Delta p)$, if (Δp) is small.

Learn the form of the p–v diagram and the p–v–T surface for a substance that contracts on freezing (not water). Learn the form of the p–v–T surface for water that expands on freezing (Frames 63 and 65).

Learn to identify the gas region on the p–v diagram.
A perfect gas is an ideal concept for which $pv = RT$, and c_v = constant.
A semi-perfect gas is an ideal concept for which $pv = RT$. Both are ideal gases.
For an ideal gas R = constant for that gas, and $pV = mRT$, where V is the volume of mass m of gas.

The volume of kmol of gas is V_M, which is constant for all gases, and the molecular mass of a gas is M, so $pV_M = MRT = R_M T$ where $R_M = MR$ = universal gas constant = 8314.5 J/kmol K. Also $pV = nR_M T$, where n = number of mols.

Programme 5

THE FIRST LAW: VAPOURS AND GASES IN CLOSED AND OPEN SYSTEMS

1

We shall begin this programme by looking at problems involving a vapour, usually steam, in closed systems. The system is initially in a known equilibrium state, and as a result of heat and work interactions at the boundary, proceeds to a second equilibrium state. We need to use the equation of the First Law applied to a closed system to solve such a problem. This was given in Frame 11 of Programme 3 as:

$$Q - W = E_2 - E_1$$

where the energy term E was shown in Frame 20 of Programme 3 to consist of internal energy, kinetic energy and potential energy. So what would be the First Law equation for problems with steam in closed systems?

2

$$\boxed{Q - W = U_2 - U_1}$$

In any problem involving a fluid in a closed system, the fluid does not have any kinetic energy and there is no significant change in potential energy between states. This equation tells us that the change of internal energy between the final and initial state is equal to the heat transfer minus the work transfer.

First, in dealing with steam, we shall think about constant volume processes. This makes any problem rather easier because one of the terms in the First Law equation immediately disappears. Which term is that?

3

$\boxed{W\text{, the work term}}$

In a constant volume process there is no displacement work. Why is this? If you are not sure look back at Programme 2, Frame 27. In the absence of any other work, the First Law equation is then:

$$Q = U_2 - U_1$$

Let us look carefully at this problem:

Steam is initially at the critical point and it is cooled at constant volume until the pressure is 0.2 MPa or 2 bar. Find the final state of the steam, (we shall need to find the *condition* and a second property), and find the heat transfer involved for 2 kg of steam.

The first diagram in Frame 37 of Programme 4 shows the conditions of wet, saturated and superheated steam. We must decide on the condition first so that we shall know how to proceed. Starting at the critical point, what is the final condition after constant volume cooling?

4

The steam is wet

The property diagram of pressure against specific volume for steam is given below; constant volume cooling from the critical point can only go into the wet steam region where the water is only partially converted into steam.

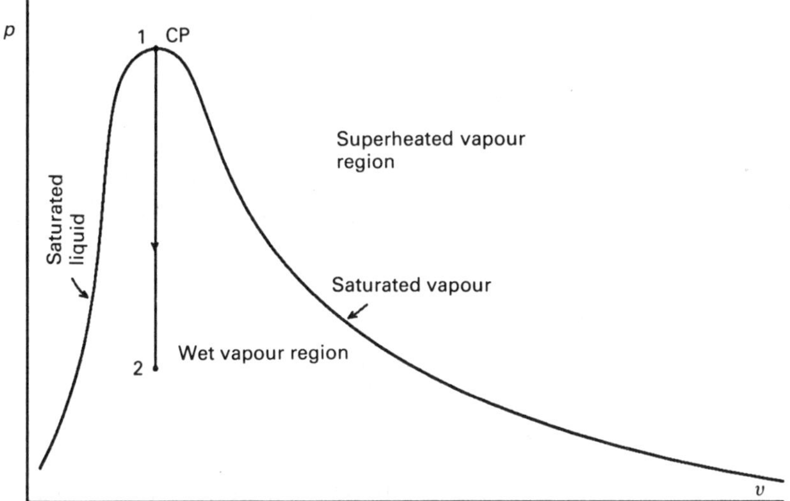

As the steam is cooled at constant volume, the final specific volume must be the same as the value at the critical point. You can find this in your steam tables, at the critical point pressure of 22.12 MN/m^2, or 221.2 bar. Remember that both v_g and v_f are equal at the critical point. Also look up u which we shall need later, again remembering that u_f and u_g are also equal at this point. (The precise values you look up will depend on which tables you are using.)

5

Rogers and Mayhew: $v = 0.003\ 17\ m^3/kg$, $u = 2014\ kJ/kg$
Cooper and Le Fevre: $v = 0.003\ 155\ m^3/kg$, $u = 2026\ kJ/kg$
Haywood: $v = 0.003\ 17\ m^3/kg$, $u = 2038\ kJ/kg$

As property values are changing very rapidly in the vicinity of the critical point, there is the least agreement between sources on some values.

We can now find U_1 which is the internal energy of 2 kg of steam at state 1, the critical point. For the Rogers and Mayhew data:

$$U_1 = 2 \times 2014 = 4028\ [kJ]$$

The other two results are 4052 and 4076 kJ, respectively.

We now know everything about state 1. At state 2 we know the specific volume is unchanged and so that we can calculate U_2 we shall need to know the dryness fraction x at state 2.

To find x we can use equation (1) of Frame 29, in the previous programme:

$$v = (1 - x)v_f + x\ v_g$$

where v is the specific volume at the critical point, and v_f and v_g are at the lower pressure of state 2. From the Rogers and Mayhew tables, page 10, we can take v_f at 1.985 bar ($0.001\ 06\ m^3/kg$), as being close enough to the value at 2 bar. On page 4 we find v_g at 2 bar is $0.8856\ m^3/kg$. Values from the other tables are:

Haywood: $v_f = 0.001\ 061\ m^3/kg$, $v_g = 0.885\ m^3/kg$ (page 12)

Cooper and Le Fevre: $v_f = 0.001\ 061\ m^3/kg$, $v_g = 0.8859\ m^3/kg$ (page 4)

Thus here we see closer agreement. Now you can calculate the dryness fraction at state 2.

The answers are $x = 0.002\ 385$, from the Rogers and Mayhew tables, $x = 0.002\ 386$ from the Haywood tables, and $x = 0.002\ 367$ from the Cooper and Le Fevre tables. The solution using the Rogers and Mayhew tables is given in the next frame.

6

We are using the equation:

$$v = (1 - x)v_f + x\ g_g$$

$$\therefore \quad 0.003\ 17 = (1 - x)\ 0.001\ 06 + 0.8856\ x$$

$$\therefore \quad 0.884\ 54\ x = 0.002\ 11$$

$$\therefore \quad x = 0.002\ 385$$

We can now find the internal energy at state 2. At 2 bar, from the Rogers and Mayhew tables, page 4, we find $u_f = 505$ kJ/kg, and $u_g = 2530$ kJ/kg. Corresponding values from the Haywood tables are $u_f = 504.5$ kJ/kg, and $u_g = 2529.2$ kJ/kg; and from the Cooper and Le Fevre tables: $u_f = 504.6$ kJ/kg, and $u_g = 2530$ kJ/kg. We can use:

$$u = (1 - x)u_f + x\ u_g$$

to find u, and then find U for 2 kg of the wet steam. (With a dryness fraction of only 0.0024, it is almost all water.) Then we can use:

$$Q = U_2 - U_1$$

to find the heat transfer. Now you can complete these three steps for the three sets of data and compare the results.

7

R and M tables: $u_2 = 509.84$ kJ/kg, $U_2 = 1019.68$ kJ, $Q = -3008.32$ kJ
C and Le F tables: $u_2 = 509.4$ kJ/kg, $U_2 = 1018.8$ kJ, $Q = -3033.2$ kJ
H tables: $u_2 = 509.48$ kJ/kg, $U_2 = 1018.96$ kJ, $Q = -3057.04$ kJ

The solution using the Haywood data is:

$$u_2 = (1 - 0.002386) \times 504.5 + 0.002386 \times 2529.2$$
$$= 509.48\ [\text{kJ/kg}]$$

Hence U_2 is $2 \times 509.48 = 1018.96$ [kJ]

We now use: $\quad Q = U_2 - U_1 = 1018.96 - 4076 = 3057.04$ [kJ]

8

The variation in the three final answers is due to the variation in the critical point data.

While all this is fresh in your mind, here are two more problems on constant volume changes with steam. Do these before moving on.

1. Steam of mass 0.25 kg is heated at constant volume from a pressure of 100 bar, 10 MN/m^2, to reach the critical point. Calculate (a) the initial dryness fraction, (b) the initial specific internal energy and (c) the heat transfer.
[R and M: (a) 0.1036, (b) 1512.35 kJ/kg, (c) +125.41 kJ to the steam;
H: (a) 0.1035, (b) 1512.92 kJ/kg, (c) +131.27 kJ to the steam;
C and Le F: (a) 0.1029, (b) 1512.34 kJ/kg, (c) +128.42 kJ to the steam]
2. Unit mass of saturated steam at 15 MPa is cooled at constant volume to a pressure of 0.3 MPa. Calculate (a) the final dryness fraction, (b) the final specific internal energy and (c) the heat transfer.
[R and M: (a) 0.0153, (b) 591.34 kJ/kg, (c) – 1864.66 kJ/kg from the steam;
H: (a) 0.0153, (b) 591.46 kJ/kg, (c) – 1868.44 kJ/kg from the steam;
C and Le F: (a) 0.0154, (b) 591.66 kJ/kg, (c) – 1865.34 kJ/kg from the steam]

9

From the diagram below you can see that if, in a constant volume heating, process (a), the specific volume is less than the critical point value, then the final state will approach the saturated water line, so x would decrease to zero.

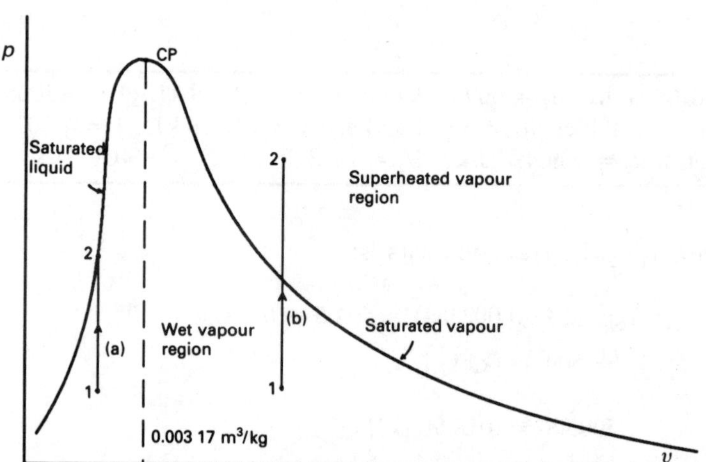

Also, for specific volumes greater than this value, there is the possibility that the final state could be superheated. This is shown in process (b). In two processes, (a) and (b), the specific volumes are 0.001 366 m^3/kg and 0.3427 m^3/kg respectively. By inspection from the tables, at what values of saturation pressure and temperature does process (a) meet the saturated water line, and process (b) meet the saturated steam line? (These figures correspond to discrete points in the Rogers and Mayhew tables.)

10

(a) 74.45 bar, 290°C
(b) 5.5 bar, 155.5°C

From page 10 of the Rogers and Mayhew tables you would see that $v_f = 0.001\ 366$ m³/kg at 74.45 bar and 290°C, and from page 4 you can see that $v_g = 0.3427$ m³/kg at 5.5 bar and 155.5°C.

In this next problem with constant volume heating, the final state is in the superheat region: wet steam at atmospheric pressure of 101.325 kN/m² with a dryness fraction of 0.4 is heated at constant volume to a pressure of 0.5 MN/m². Calculate (a) the final temperature and degree of superheat, and (b) the heat transfer for unit mass of steam.

You can start by calculating the specific volume at state 1. Using the Haywood tables, v_f is 0.001 044 m³/kg and v_g is 1.673 m³/kg.

11

$v_1 = 0.6698$ m³/kg

We have, then: $v_1 = (1 - 0.4) \times 0.001\ 044 + 0.4 \times 1.673$

$= 0.6698$ m³/kg

Why did we need to calculate the specific volume for this constant volume process?

12

This will be the specific volume at state 2, which we need to find the temperature at state 2 in the superheat region. Now we shall find the specific internal energy at state 1. From page 11 of the Haywood tables, $u_f = 419.0$ kJ/kg and $u_g = 2506.5$ kJ/kg. Hence:

$$u_1 = (1 - 0.4) \times 419.0 + 0.4 \times 2506.5$$
$$= 1254.0 \text{ [kJ/kg]}$$

Now we shall think about the final state. The specific volume is 0.6698 m³/kg, so the density is 1.493 kg/m². On page 18 of the Haywood tables we can see that this value occurs at 0.5 MN/m² between temperatures of 450° (1.506 kg/m³) and 475°C (1.455 kg/m³). We interpolate to find the actual temperature:

$$T = 450 + (475 - 450) \times \frac{1.506 - 1.493}{1.506 - 1.455}$$
$$= 450 + 6.37$$
$$= 456.37°\text{C}$$

The saturation temperature at 0.5 MN/m² is 151.8°C, so the degree of superheat is 456.37°C − 151.8°C, which is 304.57°C.

The next step is to use this temperature at 0.5 MN/m² to find u from the table on page 19. Do that. Values of u at 450°C and 475°C are 3045 and 3087 kJ/kg, respectively. The value calculated will be u_2.

13

3055.70 kJ/kg

$$u_2 = 3045 + (3087 - 3045) \times \frac{456.37 - 450.0}{475 - 450} = 3055.70 \text{ [kJ/kg]}$$

Now we can find the heat transfer for unit mass of steam:

$$Q = u_2 - u_1 = 3055.7 - 1254.0 = 1801.7 \text{ [kJ/kg]}$$

As this is positive it is heat transfer to the steam, which is expected since we are heating steam from a wet condition into the superheat region.

14

We saw from Frame 9 that if constant volume heating took place on wet steam with a specific volume less than the critical point value we would obtain saturated water. One might argue that as the pressure was increasing in the constant volume process, steam present initially would be compressed back into water. But the same argument breaks down for specific volumes greater than the critical point value when water present initially becomes saturated steam. Thus the critical point of a substance is the point at which properties are changing dramatically, and for which a physical explanation is hard to find.

To conclude our constant volume work with steam, here is an exercise to the left of the critical point constant volume: wet steam at a saturation temperature of 200°C is heated at constant volume until it becomes saturated water. The specific volume is 0.001 38 m³/kg. What are (a) the initial saturation pressure and dryness fraction, and (b) the final saturation temperature and pressure? In this exercise, Rogers and Mayhew tables are used. First check from page 10 that, at 200°C, the saturation pressure is 15.55 bar.

15

We know the initial specific volume and saturation pressure so you should be able now to find the dryness fraction. This will involve interpolation of data from page 4 to find v_g at 200°C. You then have to look at page 10 again and see at which saturation condition the specific volume of 0.001 38 m³/kg becomes equal to v_f. It will probably lie between two given states in the table. Then you can interpolate to find the exact saturation pressure and temperature. The answers are in the box below.

16

(a) 15.55 bar, x = 0.001 768
(b) 293.68°C, 78.78 bar

17

From page 10 of the tables we can see that at 200°C, the saturation pressure is 15.55 bar and v_f = 0.001 157 m³/kg. From page 4 we can see that v_g at 198.3°C is 0.1317 m³/kg, and at 201.4°C is 0.1237 m³/kg. Hence at 200°C:

$$v_g = 0.1317 - (0.1317 - 0.1237) \times \frac{(200 - 198.3)}{(201.4 - 198.3)}$$
$$= 0.1273$$

Knowing the specific volume at 200°C, we can calculate x.

Hence: $$0.001\,38 = (1 - x)v_f + xv_g$$
$$= (1 - x) \times 0.001\,157 + x \times 0.1273$$

and so: $$x = 0.001\,768$$

We can now look at page 10 of the tables again, and we see that values of v_f on either side of 0.001 38 m³/kg occur at 290°C and 300°C. Thus:

$$v_f = 0.001\,366 \text{ m}^3\text{/kg at 290°C, and 74.45 bar, and}$$
$$v_f = 0.001\,404 \text{ m}^3\text{/kg at 300°C, and 85.92 bar}$$

The saturation temperature for v_f = 0.001 38 m³/kg is given by:

$$T_s = 290 + (300 - 290) \times \frac{(0.001\,38 - 0.001\,366)}{(0.001\,404 - 0.001\,366)}$$
$$= 290 + 10 \times \frac{1.4 \times 10^{-5}}{3.8 \times 10^{-5}}$$
$$= 293.68°\text{C}$$

Since we are interpolating between the same points the saturation pressure is:

$$p_s = 74.45 + (85.92 - 74.45) \times \frac{1.4}{3.8}$$
$$= 78.68 \text{ [bar]}$$

18

Now we shall look at a constant pressure expansion of steam. In Frame 50 of Programme 2 we calculated the displacement work when steam at 5 bar expanded from a volume of 0.1 m³ to a volume of 0.15 m³. The final specific volume was 0.3748 m³/kg (Frame 54), so that the mass of the steam was 0.4002 kg. By inspection of your steam tables, what can you now say about the condition of the steam at the end of its expansion?

19

It is saturated steam

At 5 bar, or 0.5 MN/m^2, we see that $v_g = 0.375$ m^3/kg, so the steam is saturated. The expansion is shown on the p–v diagram for steam below.

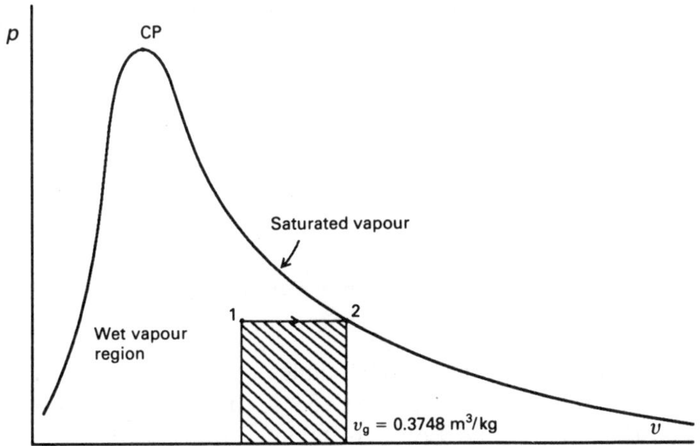

The work of expansion was found to be 25 kJ. (You may check again that this is correct.) This expansion could not take place without a heat transfer to evaporate the water content of the wet steam to give saturated steam.

From the First Law equation:

$$Q - W = U_2 - U_1$$

and

$$W = p(V_2 - V_1)$$

so

$$Q = H_2 - H_1$$

To find the heat transfer involved, we need to calculate the initial and final enthalpy. At state 1 we need to know the dryness fraction.

The specific volume at state 1 may be found from the actual volume and the mass of steam:

$$v_1 = 0.1/0.4002 = 0.2499 \text{ [m}^3\text{/kg]}$$

Then after finding v_f in your tables (we already have v_g), you can then calculate the dryness fraction at state 1.

20

$$\boxed{x_1 = 0.6658}$$

We know that $v_g = 0.375\ m^3/kg$ at 5 bar, or $0.5\ MN/m^2$, and from the Haywood and Cooper and Le Fevre tables we have that v_f is 0.001 093 m^3/kg. The same result may be obtained by interpolation between $v_f = 0.001\ 091\ m^3/kg$ at 4.76 bar, and $v_f = 0.001\ 102\ m^3/kg$ at 6.181 bar, from page 10 of the Rogers and Mayhew tables.

Then using: $v_1 = (1 - x_1)v_f + xv_g$, we get:

$$0.2499 = (1 - x_1)\ 0.001\ 093 + 0.375\ x_1$$

so that:

$$x_1 = 0.6654$$

The next thing to do is to calculate the enthalpy at the initial and final states. The initial state is for wet steam, with $x = 0.6654$ at 5 bar, and the final state is saturated steam at 5 bar. The mass of steam is 0.4002 kg. We can use $H_1 = m(h_f + x\ h_{fg})$, so what is H_1?

21

The answer now depends slightly on which tables you are using; thus it is 817.74 kJ from the Rogers and Mayhew tables; 817.35 kJ from the Haywood tables and 817.55 kJ from the Cooper and Le Fevre tables. From the Rogers and Mayhew tables:

$$H_1 = 0.4002 \times (640 + 0.6654 \times 2109)$$
$$= 817.74\ [kJ]$$

and for saturated steam at state 2 we have (Rogers and Mayhew Tables):

$$H_2 = 0.4002 \times 2749$$
$$= 1100.15\ [kJ]$$

The heat transfer, Q is then:

$$Q = H_2 - H_1$$
$$= 1100.15 - 817.74 = 282.41\ [kJ]$$

The answer is 282.20 kJ from both the Haywood and Cooper and Le Fevre tables.

22

In Programme 2 we spoke of polytropic expansions and compressions where pV^n = constant. The index n may vary from one process to another and the law is based on actual measurements of pressures and volumes during the processes. When steam expands polytropically in an engine cylinder, the process is rapid and there is not much time for heat transfer to take place. Ideally we may assume the process is adiabatic so that Q is zero in the First Law equation. The expansion index n is found to depend on the initial condition of the steam. Thus, for steam initially dry saturated, n is found to be close to 1.135. For initially superheated steam, n depends on whether or not the steam is superheated during the whole of the expansion. It is difficult to get exact values of n for an adiabatic process, since however rapid the process some heat transfer must take place.

23

In our next example, steam is initially dry saturated at 1 MPa or 10 bar, and it expands over a volume ratio of 2, according to the law $pv^{1.135} = k$. We shall calculate the work of expansion and the change of internal energy, and from the First Law see how close the process was to being adiabatic. We should find Q is close to zero. The expansion is shown in the diagram below.

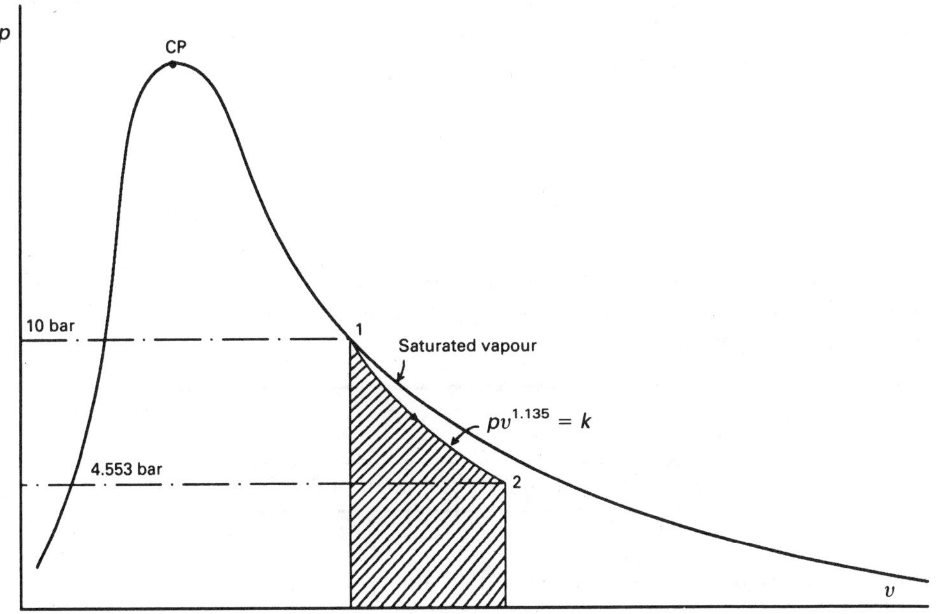

24

We may consider unit mass of steam. From the Cooper and Le Fevre tables at 1 MPa on page 5 at 10 bar, $v_1 = v_g = 194.3$ m³/Mg $= 0.1943$ m³/kg. After the expansion v_2 is twice v_1 which is 0.3886 m³/kg, and during the expansion $pv^{1.135} = k$. Remember that specific volume, as well as volume, may be used in this relationship.

We have then:

$$p_2 = 1.0\left(\frac{1}{2}\right)^{1.135}$$
$$= 0.4553 \text{ [MPa]}$$

We now know p_2 and v_2 so by inspection from the tables we can say whether the steam is wet, saturated or superheated. We cannot be completely sure since interpolation is necessary to pin-point the state of the steam after the expansion, but before we do that, what would your guess be?

25

The steam is wet

The final pressure is between 0.44 and 0.46 MPa, for which v_g is 422.8 m³/Mg and 405.4 m³/Mg, respectively, and v_2 is 388.6 m³/Mg, so we can safely say the steam is wet. For the pressure of 0.4553 MPa:

$$v_g = 422.8 - (422.8 - 405.4) \times \frac{(0.4553 - 0.44)}{(0.46 - 0.44)}$$
$$= 409.49 \text{ [m}^3\text{/Mg]} = 0.4095 \text{ [m}^3\text{/kg]}$$

Since our v_2 at 0.3886 m³/kg is smaller than this, the steam must be wet at state 2. Now find v_f at 0.4553 MPa (interpolating between the same two points), and then the dryness fraction at state 2.

26

$v_f = 0.001\,089$ m³/kg, $x_2 = 0.9488$

Interpolating between 0.44 and 0.46 MPA, for v_f, we have:

$$v_f = 1.087 + (1.089 - 1.087) \times \frac{(0.4553 - 0.44)}{(0.46 - 0.44)}$$

$$= 1.0885 \text{ m}^3/\text{Mg} = 0.001\,089 \text{ [m}^3/\text{kg]}$$

Then: $0.3886 = (1 - x_2)\, 0.001\,089 + 0.4095\, x_2$

$\therefore$ $x_2 = 0.9488$

Now interpolate between the same two points to find u_f and u_g at 0.4553 MPa, remembering that these are given as e_f and e_g in the Cooper and Le Fevre tables. Then find u_2. Also look up u_1 which is u_g at 1MPa.

27

$$\boxed{u_2 = 2458.78 \text{ kJ/kg},\ u_1 = 2583 \text{ kJ/kg}}$$

We have:

$$u_f = 619.2 + (626.2 - 619.2) \times \frac{(0.4553 - 0.44)}{(0.46 - 0.44)}$$

$$= 624.55 \text{ [kJ/kg]}$$

$$u_g = 2557 + (2558 - 2557) \times \frac{(0.4553 - 0.44)}{(0.46 - 0.44)}$$

$$= 2557.76 \text{ [kJ/kg]}$$

Then: $u_2 = (1 - 0.9488)\, 624.55 + 0.9488 \times 2557.76$

$$= 2458.78 \text{ [kJ/kg]}$$

28

The change of specific internal energy is then:

$$u_2 - u_1 = 2458.78 - 2583 \text{ kJ/kg}$$

$$= -124.22 \text{ [kJ/kg]}$$

Next we must find the work done in the polytropic expansion. The formula was derived in Programme 2. You should be able to remember it.

29

$$W = \frac{p_2v_2 - p_1v_1}{1 - n}$$

This is for unit mass of steam. The pressure is in MPa so we need the factor of 10^3 to give the work done in kJ/kg.

Hence: $$W = \frac{(0.4553 \times 0.3886 - 1.0 \times 0.1943) \times 10^3}{1 - 1.135}$$

$$= 128.67 \text{ [kJ/kg]}$$

We can now bring all the terms together to find Q, using the First Law equation for unit mass of steam:

$$Q - W = u_2 - u_1$$

$$Q - 128.67 = -124.22$$

and so: $$Q = +4.45 \text{ [kJ/kg]}$$

In an adiabatic expansion, the answer should be zero; this result indicates a small error in the value of n for these particular conditions.

For practical purposes, the rule is that $n = 1.135$ if the steam is at *any* initially dry saturated state. If the steam is *wholly superheated* in the expansion, then it is found by measuring the pressures and volumes during the process that n is close to 1.3 for an adiabatic expansion.

30

We will now work through a similar exercise starting with steam at 1 MPa and at 250°C, expanding to 0.5 MPa according to $pv^{1.3} = k$. The expansion is shown in the diagram. The saturation temperature at 1 MPa is 179.9°C, so clearly the steam is initially superheated.

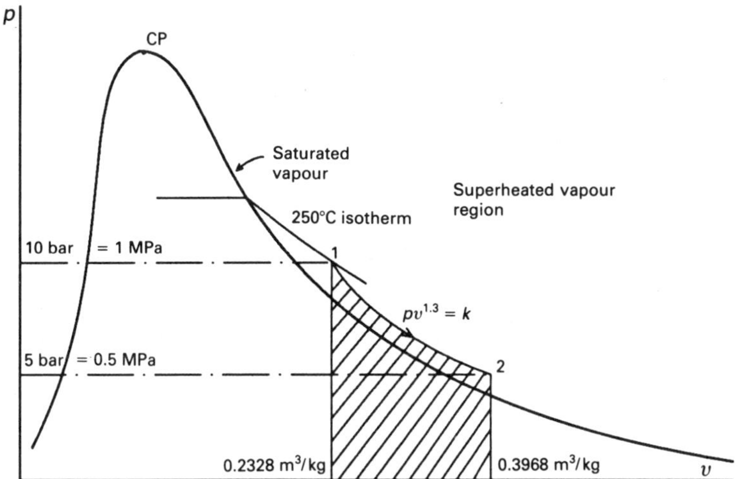

As before, we have to find the initial and final internal energies, and the work of expansion. At state 1, for steam at 1.0 MPa and 250°C, $v_1 = 232.7$ m³/Mg, or 0.2327 m³/kg, and $u_1 = 2710$ kJ/kg. These figures come from page 7 of the Cooper and Le Fevre tables.

Using $p_2 v_2^{1.3} = p_1 v_1^{1.3}$ we get:

$$v_2^{1.3} = \frac{1.0}{0.5} \times 0.2327^{1.3}$$

$$\therefore \qquad v_2 = 0.3959 \text{ [m}^3\text{/kg]}$$

Knowing v_2, we can find the condition of the steam by comparing with v_g at the same pressure.

31

From page 5 of the Cooper and Le Fevre tables we see that v_g at 0.5 MPa is 374.8 m³/Mg, or 0.3748 m³/kg, and with $v_2 = 0.3959$ m³/kg we see that the whole expansion is in the superheat region. We now have to establish the precise state point so that we can find the value of u_2.

32

From page 7 of these tables we can see that $v = 0.4250$ m^3/kg at 200°C, and at saturation $v_g = 0.3748$ m^3/kg at 151.8°C, so at state 2 the temperature is:

$$T_2 = 151.8 + (200 - 151.8) \times \frac{(0.3959 - 0.3748)}{(0.4250 - 0.3748)}$$

$$= 151.8 + 20.26 \text{ [°C]}$$

Then with $u_g = 2561$ kJ/kg, u_2 is given by:

$$u_2 = 2561 + (2643 - 2561) \times \frac{20.26}{(200 - 151.8)}$$

$$= 2595.47 \text{ [kJ/kg]}$$

We now have $u_1 = 2710$ kJ/kg, $u_2 = 2595.47$ kJ/kg, $p_1 = 1$ MPa, $v_1 = 0.2327$ m^3/kg, $p_2 = 0.5$ MPa and $v_2 = 0.3959$ m^3/kg. Now complete the solution by calculating the work of the expansion, and by using The First Law to find the value of Q.

33

$$\boxed{W = 115.83 \text{ kJ/kg}, \ Q = +1.3 \text{ kJ/kg}}$$

$$W = \frac{(0.5 \times 0.3959 - 1.0 \times 0.2327) \times 10^3}{1 - 1.3}$$

$$= 115.83 \text{ [kJ/kg]}$$

Then using the First Law equation:

$$Q - 115.83 = 2595.47 - 2710.0$$

$$Q = +1.30 \text{ [kJ/kg]}$$

As you know, during the evaporation of water at constant pressure, the temperature is also constant. **A constant temperature process is called an isothermal process.** If an isothermal expansion continues into the superheat region, the pressure starts to fall. This was shown in Frame 37 of the previous programme, and is shown again in the next frame, where we will consider the work done in an isothermal expansion.

34

Suppose wet steam, at a particular pressure and specific volume, expands isothermally to a higher specific volume in the superheat region. We wish to find the heat transfer involved. In previous problems we found the internal energy at the end states and the work of expansion in order to find Q.

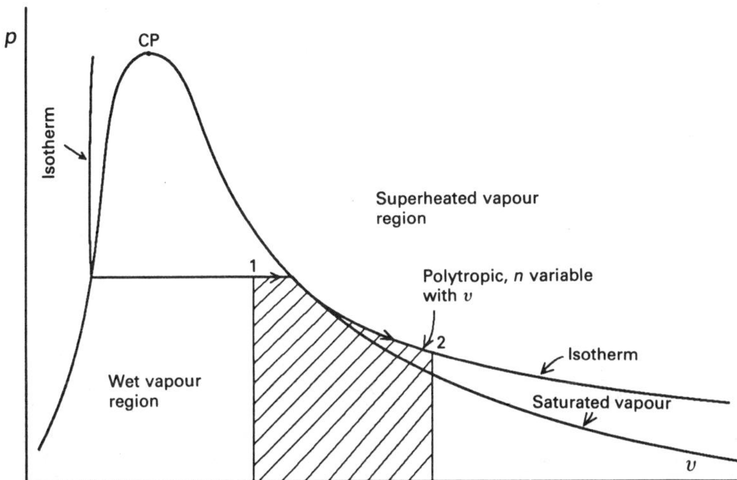

We can see from the figure that part of the expansion is at constant pressure, and the remainder follows a p–v curve that looks roughly polytropic. Unfortunately from measurements of this kind of expansion it is found that there is no single value of n that can be used in $pV^n = k$. The path function for W is too complicated. What about the path function for Q? Can we do this the other way round and calculate Q in order to find W from the First Law equation? The answer to this question is yes, because the path function for Q, with the process being isothermal, is very simple. We need to know about **entropy** before we can do this, so we shall consider a problem like this in Programme 6.

We will now think about the flow of vapours through open systems. We will start by looking at the heat transfer involved in the boiling of water as it flows through a tube.

35

We shall need the steady flow energy equation (Frame 40, Programme 3), which is the First Law applied to an open system, with no work term, and no change in potential energy:

$$\dot{Q} = \dot{m}\left(h_2 - h_1 + \frac{V_2^2 - V_1^2}{2}\right) \quad \text{[W or kW]}$$

36

Steam is generated for electricity production on a large scale in water-tube boilers. Water enters a tube at state 1 in the diagram and leaves as saturated steam at state 2. We may assume the process is at constant pressure. Heat transfer occurs through the cylindrical surface of the tube. In the steady flow energy equation, given in the previous frame, we have the heat transfer term, the enthalpy and kinetic energy terms, and the units are in W or kW. The process path is shown on the *p–v* diagram below.

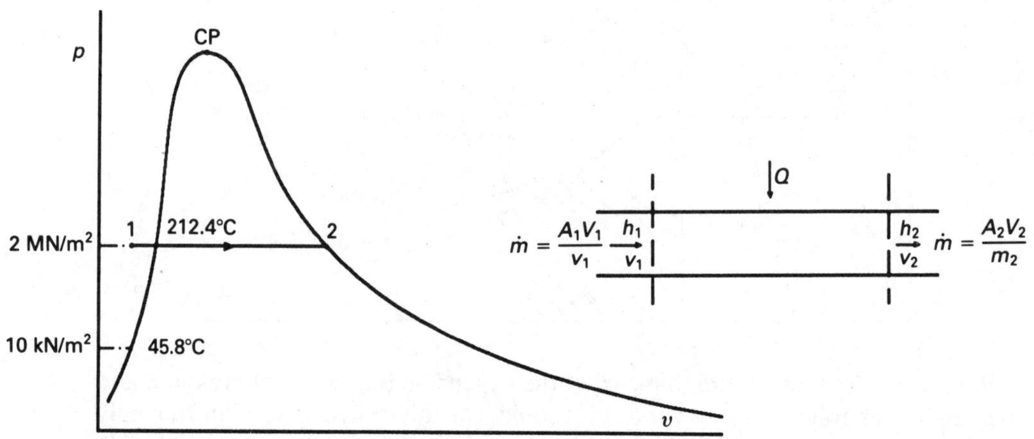

Depending on the application, $\dot{Q}$ in this equation can be very large. For a single boiler tube, $\dot{Q}$ would be in kW or possibly MW. All this heat has to flow through the wall of the tube, so the tube can be long, and boilers are physically very large compared with the size of the power-producing turbines.

The fluid expands greatly between inlet and outlet so the velocity of flow increases correspondingly. Obviously the mass flow at any point in the tube must be constant, and this may be expressed as:

$$\text{mass flow rate} = \frac{\text{volume flow rate}}{\text{specific volume}}$$
$$= \text{constant}$$

or:

$$\dot{m} = \frac{A \times V}{v} = \text{constant}$$

37

In a steam power plant, the steam that has expanded in the turbines is condensed and pumped back into the boiler. The condenser usually works at pressures below atmospheric; this allows more work to be done by the steam in the turbines.

Let us now work carefully through this problem: saturated water at 10 kN/m², 0.1 bar is extracted from a condenser and is compressed to 2 MN/m² bar at which pressure it enters a boiler tube at state 1. The flow is 10 kg/s and the tube is 50 mm diameter at entry. The flow leaves the tube as dry saturated steam at 20 bar, at state 2, and the tube diameter has increased to 70 mm. Taking into account kinetic energy terms, calculate the heat transfer to the flow from the surface of the tube.

The water at state 1 is in the compressed liquid region. At 10 kN/m² the saturation temperature is 45.8°C, so we need to find the enthalpy of water at 45.8°C and at 2 MN/m². From page 16 of the Haywood tables, h at this pressure is 106.6 kJ/kg at 25°C and is 211.0 kJ/kg at 50°C, so you can interpolate to find the value required.

38

$$\boxed{h = 193.46 \text{ kJ/kg}}$$

$$h = 106.6 + (211.0 - 106.6) \times \frac{(45.8 - 25)}{(50 - 25)}$$

$$= 193.46 \text{ [kJ/kg]}$$

This is the value of h_1 required in our energy equation. Next we need the specific volume of water in this state. From the density figures given on page 18 of the Haywood tables, we have $\rho = 998$ kg/m³ at 25°C, and $\rho = 989$ kg/m³ at 50°C, so what is v_1 at 45.8°C?

39

You should find that $v_1 = 0.001\ 0096$ m³/kg.

We have that: $\rho = 998 - (998 - 989) \times \dfrac{(45.8 - 25)}{(50 - 25)} = 990.51$ [kg/m³]

and hence $v_1 = 1/990.51 = 0.001\ 0096$ [m³/kg].

40

For the flow of 10 kg/s, with $v_1 = 0.001\ 009\ 6\ \mathrm{m^3/kg}$, and the area being that of a 50 mm diameter tube, using the formula in Frame 36, what is the velocity, V_1?

41

5.143 m/s

Using: $$\dot{m} = \frac{A \times V}{v}$$

we get: $$V = \dot{m} \times v/A$$

and with
$$\begin{aligned} A &= \pi\ D^2/4 \\ &= \pi \times 0.05^2/4 \\ &= 0.001\ 963\ [\mathrm{m^2}] \end{aligned}$$

we get:
$$\begin{aligned} V_1 &= 10 \times 0.001\ 009\ 6/0.001\ 963 \\ &= 5.143\ [\mathrm{m/s}] \end{aligned}$$

At outlet, the steam is dry saturated at 2 MN/m², which is state 2. For this condition, $v_2 = v_g = 0.0995\ \mathrm{m^3/kg}$, and $h_2 = h_g = 2797.2$ kJ/kg. The tube diameter is 70 mm, so the area is 0.003 848 m², and the velocity is:

$$\begin{aligned} V_2 &= 10 \times 0.0995/0.003\ 848 \\ &= 258.58\ [\mathrm{m/s}] \end{aligned}$$

The equation we need to use to find $\dot{Q}$ is in Frame 35, and we have $\dot{m} = 10$ kg/s, $V_1 = 5.143$ m/s, $V_2 = 258.58$ m/s, $h_1 = 193.46$ kJ/kg and $h_2 = 2797.2$ kJ/kg. What is the heat transfer required in MW? (Remember your kinetic energy terms need a factor of 10^3 in the denominator, to give units of kJ/kg.)

42

26.372 MW

We have: $$\begin{aligned} \dot{Q} &= 10 \times (2797.2 - 193.46) + 10 \times (258.58^2 - 5.143^2)/2 \times 10^3 \\ &= 26\ 037.4 + 334.19 \end{aligned}$$

$= 26\,371.6$ [kW]

$= 26.372$ [MW]

This is a positive heat transfer to the fluid.

This is heat transfer to the steam in just one boiler tube. In a power station there would be many of these tubes in parallel, carrying hundreds of MW of heat from the furnace gases to the steam. Because of the efficiency of the process, much of this heat would later be discharged in the condenser. Notice that we have not calculated the length of the tubes: we would need heat transfer theory to do that.

43

Steam is condensed by allowing the vapour to flow over a surface cooled by water. Very large amounts of cooling water are needed in the heat transfer process in power stations, so the water which has become warm is itself cooled by air in the cooling towers. It is then recirculated back to the condensers. Again we shall assume the process is at constant pressure, and the process path is shown on the p–v diagram.

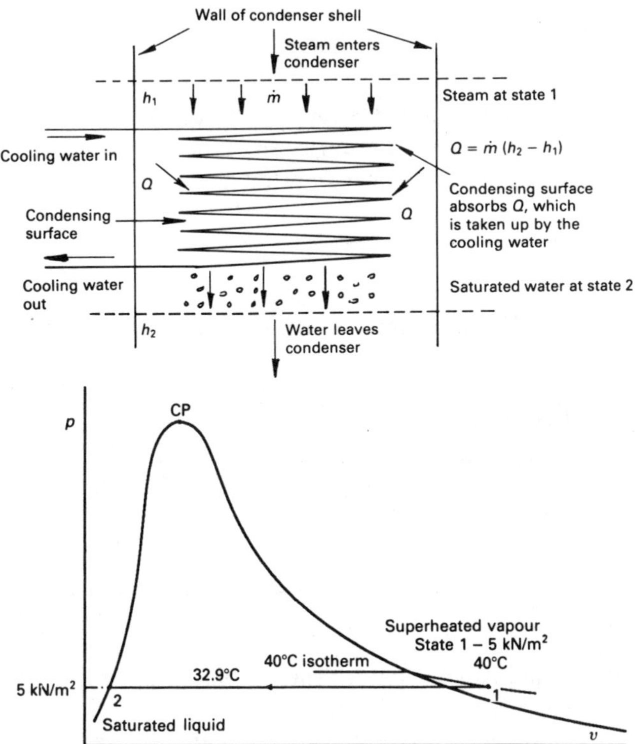

44

Suppose steam at 5 kN/m³ and 40°C is condensed to saturated water at the same pressure. We shall calculate the heat transfer for a flow of 15 kg/s, assuming the velocities are negligible. First of all, what can you say about the state of the steam at this pressure and temperature?

45

It has 7.1 K of superheat

The saturation temperature at this pressure is 32.9°C, so at 40°C there must be 7.1 K of superheat. Then if we use data from the superheat table on page 6 of the Rogers and Mayhew tables (the pressure being 0.05 bar), and remembering that column 2 gives data at saturation conditions, we can interpolate between 32.9°C and 50°C to get:

$$h_1 = 2561 + (2594 - 2561) \times \frac{(40 - 32.9)}{(50 - 32.9)}$$

$$= 2574.7 \text{ [kJ/kg]}$$

At state 2, the enthalpy is that of saturated water at 0.05 bar, which is 138 kJ/kg, from page 3 of the tables. As we are neglecting kinetic energy, you now have all you need to know to calculate the heat transfer, for a flow of 15 kg/s. What is it?

46

The answer is −36.55 MW, the negative sign indicating it is heat transfer from the steam.

We use: $\dot{Q} = \dot{m}\,(h_2 - h_1)$

$$= 15 \times (138 - 2574.7) = -36\,550.5 \text{ [kW]}$$

$$= -36.55 \text{ [MW]}$$

47

The Haywood tables and Rogers and Mayhew tables also contain data for vapours other than steam, which are used as refrigerants.

Dichlorodifluoromethane (Refrigerant 12) is commonly used, but it is one of the refrigerants which, if it escapes into the atmosphere, affects the ozone layer and its use is to be phased out. It is also known as *R*-12 or Freon 12.

In a refrigerator, the refrigerant fluid flows through coils around the ice-box, which is called the evaporator, in which it is boiled by heat transfer from the contents. At a saturation pressure of 0.1236 MN/m^2, Refrigerant 12 boils at -25°C, so the contents will be cooled to a temperature rather warmer than this.

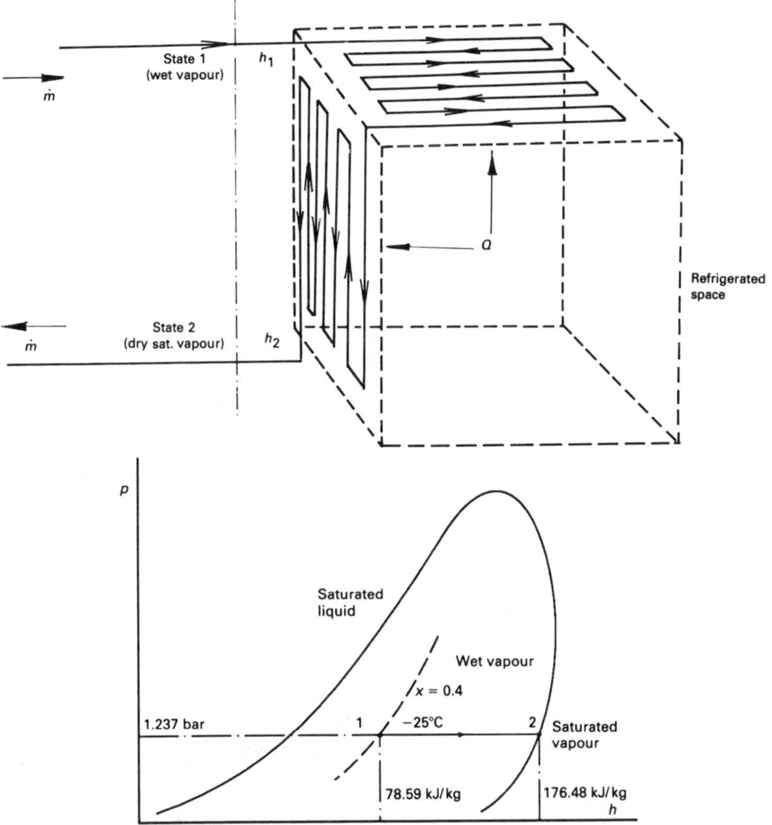

In the evaporator of a commercial refrigerator, R-12 enters at -25°C with a dryness fraction of 0.4, and it leaves as a dry saturated vapour. The mass flow is 0.1 kg/s, and kinetic energy is negligible. We want to know the rate of heat transfer to the fluid. This can be calculated by considering the enthalpy at the two end states, obtained from tables as before, and using the steady flow energy equation. The *p–h* diagram for the process path is shown above and using a *p–h* diagram for R-12 we can read off the enthalpy values at the two end states. This is quick to do and does not involve any calculation, but generally it is not so accurate as using tables.

Now find the heat transfer by both methods. The *p–h* diagram above shows the state points for which the specific enthalpy is required.

48

9.789 kW

State 1 is for fluid entering the evaporator with a dryness fraction of 0.4; using data from page 13 of the Rogers and Mayhew tables:

$$\begin{aligned} h_1 &= (1 - x)h_f + xh_g \\ &= (1 - 0.4) \times 13.33 + 0.4 \times 176.47 \\ &= 78.59 \text{ kJ/kg} \end{aligned}$$

At state 2, $h_2 = h_g = 176.48$ kJ/kg

Hence:

$$\begin{aligned} \dot{Q} &= \dot{m}\,(h_2 - h_1) \\ &= 0.1 \times (176.48 - 78.59) \\ &= 9.789 \text{ [kW]} \end{aligned}$$

From the Hickson and Taylor p–h diagram, h_1 is 79 kJ/kg and h_2 is 177 kJ/kg, so:

$$\begin{aligned} \dot{Q} &= 0.1 \times (177 - 79) \\ &= 9.80 \text{ [kW]} \end{aligned}$$

49

The refrigeration and heat pump cycles will be considered in a later programme; at the moment we can say that in part of the cycle, the refrigerant expands from a high pressure saturated liquid condition to a low pressure wet vapour condition. This is a throttling process of the type considered in Frame 64 of Programme 3, and takes place at constant enthalpy. The fluid enters the throttling process after condensation and is generally a saturated liquid, so that its state is known. Either from tables, or from a p–h chart, with the specific enthalpy constant we can find the state after throttling.

Let R-12 expand from a saturated liquid condition at 60°C, at which $p_s = 1.526$ MN/m^2, or 15.26 bar, to a wet vapour condition at -15°C and 0.1826 MN/m^2, or 1.826 bar, at constant enthalpy. From page 13 of the Rogers and Mayhew tables, h_f at 60°C is 95.74 kJ/kg. At 1.826 bar, $h_f = 22.33$ kJ/kg and $h_g = 180.97$ kJ/kg. What is the dryness fraction after the expansion? The process is shown on the p–h diagram below.

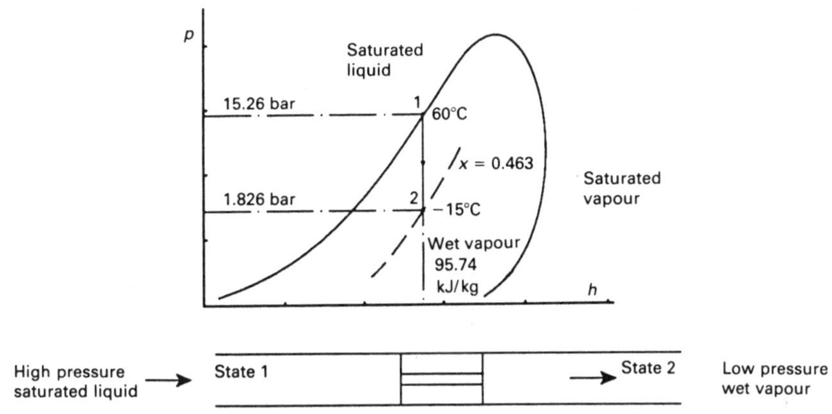

50

$$x = 0.4627$$

The specific enthalpy before and after expansion is 95.74 kJ/kg. Hence: After the expansion:

$$(1 - x)h_f + x\,h_g = 95.74$$

$$(1 - x)\,22.33 + x\,180.97 = 95.74$$

so:

$$x = 0.4627$$

From the chart, we can see that the dryness fraction is 0.46.

Adiabatic throttling also takes place in a device called the throttling calorimeter. This is shown below, and described in the next frame.

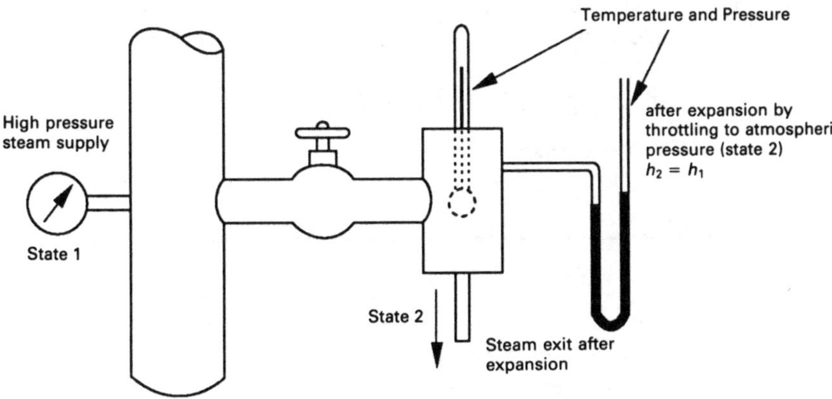

51

The throttling calorimeter is used to find the state of steam in a steam line. Steam in the line (or pipe) at a known pressure expands through a valve to atmospheric pressure, and the steam temperature is recorded. The expanded steam will be slightly superheated; knowing its temperature and pressure, we can find its enthalpy. This will be the enthalpy of steam in the line, so knowing its pressure, we can find its state.

Suppose we have steam at 1.013 25 bar and at 120°C at the outlet of the calorimeter; what is its specific enthalpy?

52

2716.4 kJ/kg

From page 6 of the Rogers and Mayhew tables, h_g at 1.013 25 bar is 2676 kJ/kg (second column), and h at 150°C is 2777 kJ/kg. Hence at 120°C:

$$h = 2676 + \frac{20}{50} \times (2777 - 2676)$$

$$= 2716.4 \text{ [kJ/kg]}$$

This is then the specific enthalpy of steam in the steam line. If the line pressure is 4 bar, we can compare the figure of 2716.4 kJ/kg with h_g at 4 bar, which is 2739 kJ/kg. We conclude that the steam is slightly wet. What is its dryness fraction?

53

The answer is the steam is 0.989 dry.

At 4 bar, from page 4, we have h_f = 605 kJ/kg, and h_{fg} = 2134 kJ/kg, so:

$$2716.4 = h_f + x\, h_{fg} = 605 + 2134\, x$$

Hence: $$x = 0.989$$

54

This concludes our work with vapours for the time being. Processes of expansion in open systems in work-producing turbines will follow in Programme 7. Before moving on to working with gases, make sure you can do all the following exercises.

The answers were obtained using Rogers and Mayhew tables; slight variations will occur using the Haywood or Cooper and Le Fevre tables.

PROBLEMS IN CLOSED AND OPEN SYSTEMS WITH VAPOURS

1. 0.5 kg of wet steam at 1.01325 bar is heated at constant volume to a pressure of 221.2 bar. What is its final state? Calculate (a) the initial dryness fraction and (b) the heat transfer. [At the critical point, (a) $x = 0.001\ 272$, (b) 796.17 kJ]
2. 4 kg of wet steam, having a specific volume of 0.001 251 m³/kg and a pressure of 0.1985 MPa, is heated at constant volume until it reaches a saturation line. What is the final state? Calculate U_1 and U_2.
[Saturated liquid at 250°C, 3.978 MPa, $U_1 = 2017.54$ kJ, $U_2 = 4321.84$ kJ]
3. Wet steam of specific volume 0.135 m³/kg is heated at constant volume until it reaches a pressure of 2.0 MN/m². Find (a) the pressure at which it passes through the saturated vapour boundary and (b) the final temperature.
[(a) 1.464 MN/m², (b) 336.26°C]
4. 2 kg of steam at a pressure of 0.1 bar and specific volume 0.8856 m³/kg is heated at constant volume until it becomes dry saturated steam. It is then heated at constant pressure until the temperature is 350°C. What is the final pressure? Calculate the heat transfer between the initial and final state.
[2 bar, 5494.25 kJ]
5. Steam at a pressure of 1.013 25 bar is heated at constant pressure from an initial dryness fraction of 0.1 to a final dryness fraction of 0.9. What is (a) the change of specific internal energy and (b) the change of enthalpy?
[(a) 1669.83 kJ/kg, (b) 1805.36 kJ/kg]
6. Unit mass of wet steam at state 1 is heated at constant pressure to a higher dryness fraction at state 2. The heat transfer is 695.4 kJ. The work transfer is 48 kJ. For the pressure of the steam, $u_{fg} = 2158$ kJ/kg. What mass of water evaporated between state 1 and state 2? [0.3 kg]
7. Wet ammonia vapour at -10°C and of specific volume 0.1926 m³/kg is heated at constant volume to the saturated vapour condition, and is then heated at constant pressure to 75 K of superheat. (a) Calculate the initial dryness fraction. (b) Find the pressure at which it becomes dry saturated. (c) Calculate the overall heat transfer per unit mass. Take v_f at -10°C as 0.001 54 m³/kg. Use ammonia data (Refrigerant 717) from your tables.
[(a) $x = 0.4582$, (b) 6.585 bar, (c) 845.07 kJ/kg]
8. Dry saturated steam at 2.0 MN/m² is expanded adiabatically over a volume ratio of 3 to 1 according to $pv^{1.135} = k$. Find the final state of the steam. For unit mass of steam, calculate the work done, the change of internal energy and the error in the adiabatic expansion assumption.
[5.748 bar, $x = 0.9068$, $W = 203.27$ kJ/kg, $U_2 - U_1 = -211.01$ kJ/kg, $Q = -7.74$ kJ/kg]
9. Steam at 3.0 MPa and 600°C is expanded to 0.1 MPa bar according to $pv^{1.3} = k$.

Find the final state of the steam. Calculate per unit mass the work done, the change of internal energy, and any heat transfer.

[124.05 K of superheat, $W = 720.03$ kJ/kg, $U_2 - U_1 = -741.97$ kJ/kg, $Q = -21.94$ kJ/kg]

10. Dry saturated steam at 1.0 MN/m^2 pressure is expanded isothermally to a pressure of 0.1 MN/m^2. The heat transfer is 522.36 kJ for unit mass. Calculate (a) the change of specific internal energy, (b) the work transfer, (c) the value of n for the expansion assuming it follows a single curve of $pv^n = k$.

[(a) +44.45 kJ/kg, (b) +480.91 kJ/kg, (c) $n = 0.9721$]

11. Compressed water at 40 bar having an enthalpy of 196 kJ/kg enters a boiler tube with negligible velocity and it leaves the tube downstream as a dry saturated vapour with a velocity of 100 m/s. Calculate (a) the heat transfer to the tube for a flow of 2 kg/s and (b) the tube diameter.

[(a) 5220 kW, (b) 35.66 mm]

12. The steam from problem **11** enters a superheater tube as dry saturated vapour at 40 bar, and it leaves the superheater at 500°C and at a pressure of 39 bar. Calculate the heat transfer to the stream for 2 kg/s flow, neglecting kinetic energy. [1290.2 kW]

13. 10 kg/s of exhaust steam at 0.075 MPa and 95°C enters a condenser from which it leaves as saturated water at the same pressure. The heat transfer in condensation passes to cooling water entering at 15°C and leaving at 70°C. Calculate (a) the rate of heat transfer and (b) the rate of flow of cooling water in kg/h. Take c_p for cooling water as 4.186 kJ/kg K.

[(a) 22.850 MW, (b) 357 298 kg/h]

14. Steam enters a desuperheater at 0.04 MN/m^2 pressure and at 125°C with a velocity of 220 m/s. It leaves at 0.039 MN/m^2 as dry saturated steam with negligible velocity. Calculate the heat transfer rate per kg of flow.

[−123.7 kJ/kg]

15. In a throttling calorimeter, stream is expanded from 5.5 bar pressure to 1.013 25 bar. The exhaust steam temperature is 117.36°C. What is the state of the steam at 5.5 bar pressure? [$x = 0.98$]

16. Refrigerant 12 expands through a throttle from a pressure of 5.2 bar as a saturated liquid to a wet vapour state at −25°C. What is the state of the wet vapour after the expansion? [By calculation, $x = 0.2364$; by chart, $x = 0.24$]

17. Superheated ammonia vapour at 100°C, and having a pressure corresponding to a saturation temperature of 50°C, is cooled and condensed to a saturated liquid state. The mass flow is 0.25 kg/s. What is the heat transfer rate?

[By calculation, $Q = -302.8$ kW; by chart, $Q = -300$ kW]

55

After working with vapours you will find working with gases relatively straightforward. We shall work with the perfect gas, which you will remember obeys the ideal gas rule, and has a constant value of c_V. Each gas has its own gas constant in the ideal gas rule, and constant value of specific heat.

Frequently we shall use air in our examples. As you know, air is a mixture of gases, and later in the programme we shall prove that mixtures of perfect gases can be treated as a pure substance.

In a constant volume process, the First Law equation for the process is, as we have already seen from Frame 3:

$$Q = U_2 - U_1$$

which, from Frame 91, Programme 4, for a perfect gas becomes:

$$Q = m\, c_V(T_2 - T_1)$$

For a constant volume process, we need to know all the properties at state 1 and one of the other properties at state 2, or information from which it can be found.

56

We are going to work through a problem involving heat transfer at constant volume in an engine cylinder. The *p–v* diagram is shown below.

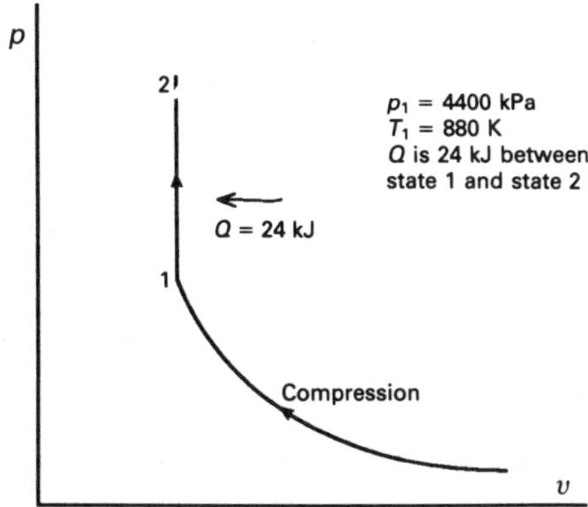

At the end of compression in an engine cylinder of 180 mm diameter, the piston is 50 mm from the closed end of the cylinder. The cylinder contents are pure air at a pressure of 4400 kPa and at a temperature of 880 K. Heat transfer at constant volume amounting to 24 kJ to the air takes place. What are the final temperature and pressure of the air? For air, $R = 0.287$ kJ/kg K and $c_V = 0.718$ kJ/kg K. We will start the solution in the next frame.

57

We need to use the First Law equation for the process, plus the ideal gas rule. First we calculate the cylinder volume:

$$V = \pi D^2 L/4 = \pi \times 0.18^2 \times 0.05/4$$
$$= 0.001\,272\ [\text{m}^3]$$

Then using the ideal gas rule at state 1 we can find the mass of air:

$$m = pV/RT = \frac{4400 \times 0.001\,272}{0.287 \times 880}\ \frac{[\text{kN/m}^2][\text{m}^3]}{[\text{kJ/kg K}][\text{K}]}$$
$$= 0.0222\ [\text{kg}]$$

With [kJ] = [kN m] you can see how the units cancel to leave mass in [kg]. In gas problems, see that you have consistent units in pressure and the gas constant.

Now we can use the First Law equation to find T_2:

$$Q = m\,c_V(T_2 - T_1)$$
$$24.0 = 0.0222 \times 0.718 \times (T_2 - 880)$$

Hence: $T_2 = 2385.7$ K

Finally, knowing T_2 we can find p_2. One way is to say for the constant volume process:

$$\frac{p_1}{T_1} = \frac{p_2}{T_2}, \text{ since } pV/T = \text{constant}$$

and so:

$$p_2 = p_1 \times (T_2/T_1)$$
$$= 4400 \times (2385.7/880)$$
$$= 11\,928.5\ [\text{kPa}]$$

Another way is to find p_2 from the ideal gas rule. Do that.

58

$$\boxed{p_2 = 11\,949.9 \text{ kPa}}$$

From the ideal gas rule we have:

$$p_2 \times 0.001\,272 = 0.0222 \times 0.287 \times 2385.7$$

$$\therefore \qquad p_2 = 11\,949.9 \text{ [kPa]}$$

The two answers agree to the first 3 figures. The calculations depend on R and c_V which, being approximate values for air, are themselves only accurate to 3 figures. In this problem, according to the two property rule, we needed to know two of the properties p, T or v at state 1 for unit mass of air, together with one property at state 2 or the value of the heat transfer. Since we were not dealing with unit mass of air, we needed to know the actual mass of air as well.

In the First Law equation we put $W = 0$ for the constant volume process since we were thinking only of displacement work. However, we might have a constant volume process involving stirring work, and/or electrical work, as well as heat transfer.

Now consider this problem: a quantity of air has a mass of 0.02 kg and its pressure and temperature are 200 kPa and 500 K. An agitator wheel in the container is driven from outside by a motor which delivers 10 kJ of work to the air. At the same time a heat transfer of 3.5 kJ occurs from the air. The volume of the air does not change. What are the final temperature and pressure? (The same work could have been done on the air by an electrically heated wire in the container.) You should be able to solve this problem, using the full First Law equation for a closed system ($c_V = 0.718$ kJ/kg K).

59

$$\boxed{T_2 = 952.65 \text{ K},\ p_2 = 381.06 \text{ kPa}}$$

Since there is a work term, we use $Q - W = U_2 - U_1$

and so: $$-3.5 - (-10) = m\, c_V(T_2 - T_1)$$

$$+6.5 = 0.02 \times 0.718 \times (T_2 - 500)$$

and hence: $$T_2 = 952.65 \text{ K}$$

and then: $p_2 = p_1 \times (T_2/T_1)$ [since the process is constant volume]

$$= 200 \times (952.65/500)$$

$$= 381.06 \text{ [kPa]}$$

60

Now we shall think about a constant pressure process. In such a process with displacement work only, the First Law equation may be written as:

$$Q - p(V_2 - V_1) = U_2 - U_1$$

or: $$Q - pm(v_2 - v_1) = mc_V(T_2 - T_1)$$

and so: $$Q = m(pv_2 - pv_1 + c_VT_2 - c_VT_1)$$

and then: $$Q = m(RT_2 - RT_1 + c_VT_2 - c_VT_1)$$

$$= m(R + c_V)(T_2 - T_1)$$

But we know that in a constant pressure process:

$$Q = mc_P\,(T_2 - T_1)$$

Hence for a perfect gas, by comparing these two equations:

$$c_P = R + c_V$$

$$\therefore \quad R = c_P - c_V$$

What is c_P for air for which c_V is 0.718 kJ/kg K and R = 0.287 kJ/kg K?

61

1.005 kJ/kg K

$$c_P = R + c_V = 0.287 + 0.718 = 1.005 \text{ kJ/kg K.}$$

The two specific heats c_P and c_V have an important ratio:

$$\gamma = \frac{c_P}{c_V}$$

Since γ (Greek gamma) is the ratio of the two specific heats which are properties, it is also a property itself, but has no units.

If you eliminate c_V between the expressions for R and γ, you can show that:

$$c_P = \frac{\gamma R}{\gamma - 1}$$

and also by eliminating c_P we can get:

$$c_V = \frac{R}{\gamma - 1}$$

For air, with $c_P = 1.005$ kJ/kg K and $c_V = 0.718$ kJ/kg K, γ is 1.3997, taken as 1.4. What is γ for a gas, for which R is 0.2081 kJ/kg K, and $c_V = 0.3122$ kJ/kg K?

62

1.667

We have that: $\gamma - 1 = R/c_V = 0.2081/0.3122 = 0.667$

Hence γ is 1.667.

A constant pressure process in a closed system automatically implies that the volume has changed, so that displacement work is done. As in a constant volume process, other forms of work could be involved as well. Consider this problem: a quantity of air has a pressure of 120 kPa, a temperature of 300 K and a volume of 0.005 m^3. It increases its volume to 0.015 m^3 at constant pressure. Calculate the mass of the air, its final temperature and the heat transfer. You can start by calculating the mass of the air. Take $R = 0.287$ kJ/kg K and $c_V = 0.718$ kJ/kg K.

63

0.006 969 kg

Using the ideal gas rule:

$$m = pV/RT$$

$$= \frac{120 \times 0.005}{0.287 \times 300}$$

$$= 0.006\,969 \text{ [kg]}$$

64

Then since the process is at constant pressure:

$$\frac{V_2}{T_2} = \frac{V_1}{T_1}$$

and so: $$T_2 = T_1 \times \frac{V_2}{V_1}$$

Hence: $$T_2 = 300 \times 0.015/0.005$$
$$= 900 \text{ K}$$

We now know both the initial and final temperatures; what is the change of internal energy?

65

3.0022 kJ

The change of internal energy is:

$$m\,c_V(T_2 - T_1) = 0.006\,969 \times 0.718 \times (900 - 300)$$
$$= 3.0022 \text{ [kJ]}$$

Then using the First Law equation $Q - W = U_2 - U_1$ we have:

$$Q - 120 \times (0.015 - 0.005) = 3.0022$$

$\therefore$ $$Q - 1.2 = 3.0022$$

$\therefore$ $$Q = 4.2022 \text{ [kJ]}$$

The air, now at 900 K and at 120 kPa, undergoes a second constant pressure process from the volume of 0.015 m^3 to 0.010 m^3, while 0.8 kJ of electrical work is done on the air. Using the First Law equation, we will find the value of Q. First, what is the temperature at the end of this process?

66

600 K

Since the process is again at constant pressure we can say:

$$T_2 = T_1 \times \frac{V_2}{V_1}$$

$$\therefore \qquad T_2 = 900 \times 0.01/0.015$$

$$= 600 \text{ K}$$

There are two work terms to be combined into the First Law equation:

$$Q - W = U_2 - U_1$$

There is electrical work done on the gas, which is negative, and a displacement work with a reduction of volume, which is also negative. The work term is:

$$\begin{aligned} W &= [-0.8 + p(V_2 - V_1)] \\ &= [-0.8 + 120 \times (0.010 - 0.015)] \\ &= [-0.8 - 0.6] \\ &= -1.4 \text{ [kJ]} \end{aligned}$$

We can now complete the First Law equation:

$$\begin{aligned} Q - [-1.4] &= m\, c_V\, (T_2 - T_1) \\ Q + 1.4 &= 0.006\,969 \times 0.718 \times (600 - 900) \\ \therefore \qquad &= -1.501 \\ Q &= -2.901 \text{ [kJ]} \end{aligned}$$

We will now think about the reversible adiabatic process for a perfect gas. You should keep in mind the fact that in all these processes we are assuming all volume changes and heat transfers to be reversible. For a vapour, you remember, we had to use approximate p–v relationships for an adiabatic expansion. What about the ideal gas? We will investigate this in the next frame.

67

We shall think of a reversible adiabatic process taking place over small changes of specific volume and pressure, dv and dp, respectively. Recalling Frame 96 of Programme 4, the ideal gas rule may be written as:

$$p\mathrm{d}v + v\mathrm{d}p = R\mathrm{d}T$$
$$= (c_P - c_V)\mathrm{d}T$$

and the First Law for unit mass of gas with no heat transfer becomes:

$$0 - p\mathrm{d}v = c_V\mathrm{d}T$$

and so:
$$\mathrm{d}T = -\frac{p\mathrm{d}v}{c_V}$$

Eliminating dT between the two equations we get:

$$p\mathrm{d}v + v\mathrm{d}p + (c_P - c_V)\frac{p\mathrm{d}v}{c_V} = 0$$

and so:
$$v\mathrm{d}p + \gamma p\mathrm{d}v = 0$$

or:
$$\gamma\frac{\mathrm{d}v}{v} + \frac{\mathrm{d}p}{p} = 0$$

and on integrating:
$$\gamma \ln v + \ln p = \text{constant}$$

or:
$$pv^{\gamma} = \text{constant}$$

Thus the index of the p–v relationship for an adiabatic process with a perfect gas is the ratio of the specific heats, γ. We use this equation in just the same way that we use pv^n = constant for a polytropic process, in order to find a pressure or specific volume at the final state, and to find the work done.

Since the adiabatic process follows a p–v relationship that is similar to the polytropic process, the integration of pdv between two end states also produces a similar result. By repeating the integration given in Frame 61 of Programme 2, using γ in place of n shows that for the adiabatic process for unit mass of gas:

$$W = \frac{p_2v_2 - p_1v_1}{1 - \gamma}$$

68

A gas for which γ is 1.4 is initially at 300 kPa pressure with a specific volume of 0.48 m^3/kg. The initial temperature is 500 K. It expands adiabatically to a pressure of

100 kPa. What is the final specific volume, the final temperature and the work done? The p–v diagram for the process is shown below:

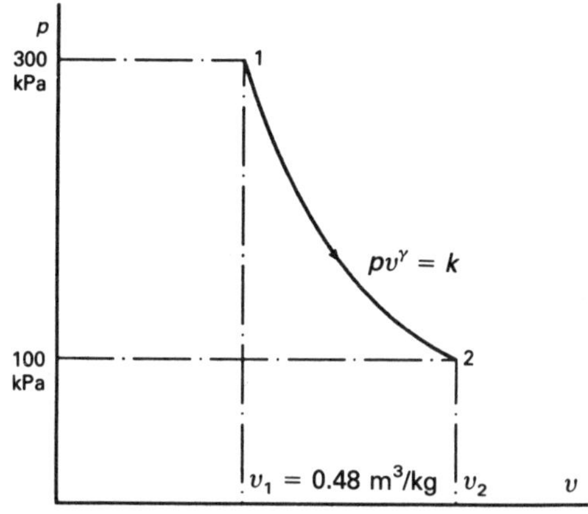

First we can find the final specific volume (at state 2). Using $pv^\gamma = k$, we have:

$$300 \times 0.48^{1.4} = 100 \times v_2^{1.4}$$

and so:

$$v_2^{1.4} = 1.0736$$

$$v_2 = 1.052 \text{ [m}^2\text{/kg]}$$

Next we have to think about how to find the temperature ratio across an adiabatic process.

69

Since for the reversible adiabatic process we have pv^γ = constant and from the ideal gas rule we have pv/T = constant, we can combine these two equations to give:

$$\frac{T_2}{T_1} = \left(\frac{p_2}{p_1}\right)^{\frac{\gamma-1}{\gamma}}$$

and:

$$\frac{T_2}{T_1} = \left(\frac{v_2}{v_1}\right)^{1-\gamma} = \left(\frac{v_1}{v_2}\right)^{\gamma-1}$$

So we get two useful relationships between temperature and pressure, and between temperature and specific volume in an adiabatic process. This may offer an element of choice in solving a problem but generally there is only one relationship appropriate to the properties you already know.

70

We are now in a position to calculate the remaining property, temperature, at state 2, in the example in Frame 68.
Thus:

$$\frac{T_2}{T_1} = \left(\frac{p_2}{p_1}\right)^{(\gamma-1)/\gamma} = \left(\frac{100}{300}\right)^{0.286}$$

$$= 0.7304$$

Hence: $$T_2 = 500 \times 0.7304$$

$$= 365.2 \text{ K}$$

In the example we have been following, with p_1 = 300 kPa, p_2 = 100 kPa, v_1 = 0.48 m³/kg, v_2 = 1.052 m³/kg and γ = 1.4, what is the work term? (Use the formula given in Frame 67.)

71

+97.0 kJ/kg

Using $W = (p_2v_2 - p_1v_1)/(1 - \gamma)$, we get:

$$W = (100 \times 1.052 - 300 \times 0.48)/(-0.4)$$

$$= +97.0 \text{ [kJ/kg]}$$

This is the area under the process path in the figure in Frame 68.

The process is adiabatic, with $Q = 0$ remember, so this work of expansion is at the expense of the internal energy. This will give us another route to the final temperature:

For unit mass: $$Q - W = u_2 - u_1$$

$$-97.0 = 0.718 \times (T_2 - T_1)$$

$$= 0.718 \times (T_2 - 500)$$

Hence: $$T_2 = 364.9 \text{ K}$$

The small difference is again due to rounding-off errors in the two methods.

Now do this example: a petrol vapour–air mixture is compressed adiabatically in an engine cylinder through a volume ratio of 10 to 1, with p_1 = 100 kPa and T_1 =

300 K. The mixture may be treated as a perfect gas for which $\gamma = 1.32$. (Remember that γ is 1.4 only for pure air.) Calculate the pressure and temperature at the end of compression.

72

2089.3 kPa, 626.8 K

We have:

$$\frac{p_2}{p_1} = \left(\frac{v_1}{v_2}\right)^{\gamma} \text{ and } \frac{T_2}{T_1} = \left(\frac{v_1}{v_2}\right)^{\gamma-1}$$

and so: $p_2 = 100 \times 10^{1.32} = 100 \times 20.893$

$= 2089.3$ [kPa]

$\therefore$ $T_2 = 300 \times 10^{0.32} = 300 \times 2.089$

$= 626.79$ K

You may have thought of using the ideal gas rule in this problem to find T_2 after finding p_2 from $pv^{\gamma} = k$. You would not have been able to do so because something was missing. What was that? (Remember we are dealing with a petrol–air mixture, not pure air.)

73

R for the mixture

We had to use the ratio formulae since we did not have the gas constant.

Next we will look at polytropic processes with perfect gases. We have already met the work term, (Frame 61, Programme 2), and since the process is not adiabatic, a heat transfer term must be involved.

74

For the polytropic process we have $pv^n = k$, and $pv = RT$ as the equation of state, hence we have:

$$\frac{T_2}{T_1} = \left(\frac{p_2}{p_1}\right)^{\frac{n-1}{n}}$$

and

$$\frac{T_2}{T_1} = \left(\frac{v_2}{v_1}\right)^{1-n} = \left(\frac{v_1}{v_2}\right)^{n-1}$$

These relationships are identical to those for the adiabatic process, except for n replacing γ, and again are useful for dealing with problems when we do not know the gas constant. The figure below shows p–v paths for various values of n in polytropic processes in comparison with that of the adiabatic process.

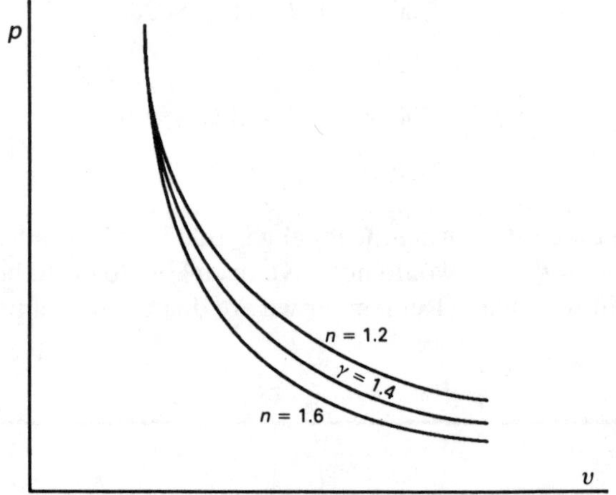

If in a polytropic compression $(v_1/v_2) = 10$, and $n = 1.25$, what are (p_2/p_1) and (T_2/T_1)?

75

17.783, 1.778

We have $v_1/v_2 = 10$ and $n = 1.25$, so:

$$\frac{p_2}{p_1} = 10^{1.25} = 17.783$$

and: $$\frac{T_2}{T_1} = \left(\frac{v_1}{v_2}\right)^{n-1}$$

$$= 10^{0.25} = 1.778$$

The displacement work for unit mass in a polytropic process is:

$$W = \frac{p_2v_2 - p_1v_1}{1 - n}$$

We may combine this with the First Law equation for unit mass to give:

$$Q = (u_2 - u_1) + \frac{p_2v_2 - p_1v_1}{1 - n}$$

and substituting $c_V T$ for u, and RT for pv, we get:

$$Q = c_V(T_2 - T_1) + \frac{R}{1-n}(T_2 - T_1)$$

$$= \left(c_V + \frac{R}{1-n}\right)(T_2 - T_1)$$

This is the heat transfer in a polytropic process of index n. The group ($c_V + R/[1-n]$) is, if you like to think of it that way, a specific heat for a polytropic of index n. But n is related to the graph of p against v, and is not a property of a system, so the group as a whole is not a property. This also follows since the group has a different value for each path value of n.

76

Next we are going to work through this exercise:

An expansion takes place in a cylinder, from 12 000 kPa and 2000 K over a volume ratio of 1 to 9.5. The polytropic index is 1.56. Calculate the final pressure and temperature, and the work and heat transfers per unit mass. The properties of air may be used: c_V = 0.718 kJ/kg K and R = 0.287 kJ/kg K.

For p_2 we have:

$$\frac{p_2}{p_1} = \left(\frac{v_1}{v_2}\right)^{1.56} \quad \text{hence: } p_2 = 12\,000 \left(\frac{1}{9.5}\right)^{1.56} = 358.04 \text{ [kPa]}$$

Now find T_2.

77

$$\boxed{T_2 = 566.85 \text{ K}}$$

We have that: $$\frac{T_2}{T_1} = \left(\frac{p_2}{p_1}\right)^{\frac{0.56}{1.56}}$$

$$\therefore \quad T_2 = 2000 \times \left(\frac{358.04}{12\,000}\right)^{0.359}$$

$$= 566.85 \text{ K}$$

We need to know v_2 and v_1 as well as the ratio between them. From the ideal gas rule:

$$v_1 = \frac{RT_1}{p_1}$$

$$= 0.287 \times 2000/12\,000$$

$$= 0.0478 \text{ [m}^3\text{/kg]}$$

and $$v_2 = 0.0478 \times 9.5$$

$$= 0.4541 \text{ [m}^3\text{/kg]}$$

Now you can calculate the work per unit mass.

78

$$\boxed{+733.95 \text{ kJ/kg}}$$

The work term is:

$$W = \frac{p_2 v_2 - p_1 v_1}{1 - n}$$

$$= \frac{358.04 \times 0.4541 - 12\,000 \times 0.0478}{1 - 1.56}$$

$$= \frac{162.59 - 573.6}{-0.56}$$

$$= +733.95 \text{ [kJ/kg]}$$

There is no need to find the internal energies; we can find Q directly from the formula in Frame 75. You should be able to do that.

79

-294.51 kJ/kg

The formula to use is:

$$Q = \left(c_V + \frac{R}{1-n}\right)(T_2 - T_1)$$

Hence:

$$Q = (0.718 + 0.287/[-0.56])(566.85 - 2000)$$

$$= 0.2055(-1433.15)$$

$$= -294.51 \text{ [kJ/kg]}$$

The negative heat transfer is from the hot expanding gases. This is as we would expect. It is the heat transfer which occurs naturally in the engine cylinder, that controls the form of the pressure–volume graph of the expansion. This gives rise to the particular value of n. For whatever the working fluid in the engine cylinder, we find that with heat transfer *from* the cylinder contents in both the compression and expansion strokes that n is less than γ in compression and greater than γ in expansion. Then if both processes are adiabatic, n becomes equal to γ. You can confirm these results by inspection of the formula for polytropic heat transfer given above.

A special case of the polytropic process is when $n = 1$. This gives $pv = k$. For a perfect gas this means T is constant along the p–v graph, and the line of the p–v function is called an **isothermal**. We met an isothermal expansion for steam in Frame 33. Since $u = c_V T$ for a gas and with the temperature constant there is no change of internal energy during the process, so in applying the First Law it must follow that $Q = W$. To find Q we have to calculate W. What is the work term when $n = 1$? (If you need to look it up, you will find the answer in Programme 2, Frame 75.)

80

$$W = p_1 v_1 \ln \frac{v_2}{v_1}$$

This is one of a number of possible results for the work term, given in Frame 75 of Programme 2, for a process for which $pv = k$.

In an isothermal compression of air, $p_1 = 100$ kPa, $V_1 = 0.015$ m³ and $T_1 = 300$ K. During the compression the heat transfer is -1.648 kJ. What is the mass of air, and what is its final state? (Take $R = 0.287$ kJ/kg K.)

Since $W = Q$, we have:

$$-1.648 = p_1 V_1 \ln \frac{V_2}{V_1}$$

$$= 100 \times 0.015 \ln \frac{V_2}{V_1}$$

Hence: $\ln(V_2/V_1) = -1.0987$

and so: $V_2/V_1 = 0.3333$

which gives: $V_2 = 0.005$ [m³]

With $V_2 = 0.005$ m³, the ratio $V_1/V_2 = 3$, so p_2 will be 300 kPa and T_2 will be unchanged.

What is the mass of the air?

81

$$m = 0.0174 \text{ kg}$$

The mass of air comes from the ideal gas rule using properties at either state 1 or state 2:

$$m = \frac{pV}{RT} = \frac{100 \times 0.015}{0.287 \times 300} \quad \text{(at state 1)}$$

$$= 0.0174 \text{ [kg]}$$

This brings us to the end of our work on single gases in closed systems. In the examples we have treated air, and, on one occasion, a petrol vapour–air mixture, as single gases whose state can be determined from the ideal gas rule. Both of these are gas mixtures, and we now have to establish that a mixture of perfect gases can be treated as a pure substance.

82

When two or more gases are mixed together, the molecules of the different gases are intimately mixed, and samples of the mixture at different points will have identical compositions. When each constituent is a pure substance and perfect gas, the mixture is also a pure substance and perfect gas. We need to establish the properties of the mixture, in terms of the properties of each constituent gas.

Consider a closed system consisting of a volume, V, of a mixture of three different gases, gas a, gas b and gas c. The molecules of any of the three gases will be found anywhere in the volume V, so for the gases making up the mixture we can say:

$$V = V_a = V_b = V_c \qquad (1)$$

We are thinking of three gases, but this and following equations may be written for any number of gases in the mixture.

83

The mass of the mixture is m, and the masses of each gas in the mixture are m_a, m_b and m_c respectively, so:

$$m = m_a + m_b + m_c \qquad (2)$$

The volume V of the closed system in an equilibrium state will have a single intensive property of temperature throughout, so we can say:

$$T = T_a = T_b = T_c \qquad (3)$$

The intensive property of pressure, p, will also be uniform throughout, but the value of p will be due to the molecular bombardment of the surface by molecules of each constituent gas in the mixture. The way in which each gas contributes to the pressure of the mixture is given by Dalton's Law in the next frame.

84

Dalton's Law of Partial Pressures states:

The pressure of a gaseous mixture is the sum of the pressures which each constituent gas would exert if it alone occupied the volume of the mixture at the temperature of the mixture.

The pressure which each gas would exert is called the **partial pressure** of the gas. Thus we have:

$$p = p_a + p_b + p_c \tag{4}$$

To find the pressure of a gas mixture we calculate the partial pressure of each gas, of known mass occupying a known volume at a known temperature, and then add together all the partial pressures.

Here is an example you ought to be able to do: 1 kg of air occupies a volume of 1 m^3 at 298 K. We may take the composition of the air to be 0.233 kg of oxygen (R = 259.8 J/kg K) and 0.767 kg of nitrogen (R = 296.8 J/kg K). What is the pressure of the air? Do this by finding the partial pressures of the oxygen and nitrogen as if each gas occupied the total volume at 298 K, and add the results together. Your answer will be in Pa, or N/m^2.

85

p_{O_2} = 18 038.95 Pa, p_{N_2} = 67 838.39 Pa, p = 85 877.34 Pa

For the oxygen:
$$p_{O_2} = \frac{mRT}{V} = \frac{0.233 \times 259.8 \times 298}{1}$$
$$= 18\,038.95\ [\text{Pa}]$$

For the nitrogen:
$$p_{N_2} = \frac{mRT}{V} = \frac{0.767 \times 296.8 \times 298}{1}$$
$$= 67\,838.39\ [\text{Pa}]$$

Hence the pressure of the mixture is:

$$p = p_{O_2} + p_{N_2}$$
$$= 18\,038.95 + 67\,838.39$$
$$= 85\,877.34\ [\text{Pa}] = 85.88\ [\text{kPa}]$$

We can go on to show that the density of the mixture is the sum of the densities of the individual gases. By expressing volume as the product of mass and specific volume, we can write equation (1) in Frame 82 as:

$$mv = m_a v_a = m_b v_b = m_c v_c \tag{5}$$

and then using equation (2) from Frame 83 we get:

$$\frac{m}{mv} = \frac{m_a}{m_a v_a} + \frac{m_b}{m_b v_b} + \frac{m_c}{m_c v_c}$$

or:

$$\frac{1}{v} = \frac{1}{v_a} + \frac{1}{v_b} + \frac{1}{v_c} \tag{6}$$

or:

$$\rho = \rho_a + \rho_b + \rho_c \tag{7}$$

The Greek rho, ρ, is the usual symbol for density.

86

From the previous example we can calculate the individual specific volumes of the oxygen and nitrogen components:

$$v_{O_2} = \frac{RT}{p_{O_2}} = \frac{259.8 \times 298}{18\,038.95}$$

$$= 4.2918\ [\mathrm{m^3/kg}]$$

$$v_{N_2} = \frac{RT}{p_{N_2}} = \frac{296.8 \times 298}{67\,838.39}$$

$$= 1.3038\ [\mathrm{m^3/kg}]$$

Remembering that density is the reciprocal of specific volume, you can now calculate the densities of oxygen and nitrogen present, and confirm that the sum of these two densities is the density of the air in the example in Frame 84.

87

$\rho_{O_2} = 0.233\ \mathrm{kg/m^3}$, $\rho_{N_2} = 0.767\ \mathrm{m^3/kg}$, $\rho = 1\ \mathrm{kg/m^3}$

88

We have, then:

$$\rho_{O_2} = 1/4.2918 = 0.233 \text{ [kg/m}^3\text{]}$$

$$\rho_{N_2} = 1/1.3038 = 0.767 \text{ [kg/m}^3\text{]}$$

and:

$$\rho = 0.233 + 0.767$$

$$= 1.0 \text{ [kg/m}^3\text{]}$$

Thus the density of the mixture is the sum of these densities which is 1 kg/m^3, which is correct as we had a total of 1 kg of air occupying a volume of 1 m^3.

Next we shall think about the internal energy and enthalpy of a gas mixture.

89

Dalton's Law of Partial Pressures is part of a more general law about gas mixtures, the Gibbs–Dalton Law, which extends the additive law of pressures to include additive behaviour of internal energy and enthalpy.

The pressure, internal energy and enthalpy of a mixture of gases are respectively equal to the sums of the pressures, internal energies and enthalpies that each constituent gas would have if it alone occupied the volume of the mixture at the temperature of the mixture.

This means that in addition to equation (4):

$$U = U_a + U_b + U_c \quad (8)$$

and

$$H = H_a + H_b + H_c \quad (9)$$

In terms of specific properties:

$$mu = m_a u_a + m_b u_b + m_c u_c \quad (10)$$

and

$$mh = m_a h_a + m_b h_b + m_c h_c \quad (11)$$

90

In Frame 27 of Programme 3 we defined specific heat at constant volume as the rate of change of specific internal energy with temperature, with the volume held constant. If we differentiate equation (10) with respect to temperature with the volume held constant, we shall obtain:

$$m\left(\frac{\partial u}{\partial T}\right)_V = m_a\left(\frac{\partial u_a}{\partial T}\right)_V + m_b\left(\frac{\partial u_b}{\partial T}\right)_V + m_c\left(\frac{\partial u_c}{\partial T}\right)_V \qquad (12)$$

or:

$$m\, c_V = m_a\, c_{V_a} + m_b\, c_{V_b} + m_c\, c_{V_c} \qquad (13)$$

so that:

$$c_V = \frac{m_a\, c_{V_a} + m_b\, c_{V_b} + m_c\, c_{V_c}}{m} \qquad (14)$$

Similarly, from Frame 31 of Programme 3, we can write:

$$m\left(\frac{\partial h}{\partial T}\right)_P = m_a\left(\frac{\partial h_a}{\partial T}\right)_P + m_b\left(\frac{\partial h_b}{\partial T}\right)_P + m_c\left(\frac{\partial h_c}{\partial T}\right)_P \qquad (15)$$

or:

$$m\, c_P = m_a\, c_{P_a} + m_b\, c_{P_b} + m_c\, c_{P_c} \qquad (16)$$

so that:

$$c_P = \frac{m_a\, c_{P_a} + m_b\, c_{P_b} + m_c\, c_{P_c}}{m} \qquad (17)$$

Equations (14) and (17) give simple results for the two specific heats of a gas mixture, involving the specific heats and masses of the constituent gases. We can then calculate γ for the mixture as c_P/c_V.

91

In a sample of air, we have 0.233 kg of oxygen and 0.767 kg of nitrogen, and the specific heats at constant pressure respectively are 0.918 kJ/kg K and 1.040 kJ/kg K. What is c_P for air?

Using equation (17) we have:

$$c_P = \frac{0.233 \times 0.918 + 0.767 \times 1.040}{1}$$

$$= \frac{0.2139 + 0.7977}{1}$$

$$= 1.0116 \text{ kJ/kg K}$$

Now do this exercise, using equation (14):

A gas mixture consists of 2 kg of nitrogen, 3 kg of carbon dioxide and 0.8 kg of hydrogen. The specific heats at constant volume, respectively, are 0.743, 0.657 and 10.153 kJ/kg K. What is c_V for the mixture?

$$\boxed{1.9964\ \text{kJ/kg K}}$$

Using equation (14) we have:

$$c_V = \frac{2 \times 0.743 + 3 \times 0.657 + 0.8 \times 10.153}{5.8}$$

in which the mass of the mixture is 5.8 kg. This gives:

$$c_V = \frac{1.486 + 1.971 + 8.122}{5.8}$$

$$= 1.9964\ [\text{kJ/kg K}]$$

You will recall that the difference between the two specific heats of a gas is equal to the gas constant for that gas. We have c_P and c_V for a gas mixture from equations (17) and (14), so we could expect a similar formula for R for a gas mixture.

Using equation (4):

$$p = p_a + p_b + p_c$$

we can insert each partial pressure in terms of T and V for the mixture:

$$p = \frac{T}{V}(m_aR_a + m_bR_b + m_cR_c) \tag{18}$$

This is equivalent to:

$$p = \frac{T}{V}(mR)$$

where m and R are the mass and gas constant for the mixture,

so: $$mR = m_aR_a + m_bR_b + m_cR_c$$

and so: $$R = \frac{m_aR_a + m_bR_b + m_cR_c}{m} \tag{19}$$

The relation between R, c_P and c_V will also be true for a gas mixture:

$$R = c_P - c_V$$

93

We can now calculate the gas constant for air, using the values of R = 259.8 J/kg K for oxygen, and R = 296.8 J/kg K for nitrogen. We shall assume we have 0.233 kg of oxygen and 0.767 kg of nitrogen:

$$R = \frac{0.233 \times 259.8 + 0.767 \times 296.8}{1} = 288.18 \text{ [J/kg K]}$$

This compares with 287.0 J/kg K given in Programme 4 and used in examples earlier in this programme. The value we have just obtained is approximate since we have taken only an approximate composition for air.

You can now confirm the more accurate value, using a more complete analysis for air, which consists of (per kg of air): 0.7553 kg of nitrogen, 0.2314 kg of oxygen, 0.0128 kg of argon (R = 208.1 J/kg K) and 0.0005 kg of carbon dioxide (R = 188.9 J/kg K).

94

$$\boxed{R = 287.04 \text{ J/kg K}}$$

Inserting data for the gases in the order given:
For air,

$$R = \frac{0.7553 \times 296.8 + 0.2314 \times 259.8 + 0.0128 \times 208.1 + 0.0005 \times 188.9}{1}$$

$$= \frac{224.17 + 60.12 + 2.66 + 0.094}{1} = 287.04 \text{ [J/kg K]}$$

Now, from Programme 4, Frame 82, we know that $n\,R_M = m\,R$ for any perfect gas where n is the number of moles and R_M is the universal gas constant, so we can use this for each gas in equation (18) of Frame 92:

$$p = \frac{T}{mv}(n_a R_M + n_b R_M + n_c R_M)$$

The sum of the moles of each gas must be the total moles in the mixture:

$$n = n_a + n_b + n_c$$

so for the mixture:

$$pV = n\,R_M T \qquad (20)$$

In the next frame we shall think about the idea of molecular mass for a mixture.

95

Each individual gas has its own molecular mass which is based on the actual mass of a molecule of that gas relative to the mass of the carbon-12 atom. Consequently a mixture of gases is a mixture of molecules of differing molecular masses. Taking an overall, or macroscopic, view, it is possible to find a molecular mass for the mixture, which although it does not represent the mass of any actual molecule in the mixture, it can be used to calculate properties of the mixture. Equation (20) in the previous frame can be expressed for one mole, so

$$pV = nR_MT \qquad (20)$$

for $n = 1$ becomes:

$$pV_M = R_MT \qquad (21)$$

where V_M is the molar volume, and then:

$$pV_M = MRT \qquad (22)$$

where M is defined as the molecular mass of the mixture.

96

We need to be able to calculate the molecular mass of a mixture of gases from the individual molecular masses of the constituent gases.

From equation (18) of Frame 92 we have:

$$p = \frac{T}{mv}(m_aR_a + m_bR_b + m_cR_c)$$

which may be expressed as:

$$p = \frac{T}{mv}(n_aR_M + n_bR_M + n_cR_M) \qquad (23)$$

$$= \frac{n\,R_MT}{mv}$$

$$= \frac{R_MT}{V_M} \text{ using equation (21) from the previous frame}$$

The molecular volume V_M is the product of mass, which is the molecular mass, and specific volume, so the previous result becomes:

$$p = \frac{R_MT}{Mv} \qquad (24)$$

Now we may compare the right-hand sides of equations (23) and (24):

$$\frac{T}{mv}(n_a R_M + n_b R_M + n_c R_M) = \frac{R_M T}{Mv}$$

and so by cancelling T, v and R_M we get:

$$M = \frac{m}{n_a + n_b + n_c}$$

and finally:

$$M = \frac{m}{\dfrac{m_a}{M_a} + \dfrac{m_b}{M_b} + \dfrac{m_c}{M_c}} \qquad (25)$$

Thus, the molecular mass of a mixture is the mass of the mixture, divided by the total number of moles of mixture, which is the sum of the individual masses of the constituents divided respectively by their own molecular masses.

97

We can now use equation (25) to calculate the molecular mass of air. Per kg we have 0.7553 kg of nitrogen ($M = 28$), 0.2314 kg of oxygen ($M = 32$), 0.0128 kg of argon ($M = 40$), and 0.0005 kg of carbon dioxide ($M = 44$).

Hence:

$$M = \frac{1}{\dfrac{0.7553}{28} + \dfrac{0.2314}{32} + \dfrac{0.0128}{40} + \dfrac{0.0005}{44}}$$

$$= \frac{1}{0.026\,975 + 0.007\,23 + 0.000\,32 + 0.000\,011}$$

$$= \frac{1}{0.034\,536}$$

$$= 28.955$$

The rounded-off value of the molecular mass of air is 29.

So far we have considered gas mixtures in terms of the partial pressures and masses of the constituent gases. However, the contents of a gas mixture are often expressed in terms of percentage composition by volume. Next we need to see how this may be done.

We can look again at our mixture of three gases occupying a volume V at a pressure p, with the individual gases having partial pressures of p_a, p_b and p_c. The same three gases can be considered in a closed system, but with each gas separated into its own volume, V_a, V_b and V_c. These can be thought of as partial volumes, and are shown in the figure in the next frame.

98

Previously we have had, from Frame 84:

$$p = p_a + p_b + p_c \tag{4}$$

where p_a, p_b and p_c are the partial pressures each gas would exert if each alone occupied the whole volume.

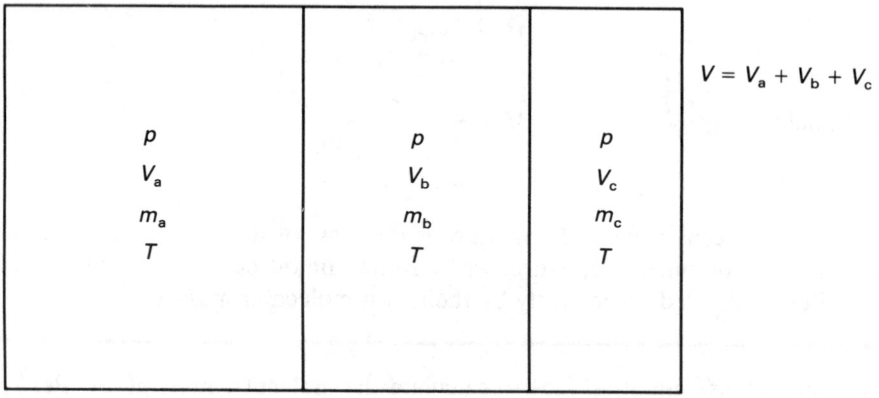

p	p	p
V_a	V_b	V_c
m_a	m_b	m_c
T	T	T

$V = V_a + V_b + V_c$

This gave us, from Frame 92:

$$p = \frac{T}{V}(m_aR_a + m_bR_b + m_cR_c) \tag{18}$$

Now we can say, using the ideal gas rule:

$$pV_a = m_aR_aT$$

so:

$$V_a = \frac{m_aR_aT}{p} \tag{26}$$

We will get similar results for V_b and V_c, and then we may add all three volumes together. This will give us:

$$V = V_a + V_b + V_c$$

$$= \frac{T}{p}(m_aR_a + m_bR_b + m_cR_c)$$

or:

$$p = \frac{T}{V}(m_aR_a + m_bR_b + m_cR_c) \tag{27}$$

which is the same result as equation (18).

99

We can now consider this problem:

We have 100 m^3 of gas at a pressure of 100 kPa, and at a temperature of 298 K. The composition by volume of the gas is: 30% carbon dioxide (CO_2), 12% hydrogen (H_2), 52% nitrogen (N_2) and 6% oxygen (O_2). Calculate (a) the masses of each gas present and the percentage composition by mass, (b) the molecular mass of the mixture, (c) the gas constant and (d) c_P, c_V and γ for the mixture. For the gases:

	M	R	c_P
Carbon dioxide	44	0.1889 [kJ/kg K]	0.846 [kJ/kg K]
Hydrogen	2	4.124	14.31
Nitrogen	28	0.2968	1.040
Oxygen	32	0.2598	0.918

For each gas:

$$pV = mRT$$

so, for the carbon dioxide, the volume is 30% of 100 m^3 which is 30 m^3, and:

$$\begin{aligned} m &= \frac{pV}{RT} \\ &= \frac{100 \times 30}{0.1889 \times 298} \\ &= 53.293 \text{ kg} \end{aligned}$$

Now calculate the masses of the remaining three gases.

100

The answers are: H_2 = 0.976 kg, N_2 = 58.793 kg and O_2 = 7.750 kg.

For the hydrogen the volume is 12 m^3, so:

$$m = \frac{100 \times 12}{4.124 \times 298} = 0.976 \text{ [kg]}$$

For the nitrogen and oxygen the volumes are 52 m^3 and 6 m^3, respectively, so the masses are:

$$\text{nitrogen: } m = \frac{100 \times 52}{0.2968 \times 298} = 58.79 \text{ [kg]}$$

$$\text{oxygen: } m = \frac{100 \times 6}{0.2598 \times 298} = 7.75 \text{ [kg]}$$

101

The sum of the four masses is 120.812 kg, so we divide each mass by the total and express as a percentage:

$$\% \text{ Carbon dioxide} = \frac{53.293}{120.812} \times 100 = 44.11\%$$

The remaining percentages are: 0.81% hydrogen, 48.67% nitrogen and 6.41% oxygen.

The molecular mass is found from equation (25) in Frame 96. Thus:

$$M = \frac{120.812}{\frac{53.293}{44} + \frac{0.976}{2} + \frac{58.793}{28} + \frac{7.750}{32}}$$

$$= \frac{120.812}{1.211 + 0.488 + 2.10 + 0.242}$$

$$= 120.812/4.041$$

$$= 29.897$$

Next you may calculate R and c_P from equations (19) and (17).

102

$$\boxed{R = 0.2777 \text{ kJ/kg K}, \ c_P = 1.0538 \text{ kJ/kg K}}$$

From equation (19) in Frame 92:

$$R = \frac{53.293 \times 0.1889 + 0.976 \times 4.124 + 58.793 \times 0.2968 + 7.75 \times 0.2598}{120.812}$$

$$= \frac{10.067 + 4.025 + 17.45 + 2.013}{120.812}$$

$$= 0.2777 \text{ [kJ/kg K]}$$

From equation (17) in Frame 90:

$$c_P = \frac{53.293 \times 0.846 + 0.976 \times 14.31 + 58.793 \times 1.04 + 7.75 \times 0.918}{120.812}$$

$$= \frac{45.086 + 13.967 + 61.145 + 7.115}{120.812}$$

$$= 1.0538 \text{ [kJ/kg K]}$$

As we now have R and c_P we can easily find c_V and γ:

$$c_V = c_P - R = 1.0538 - 0.2777$$
$$= 0.7761 \text{ [kJ/kg K]}$$

Then
$$\gamma = c_P/c_V = 1.0538/0.7761$$
$$= 1.3578$$

103

In dealing with gas mixtures we have met 27 separate equations, some of which were steps in developing the required relationships. The following is a list of 12 of those results which should be memorised:

Frame 82:
$$V = V_a = V_b = V_c \quad (1)$$

Frame 83
$$m = m_a + m_b + m_c \quad (2)$$

$$T = T_a = T_b = T_c \quad (3)$$

Frame 84
$$p = p_a + p_b + p_c \quad (4)$$

Frame 85
$$\frac{1}{v} = \frac{1}{v_a} + \frac{1}{v_b} + \frac{1}{v_c} \quad (6)$$

$$\rho = \rho_a + \rho_b + \rho_c \quad (7)$$

Frame 89
$$U = U_a + U_b + U_c \quad (8)$$

$$H = H_a + H_b + H_c \quad (9)$$

Frame 90
$$c_V = \frac{m_a\, c_{V_a} + m_b\, c_{V_b} + m_c\, c_{V_c}}{m} \quad (14)$$

$$c_P = \frac{m_a\, c_{P_a} + m_b\, c_{P_b} + m_c\, c_{P_c}}{m} \quad (17)$$

Frame 92
$$R = \frac{m_a\, R_a + m_b\, R_b + m_c\, R_c}{m} \quad (19)$$

Frame 96
$$M = \frac{m}{\dfrac{m_a}{M_a} + \dfrac{m_b}{M_b} + \dfrac{m_c}{M_c}} \quad (25)$$

104

There is also one more equation which reminds us that in addition to the concept of partial pressure, a gas mixture can be expressed in terms of the percentage mass and percentage volume of each constituent:

Frame 98: $$p = \frac{T}{V}(m_a R_a + m_b R_b + m_c R_c) \quad (27)$$

From equations (14), (17) and (19) the following relationships for single gases also apply for gas mixtures:

$$R = c_P - c_V$$

$$\gamma = c_P/c_V$$

$$c_P = \frac{\gamma R}{\gamma - 1}$$

$$c_V = \frac{R}{\gamma - 1}$$

105

Finally for this programme, we shall think about gas mixtures in steady flow. Gases can be made to mix adiabatically in steady flow in open systems. The figure shows a control surface through which pass two inlet streams, and from which is a single outlet stream.

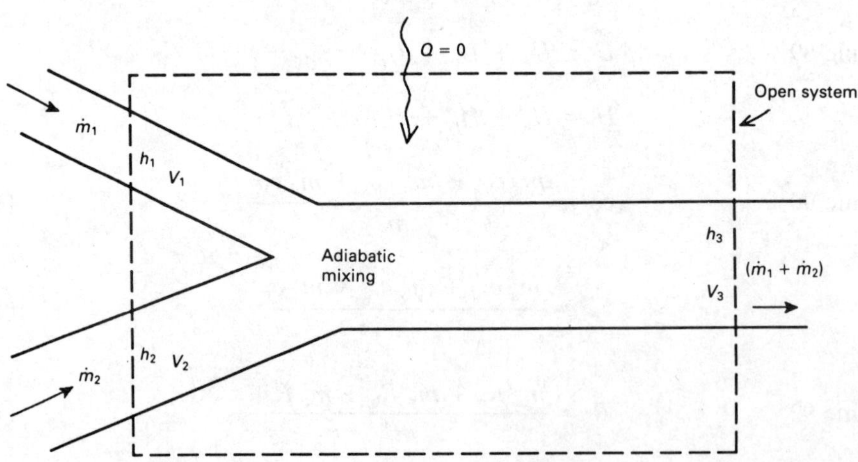

The steady flow energy equation, Frame 40, Programme 3, may be extended to cover this case. The process is adiabatic and there is no work term, in addition we may assume no change in potential energy, so the equation must say:

0 = enthalpy flow out − enthalpy flow in + kinetic energy of outlet stream − kinetic energy of inlet stream

In this problem we have two inlet streams, 1 and 2, and an outlet stream 3, so the enthalpy flow in is $(\dot{m}_1h_1 + \dot{m}_2h_2)$, where $\dot{m}$ is mass flow rate, and the enthalpy flow out is $(\dot{m}_1 + \dot{m}_2)h_3$, the kinetic energy of the inlet streams is $(\dot{m}_1V_1^2/2 + \dot{m}_2V_2^2/2)$, and the kinetic energy of the outlet stream is $(\dot{m}_1 + \dot{m}_2)V_3^2/2$ Putting these terms together we get:

$$0 = (\dot{m}_1 + \dot{m}_2)h_3 - (\dot{m}_1h_1 + \dot{m}_2h_2) + (\dot{m}_1 + \dot{m}_2)\frac{V_3^2}{2} - \left(\dot{m}_1\frac{V_1^2}{2} + \dot{m}_2\frac{V_2^2}{2}\right)$$

We will illustrate the use of this equation in a problem in the next frame.

106

In a gas-burning boiler, methane gas and air are supplied in separate ducts which enter the burner. The air flow is 20 kg/s at 450 K and 220 kPa pressure in a duct 400 mm diameter, and the methane flow is 1.5 kg/s at 300 K and 400 kPa pressure in a duct 70 mm diameter. Immediately after mixing the pressure is 210 kPa, in a duct 580 mm diameter. What is the temperature of the mixed gas stream?

For air, take $R = 0.287$ kJ/kg K, and $c_P = 1.0128$ kJ/kg K; and for methane take $R = 0.5197$ kJ/kg K, and $c_P = 2.226$ kJ/kg K.

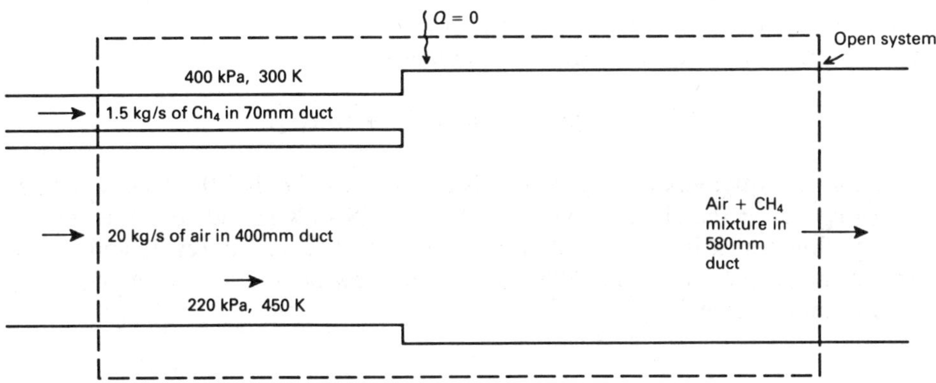

As we shall see in the next frame, this problem involves some ‘trial and error’.

107

This is because we do not know the outlet velocity V_3 for the outlet kinetic energy term. We need to know the temperature of the mixture at outlet, T_3 (this is the answer we want), in order to calculate V_3, so we first find T_3 neglecting the kinetic energy terms, then find V_3 and recalculate T_3 including the kinetic energy terms.

You can start to solve this problem by finding c_P and R (equation (17), Frame 90, and equation (19), Frame 92) for the mixture of 20 kg of air and 1.5 kg of methane. This quantity of mixture (21.5 kg) will leave the open system every second.

108

$$\boxed{c_P = 1.0974 \text{ kJ/kg K}, R = 0.3033 \text{ kJ/kg K}}$$

From equation (17): $$c_P = \frac{20 \times 1.0128 + 1.5 \times 2.226}{21.5} = 1.0974 \text{ [kJ/kg K]}$$

From equation (19): $$R = \frac{20 \times 0.287 + 1.5 \times 0.5197}{21.5} = 0.3033 \text{ [kJ/kg K]}$$

Now we will use the equation in Frame 105 to find the outlet temperature, T_3. Remember that in this equation, for a gas or gas mixture, $h = c_PT$, so after neglecting the kinetic energy terms we are left with:

$$(\dot{m}_1 + \dot{m}_2)c_PT_3 = (\dot{m}c_P)_1T_1 + (\dot{m}c_P)_2T_2$$

where for the outlet mixture $(\dot{m}_1 + \dot{m}_2)$ is 21.5 kg/s, and c_P is 1.0974 kJ/kg; and at inlet $(\dot{m}c_P)_1$ is for the air flow and is 20.0×1.0128 kJ/K s; and $(\dot{m}c_P)_2$ is for the methane flow which is 1.5×2.226 kJ/K s. The inlet temperatures T_1 and T_2 are given in the problem in Frame 106. You should now be able to use the equation above to find a value for T_3.

$$\boxed{T_3 = 428.79 \text{ K}}$$

Putting all the terms together we get:

$$21.5 \times 1.0974 \times T_3 = 20 \times 1.0128 \times 450 + 1.5 \times 2.226 \times 300$$

$$23.5941 \times T_3 = 9115.2 + 1001.7$$

so:

$$T_3 = 428.79 \text{ K}$$

This is the outlet temperature if we neglect the kinetic energy terms. Now we shall see the effect of including these terms. First we will calculate the velocity of the air stream. We have to use the continuity equation:

$$\dot{m} = \frac{A\,V}{v}$$

so we need the specific volume v and the cross-sectional area of flow A.

For air at 450 K and 220 kPa:

$$v = \frac{RT}{p} = \frac{0.287 \times 450}{220}$$

$$= 0.585 \text{ [m}^3\text{/kg]}$$

The cross-sectional area of the duct is:

$$A = \frac{\pi\,D^2}{4} = \frac{\pi \times 0.4^2}{4}$$

$$= 0.1257 \text{ [m}^2\text{]}$$

Then we have:

$$20 \text{ [kg/s]} = \frac{0.1257 \times V}{0.585}$$

$\therefore$

$$V = 93.4 \text{ [m/s]}$$

Following the same procedure, you can now find the specific volume and velocity of the methane.

110

$$\boxed{v = 0.3898 \text{ m}^3\text{/kg},\ V = 151.87 \text{ m/s}}$$

For the methane, $$v = \frac{0.5197 \times 300}{400}$$

$$= 0.3898 \text{ [m}^3\text{/kg]}$$

The duct area is: $$\frac{\pi\ 0.07^2}{4} = 0.003\ 85 \text{ [m}^2\text{]}$$

and for continuity: $$1.5 \text{ [kg/s]} = \frac{0.003\ 85 \times V}{0.3898}$$

so: $$V = 151.87 \text{ [m/s]}$$

We know the temperature of the mixture if we neglect the kinetic energy terms. Now you can use this temperature to find values for the specific volume and velocity of the mixture.

111

$$\boxed{v = 0.6193 \text{ m}^3\text{/kg},\ V = 50.4 \text{ m/s}}$$

For the mixture: $$v = \frac{RT}{p} = \frac{0.3033 \times 428.79}{210}$$

$$= 0.6193 \text{ [m}^3\text{/kg]}$$

The area for flow is: $$\frac{\pi\ D^2}{4} = \frac{\pi \times 0.58^2}{4}$$

$$= 0.2642 \text{ [m}^2\text{]}$$

Then, for continuity: $$\dot{m} = 21.5 = \frac{A\ V}{v} = \frac{0.2642\ V}{0.6193}$$

so: $$V = 50.4 \text{ [m/s]}$$

112

We can now take the full equation of Frame 105 to find a value for T_3 taking into account the kinetic energy terms. Remembering that the kinetic energy terms require 10^3 in the denominator, we get:

$$0 = 21.5 \times 1.0974 \times T_3 - (20 \times 1.0128 \times 450 + 1.5 \times 2.226 \times 300)$$

$$+ 21.5 \times \frac{50.4^2}{2000} - \left(20 \times \frac{93.4^2}{2000} + 1.5 \times \frac{151.87^2}{2000}\right)$$

so: $$0 = 23.594 \times T_3 - (9115.2 + 1001.7) + 27.307 - (87.236 + 17.298)$$

and: $$23.594 \times T_3 = 10\,194.127$$

$$T_3 = 432.06 \text{ K}$$

The value we got by neglecting kinetic energy was 428.79 K, in Frame 109; we will regard the value of 432.06 K as being the correct answer. To take the solution further we would need to recalculate the outlet specific volume and velocity, V_3, using $T_3 = 432.06$ K, and then find another value of T_3, which you would find to be very close to the present answer.

113

Another example of gas mixtures in steady flow is to be found in air conditioning. Atmospheric air contains water vapour in varying amounts. The water content is called the **humidity** of the air; this may be controlled in steady flow in air conditioning plant to provide a comfortable environment. Humid air is treated as a mixture of dry air and water vapour.

The ratio of the mass of water vapour, m_w, to mass of dry air, m_{air}, in any given volume of mixture is called the **specific humidity**, ω (Greek 'omega').

Thus: $$\omega = \frac{m_w}{m_{air}}$$

Air at any given temperature can hold only a certain amount of water before it becomes saturated.

114

The degree of saturation is defined by the **relative humidity**, ϕ (Greek 'phi'):

$$\phi = \frac{p_w}{p_s}$$

where p_s is the saturation pressure of water vapour at the prevailing air temperature, and p_w is the actual partial pressure of the water vapour. ω is also related to the actual partial pressure of the vapour. We have:

$$p_w V = m_w R_w T$$

and

$$p_{air} V = m_{air} R_{air} T$$

so:

$$\omega = \frac{m_w}{m_{air}} = \frac{R_{air} p_w}{R_w p_{air}}$$

Since $R_{air} = 0.2871$ kJ/kg K, $R_w = 0.4619$ kJ/kg K and $p_{air} = p - p_w$, where p is atmospheric pressure of the humid air, we have:

$$\omega = 0.622 \frac{p_w}{p - p_w}$$

An example in the next frame illustrates the use of these results.

115

Air at 1.013 25 bar pressure at 28°C has a relative humidity of 70%. What is the specific humidity?

At 28°C, $p_s = 0.037\ 78$ bar from steam tables.

$$\therefore \quad p_w = 0.7 \times 0.037\ 78 = 0.026\ 45 \text{ [bar]}$$

Then:

$$\omega = 0.622 \times \frac{0.026\ 45}{1.013\ 25 - 0.026\ 45}$$

$$= 0.016\ 67 \text{ [kg vapour/kg air]}$$

The partial pressure of water vapour in air cannot exceed the saturation pressure of water vapour at the temperature of the air. Thus for air at 28°C, p_w cannot exceed p_s which is 0.037 78 bar. If it has this value, then the relative humidity is 100%. As humid air is cooled at constant pressure, p_s drops and approaches p_w. The temperature at which they are equal is the **dew point temperature**, and further cooling leads to the condensation of vapour as water, or dew.

Humid air at a pressure of 1.0124 bar and a temperature of 30°C has a specific humidity of 0.020 08 [kg vapour/kg dry air]. What are the relative humidity and the dew point temperature?

116

74.63%, 25°C

We can find the partial pressure of vapour from the formula in Frame 114:

$$0.020\,08 = 0.622 \times \frac{p_w}{1.0124 - p_w}$$

Hence: $$p_w = 0.031\,66 \text{ [bar]}$$

At 30°C, $p_s = 0.042\,42$ bar, so $\phi = 0.031\,66/0.042\,42 = 74.73\%$
From your tables you can find that $p_s = 0.031\,66$ bar at 25°C, so the dew point temperature is 25°C.

117

To illustrate the use of these ideas we shall consider an air conditioning plant. Examples of this type are rather long, but if we take it step by step, you should find there is nothing very difficult about it.

Atmospheric air at 1.013 25 bar and at 35°C with 85% relative humidity passes into an air conditioning plant to be delivered at 25°C with 55% relative humidity. This is achieved in two stages: (a) by cooling the air to the dew point temperature of the output, and then (b) by reheating to 25°C. The fan power input is 0.3 kJ/kg of air flow. Calculate the heat transfers in the cooling and reheating.

This method of cooling followed by reheating means that the required vapour content in the output is achieved in the cooling process, and the vapour content is constant in the second stage reheating. A diagram of the plant is given in the next frame.

118

A diagram of the plant showing the two stages of cooling and reheating is given below.

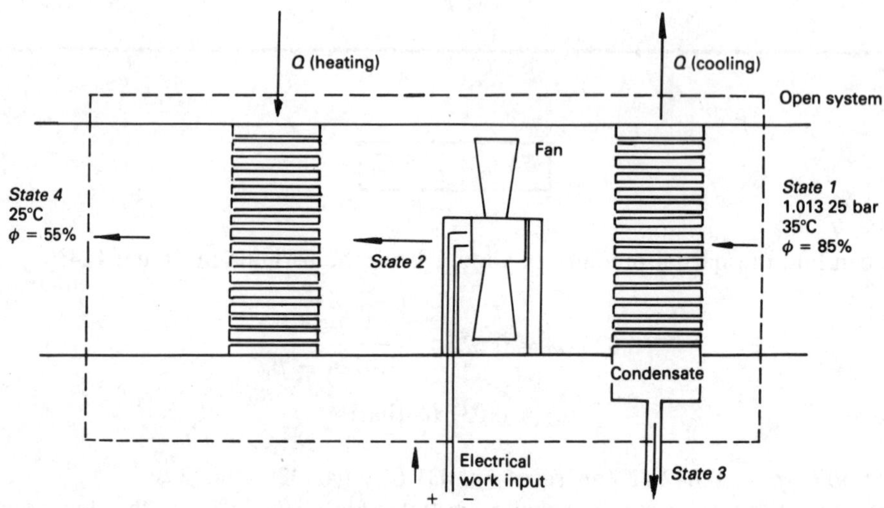

We start by finding the partial pressure of the water content, p_w, at the outlet at position [4]. This will also be p_w at position [2]. At the saturation temperature of 25°C, p_s is 0.031 66 bar, so p_w at [4] is 0.55 × 0.031 66 = 0.017 41 bar. What is the saturation temperature corresponding to this pressure?

119

$$\boxed{T_s = 15.33°\text{C}}$$

The figure for T_s is found by interpolation. From the Rogers and Mayhew tables:

$$T_s = 15 + (16 - 15) \times \frac{(0.017\,41 - 0.017\,04)}{(0.018\,17 - 0.017\,04)}$$

$$= 15.33°\text{C}$$

This is the temperature to which the air is cooled at position [2]. Between positions [1] and [2] the air is cooled from 35°C, with ϕ = 85%, to 15.33°C, with ϕ = 100%. There will be two outlets for this open system: the saturated air at

15.33°C, at [2], and the water that has condensed out in the cooling process, at [3]. Next we need to calculate the vapour content, ω, at positions [1] and [2].

At [1] $\quad \phi = 85\%$, and p_S at 35°C is 0.056 22 bar

so $\quad p_W = 0.85 \times 0.056\,22$

$= 0.047\,79$ [bar]

Then, $\omega = 0.622 \times \dfrac{0.047\,79}{1.013\,25 - 0.047\,79}$ (using the equation for ω from Frame 114)

$= 0.030\,79$ [kg vapour/kg dry air]

Using the same procedure, what is ω at position [2]?

120

$\omega = 0.010\,87$ [kg vapour/kg dry air]

At [2], $p_W = 0.017\,41$ bar, from Frame 118, so:

$$\omega = 0.622 \times \frac{0.017\,41}{1.013\,25 - 0.017\,41}$$

$= 0.010\,87$ [kg vapour/kg dry air]

Hence between these points 0.030 79 − 0.010 87 = 0.019 92 kg of vapour is removed as water per kg of dry air. This is the condensate at position [3].

We can write the steady flow energy equation for the open system bounded by positions [1], [2] and [3]:

$$\dot{Q} - \dot{W} = (\dot{m}_W)_2\, h_g + \dot{m}_{air}\, c_P\, T_2 + (\dot{m}_W)_3\, h_f - (\dot{m}_W)_1\, h_1 - \dot{m}_{air}\, c_p\, T_1$$

where for unit mass flow of dry air, $\dot{W}$ = work input = −0.3 kJ/kg, $\dot{m}_{air}$ = 1 kg/s, T_1 = 35°C = 308.15 K, T_2 = 15.33°C = 288.48 K, $(\dot{m}_W)_1$ = 0.030 79 kg/s, $(\dot{m}_W)_2$ = 0.010 87 kg/s, and $(\dot{m}_W)_3$ = 0.019 92 kg/s. In addition, h_g at [2] and h_f at [3] are both at 15.33°C. At [1], p_W = 0.047 79 bar which is the saturation pressure for what temperature?

121

$$\boxed{T_s = 32.09°C}$$

Interpolating between T_s of 32°C and 34°C in the Rogers and Mayhew tables, we have:

$$T_s = 32 + 2 \times \frac{(0.047\,79 - 0.047\,54)}{(0.053\,18 - 0.047\,54)}$$

$$= 32.09°C$$

At [1], therefore, we have water vapour of pressure 0.047 79 bar with (35 − 32.09) = 2.91 K of superheat. The specific enthalpy at this state is h_g at 32.09°C plus enthalpy due to 2.91 K of superheat. From page 2 of the Rogers and Mayhew tables:

$$h_g \text{ at } 32.09°C = 2559.3 + (2562.9 - 2559.3) \times \frac{0.09}{2} = 2559.5 \text{ kJ/kg.}$$

To obtain the superheat h we may add on $c_P \times (35 - 32.09)$, where c_P for water vapour is 1.886 kJ/kg K. Hence at [1]:

$$h = 2559.5 + 1.886 \times 2.91 = 2564.99 \text{ kJ/kg.}$$

At 15.33°C at position [2] $h = h_g$, and at [3] $h = h_f$ at 15.33°C. Using data from page 2 we have:

$$h_f = 62.9 + 0.33 \times (67.1 - 62.9) = 64.29 \text{ kJ/kg, and}$$
$$h_g = 2528.4 + 0.33 \times (2530.2 - 2528.4) = 2528.99 \text{ kJ/kg}$$

Finally, for air c_P = 1.0052 kJ/kg K. Next we can find Q from the open system bounded by [1], [2] and [3].

122

By putting all the terms into the steady flow energy equation we have:

$$\dot{Q} - (-0.3) = 0.010\,87 \times 2528.99 + 1.0 \times 1.0052 \times 288.48 + 0.019\,92 \times 64.29 - 0.030\,79 \times 2564.99 - 1.0 \times 1.0052 \times 308.15$$

$$= 27.490 + 289.980 + 1.281 - 78.976 - 309.752$$

$= -69.977$ [kJ/kg of dry air]

Hence: $\dot{Q} = -70.277$ [kJ/kg of dry air]

Between [1] and [2] this is a heat transfer from the open system, so that as we expect, cooling has taken place.

The energy equation for the open system between positions [2] and [4] is:

$$\dot{Q} = (\dot{m}_W)_4\, h_4 + \dot{m}_{air}\, c_P\, T_4 - (\dot{m}_W)_2 h_g - \dot{m}_{air}\, c_P\, T_2$$

where $(\dot{m}_W)_4 = (\dot{m}_W)_2$, and h_4 is (25 − 15.33) K of superheat above h_2 which is h_g at 15.33°C, and T_4 is 25°C which is 298.15 K.

Hence: $\dot{Q} = 0.010\,87 \times (2528.99 + 1.866 \times [25 - 15.33]) + 1.0 \times 1.0052 \times 298.15$

$- 0.010\,87 \times 2528.99 - 1.0 \times 1.0052 \times 288.48$

so: $\dot{Q} = 27.686 + 299.70 - 27.490 - 289.980$

$\dot{Q} = 9.916$ [kJ/kg of dry air]

This positive heat transfer to the air is the second heat transfer required per kg of dry air to condition the air to the state specified. For each kg of air, 0.019 92 kg of water is removed. A problem similar to the one we have just completed may be found at problem **14** in the following frame.

123

PROBLEMS INVOLVING PERFECT GASES IN CLOSED AND OPEN SYSTEMS

In the following problems, when air is involved, you are given that $c_P = 1.0052$ kJ/kg K and $c_v = 0.718$ kJ/kg K. R and γ, if needed, may be calculated from these values.

1. Air at 3000 kPa pressure and at 1400 K temperature undergoes a constant volume process so that the final temperature is 650 K. Calculate (a) the final pressure and (b) the heat transfer per kg of air.

[(a) 1392.86 kPa, (b) −538.5 kJ/kg]

2. A quantity of air has a mass of 0.05 kg and the pressure and temperature are 100 kPa and 310 K respectively. Stirring work is done on the gas by an impeller on a shaft passing through the system boundary. The torque on the shaft is 20 Nm. The shaft rotates at 500 rev/min for 35 seconds. The volume of the gas does not change and a heat transfer of 25 kJ from the gas takes place. Calculate

(a) the volume of the gas, (b) the final temperature and (c) the final pressure. [(a) 0.0445 m^3, (b) 634.57 K, (c) 204.7 kPa]

3. Air having a volume of 0.02 m^3 is expanded at constant pressure to a volume of 0.05 m^3. The pressure is 90 kPa and the initial temperature is 300 K. Calculate (a) the mass of air, (b) the displacement work, (c) the final temperature and (d) the heat transfer. [(a) 0.0209 kg, (b) +2.7 kJ, (c) 750 K, (d) +9.453 kJ]
4. Air having a volume of 0.04 m^3 , a mass of 0.0961 kg and a temperature of 290 K expands at constant pressure to a volume of 0.12 m^3. An electric coil passes through the system boundary, and the coil receives a current of 5 amps at 12 volts for 2 minutes so that electrical work is done on the air during the expansion, while there is also a heat transfer. Calculate (a) the value of the constant pressure and (b) the value of the heat transfer. [(a) 200 kPa, (b) +48.82 kJ]
5. A quantity of air having a volume of 0.08 m^3, a mass of 0.0989 kg and a pressure of 110 kPa is compressed adiabatically through a volume ratio of 10 to 1. Calculate (a) the initial temperature, (b) the final pressure and temperature and (c) the work done on the air. [(a) 310 K, (b) 2763.17 kPa, 778.7 K, (c) −33.263 kJ]
6. The same mass of air at the same initial conditions in problem **5** is compressed polytropically with an index of $n = 1.3$ through the same volume ratio of 10 to 1. Calculate (a) the final pressure and temperature, (b) the work done and (c) the heat transfer. [(a) 2194.85 kPa, 618.5 K, (b) −29.196 kJ, (c) −7.289 kJ]
7. Again the same mass of air at the same initial conditions is now compressed isothermally through the same volume ratio of 10 to 1. Calculate (a) the final pressure and (b) the work and heat transfers. [(a) 1100 kPa, (b) $W = -20.263$ kJ, $Q = -20.263$ kJ]
8. Products of combustion in an engine cylinder consist of 9.565% CO_2, 5.783% O_2, 75.31% N_2 and 9.342% H_2O vapour, by volume. Treating as a mixture of perfect gases, calculate (a) the percentage composition by mass and (b) R and c_v for the mixture. An expansion takes place over a volume ratio of 1 to 10 with a polytropic index of 1.52, from an initial condition of a pressure of 12 500 kPa and a temperature of 2150 K. Calculate (c) the final pressure and temperature and (d) the work and heat transfers per unit mass of gas mixture. For the gases, in kJ/kg K units:

	R	c_V	
CO_2	$R = 0.1889$,	$c_V = 1.0451$	
O_2	0.2598	0.8302	
N_2	0.2968	0.8702	
H_2O	0.4619	1.8261	(treating as a gas)

[(a) CO_2 14.6%, O_2 6.42%, N_2 73.15%, H_2O 5.83%, (b) $R = 0.2883$, c_V = 0.9489 kJ/kg K, (c) 377.49 kPa, 649.29 K, (d) $W = +832.07$ kJ/kg, $Q = -591$ kJ/kg]

9. A gas expands against a spring-loaded piston and the pressure–volume relationship is $pV^{-1} = k$. Initially $p_1 = 110$ kPa, $V_1 = 0.004$ m^3 and $T_1 = 320$ K. During the expansion, stirring work is done on the gas, amounting to 2.5 kJ. The final volume is 0.008 m^3. Calculate (a) the mass of gas, (b) the final pressure and temperature, (c) the net work done by the gas and (d) the heat transfer. For the gas, $R = 0.189$ kJ/kg K, $c_V = 0.657$ kJ/kg K.
[(a) 0.007 275 kg, (b) 220 kPa, 1280 K, (c) -1.84 kJ, (d) $+2.7485$ kJ]

10. Gas of mass 2 kg consists of 0.5 kg of hydrogen, 1.2 kg of carbon monoxide and 0.3 kg of nitrogen. The total volume is 5 m^3, and the temperature is 350 K. Calculate (a) the partial pressures of each gas and the pressure of the mixture and (b) the specific volume and density of each gas present. Values of R are: H_2 4.124, CO 0.2968, N_2 0.2968 kJ/kg K.
[(a) $p_{H_2} = 144.34$ kPa, $p_{CO} = 24.93$ kPa, $p_{N_2} = 6.23$ kPa, $p = 175.5$ kPa. (b) For H_2 $v = 10$ m^3/kg, $\rho = 0.1$ kg/m^3; for CO $v = 4.167$ m^3/kg, $\rho = 0.24$ kg/m^3; for N_2 $v = 16.674$ m^3/kg, $\rho = 0.06$ kg/m^3]

11. Calculate c_P, c_V and γ for the mixture of question **10**, and the change of enthalpy and internal energy for a temperature rise of 150 K. Values of specific heats are : H_2, $c_P = 14.43$, $c_V = 10.306$; CO, $c_P = 1.043$, $c_V = 0.746$; N_2, $c_P = 1.043$, $c_V = 0.746$ kJ/kg K.
[$c_P = 4.3895$ kJ/kg K, $c_V = 3.1357$ kJ/kg K, $\gamma = 1.3998$; $\Delta H = 1316.85$ kJ, $\Delta U = 940.71$ kJ]

12. A gas mixture consists by mass of 70% CO_2, 8% H_2, 12% CH_4 and 10% O_2. Calculate (a) the percentage composition by moles and (b) the molecular mass of the mixture. Values of molecular mass are: CO_2 44, H_2 2, CH_4 16 and O_2 32.
[(a) CO_2 23.91%, H_2 60.12%, CH_4 11.27%, O_2 4.7%, (b) $M = 15.03$]

13. Humid air of pressure 1.0125 bar and at 24°C has a relative humidity of 73.64%. What are (a) the specific humidity and (b) the dew point temperature?
[(a) 0.013 79 kg/kg, (b) 19°C]

14. Atmospheric air at 1.0133 bar pressure and at 30°C has a 90% relative humidity. An air conditioning requirement involves 4 kg/s of air at 20°C with 60% relative humidity. The air flow is to be cooled to the dew point temperature of the stated output and then reheated to 20°C. Fan power is 1.5 kW. Calculate the magnitude of the two heat transfer rates, and the rate of condensation extraction. For water vapour, $c_P = 1.866$ kJ/kg K, for air $c_P = 1.0052$ kJ/kg K.
[-231.19 kW, $+32.63$ kW, 0.0625 kg/s]

124

Points to remember

In the expansion of steam, n in the relationship $pV^n = k$ for an **adiabatic process** depends on the initial condition of the steam.

For initially dry saturated steam, $n = 1.135$

For initially superheated steam, $n = 1.3$

For **perfect gases:** $R = c_P - c_V$ and $\gamma = c_P/c_V$

$$c_P = \frac{\gamma R}{\gamma - 1} \qquad c_V = \frac{R}{\gamma - 1}$$

For **adiabatic processes:** $pv^\gamma = k$

Work of expansion or compression $= \dfrac{p_2v_2 - p_1v_1}{1 - \gamma}$ where v = specific volume

$$\frac{T_2}{T_1} = \left(\frac{p_2}{p_1}\right)^{\frac{\gamma - 1}{\gamma}}; \quad \frac{T_2}{T_1} = \left(\frac{v_2}{v_1}\right)^{1-\gamma} = \left(\frac{v_1}{v_2}\right)^{\gamma - 1}$$

For **polytropic processes:** $pv^n = k$

Work of expansion or compression $= \dfrac{p_2v_2 - p_1v_1}{1 - n}$

$$\frac{T_2}{T_1} = \left(\frac{p_2}{p_1}\right)^{\frac{n - 1}{n}}; \quad \frac{T_2}{T_1} = \left(\frac{v_2}{v_1}\right)^{1-n} = \left(\frac{v_1}{v_2}\right)^{n - 1}$$

$$Q = \left(c_V + \frac{R}{1 - n}\right)(T_2 - T_1) \text{ for unit mass}$$

For **isothermal processes:** $pv = k$

$$Q = W = p_1v_1 \ln \frac{v_2}{v_1}$$

See Frame 103 for a summary of relationships for **gas mixtures**.

Specific humidity: $\omega = \dfrac{m_W}{m_{air}} = 0.622 \dfrac{p_W}{p - p_W}$ [kg water vapour/kg dry air]

Relative humidity: $\phi = p_W/p_S$, where p_W = partial pressure of water vapour; p_S = saturation pressure of water vapour at the prevailing temperature of the humid air; and p = atmospheric pressure of the humid air.

Programme 6

THE SECOND LAW: THERMODYNAMIC TEMPERATURE AND ENTROPY

1

In Programme 1 we said that in studying Thermodynamics we have to take a macroscopic view. This is because some of the forms of energy we have to consider can only be treated in terms of macroscopic ideas. Internal energy of a fluid depends on the translational and vibrational energy of countless myriads of tiny molecules, and we can only get a picture of the state of that fluid by using the overall concepts of temperature and pressure. Since internal energy is that mode of energy directly affected by a heat interaction, heat itself is an energy transfer that depends on random molecular activity, which again can only be viewed with macroscopic ideas. Even though similar macroscopic ideas, such as force (being pressure × area) are used describe a work transfer, heat and work, as interactions at a system boundary, are fundamentally different.

2

Consider a bucket of water. It would be possible in two separate experiments to arrange heat and work interactions of equal magnitude. Suppose we have water of mass 4 kg at 20°C at ground level. We allow a heat transfer of 30 kJ to take place. Neglecting the bucket, and taking the specific heat of water as 4.186 kJ/kg K, we have:

$$30\ [\text{kJ}] = 4\ [\text{kg}] \times 4.186\ [\text{kJ/kg K}] \times (\Delta T)\ [\text{K}]$$

$$\therefore \quad (\Delta T) = 1.792\ [\text{K}]$$

So the water is warmed by 1.792 degrees. Now imagine that a work transfer of 30 kJ takes place; to what height above ground level would this raise the water? (Again we may neglect the bucket.)

3

764 m

We have: $30\ [\text{kJ}] = 4\ [\text{kg}] \times 9.81\ [\text{m/s}^2] \times \text{H}\ [\text{m}] \times 10^{-3}\ [\text{kJ/J}]$

$$\therefore \quad H = 764\ \text{m}$$

The two separate interactions of equal magnitude have produced strikingly different results. We can say that the work transfer which has given an increase in potential energy of the water was an **ordered** energy transfer, while the heat transfer which has only increased the internal energy of the water was a **disordered** energy transfer. The water will cool down and slightly warm the surrounding air; in this way the **disorder will increase**. If the bucket of water is secure at a height of 764 m it will stay there until some further action disturbs it. It has potential energy which is an **ordered** form of energy. If the water is allowed to fall or flow back to ground level it will acquire kinetic energy which is again an **ordered** form of energy; both forms are capable of producing work directly, as in turning a waterwheel or turbine. If the water simply falls to the ground, its original potential energy will become internal energy as the water and its surroundings are slightly warmed, so the **ordered** energy form has become **disordered**.

4

In Thermodynamics we are concerned with how to obtain the **ordered** interaction of work from the **disordered** interaction of heat. In the environment, work and ordered forms of energy are produced in thermodynamic processes from the disordered energy form of heat from the sun. Thus water is transfered from the sea to the tops of hills and mountains through the processes of evaporation, condensation and rain. Work has been done on that water and its potential energy has increased; it is now capable itself of doing work in flowing back to the sea in rivers or in man-made ducts. Other water has remained in the sea; it has simply become warm from the heat of the sun and it is not capable of doing work. In man-made machines, work is done and both the ordered and disordered forms of energy flow in a controlled way. In nature these things are uncontrolled and just arise from the particular set of conditions at the time. Can you think of other examples of potential and kinetic energy present in the environment, in an uncontrolled way?

5

The wind – kinetic energy
Waves breaking on the beach – potential and kinetic energy
Moving glaciers – potential and kinetic energy
Rocks and lava thrown from a volcano – kinetic energy

6

The first two examples have been used to obtain work; there is also the potential energy of the tide, harnessed as tidal energy, but this is controlled and ordered by natural forces in a way that other examples are not.

There is a natural degradation of the ordered energy forms to the disordered. As we have seen, as falling water comes to rest its original potential energy has become internal energy. (In Joule's studies of heat, work and energy, he actually tried to measure a temperature rise at the base of a waterfall, but the cooling effect of the evaporation of spray prevented him getting meaningful results.) Whenever work is done in driving machinery, or in powering cars, trains or aircraft, the energy expended ultimately becomes internal energy of some part of a system or its surroundings.

The family car is taken from its garage on a round trip of some miles and after the expenditure of several gallons of petrol finds itself back where it started in the garage. What has happened to the chemical energy of the petrol that has been burnt?

7

Some of the chemical energy in the petrol has enabled the car's engine to do work to overcome inertia and friction to make the car move. Ultimately all the energy of the fuel has become internal energy of parts of the car, the road surface and the air through which the car has passed.

Since there is a natural degradation of ordered forms of energy to the disordered, in trying to obtain work from heat we are working against this natural process. The natural degradation also means that in the disordered forms of energy the 'level of disorder' increases as temperatures fall. Thus a block of metal at a low temperature is 'more disordered' than one having a higher temperature. There is a natural tendency for the 'less disordered state' to proceed spontaneously to the 'more disordered state'. All real processes are irreversible and in any process there is always an increase in disordered energy. In this programme we are going to see how, in some measure, we can work against the natural increase of disorder, and obtain work from heat.

8

As we have seen, obtaining heat from work is very easy. Joule showed the equivalence of these two interactions in his famous experiment, which was described in Frame 100 of Programme 2. Work performed on water by stirring it will raise the temperature of the water; this is followed by a heat transfer to restore the water to its original temperature. Work could be performed on water by any number of different and complicated means (even by shaking it), but the effect would be the same.

In previous programmes we have seen how work may be obtained from the expansion of a fluid, say steam, in an engine cylinder. To obtain work in a continuous way we have to keep on repeating this process with each revolution of the engine crankshaft. A device that enables us to obtain work in a continuous way is called a **heat engine**. A heat engine may be defined:

A heat engine is a continuously operating system at the boundary of which there are continuous heat and work interactions.

Notice that this definition does not specifically say the work interaction is a positive work output. This allows the possibility of a heat engine on which negative work is done.

To achieve a continuous repetition of an expansion process of steam in an engine, we need a supply of steam from a boiler. After each expansion process, what happens to the steam?

9

It either goes to waste or is condensed to water to be returned to the boiler.

In the reciprocating steam engine (mostly preserved examples nowadays), the expanded steam usually goes to a condenser, but in the case of a locomotive it goes to waste. If in a modern machine an alternative vapour is being used, it will be carefully contained. In steam turbine power plant for the generation of electricity, the expansion of steam is a continuous rather than a repetitive process and the expanded steam goes to a condenser. We first spoke of condensers in Frame 43 of Programme 5. The steam is condensed to water which is returned to the boiler; it transfers its enthalpy to a supply of cold water which is heated in the process. The quantity of energy involved can be very large, and increasingly today combined heat and power (CHP) plant is being built so that this 'waste heat' can be usefully employed in, for example, a district heating scheme.

10

But isn't all this rather unsatisfactory from the point of view of producing work? Could we not take the expanded steam from the engine or turbine and instead of condensing it, reheat it to the required condition to be used again, and so save a lot of fuel?

11

Suppose steam is supplied from a boiler to an engine cylinder at state 1 in the diagram. It expands to state 2 doing useful work. This is repeated many times a second and power is produced.

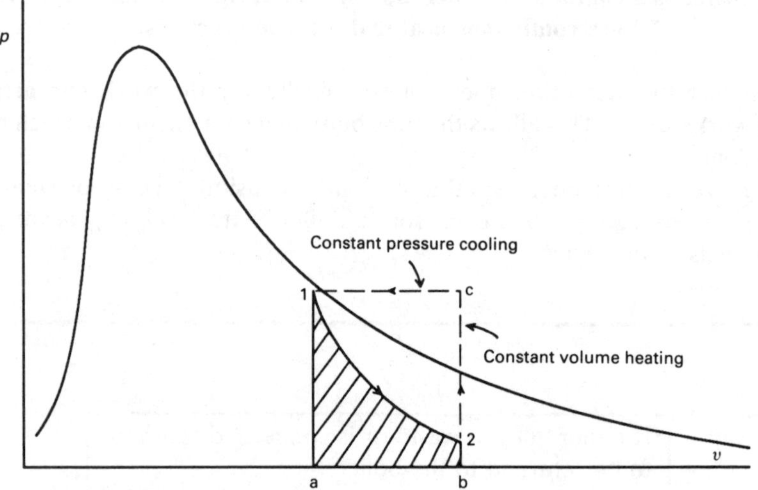

The used steam is at low pressure and its specific volume is relatively high. Suppose we arrange to heat it at constant volume to raise the pressure to the original value as shown as process 2–c, and then cool it at constant pressure in process c–1 to reach the original specific volume. The steam is then back in the state required to enter the cylinder again for another power-producing expansion. In the constant pressure process you can see that displacement work is done which is the area c–1–a–b–c, and this is larger than the work obtained in the expansion which is 1–2–b–a–1. If we consider the steam at state 1 as a closed system undergoing the cycle 1–2–c–1, we would find that we had performed on it negative work given by the area 1–2–c–1 followed by a negative heat transfer of magnitude equal to the constant pressure cooling less the constant volume heating. This whole sequence would be equivalent to a Joule experiment in which heat is produced from work.

Thus we would have **done** work; we would not have **obtained** any work with which we could do something useful.

12

We could try and invent other ideas to re-use expanded steam or other working fluids in engines, but however clever we thought we were being, we would always find in the end that our machine required net work to be done on it from which it produced a supply of heat. This of course is the opposite of what we are trying to do. In a steam engine we cannot escape the necessity of having to condense the steam and then to pump the condensate back into the boiler. This way very little negative work is involved so we then have a device for producing positive work from the heat from the boiler. The cost of doing this is the necessity of having to throw away a lot of heat from the condensation of the steam (or of having to find another use for it). This fact is embodied in a statement of the **Second Law of Thermodynamics**. There are other statements one of which we shall meet later; this one is known as the Kelvin–Planck statement:

It is impossible to construct a cyclic process whose sole result is the flow of heat from a single heat reservoir and the performance of an equivalent amount of work.

13

The 'cyclic process' takes place in a heat engine, and this law states that it is impossible to build such an engine that has an efficiency of 100%. We deduce this because, in Programme 1 we spoke of efficiency as being a measure of the net work obtained as a fraction of the heat that is supplied. As only one heat reservoir is involved (the hot gases from a boiler furnace, for example), it is implied that all the heat is converted into work, since in a cyclic process $Q = W$ for the First Law to be satisfied. Thus we cannot do away with a second heat reservoir such as cooling water from a river. Whatever type of engine we have in mind, it must reject heat as well as receive heat.

A heat engine operating between two heat reservoirs is shown below.

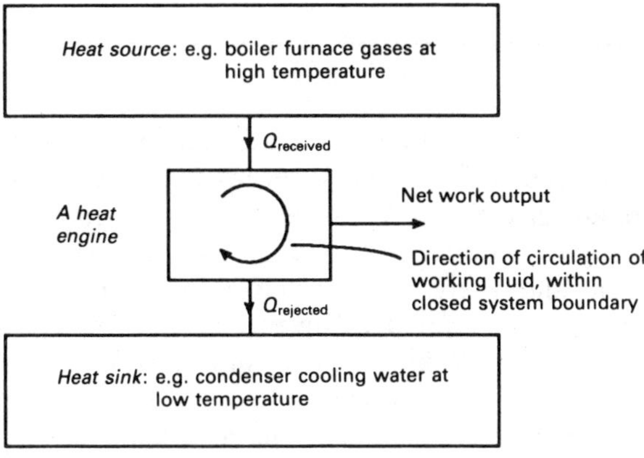

14

The statement of the Second Law in Frame 12 tells us that a heat engine has to operate between a high temperature heat source from which it receives heat, and a low temperature heat sink to which it rejects heat. Within the closed system boundary of this representation of a work-producing heat engine the working fluid is circulating, by convention, in a clockwise direction. As a result of the interactions, net positive work is obtained.

For a steam power plant, the contents of the closed system boundary are shown in the figure below.

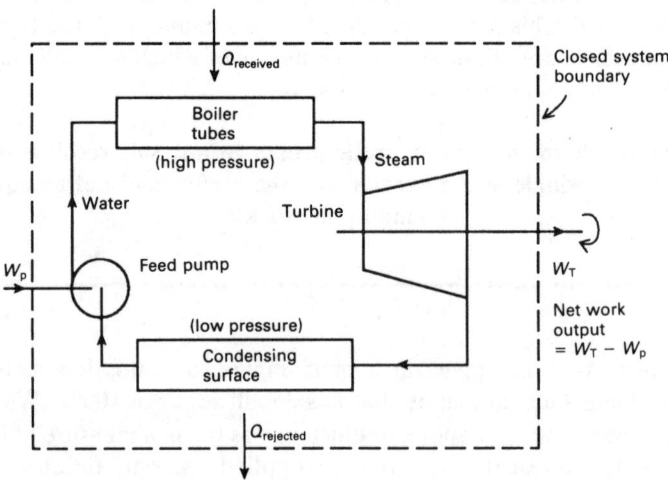

The working fluid flows round within the boundary passing from the boiler to the turbines (or to a reciprocating steam engine for a small output), to the condenser, to the feed pump, and back to the boiler. Heat transfers are *to* the boiler at high pressure, and *from* the condenser at low pressure, and the work output is the difference between the positive work from the turbines and the small negative work of the feed pump.

15

It is equally possible for a heat engine to operate in the reverse direction. It is then called a **reversed heat engine**. Everything is reversed: the net work is negative, and both heat transfers are in the opposite direction. This means that where heat is rejected, the temperature within the closed system must be hotter than the high temperature reservoir, and where heat is received the temperature must be colder than the low temperature reservoir. By convention, in the representation below of a reversed heat engine, the working fluid is shown flowing in an anticlockwise direction. When work is done on a reversed heat engine there is a net transfer of heat from the low temperature reservoir to the high temperature reservoir.

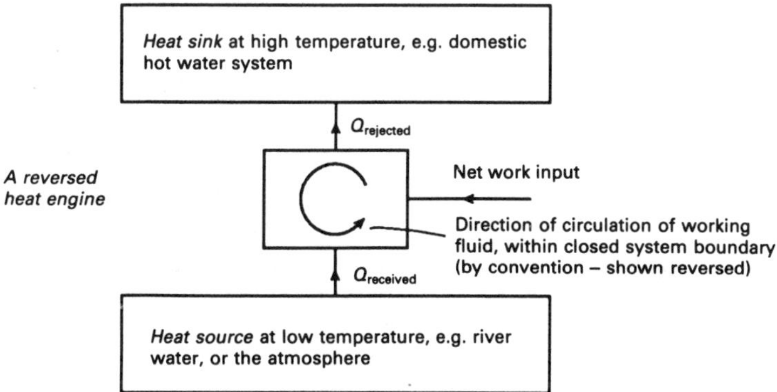

If everything is reversed in a **reversed heat engine**, will boiling and condensation in the cycle take place at high or low pressure?

16

Boiling – low pressure
Condensation – high pressure

Heat transfer to the cycle is at low temperature and pressure; this is where boiling occurs; and heat transfer is from the cycle at the high temperature and pressure condensation.

Heat pumps and **refrigerators** are examples of a reversed heat engine. The principle of each is the same. In a heat pump, heat extracted from a river or the atmosphere may be used to heat a house. In some food supermarkets, heat extracted from freezers is used to supply a heating requirement.

In comparison with the figure in Frame 14, the contents of the closed system boundary of a reversed heat engine are shown below.

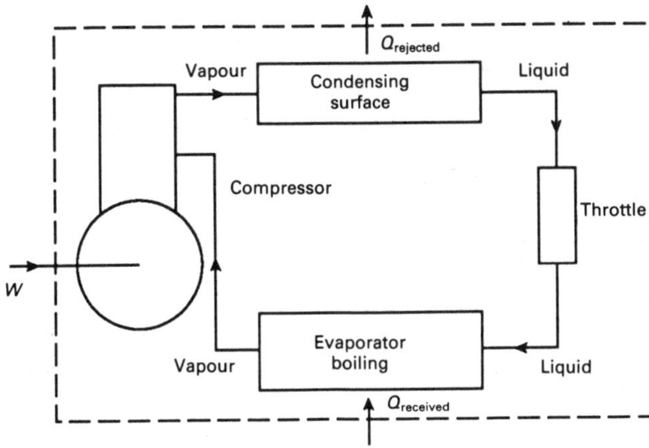

17

In a reversed heat engine we see that the net result of work being done is for heat to flow from a lower temperature to a higher temperature. The fact that heat will not do this unaided is the basis of the **Clausius statement of the Second Law**:

It is impossible for any device to cause heat to flow from one region to another at a higher temperature, except by performing work on that device.

18

We now need to look at some of the characteristics of heat engines, in order to establish how efficient they can be. We now know that we have to reject some heat, but how much?

In Programme 2 we discussed some ideas about reversible work transfers in the form of expansions and compressions, and also how in an idealisation we might achieve reversible heat transfers. Thus, in Frame 35 of Programme 2 we saw that reversible expansions and compressions are superior to irreversible ones, and in Frame 118 we saw that a reversible heat transfer is 100% efficient. Remember though, that to be truly reversible these processes would be infinitely slow and of no practical use. However, we can postulate that we have designs for some heat engines that operate entirely from reversible processes, though it is not actually possible to make them.

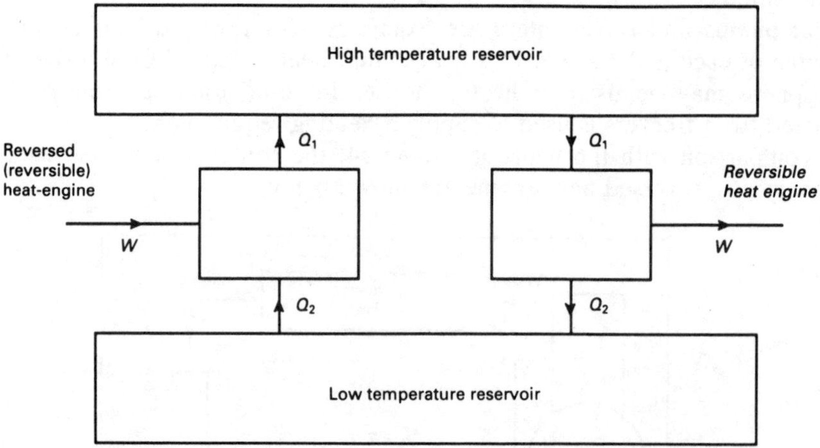

The diagram shows a reversible heat engine operating between two heat reservoirs, each at a constant temperature, with heat transfers Q_1 to the engine, Q_2 from the engine, and a work output of W which by the First Law must be $Q_1 - Q_2$. If a reversible heat engine is operated backwards as a reversed heat engine, which is also shown, between the same temperatures, then the heat and work transfers must have the same values as before, but be of opposite sign.

The efficiency of a heat engine, whether it is reversible or not, is the ratio of the work output, which is what we are seeking, to the heat input, which is what we have to pay for. What is the efficiency of the heat engine in terms of Q_1 and Q_2?

19

$$\boxed{\frac{Q_1 - Q_2}{Q_1}}$$

By the First Law the work output is $Q_1 - Q_2$, and the heat supplied is Q_1, so the **efficiency** must be $(Q_1 - Q_2)/Q_1$. We cannot yet evaluate the efficiency because we do not know how large Q_2 has got to be in relation to Q_1, but we know the efficiency will be less than 100%. When a heat engine is operating in reverse as a heat pump or refrigerator, we need another expression for a measure of its performance. 'What we are seeking' is now either the heat output, Q_1, for the heat pump, or the heat input, Q_2, for a refrigerator. 'What we have to pay for' in either case is the work input W, so we have two expressions which are known as the **coefficient of performance**, or **COP**. Thus:

$$\text{COP (refrigerator)} = Q_2/W = Q_2/(Q_1 - Q_2)$$

$$\text{COP (heat pump)} = Q_1/W = Q_1/(Q_1 - Q_2)$$

We also have that:

$$\begin{aligned}\text{COP (heat pump)} &= Q_1/W = (W + Q_2)/W = 1 + Q_2/W \\ &= 1 + \text{COP (refrigerator)}\end{aligned}$$

If a reversible heat engine has an efficiency of 0.5 or 50%, what would be the COP of a heat pump which involved this engine being reversed?

20

2.0

From the previous frame we have seen that the efficiency is W/Q_1, and when operating in reverse the COP as a heat pump is Q_1/W, so the COP = 1/0.5 = 2.0.

This is another important result. For a **reversible engine** operating forwards as an engine and in reverse as a heat pump:

$$\text{Efficiency as an engine} = \frac{1}{\text{COP (heat pump)}}$$

In Frame 18 we said that for a reversible engine, operating normally and in reverse between the same temperatures, the heat and work transfers for the reversed engine would be numerically the same as for the engine running normally, but of opposite sign. If we let the engine drive the reversed engine, what would be the overall effect on each heat reservoir?

21

Zero net gain or loss of energy from each reservoir

Since the reservoirs are each giving and receiving the same amount of heat, the net gain or loss is zero for each reservoir. This has the overall effect of producing an adiabatic barrier between the reservoirs with no heat transfer. But we know in fact that even with the best insulation available no boundary is ever adiabatic and **we may argue that an adiabatic surface cannot exist because reversible heat engines do not exist**.

We can go on to think about what would be the net result of having a real engine driving a reversed real engine, with all the work output of the engine absorbed by the reversed engine, between a high and low temperature heat reservoir.

If the engines are not reversible, the only other possibility is a net heat transfer from the high to the low temperature reservoir.

We see that: $Q_1 - Q_2 = W = Q_{1a} - Q_{2a}$

and so: $Q_1 - Q_{1a} = Q_2 - Q_{2a} = W$

Thus there is a net heat flow between the reservoirs equal in magnitude to W.

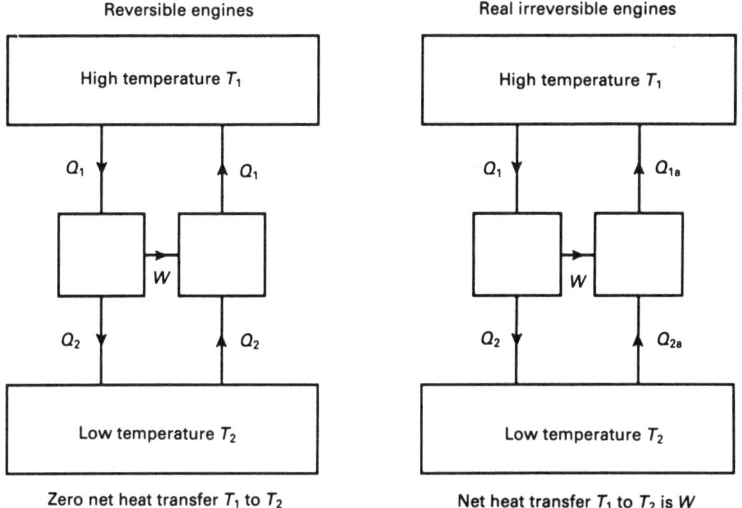

22

We will now seek to find out whether the efficiencies of all reversible engines of different designs operating between the same two heat reservoirs are the same. Even though reversible engines can only exist in theory, it is important to see what the implications are.

The figure shows a reversible heat engine driving a reversible engine operating in reverse as a heat pump. **We shall assume that the two machines are of different design** and that:

$$\text{Reversible engine efficiency} > 1/\text{COP(heat pump)}$$

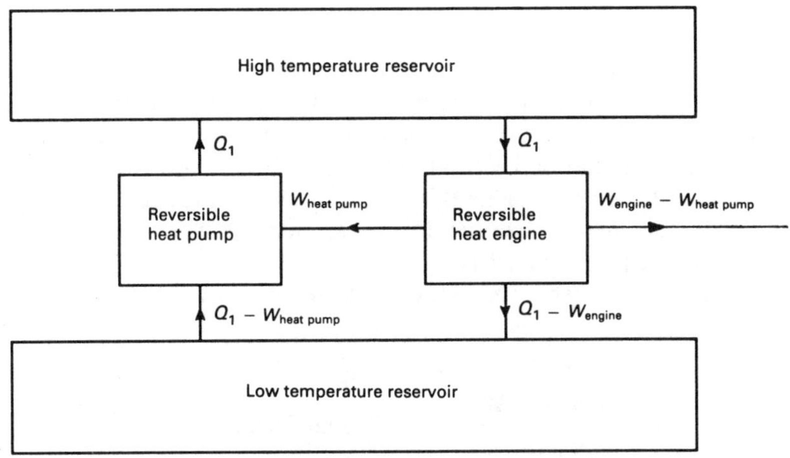

We shall investigate the consequences of this assumption.

23

We have assumed: $(W/Q_1)_{\text{engine}} > (W/Q_1)_{\text{heat pump}}$

so for the same Q_1 we have: $(W) > (W)_{\text{heat pump}}$

There is therefore surplus work of $(W)_{\text{engine}} - (W)_{\text{heat pump}}$

and also for the low temperature reservoir the heat received from the engine is $Q_1 - (W)_{\text{engine}}$ and the heat taken by the heat pump is $Q_1 - (W)_{\text{heat pump}}$.

The net heat flow from the low temperature reservoir is:

$$Q_1 - (W)_{\text{heat pump}} - [Q_1 - (W)_{\text{engine}}]$$
$$= (W)_{\text{engine}} - (W)_{\text{heat pump}}$$

There is thus a net flow of heat from the low temperature reservoir, and an equivalent output of work. We know that this does not happen (if it did, we would not need to use fuel at all to obtain work!), so we conclude that the COP of the heat pump must be the reciprocal of the reversible engine efficiency. Then as before, since we are dealing with reversible engines, we have an adiabatic boundary.

We could then swap the two engines round and let the second one now drive the first one as a heat pump. We must obtain the same result as before.

Does this mean that two reversible engines of different design, operating between the same two temperatures, must have the same efficiencies?

24

Yes

We can explore the implication of a 'no' answer, which means that one reversible engine is more efficient than another, between the same two temperatures.

In the diagram we have a reversible heat engine A that is supposed to have an efficiency greater than that of a second heat engine B operating between the same two temperatures T_1 and T_2. Engine B is operating as a reversed engine and is being driven by engine A, and we size the engines so that the work output of A is the same as the work requirement of B.

The two engines together form a closed system. We have then that:

$$\eta_A > \eta_B$$

Therefore: $\dfrac{W}{Q_{1_A}} > \dfrac{W}{Q_{1_B}}$ with W equal for both

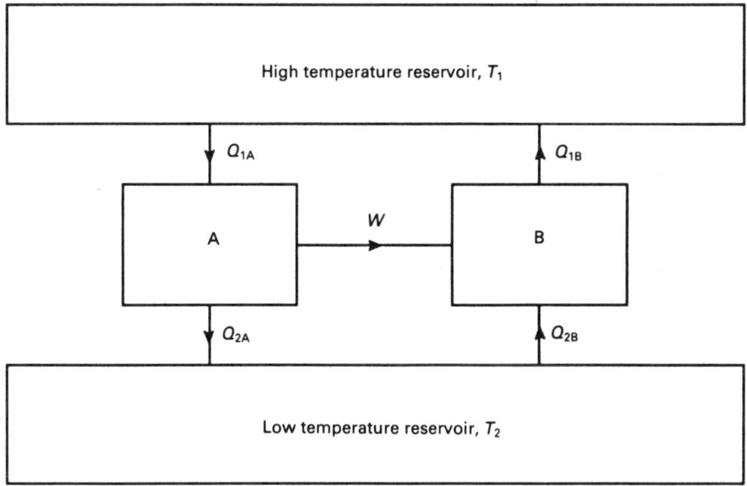

and so: $$\frac{Q_{1_A}}{Q_{1_B}} < 1$$

And then because $Q_{2_A} = Q_{1_A} - W$

and $Q_{2_B} = Q_{1_B} - W$

with W equal for both,

we have: $$\frac{Q_{2_B}}{Q_{2_A}} = \frac{Q_{1_B} - W}{Q_{1_A} - W} > 1$$

These results mean that there is a net flow of heat from the lower temperature reservoir to the higher temperature reservoir, without any work being done, and again we know that this cannot happen. So we conclude that any two reversible engines operating between the same two temperatures must have the same efficiency, irrespective of the design of the engine or of the working fluid being used in the engine.

25

In the next frame we shall think about how a reversible heat engine operating between two constant temperature heat reservoirs, a source and sink, might be conceived.

26

Since all processes used in this reversible engine must be reversible, there cannot be any temperature drop between the source temperature and the process in the engine receiving the heat transfer; likewise there cannot be a temperature drop between the process in the engine which rejects heat and the sink temperature. This means both heat transfers must be **reversible isothermal heat transfers**. For this to be so there must be an infinitesimally small temperature difference in the required direction for the heat to flow. Such a heat transfer is infinitely slow in practice, but it remains a useful theoretical idea.

The **Carnot cycle** is a reversible engine cycle that uses reversible isothermal heat transfers, so it can operate between two constant temperature heat reservoirs. The cycle also uses two reversible adiabatic processes.

27

Since the Carnot cycle is a reversible engine cycle, we know that if we consider such a cycle with a perfect gas as the working fluid, and secondly with steam, then between the same two temperatures both cycles must have the same efficiency. We established this in Frame 24. A Carnot cycle with a perfect gas as the working fluid is shown in the figure.

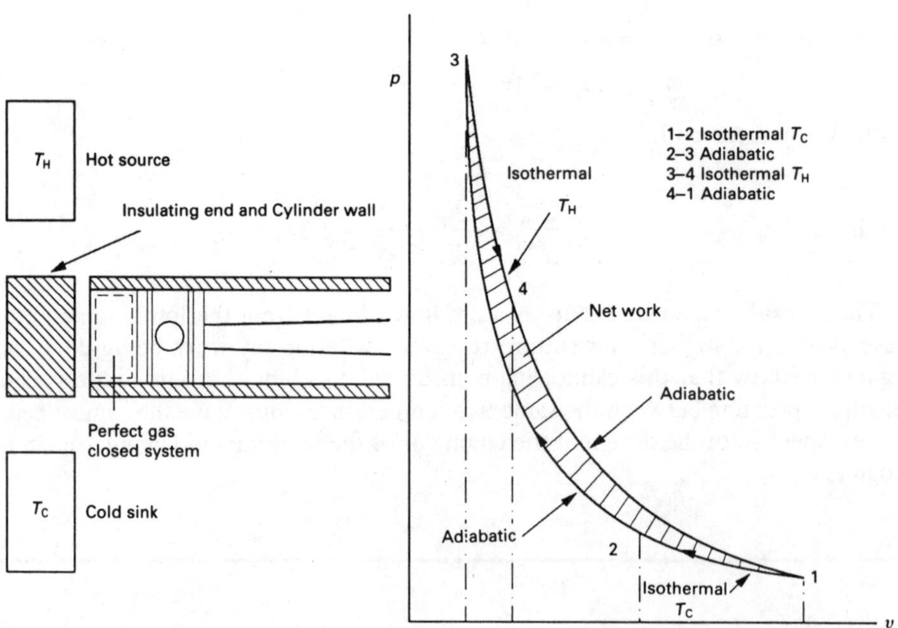

Imagine the cycle to be performed by a reciprocating engine using a cylinder with insulated walls, but with a closed end that may be insulating or conducting according to the point in the cycle. We may start at state 1, with the piston at the outer limit of its stroke. A normal compression would be rapid and the temperature of the gas would rise, but in this cycle the gas is compressed slowly so that heat transfer can take place and the gas temperature remains constant. Thus the gas is compressed **isothermally** at the lower temperature, T_C, and heat transfer occurs to the heat sink at T_C with the closed end of the cylinder conducting. At state 2, the insulating end is put in place, and compression continues **adiabatically** to state 3, where the gas temperature is now T_H which is the temperature of the heat source. At state 3, the piston is at the inner limit of its stroke, the gas is now expanded **isothermally** to state 4, by allowing heat to flow to the expanding gas with the conducting end in place. Finally, at state 4 the insulating end is replaced and the expansion continues **adiabatically** so that we return to state 1 at the lower temperature with the piston again at the outer limit of its stroke.

If the heat supplied is Q_H at T_H and the heat rejected is Q_C at T_C, what is the thermal efficiency of the cycle?

28

$$\eta_R = (Q_H - Q_C)/Q_H$$

The work done in the cycle is $Q_H - Q_C$ by the First Law. (In the cycle, net work = net heat, and net heat = $Q_H + Q_C$. But as Q_C is negative, net work = $Q_H - Q_C$.)

Hence the cycle efficiency is $(Q_H - Q_C)/Q_H$.

Since in an isothermal compression or expansion the heat transfer is equal to the work term, for the isothermal expansion 3–4 (Frame 80, Programme 5), we can say that:

$$Q_H = p_3V_3 \ln \frac{V_4}{V_3} = m\ R\ T_H \ln \frac{V_4}{V_3}$$

and for the isothermal compression 1–2 we can say:

$$Q_C = p_1V_2 \ln \frac{V_2}{V_1} = m\ R\ T_C \ln \frac{V_2}{V_1}$$

This is continued in the next frame.

29

Because the two isothermal processes are linked by two adiabatic processes (Frame 69, Programme 5), we can see that:

$$\frac{T_H}{T_C} = \left(\frac{V_2}{V_3}\right)^{\gamma - 1} = \left(\frac{V_1}{V_4}\right)^{\gamma - 1}$$

and so:

$$\frac{V_2}{V_3} = \frac{V_1}{V_4}$$

and by cross-multiplying:

$$\frac{V_4}{V_3} = \frac{V_1}{V_2}$$

Thus in the terms for Q_H and Q_C, the volume ratios V_4/V_3 and V_1/V_2 are the same. Hence we may express the efficiency of the Carnot cycle as:

$$\eta_R = \frac{Q_H - Q_C}{Q_H} = \frac{m\,R\,T_H - m\,R\,T_C}{m\,R\,T_H}$$

$$= \frac{T_H - T_C}{T_H}$$

We see, therefore, that the temperatures of the two heat reservoirs are related to the efficiency of a reversible engine operating between those reservoirs. We have already seen in Programme 4 that by using a constant volume ideal gas thermometer that an absolute temperature scale is predicted such that 0°C and 100°C become 273.15° and 373.15° on the absolute scale. These are the absolute temperatures which are used in the ideal gas rule and temperatures on this scale now appear in the equation above.

30

However, even though the constant volume gas thermometer predicts an absolute zero of temperature at zero pressure, it is just another example of a device which gives a thermometric scale depending on the physical properties of a substance. The question now arises: can there be a theoretical absolute temperature scale which is independent of the physical properties of actual substances, and what would its relation be to the scale of the gas thermometer? Also, can an absolute zero of temperature really exist?

We may argue that it is not possible to attain a temperature of absolute zero because it is not possible to create an adiabatic enclosure; heat transfer must always take place to any system at a temperature lower than its surroundings. Nevertheless, absolute zero temperature, like the reversible heat engine, is seen to be a useful theoretical concept. It was Lord Kelvin, who, in 1851, foresaw that an absolute temperature scale was possible so that absolute temperatures would be related to the efficiencies of reversible engines.

The **Kelvin scale of thermodynamic temperature** is defined in terms of the efficiency of reversible engines, such that:

$$\eta_R \equiv \frac{T_H - T_C}{T_H}$$

We shall have to examine this carefully in the next few frames to see what this means and to see whether the temperatures T_H and T_C in this equation are the same as those in Frame 29.

31

We have seen that the efficiencies of reversible engines operating between heat reservoirs depend on the heat transfers involved at those reservoirs; **Kelvin proposed that these heat transfers depended only on the absolute temperatures of those reservoirs**. There was no other identifiable property of a heat reservoir that could affect the magnitude of a heat transfer to or from a reversible engine.

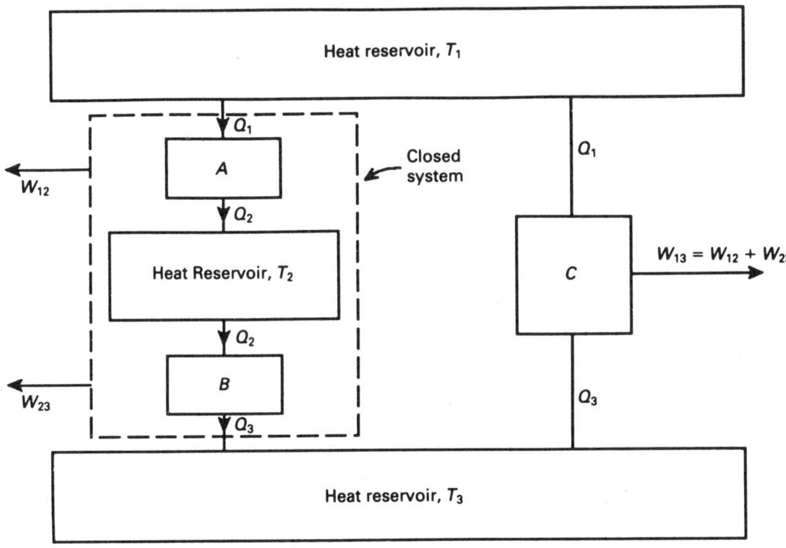

We shall consider this diagram in the next frame.

32

Between the heat reservoirs at T_1 and T_3 we have a closed system comprising reversible engines A and B and an intermediate heat reservoir at T_2. This closed system constitutes a single reversible heat engine which must have the same efficiency as engine C which is also operating between temperatures T_1 and T_3, since we have seen that all reversible engines operating between the same reservoirs must have the same efficiency which is a function only of the temperatures of the reservoirs.

For engine A we may express the efficiency as a function of T_1 and T_2 as:

$$\eta_{R_{12}} = f(T_1, T_2)$$

and we may also write the efficiency as:

$$\eta_{R_{12}} = 1 - \frac{Q_2}{Q_1}$$

In the next frame we shall compare these two equations.

33

Since the right-hand sides of these equations are equal, there must be a functional relationship between T and Q, so we may write:

$$\frac{Q_1}{Q_2} = \phi(T_1, T_2)$$

By similar reasoning we can deduce that:

$$\frac{Q_2}{Q_3} = \phi(T_2, T_3), \text{ and } \quad \frac{Q_1}{Q_3} = \phi(T_1, T_3)$$

Next we can combine the three heat ratios:

$$\frac{Q_1}{Q_2} = \frac{Q_1}{Q_3} \times \frac{Q_3}{Q_2} = \frac{Q_1/Q_3}{Q_2/Q_3}$$

and by introducing the temperature functions we get:

$$\phi(T_1, T_2) = \frac{\phi(T_1, T_3)}{\phi(T_2, T_3)}$$

which means that:

$$\phi(T_1, T_2) = \frac{\psi(T_1)}{\psi(T_2)}$$

where ψ is a new functional relationship arising after the cancellation of T_3 in the previous equation. The simplest functional relationship of temperature to satisfy this equation was adopted by Kelvin, which is:

$$\psi(T) = T$$

This means therefore that:

$$\frac{Q_1}{Q_2} = \frac{T_1}{T_2}$$

where Q_1 and Q_2 are the reversible heat transfers from any two reservoirs at T_1 and T_2.

The efficiency of a reversible engine is therefore:

$$\eta_{R_{12}} = \frac{Q_1 - Q_2}{Q_1}$$

$$= 1 - \frac{Q_2}{Q_1} = 1 - \frac{T_2}{T_1}$$

This is seen to be similar to the relationship of Frame 29 in which T_H would be T_1 and T_C would be T_2, which was obtained by using the absolute temperatures associated with the ideal gas rule and the constant volume ideal gas thermometer of Programme 4.

34

We have now established that the efficiency of a reversible heat engine is given by:

$$\eta_R = \frac{T_H - T_C}{T_H}$$

where in the Carnot cycle, using an ideal gas, the temperatures are absolute values from the constant volume gas thermometer, and also independently where T_H and T_C are absolute thermodynamic temperatures.

35

The problem now facing us is how do we attach values to T_H and T_C, and thus calibrate the thermodynamic temperature scale, if we cannot build reversible engines? Surprisingly, perhaps, this can be done with relatively simple experiments in the laboratory, as we shall see in Frames 39–41. We shall also see that both the thermodynamic temperature scale and the ideal gas temperature scale predict the same value for absolute zero temperature, starting from 0°C and 100°C as the fixed points on the Celsius scale, so that there is just the one absolute temperature scale. The fixed points on the Celsius scale of 0°C and 100°C, have absolute values of 273.15° and 373.15° on the ideal gas scale; what is the reversible engine efficiency between these temperatures?

36

0.268

$$\eta_R = (373.15 - 273.15)/373.15 = 100.0/373.15$$

hence: $\eta_R = 0.268$

Before we can be completely sure that this and similar calculations are correct, we must check that ideal gas absolute temperatures have the same values as absolute thermodynamic temperatures.

The route to giving numbers to Kelvin temperatures is via the Carnot cycle using a vapour such as steam. Consequently we must now consider the Carnot cycle again, using steam, as in addition we must also confirm that the efficiency is the same as with an ideal gas when operating between the same two temperatures.

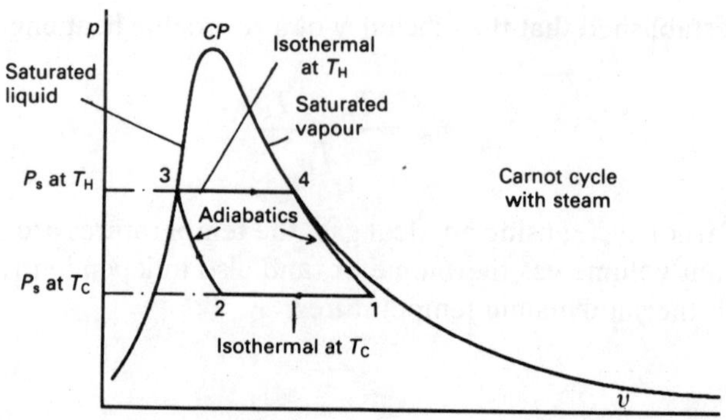

The p–v diagram for the Carnot cycle using steam is shown in the figure.

37

The cycle operates between saturation pressures corresponding to T_H and T_C, and as before we have two isothermal and two adiabatic processes. We may start the cycle at the outer position of the piston at state 1, which is a wet steam condition achieved by an adiabatic expansion from the saturated vapour condition at T_H at state 4. From state 1 the steam is compressed **isothermally** to state 2, and the heat rejected is Q_C at T_C. Then after an **adiabatic** compression to the inner limit of the piston stroke, the fluid is saturated water at state 3. This is followed by an **isothermal** expansion to the saturated vapour condition at state 4, so that the heat received is Q_H at T_H. The final adiabatic expansion completes the cycle.

The cycle efficiency is then:

$$\eta_R = (Q_H - Q_C)/Q_H$$

with the heat transfers taking place at T_H and T_C respectively. If these temperatures can be shown to be absolute thermodynamic temperatures, then the cycle efficiency again becomes:

$$\eta_R = (T_H - T_C)/T_H$$

and provided that we can show these temperatures to be the same as ideal gas absolute temperatures, we will have shown that both Carnot cycles between the same two temperatures have the same efficiency.

38

We must always remember that whether we use a perfect gas or steam, the Carnot cycle is only a theoretical idea, developed to explain in an idealised way the workings of real engines, which at the time had very low efficiencies. It was Kelvin in 1851 who embodied the ideas of Sadi Carnot into a theoretical reversible engine cycle, the efficiency of which would define the thermodynamic temperature scale. Carnot's work, entitled *Reflections on the motive power of fire*, was published some time before in 1824, when the Stockton and Darlington Railway was being built.

Later we shall consider other ideal engine cycles, which are more practical, and more capable of being realised. However, in these cycles the heat transfers are not necessarily isothermal and we shall see that the efficiencies will be less than that of Carnot cycles operating between the same temperature limits.

Practical measurements of thermodynamic temperature may be made from experiments with steam, by considering the Carnot cycle operating between two temperatures only incrementally different.

The cycle is shown in the figure. The lower pressure in the cycle is the saturation pressure corresponding to absolute temperature T, and the upper pressure corresponds to absolute temperature $T + \delta T$.

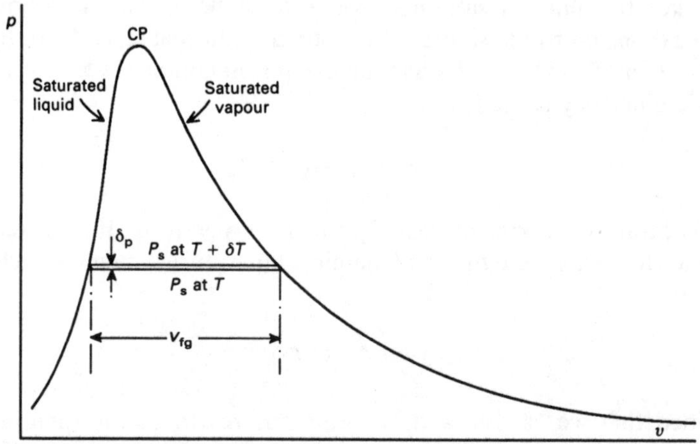

The pressure difference is δp. The work done in the cycle for unit mass of steam is, since δp is very small, the area of the rectangle of height δp and width v_{fg}, so:

$$W = \delta p \; v_{fg}$$

The heat supplied in the cycle is the heat required to evaporate unit mass of saturated water to saturated steam at constant pressure, h_{fg}, so:

$$\eta_R = \frac{\delta p \; v_{fg}}{h_{fg}}$$

and also:

$$\eta_R = \frac{T_H - T_C}{T_H} = \frac{\delta T}{T}$$

Hence we can write:

$$T = \frac{h_{hg}}{v_{hg}} \frac{\delta T}{\delta p}$$

This equation is known as the **Clausius–Clapeyron equation**, after its originators. It is possible to measure h_{fg} and v_{fg} together with the rate of change of temperature with pressure at a particular saturation pressure, and then the corresponding absolute temperature may be calculated.

40

Unfortunately the increments of pressure and temperature in the steam tables are a little too large for this formula to be used to check accurately the consistency of the data in the tables. Thus suppose we consider steam at 0.2 MN/m^2, from the Haywood tables we have that h_{fg} = 2201.6 kJ/kg, v_g = 0.885 m^3/kg and v_f = 0.001 061 m^3/kg. We can take δp as (0.22 − 0.19) MN/m^2 and δT as (123.3−118.6) K. The Clausius–Clapeyron equation then gives us:

$$T = \frac{2201.6}{(0.885 - 0.001\,061)} \times \frac{(123.3 - 118.6)}{0.03 \times 10^3}$$

$$= 390.2 \text{ K}$$

The actual temperature at 0.2 MN/m^2 is 120.2°C which is 393.35 K. Thus we see that there is an error of 3.15 K.

Using more accurate data not available in your steam tables, at the saturation pressure of 1.013 25 bar, h_{fg} = 2256.7 kJ/kg, v_{fg} = 1.671 96 m^3/kg and $\delta T/\delta p$ = 27.646 K/bar. What is the absolute temperature at the pressure of 1.013 25 bar?

41

373.147 K

$\delta T/\delta p$ = 27.646 K/bar = 0.276 46 K/kPa, for consistent units, so we have:

$$T = \frac{2256.7}{1.67196} \times 0.276\,46 = 373.147 \text{ K}$$

As water at a pressure of 1.013 25 bar boils at 100°C, we have shown that the absolute zero of temperature on the thermodynamic scale is −273.147 K. This is the same as the result given by the ideal gas thermometer. A whole range of absolute temperatures for steam boiling at different pressures may be found by repeating this experiment. Thus the absolute zero of temperature relative to the 0°C and 100°C fixed points on the Celsius scale has been established by two routes: (a) a practical route using the ideal gas thermometer (see Frames 85–89 in Programme 4) and (b) a theoretical route using the efficiency of a reversible engine backed by experimental measurements, with both routes using Celsius thermometers.

42

Both approaches give the same answer. The international scale of temperature, discussed in Programme 1, is the result of much careful experimentation to provide a means of measuring temperatures in °C which when added to 273.15 K gives as near as possible temperatures on an absolute scale having an equal K unit between values. The near linear scale of the ideal gas thermometer plays a part in the procedures, even though other fixed points and interpolation procedures are specified.

Thus we have shown that the Carnot cycle, using either an ideal gas or steam, operating between the same two temperatures, has the same efficiency. We also established that this must be so in Frame 24.

43

Now that we have firmly established values of thermodynamic temperature, we are in a position to see what sort of efficiencies reversible engines operating between known temperature levels could have. We can now see that the calculation in Frame 36 was correct.

A reversible engine receives heat from a reservoir at 600 K and rejects heat to a reservoir at 300 K. Calculate the efficiency of the engine and the amount of work the engine could do when receiving 1000 kJ at 600 K.

The efficiency is:
$$\eta = \frac{600 - 300}{600}$$
$$= 0.5 \text{ or } 50\%$$

The work output is therefore 0.5 of the heat input, which is 500 kJ. The actual efficiency of a real steam engine operating between these temperatures would probably be no more than about 25%.

A reversible engine on a spacecraft receives heat from a solar panel maintained at 400 K and heat is rejected on the shaded side of the craft by a further panel at 50 K which radiates to space. What is the efficiency of this reversible engine?

44

87.5%

The efficiency is:
$$\eta = \frac{400 - 50}{400} = 0.875 = 87.5\%$$

This example shows that the efficiencies of reversible engines are limited by the temperature of the earth's environment, where the minimum heat rejection temperature is around 273 K or more. In the space environment it is theoretically possible to reject heat at much lower temperatures. Again the real engine efficiency would be much less.

Now calculate the efficiencies of a reversible engine rejecting heat at 300 K and receiving heat at 500 K and 400 K respectively, and compare with the result of the example in Frame 43.

45

T_1 =	600 K	500 K	400 K
η_R =	0.5	0.4	0.25

The results are:

$$\eta_R = (500 - 300)/500 = 0.4$$

and

$$\eta_R = (400 - 300)/400 = 0.25$$

We notice a definite trend in these results which is that the efficiency increases as the temperature of the heat source increases. This was realised quite early on in the development of the steam engine. To do this required either the use of superheat at lower pressures, or higher pressures to increase the saturation temperature. Progress in both developments was limited by metallurgical constraints of the time, sometimes resulting in tragically fatal boiler explosions.

We can also look at the upper limit of performance of a reversed (reversible) heat engine. A reversible heat pump extracts heat from a river of temperature 5°C and delivers warm air to a house at 45°C. Find the COP and the rate of heat delivery if it is powered by a motor of 0.8 kW. (The COP of a heat pump was given in terms of Q_1 and Q_2 in Frame 19, and you should be able to express the COP in terms of the absolute temperatures of the reservoirs – see Frame 33.)

46

7.954, 6.363 kW

The COP of a reversible heat pump was given in Frame 19 as $Q_1/(Q_1 - Q_2)$, and we now know that this may be expressed in terms of the absolute temperatures of the reservoirs as $T_1/(T_1 - T_2)$, so with $T_1 = 318.15$ K and $T_2 = 278.15$ K, we have:

$$\begin{aligned} \text{COP} &= 318.15/(318.15 - 278.15) \\ &= 7.954 \end{aligned}$$

The rate of work input is 0.8 kW, so the rate of heat output is 0.8×7.954 kW, which is 6.363 kW. The heat extracted from the river is then $6.363 - 0.8 = 5.563$ kW.

The performance of a real heat pump would again be very much less than this, having a COP of about 3.5–4 between these temperatures.

We referred to other possible engine cycles in Frame 38. We shall look at examples of these cycles in later programmes. For the moment we can work with these other ideal cycles, and with real cycles in which irreversibilities occur, by specifying their efficiency in terms of a fraction of the corresponding Carnot efficiency.

47

For example, a heat engine operates between a maximum temperature of 2000 K and a minimum temperature of 400 K with 0.6 of the Carnot efficiency between these temperatures. Calculate the work output per 100 kJ of heat input.

The Carnot efficiency is $1 - T_2/T_1 = 1 - 400/2000 = 0.8$

The actual efficiency is $0.8 \times 0.6 = 0.48$

The work output is the actual efficiency × heat input,

$$\begin{aligned} &= 0.48 \times 100 \\ &= 48 \text{ [kJ]} \end{aligned}$$

A heat pump uses water in a river at 6°C as an energy source and it delivers heat at 65°C to a building. It operates at 65% of its maximum possible COP between these temperatures and is powered by a 1.5 kW motor. What is the rate of heat output to the building?

48

5.588 kW

The thermodynamic temperatures are:

$$T_1 = 273.15 + 65.0 = 338.15 \text{ K}$$

$$T_2 = 273.15 + 6.0 = 279.15 \text{ K}$$

The maximum possible COP for the heat pump $= T_1/(T_1 - T_2)$

$$= 338.15/(338.15 - 279.15) = 5.731$$

The actual COP for the heat pump $= 0.65 \times 5.731 = 3.725$

The rate of heat transfer output $= 3.725 \times$ power input

$$= 3.725 \times 1.5$$

$$= 5.588 \text{ [kW]}$$

As refrigerators and heat pumps require to be driven, it is not unusual for heat engines to be used for this purpose. By using electric power generated by steam power plant one can argue that this is done anyway, but it is interesting to look at the possibilities of a small-scale power source being used to drive a reversed heat engine.

In a sports complex, there is an ice-rink and an indoor heated swimming pool. A small power plant drives a reversed heat engine, which acts as a refrigerator for the ice-rink and also as a heat pump to heat both the swimming pool, and to some extent the buildings of the complex. Heat is extracted from the ice-rink at −5°C and is delivered by the heat pump at 65°C for its various heating purposes. The power plant operates between a maximum temperature of 300°C and a heat rejection temperature of 65°C. The waste heat from the power plant is added to the output from the reversed heat engine. Both machines operate at 68% of their respective Carnot performance. The heat extraction rate from the ice-rink is 50 kW. In the next frame we shall calculate the rate of heat supply to the power plant, and the heat output to the swimming pool. But suppose the first answer was X kW, what would be the second answer? (Consider the heat pump and heat engine together as a closed system.)

49

$X + 50$ kW

If you consider the whole plant as a closed system, it has three heat transfers: heat from the ice-rink, heat from the source at 300°C and heat to the complex at 65°C. The last of these must be the sum of the first two. The plant is shown below.

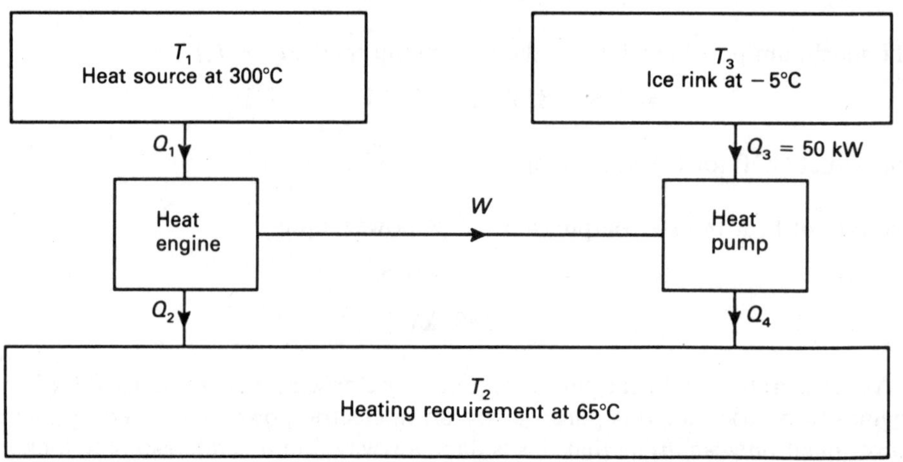

The heat reservoirs are shown as outline blocks at their respective temperatures. We may assume the heat engine is a vapour power cycle boiling at 300°C and condensing at 65°C, driving directly the heat pump compressor. The refrigerant in this cycle is boiling at −5°C at the ice-rink temperature, and condensing at 65°C, the heating load temperature. The sort of hardware involved in a refrigerant or heat pump cycle was shown in Frame 16.

The thermodynamic temperatures are:

Heat source to power plant:	$300 + 273.15 = 573.15$ K, T_1
Ice-rink source to the heat pump:	$-5 + 273.15 = 268.15$ K, T_3
Heat sink at 65°C:	$65 + 273.15 = 338.15$ K, T_2

The efficiency of the power plant is $(1 - T_2/T_1) \times 0.68$

$$= (1 - 338.15/573.15) \times 0.68$$

$$= 0.28$$

What is the COP of the heat pump?

3.285

The COP of the heat pump is $T_2/(T_2 - T_3) \times 0.68$

$$= 338.15/(338.15 - 268.15) \times 0.68$$

$$= 3.285$$

We know that Q_3 for the heat pump is 50 kW, and $Q_4 = W + Q_3$

hence for the heat pump $(W + Q_3)/W = \text{COP} = 3.285$

$\therefore \quad 3.285 = (W + 50)/W$

hence: $\quad W = 21.88$ [kW]

and: $\quad Q_4 = 50 + 21.88 = 71.88$ [kW]

For the power plant: $W/Q_1 = 0.28$

$\therefore \quad Q_1 = W/0.28 = 21.88/0.28 = 78.14$ [kW]

and also: $\quad Q_2 = Q_1 - W = 78.14 - 21.88 = 56.26$ [kW]

The total heat rejected is $Q_2 + Q_4 = 56.26 + 71.88 = 128.14$ [kW]

This must also be the sum of Q_1 and $Q_3 = 78.14 + 50 = 128.14$ [kW]

The overall performance is that we use heat at the rate of 78.14 kW, and obtain 128.14 kW of heating and maintain the ice-rink at −5°C. The 50 kW from the ice-rink is heat from the atmosphere and the skaters, and from heat leakage into the ice-rink enclosure from the environment. We can say the overall COP for the system is 128.14/78.14 = 1.64.

51

Our next topic is the new property called **entropy**. Before we move on, make sure you can do the following problems.

1. A Carnot engine receives heat from a source at 200°C and rejects to a sink at 50°C. Calculate (a) the engine efficiency and (b) the power output for a heat input of 150 W. [(a) 0.317, (b) 47.55 W]
2. A reversed Carnot engine receives heat at 0°C and rejects heat at 60°C. Calculate (a) the COP as a heat pump and (b) the COP as a refrigerator. [(a) 5.553, (b) 4.553]
3. A reversible Carnot engine operates at 30% efficiency and rejects heat to a sink at 300 K. What is the temperature of the heat source? [428.57 K]
4. A heat engine operates between maximum and minimum temperatures of 300°C and 80°C at 0.6 of the reversible engine efficiency. What is the work output for a heat input of 800 kJ? [184.245 kJ]
5. A heat pump operates between a river of mean temperature 10°C as a heat source and delivers heat at 65°C to a hot water system. It has a COP of 50% of a reversible heat pump COP. Power input to the heat pump is 2 kW. What is the heat output? [6.148 kW]
6. A heat engine of thermal efficiency 35% drives a reversed engine operating, both as a refrigerator and heat pump between − 5°C and 50°C. The COP as a refrigerator is 70% of the reversible maximum. The heat input to the engine is 5 kW. Calculate (a) the heat extraction rate as a refrigerator and (b) the heat output as a heat pump. [(a) 5.448 kW, (b) 7.198 kW]
7. A heat engine having 55% of the Carnot efficiency operates between 250°C and 60°C. It drives a heat pump having 65% of its maximum possible COP operating between 10°C and 60°C. The waste heat of the heat engine is added to the heat pump output. Calculate the total heat delivery at 60°C for every 100 kJ supplied to the heat engine. [166.552 kJ]
8. OTEC stands for ocean thermal energy conversion. The principle involves a heat engine operating between water at say 30°C at the surface of the sea, and water at say 5°C at a depth of 300 m. Calculate the Carnot efficiency of such a plant operating at these temperatures. [8.25%]
9. An OTEC plant is operating at 30% of its Carnot efficiency between the temperatures given in **8**. Calculate the work output of the plant in kWh for every cubic km of sea water at 30°C passing through it, if the water drops in temperature by 2 K in the plant. (Specific heat of water is 4.186 kJ/kg K.) [57.44×10^6 kWh]
10. You have been approached to invest your money in a company making heat pumps which are claimed to deliver 1 kW of heat using a 150 W motor, taking heat from the atmosphere at 10°C and delivering at 65°C to the house. Would you do so, and if not, why not? [No. Claimed COP is 6.66; maximum theoretical COP is 6.15]
11. At 0.8 bar, in the wet steam region, the rate of change of temperature with pressure is 33.649 K/bar; using the Clausius–Clapeyron equation, calculate the absolute temperature. [366.65 K]

52

The property of **entropy** is associated with the Second Law, as energy is associated with the First Law, but there are differences. We shall see that the entropy change round a *reversible* cycle is zero, just as the energy change round *any* cycle is zero, but the Second Law is not a law of conservation of entropy as the First Law is a law of conservation of energy.

So what is entropy and what are its uses? Except to say that entropy is a property which increases when isolated systems go into equilibrium, entropy does not have such a readily grasped physical meaning as energy has, but we can begin to get an understanding of entropy by looking at its uses and by seeing what changes in entropy values mean. Clausius invented the concept of entropy around 1865, and the word means 'transformation' or 'change'. In his writings, Clausius extended his ideas to the whole universe as an isolated system; from the First Law the energy of the isolated system remains constant, but the entropy must tend towards a maximum.

53

We have already met two examples of isolated systems going into equilibrium. These are shown again below.

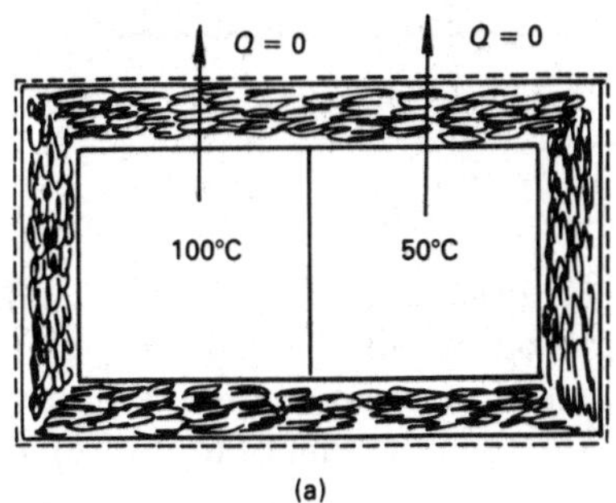

(a)

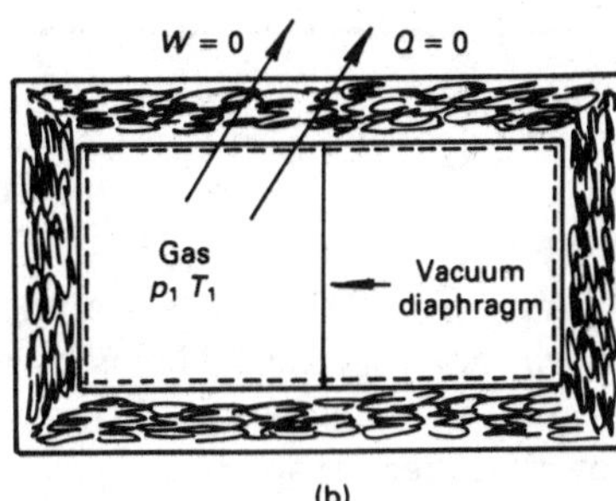

(b)

In case (a), from Frame 24, Programme 3, we have two equal blocks at different temperatures in an isolated system (for which $Q = 0$ and $W = 0$). We all know from experience that this system is not in equilibrium and given time the two blocks will have the same temperature as internal energy within the system is redistributed. In case (b), from Frame 37, Programme 2, we have gas held by a thin diaphragm in one part of a rigid insulated vessel, with a vacuum in the remaining volume of the vessel. The gas is being held back artificially from the rest of the vessel, and we know that without the diaphragm the gas would immediately fill the whole vessel. In each case we have initially a non-equilibrium state, or a state of lower disorder, followed naturally by an equilibrium state, or a state of higher disorder, and in each case the entropy finally may be shown to be more than it was initially. Thus an entropy increase in an isolated system is linked to the increase of disorder in the system.

54

The change from a non-equilibrium state to an equilibrium state in an isolated system is accompanied by an entropy increase.

The entropy of an isolated system cannot decrease. The reverse of the processes in (a) and (b) does not happen. Now consider the following example:

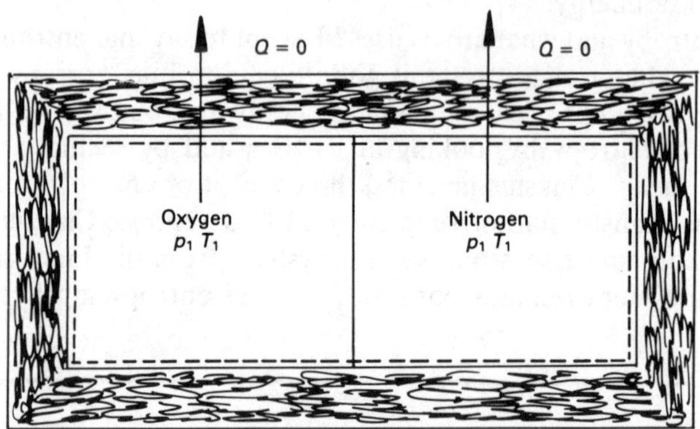

An insulated rigid vessel contains oxygen and nitrogen separately in two parts of the vessel at the same pressure, separated by a thin diaphragm. The diaphragm is removed; what will happen?

55

The gases will mix

We know from experience that the gases will not stay in separate parts of the vessel, but will mix together. This is by a process of diffusion. Again the isolated system has gone from a non-equilibrium state to an equilibrium state. We would be able to show that the entropy of the gas mixture will be more than the sum of the entropies of the two separate gases.

These are simple examples and we know what will happen from experience. The entropy function is vital in other situations where we cannot say from common experience what is going to happen.

56

Suppose we have a reaction vessel containing carbon dioxide gas, oxygen and nitrogen at room temperature and pressure. We would say that probably it was in equilibrium as we would not expect any reaction to take place. We then heat the contents at constant volume to say 2000 K, and again isolate the closed system from its surroundings. Is it still in equilibrium?

We do not know. But we could find out by calculating the entropy of the original mixture, and by recalculating the entropy for other possible states of the mixture.

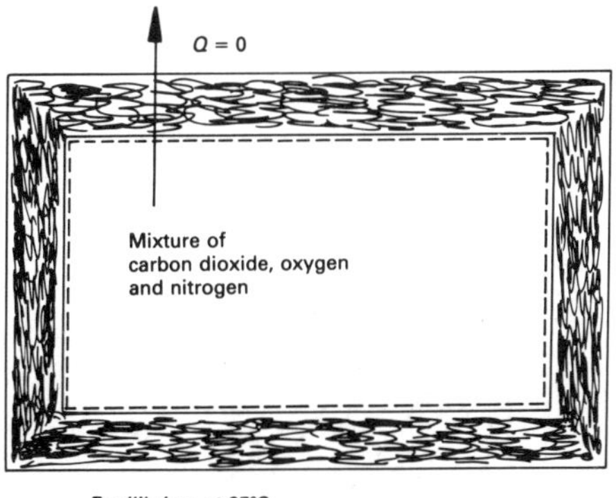

Equilibrium at 25°C

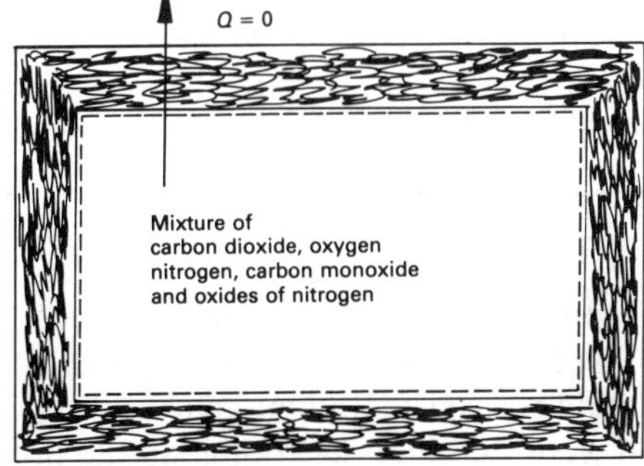

Equilibrium at 2000°C

Thus, in fact, carbon dioxide at high temperature *dissociates* partially into carbon monoxide and a single oxygen atom, and certain oxides of nitrogen are also formed. The equilibrium state would be a mixture derived from the whole of the original stated contents **that had the maximum possible entropy for that isolated system**. We could not find this state by any other way, except by using the entropy function.

57

A second valuable use of entropy is that it gives us a new property of a system which is related to the path function of Q_R, where Q_R is a reversible heat transfer in a process. You are familiar with drawing graphs of pressure against volume or specific volume for a process, and with integrating under the curve to find the value of the displacement work. Since work depends on the path of pressure with volume, we said work is a path function. Soon you will be drawing graphs of temperature against entropy and by integrating under the curve finding the value of the reversible heat transfer. Again this will depend on the temperature–entropy path, so for this reason a reversible heat transfer is said to be a path function.

58

In Frame 34 of Programme 5 we got stuck on a problem involving the isothermal expansion of steam because we could not calculate the work term.

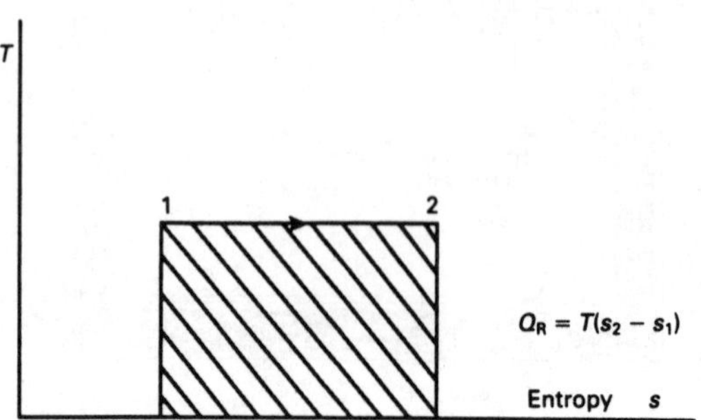

If we had known about entropy we could have plotted T against entropy (S), and easily determined the heat transfer, Q_R, so that we could then have completed the problem. Since the process was isothermal, the T–s plot is simply a rectangle.

59

A third valuable use of entropy is to be found when we need to find out how much work can be obtained in an expansion of a working fluid in a turbine or reciprocating engine. In Programme 5 we considered an adiabatic expansion of steam in an engine cylinder, and you were told that depending on the initial state of the steam the index of expansion was either 1.135 or 1.3. How did we know this? Could we not arrange for the expansion to take place in such a way that much of the subsequent condensation could be avoided and so increase the cycle efficiency? Various hypothetical expansions are shown in the figure; which are possible and which are not?

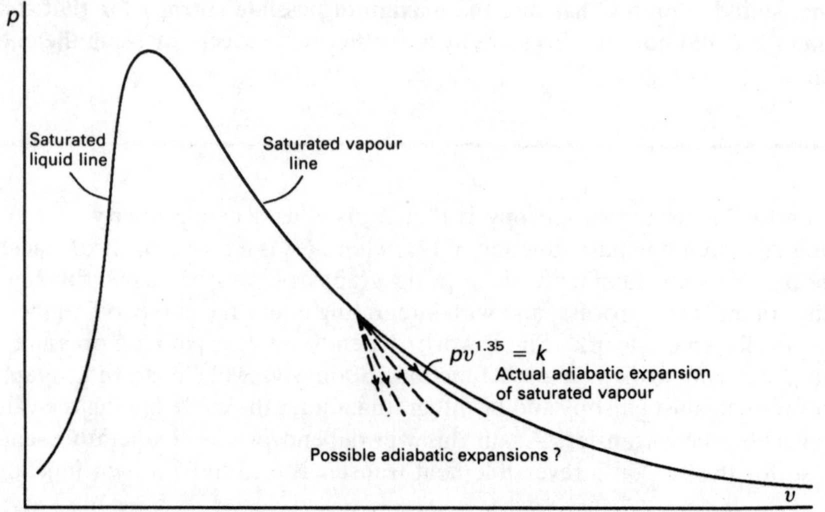

The answer lies in the fact that a reversible adiabatic expansion is a constant entropy expansion, or an **isentropic** process. We shall see that an isentropic process is superior in giving the maximum possible work in an expansion. **The entropy of a system remains constant when a reversible adiabatic process takes place**. This gives us a vertical straight line on the temperature–entropy diagram so it enables us to find the state after the expansion, which tells us what is possible and what is impossible. Knowing the final state from entropy considerations, we would be able to find an index, n, for the p–v relation in the expansion.

60

At the beginning of this programme we spoke of processes going from less to more disordered states. In Frame 53 we saw how a gas released from half of an available space would quickly fill the whole space. We would not expect otherwise; the final state is **more disordered**. The **more ordered** state (or, less disordered state) of having the gas in half the space could be created by means of a vacuum pump. It is interesting to look at this example from the microscopic point of view, in which we consider individual molecules.

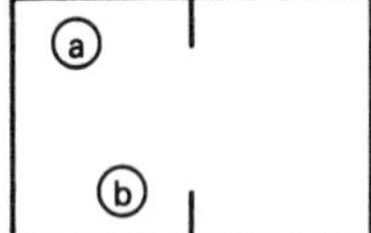

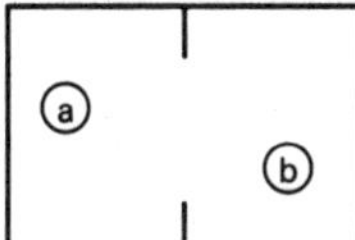

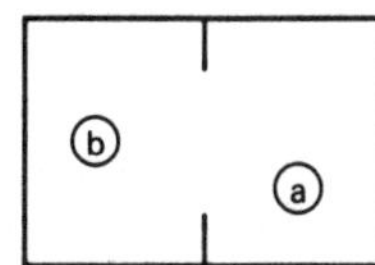

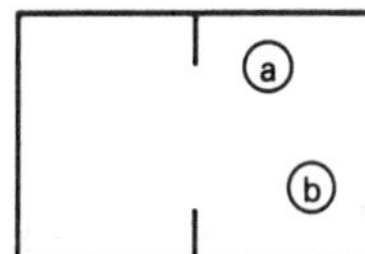

Four possible microstates with two molecules

Suppose the gas in the space consisted of only two molecules, like table-tennis balls, which we can identify as [a] and [b]. The figure shows there are 4 possible microscopic states for the arrangement of these molecules in the two halves of the space. There are two chances in four that there will be a molecule in each half, and two chances in four that both molecules will be in one half. These possible microscopic states would be changing exceedingly rapidly as the molecules moved about, but with only two molecules the chances of having one molecule in each half are the same as having both molecules in one half. The **probability** of either result is the same. We could not therefore say that disorder had increased when the gas was released from one half of the space.

61

We may now suppose that there are four molecules in the space, and we would find 16 possible microscopic states. We would find that there are 6 chances in 16 that we had two molecules in each half of the space, and 2 chances in 16 that we had all the molecules in one or other half. In the remaining 8 possibilities there would be at least one molecule in each half. Since we have 6 chances of an equal distribution and only 2 chances of all the molecules being in one half, we can say that the **probability** of an equal distribution has increased, whereas with only two molecules the probability of either result was the same. Thus we would find that with increasing the number of molecules, the **probability** of an equal distribution of molecules is increasing, and the **probability** of having the original distribution (in either half) is decreasing.

These trends increase, so that for the vast number of molecules that really exist in any space, the probability of an equal distribution becomes exceedingly close to 100% and the probability of having all the molecules in one half is exceedingly close to zero. In a study of Thermodynamics from the **microscopic** point of view, **entropy is directly related to probability**: an entropy increase in an isolated system occurs as the system proceeds to equilibrium, or to a more disordered or **more probable state**.

62

We shall not develop these ideas any further, but by looking at things from a microscopic viewpoint we do begin to appreciate why matter behaves in the way it does. It is interesting to see in the example that in theory the **possibility** of all the gas in the space collecting in one half is not ruled out, but the **probability** is so incredibly small that we know it will never happen. One could go on to try and picture other events: for example, for the Joule experiment to go in reverse all the molecules in the warm water would have to push simultaneously on the impeller in an ordered way to do work at the expense of the internal energy of the molecules, but the laws of probability tell us that this would never happen.

63

We shall now begin our detailed study of the concept of entropy. First we may recall from Frame 4 in Programme 3 that in a cycle the First Law tells us that the sum of the heat interactions round the cycle equals the sum of the work interactions round the cycle:

$$\sum_{\text{cycle}} \delta Q = \sum_{\text{cycle}} \delta W$$

and for a cycle in which the work is done by movement of the system boundary:

$$\sum_{\text{cycle}} \delta W = \sum_{\text{cycle}} p\mathrm{d}V$$

We thus remind ourselves that displacement work is the cyclic integral of the product of the intensive property p and the change in the extensive property V. In a similar way, reversible heat can be shown to be the cyclic integral of the product of an intensive property and the change in an extensive property. The new extensive property is **entropy**, S, from which follows the **specific entropy**, s, per unit mass, and the intensive property is T, absolute or thermodynamic temperature.

64

The proof that entropy is a property of a system is rather lengthy, and not very easy. You will probably find you need to read through the next few frames a few times. There are three stages in the proof, and these are summarised in this and the next frame. Do not be too despondent if you cannot fully understand all the arguments in one go; the most important thing is for you to learn how the property entropy is used in solving problems, and you will probably find that as you gain experience your understanding of entropy will increase. Remember these ideas did not appear overnight: 25 or more years elapsed between the published work of Carnot and statements of thermodynamic temperature by Kelvin, and of entropy by Clausius. We shall start to explore the use of entropy in Frame 77.

(1) The first stage is to establish what is known as the **Clausius inequality**. This is that in *any* cycle:

$$\sum_{\text{cycle}} \frac{\delta Q}{T} \leqslant 0$$

This statement says that, on integrating $\delta Q/T$ round the cycle, the result must be equal to or less than zero. [In comparison with the First Law, the integration of $(\delta Q - \delta W)$ round a cycle *is* zero]. The Clausius inequality enables us to look at any real cycle and calculate the degree to which irreversibilities are taking place. An example on the use of the Clausius inequality is given in Frame 69.

65

(2) In this second stage we show that if the cycle is *reversible*:

$$\sum_{\text{cycle}} \frac{\delta Q_R}{T} = 0$$

The heat transfer is now reversible and when $\delta Q_R/T$ is integrated round the cycle the result *is* zero. We then deduce that it is less than zero if the processes making up the cycle are not reversible, which happens in real cycles. In this and the third stage, we again consider properties which do not change at state points and different paths between these state points. You first met these ideas in Frame 13 of Programme 3.

(3) In the third stage we show that:

$$S_2 - S_1 = \int_1^2 \frac{\delta Q_R}{T}$$

for a reversible process between states 1 and 2. We show that the integral is independent of the path of the process, and therefore establishes itself as a change in the value of a property of the system. Bear in mind that while a reversible heat transfer Q_R **is a path function**, which depends on the path, $\delta Q_R/T$ **is not a path function**, and is independent of the path.

66

Stage (1): the Clausius inequality

Clausius established the equation from Frame 64 by thinking about systems not in equilibrium, in which as we have seen, irreversible processes take place. Imagine any closed system, such as gas in a cylinder contained by a piston. The temperature may vary, and heat and work may occur.

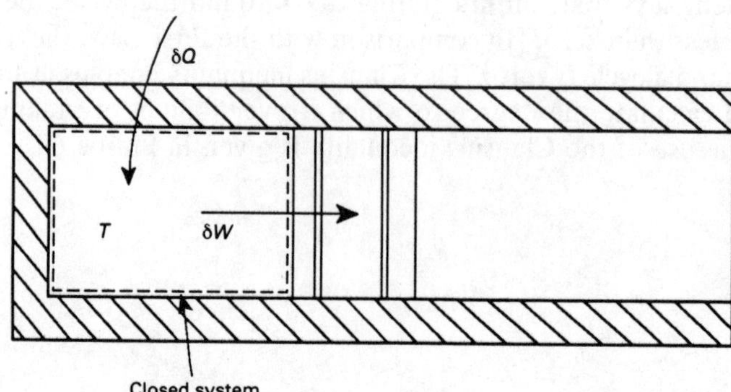

At a particular instant the temperature is T and the system receives heat δQ from the surroundings. At the same time the system does δW of work on the surroundings. Imagine that the system is able to receive heat and do work in a cyclic manner, like gas in a cylinder, so we have, by the First Law:

$$\sum_{\text{cycle}} (\delta Q - \delta W) = 0 \tag{1}$$

67

Now imagine that the system at temperature T received its heat δQ as a rejection from a reversible engine operating between a reservoir at T_H and the system at T. This is shown in the figure below. The direction of δQ_H, δW_R and δQ will change as the closed system receives or rejects heat.

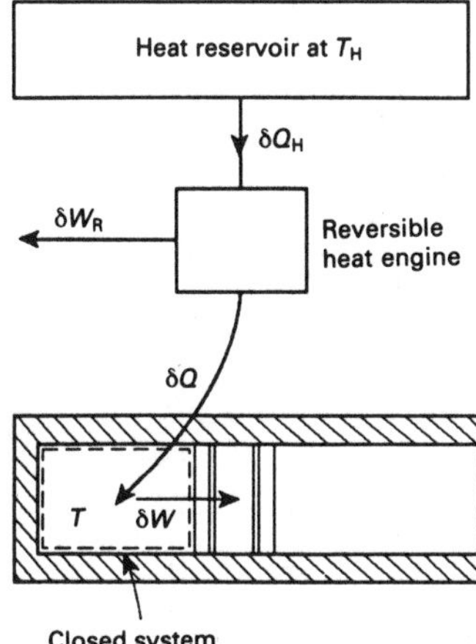

The system and reversible engine combined receive heat from the reservoir at T_H and do work of $(\delta W + \delta W_R)$ on the surroundings. The heat engine is sized so that it executes a complete number of cycles for a single cycle of the system. The work sum of $(\delta W + \delta W_R)$ in a complete cycle of the system resulting from heat from a single reservoir at T_H cannot be positive by the Second Law, so:

$$\sum_{\text{cycle}} (\delta W + \delta W_R) \leqslant 0 \tag{2}$$

From the relationships between thermodynamic temperature and heat to reversible engines we have:

$$\frac{\delta Q_H}{T_H} = \frac{\delta Q}{T} \tag{3}$$

68

Now we apply this equation to the reversible engine in the figure:

$$\frac{\delta W_R}{\delta Q_H} = \frac{\delta Q_H - \delta Q}{\delta Q_H} = \frac{T_H - T}{T_H} \tag{4}$$

By cross-multiplying in (4) and introducing (3) we get:

$$\frac{\delta W_R}{T_H - T} = \frac{\delta Q}{T} \tag{5}$$

We now take equation (2) and use equations (1) and (5) to eliminate dW and dW_R. First substitute for dW_R in (2) using (5):

$$\sum_{\text{cycle}} \left(\delta W + \frac{T_H - T}{T} \delta Q \right) \leqslant 0$$

and then using (1) to eliminate dW we get:

$$\sum_{\text{cycle}} \left(\delta Q + \frac{T_H - T}{T} \delta Q \right) \leqslant 0$$

or: $$\sum_{\text{cycle}} \frac{T_H \delta Q}{T} \leqslant 0 \text{ which means } T_H \sum_{\text{cycle}} \frac{\delta Q}{T} \leqslant 0$$

and so it follows that: $$\sum_{\text{cycle}} \frac{\delta Q}{T} \leqslant 0$$

This completes the proof of the Clausius inequality.

69

Before moving on to the second stage, we may now consider a simple steam engine cycle to demonstrate the use of the Clausius inequality. We have to show that in the cycle the integral of $\delta Q/T$ round the cycle is less than zero if the cycle involves any irreversible processes, or is zero if all processes are reversible. The cycle takes place in an engine cylinder, with reversible isothermal heat transfers to and from the steam, and with an adiabatic expansion and compression.

However, owing to turbulence in the steam, the expansion and compression, though adiabatic, are not reversible. Heat transfer to the steam takes place at 1.2 MN/m^2 or 12 bar and 188°C, and from the steam at 101.325 kN/m^2, or 1.013 25 bar and 100.0°C. State 1 is saturated water, state 2 is saturated steam, state 3 has a

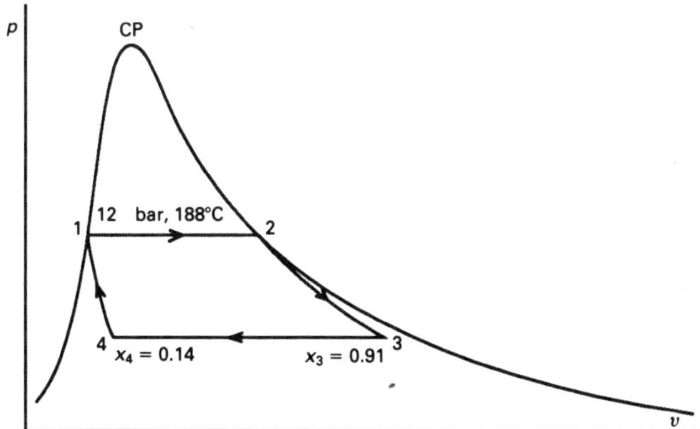

dryness fraction of 0.91 and state 4 has a dryness fraction of 0.14. If the adiabatic expansion and compression had been reversible, the entropies at states 2 and 3 would be equal, and again at states 1 and 4. We shall be able to return to this problem more fully later. For the constant pressure process 1–2:

$$Q - p(V_2 - V_1) = U_2 - U_1$$

or:
$$Q = H_2 - H_1$$

For unit mass of steam:
$$Q = h_2 - h_1 = h_g - h_f$$

So for heat transfer to the steam: **70**

$$\delta Q/T = \frac{h_g - h_f}{T}$$

so:
$$\delta Q/T = \frac{(2784 - 798)}{(188 + 273.15)} = +4.307 \text{ [kJ/kg K]}$$

Similarly, heat transfer from the steam, per unit mass, is

$$\delta Q = h_4 - h_3$$

where
$$\begin{aligned} h_4 &= (1 - x_4)h_f + x_4 h_g \\ &= (1 - 0.14) \times 419.1 + 0.14 \times 2675.8 \\ &= 735.04 \text{ [kJ/kg]} \end{aligned}$$

Now calculate h_3 at state point 3.

71

$$\boxed{h_3 = 2472.7 \text{ kJ/kg}}$$

$$\begin{aligned} h_3 &= (1 - x_3)h_f + x_3 h_g \\ &= (1 - 0.91) \times 419.1 + 0.91 \times 2675.8 \\ &= 2472.7 \text{ [kJ/kg]} \end{aligned}$$

For heat transfer from the steam:

$$\begin{aligned} \delta Q/T &= \frac{(735.04 - 2472.7)}{(100.0 + 273.15)} \\ &= -4.657 \text{ [kJ/kg K]} \end{aligned}$$

For the complete cycle, therefore:

$$\begin{aligned} \Sigma \frac{\delta Q}{T} &= +4.307 - 4.657 \\ &= -0.350 \text{ [kJ/kg K]} \end{aligned}$$

This negative result shows that irreversible processes are taking place in the cycle. A larger negative result would mean the processes have greater irreversibilities. If the adiabatic expansions had been reversible, the integral of $\delta Q/T$ for the cycle would have been zero. We will be able to check this later. If you apply the Clausius inequality to the Carnot cycle, what result would you expect to get?

72

$$\boxed{\sum_{\text{cycle}} \frac{\delta Q}{T} = 0}$$

Since the Carnot cycle is entirely reversible, we would expect to get that $\Sigma\, \delta Q/T = 0$. If the adiabatic expansions in the example had been reversible, we would have had a Carnot Cycle.

We now come to the second stage in establishing that entropy is a property of a system. This time the argument is quite brief. We have to show that for a reversible cycle:

$$\sum_{\text{cycle}} \frac{\delta Q_R}{T} = 0$$

We can prove this by considering any reversible cycle and by applying the Clausius inequality to the cycle, first in a clockwise and then in an anticlockwise direction.

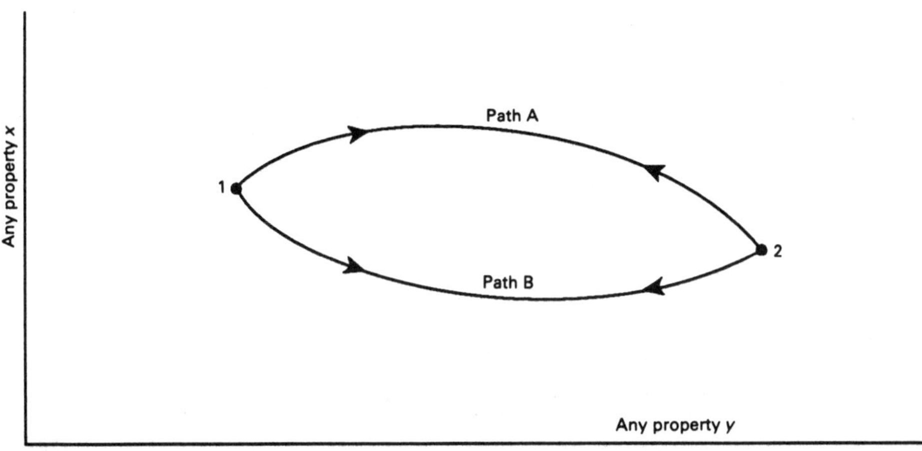

For the clockwise cycle 1A2B1 we can write:

$$\sum_{\text{cycle}} \frac{\delta Q_{\text{Rab}}}{T} \leqslant 0$$

For the anticlockwise cycle 1B2A1 we can say:

$$\sum_{\text{cycle}} \frac{\delta Q_{\text{Rba}}}{T} \leqslant 0$$

Since both cycles are reversible, and the second is the first in the reversed direction, the heat transfers in the two cycles are numerically equal but of opposite sign. The Clausius inequality can only be true for both cases above if the integral round the cycle of $\delta Q_{\text{R}}/T = 0$ for both cases, and so:

$$\sum_{\text{cycle}} \frac{\delta Q_{\text{R}}}{T} = 0$$

for a reversible cycle.

73

We now have to show that for a reversible process, $\int \delta Q_R/T$ between the fixed end states of the process is independent of the path of the process.

Consider a closed system that executes a reversible cycle between two state points only. There could be any number of different paths between these two points. The diagram shows two cycles, one consisting of path A and path B; the other consisting of path A and path C. Since the cycle is reversible, all paths are reversible.

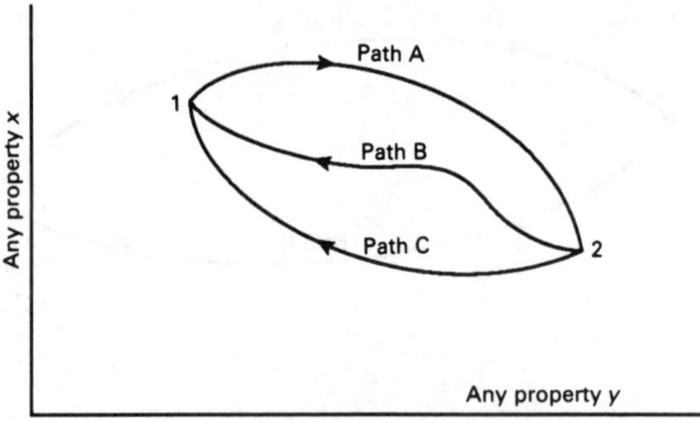

For the cycle of paths A and B, using the result from Frame 72:

$$\int_{A\,1}^{2} \frac{\delta Q_R}{T} + \int_{B\,2}^{1} \frac{\delta Q_R}{T} = 0$$

and for the cycle of paths A and C we have:

$$\int_{A\,1}^{2} \frac{\delta Q_R}{T} + \int_{C\,2}^{1} \frac{\delta Q_R}{T} = 0$$

From these two equations it must follow that:

$$\int_{B\,2}^{1} \frac{\delta Q_R}{T} = \int_{C\,2}^{1} \frac{\delta Q_R}{T}$$

Paths B and C are any two arbitrary paths between the fixed points 1 and 2, so we may conclude that the integral of $\delta Q_R/T$ between two state points is independent of the path, and therefore this integral is equal to a change of a property value between these points. This property is called **entropy**, S. Thus:

$$S_2 - S_1 = \int_{1}^{2} \frac{\delta Q_R}{T}$$

74

This is an equation relating properties; what would it be in differential form?

$$\boxed{dS = \delta Q_R/T}$$

This may also be written as $\delta Q_R = T dS$, where dS is the change of entropy of a system of temperature T having received a boundary interaction of δQ_R. In all real processes heat transfer is not reversible, so is there a relationship between dS, δQ and T? As a precedent we know that $\delta W_R = p dV$, and in compression δW is greater than $p dV$, and in expansion is less (see Frame 35 of Programme 2), and in the next two frames we shall show that $\delta Q \leqslant T dS$.

75

Consider a cycle involving two state points 1 and 2. Between 1 and 2 there is a path A that involves an irreversible heat transfer, and between 2 and 1 there is a reversible heat transfer in path B. The cycle is shown below.

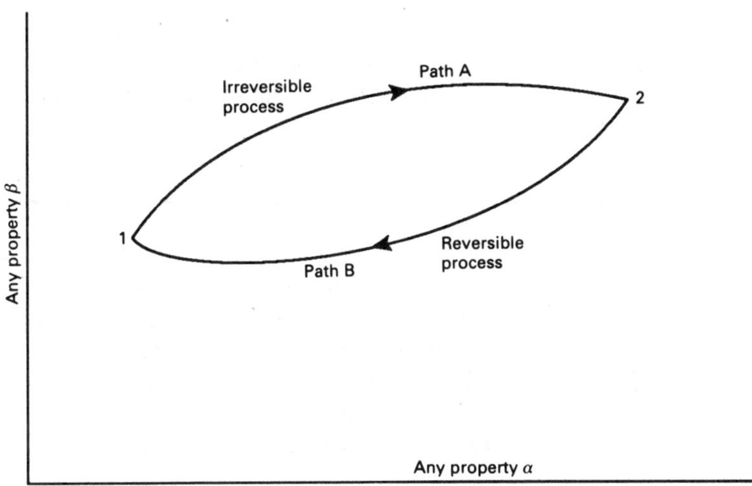

Suppose as an example there was ice on a pond in equilibrium with the atmosphere, all at 0°C. Some of the ice on the top surface is melted by the sun. This is the irreversible heat transfer to the ice; the atmosphere at 0°C is not affected by solar radiation passing through it. Then cloud covers the sun and reversible isothermal heat transfer with the ice and atmosphere freezes the water film again. This completes the cycle. The proof follows in the next frame.

76

By the Clausius inequality, Frame 68, we can say:

$$\int_{A\,1}^{2}\frac{\delta Q}{T}+\int_{B\,2}^{1}\frac{\delta Q_R}{T}<0$$

and from Frame 73 we can say:

$$S_1 - S_2 = \int_{B\,2}^{1}\frac{\delta Q_R}{T}$$

$$\therefore \quad \int_{A\,1}^{2}\frac{\delta Q}{T} + S_1 - S_2 < 0$$

$$\therefore \quad S_2 - S_1 > \int_{1}^{2}\frac{\delta Q}{T}$$

for any irreversible path between state 1 and state 2. Therefore in differential form:

$$\mathrm{d}Q \leqslant T\mathrm{d}S$$

with the equality holding in the case of the reversible heat transfer. We shall use this result in a later Programme to investigate the maximum work obtainable in real processes.

We are now in a position to explore the use of entropy.

77

So that the property entropy may be used in the ways indicated earlier, we need to know the relationships between entropy and other properties with which we are familiar. To start with we can derive two important equations.

In the absence of potential and kinetic energy terms, from the First Law we have:

$$Q - W = U_2 - U_1$$

which, for a reversible process with displacement work, becomes, in differential form in terms of specific properties:

$$\mathrm{d}Q_R - p\mathrm{d}v = \mathrm{d}u$$

From Frame 74, in terms of specific entropy, we have:

$$\mathrm{d}Q_R = T\,\mathrm{d}s \qquad (1)$$

and so from the last two equations we get:

$$T\,\mathrm{d}s = \mathrm{d}u + p\mathrm{d}v \qquad (2)$$

78

This second equation gives us a relation between thermodynamic temperature, pressure and the specific properties of entropy, internal energy and volume.

Using the relation between enthalpy, internal energy, volume and pressure in differential form (Programme 4, Frame 57), we have:

$$\mathrm{d}h = \mathrm{d}u + p\mathrm{d}v + v\mathrm{d}p$$

and:

$$\mathrm{d}u = \mathrm{d}h - p\mathrm{d}v - v\mathrm{d}p$$

Using du from this equation in equation (2) we get:

$$T\,\mathrm{d}s = \mathrm{d}h - v\mathrm{d}p \qquad (3)$$

Equations (2) and (3) have been obtained by considering a particular process, or path, which was a reversible process without any potential or kinetic energy terms present. The equations obtained are simply relations between properties which depend only on end states and not on any path, so these equations may be used to find the change of entropy of a pure substance **in any process, reversible or not**, provided the end states are known.

One very important process is the reversible adiabatic process in which there is no heat transfer. Considering equation (1) in Frame 77:

$$\mathrm{d}Q_{\mathrm{R}} = T\,\mathrm{d}s$$

what is the entropy change in a reversible adiabatic process?

79

$$\boxed{\mathrm{d}s = 0}$$

In equation (1), since $\mathrm{d}Q_{\mathrm{R}} = 0$ in an adiabatic process, and T must have a finite value, it follows that $\mathrm{d}s = 0$.

A reversible adiabatic process in which there is no entropy change is an isentropic process.

80

However, an isentropic process is not necessarily a reversible adiabatic process: a process with irreversibilities and heat transfer could follow an isentropic path.

We have seen how pressure, volume, internal energy and thermodynamic temperature may be measured or calculated, so entropy of a pure substance may be found from these equations. For gases, these equations are combined with the ideal gas rule in differential form to give formulae to use for entropy change calculations. For steam, entropy values are given in the tables.

If temperature is plotted against entropy for steam, the following diagram is obtained:

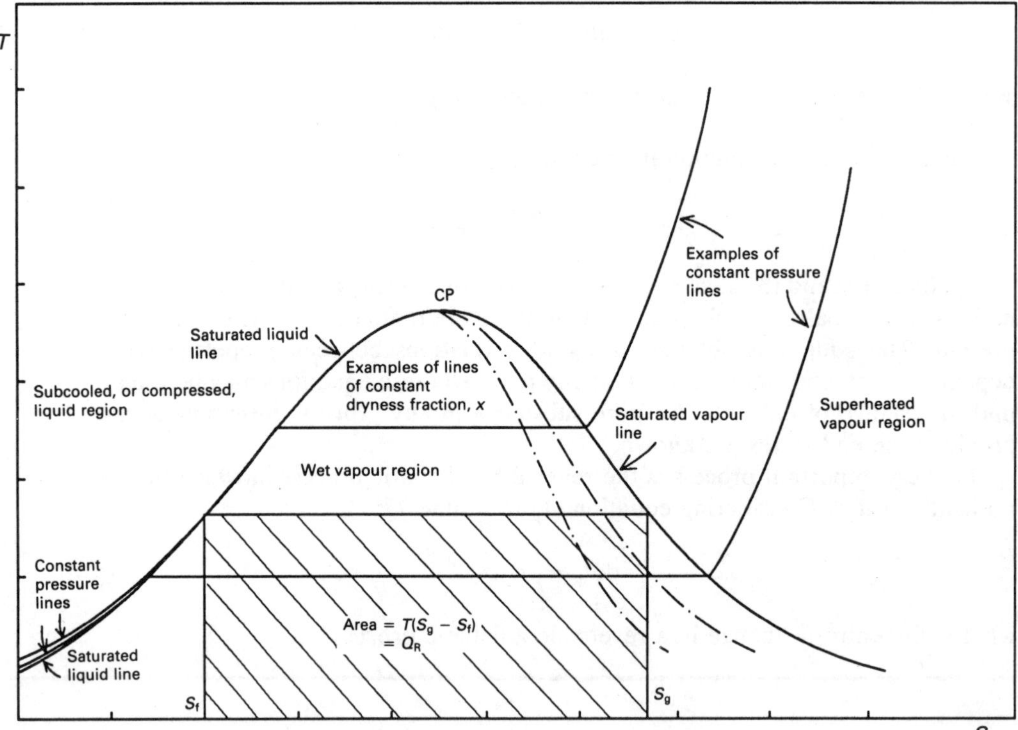

As in previous diagrams, we have the loci of phase transition states giving us the saturated liquid and saturated vapour lines which meet at the critical point. The diagram is then divided into regions of compressed liquid, wet vapour and superheated vapour. Constant pressure lines are also shown. These coincide with isotherms in the wet vapour region.

If the path of a process is shown on the *T–s* diagram, the area under that path line is $\int T\,ds$. If the process is reversible, what does this area represent? (Refer back to equation (1) in Frame 77 if you are in doubt.)

81

Reversible heat transfer, Q_R

We now have that areas under path lines on *T–s* diagrams are reversible heat transfers, in a similar way that areas under path lines on *P–v* diagrams are reversible work transfers. Thus the rectangular area shown under the isotherm joining the saturated liquid and vapour states is the reversible heat transfer for that process between those states.

It will now be obvious that a reversible adiabatic, or isentropic, process will simply be a vertical straight line on the *T–s* property diagram. A Carnot cycle for steam was shown on a *P–v* diagram in Frame 36. You remember that the Carnot cycle consists of two isothermal processes and two adiabatic processes, which with all processes being reversible, become isentropic processes. You should then be able to sketch a Carnot cycle for steam on a *T–s* diagram, with the isothermal process at the higher temperature being from the saturated liquid to the saturated vapour line.

82

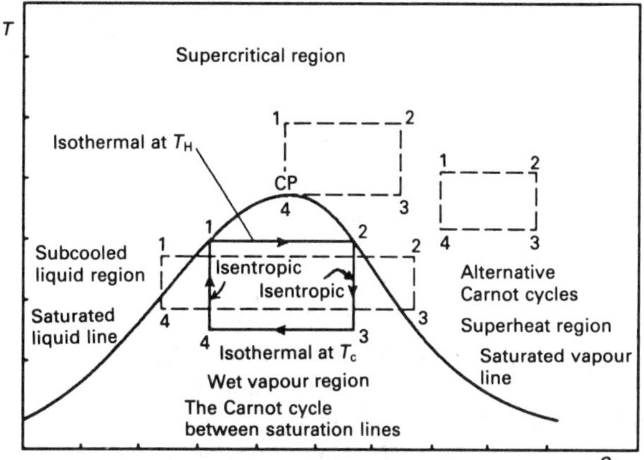

The solid lines in the diagram above show the Carnot cycle *between the saturation lines* that you were asked to draw. In fact, any rectangle with horizontal and vertical sides on a *T–s* diagram represents a Carnot cycle; other examples are shown as dotted lines.

In any of the cycles shown we have that $s_1 = s_4$, and $s_2 = s_3$; also the higher temperature is T_H and the lower temperature is T_C.

83

For the reversible processes:

$$Q_H = T_H(s_2 - s_1)$$

and $$Q_C = T_C(s_4 - s_3) = T_C(s_1 - s_2)$$

Notice that Q_C is a negative quantity, so the net heat in the cycle is:

$$\begin{aligned}\sum\delta Q_R &= T_H(s_2 - s_1) + T_C(s_1 - s_2)\\ &= (T_H - T_C)(s_2 - s_1)\end{aligned}$$

By the First Law this is also the net work of the cycle. The cycle efficiency is therefore the net work over the heat supplied Q_H:

$$\eta_R = \frac{(T_H - T_C)(s_2 - s_1)}{T_H(s_2 - s_1)}$$

$$\eta_R = \frac{(T_H - T_C)}{T_H}$$

We are thus able to confirm the relationship for the efficiency of the Carnot cycle in terms of the temperatures T_H and T_C of the high and low temperature reservoirs.

Another very useful property diagram involving entropy is the enthalpy–entropy, *h–s*, or Mollier diagram. The diagram is named after the German scientist who introduced its use. It is shown opposite.

84

The figure shows the loci of the saturated liquid and vapour states, the critical point, regions of compressed liquid, wet vapour and superheated vapour, and lines of constant pressure and temperature. Lines of constant quality or dryness fraction are also shown. The *h–s* chart is useful for quick (but less accurate) determination of properties for calculations. Thus, if steam expands isentropically from 2 MN/m^2, or 20 bar, and 400°C to a pressure of 20 kN/m^2, or 0.2 bar, what is the difference in value of specific enthalpy between the initial and final states, and what is the final state of the steam? We find the first point on the chart at 20 bar and 400°C, and draw a vertical line to meet the 0.2 bar pressure line. The point of intersection represents the final state and we can see that the steam is 0.89 dry at 0.2 bar. Then we can read off the specific enthalpy values at the two points as 3245 and 2350 kJ/kg, so the difference is 895 kJ/kg. We shall be using this chart in the next programme.

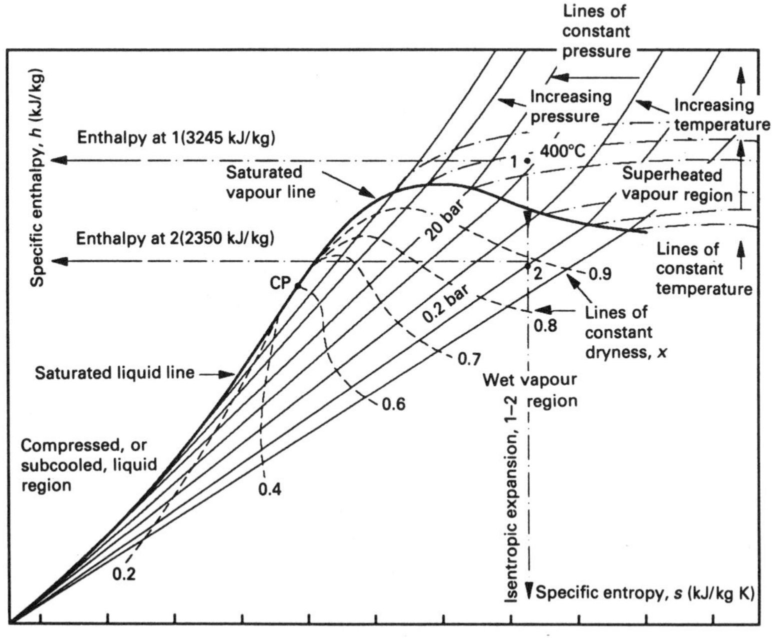

Values of entropy are given in the steam tables. Being an extensive property like volume, internal energy and enthalpy, entropy can be expressed specifically, and for the wet vapour region we have relationships in terms of the properties for saturated liquid, saturated vapour and dryness fraction, of similar form to those we met in Programme 4. Thus:

$$s = (1 - x)\, s_1 + x s_g$$
$$= s_f + x s_{fg}$$
$$= s_g - (1 - x) s_{fg}$$

where: $$s_{fg} = s_g - s_f$$

and where s_f = specific entropy of saturated liquid, and s_g = specific entropy of saturated vapour. The units are kJ/kg K. For this region the Rogers and Mayhew tables and the Cooper and Le Fevre tables give values of s_f, s_{fg} and s_g, while the Haywood tables give only s_f and s_g, from which s_{fg} may be found.

Entropy values for steam are based on a datum for saturated liquid at the triple point. All values are calculated from this datum using the equations given in Frames 77 and 78.

In the next frame we shall see how we can check the self-consistency of the data.

86

Thus from equation (3) of Frame 78, for a constant pressure process, with dp = 0, we have:

$$T\,\mathrm{d}s = \mathrm{d}h$$

For the change of state between saturated liquid and saturated vapour, this gives:

$$T\,s_{fg} = h_{fg}$$

and using Rogers and Mayhew data, for saturated steam at 100°C:

$$(100 + 273.15)\,s_{fg} = h_{fg} = 2256.7$$

Hence:

$$s_{fg} = 2256.7/373.15$$
$$= 6.0477\ [\text{kJ/kg K}]$$

This is consistent with the tabulated value of 6.048 [kJ/kg K].

At 100 bar and 311°C, s_f = 3.360 kJ/kg K and h_{fg} = 1317 kJ/kg. What is the value of s_g?

87

5.615 kJ/kg K

We have s_f = 3.360 kJ/kg K and s_{fg} = 1317/(273.15 + 311) for the constant pressure process, so:

$$s_g = s_f + s_{fg} = 3.360 + 1317/(273.15 + 311)$$
$$= 3.360 + 2.255$$
$$= 5.615\ [\text{kJ/kg K}]$$

You can check that these figures are consistent in the Rogers and Mayhew tables, and similar checks may be made with the other tables.

In the superheat states the values of specific entropy, s, are tabulated along with values of v, u and h. For the refrigerant fluids, values of s_f and s_g of the saturated states, and s of the superheat states, are given.

We shall explore the use of entropy values for vapours, both steam and refrigerant fluids, in the next programme.

88

In this frame we shall develop property relationships involving entropy for perfect gases.

We may start with equation (2) of Frame 77:

$$T\,ds = du + p dv$$

so

$$ds = \frac{du}{T} + \frac{p dv}{T} \tag{4}$$

and then using $du = c_V dT$ and the ideal gas rule $pv = RT$, we have:

$$ds = c_V \frac{dT}{T} + \frac{R dv}{v}$$

We may then integrate to find the change of entropy between two states, with c_v constant for a perfect gas:

$$s_2 - s_1 = c_V \ln\left(\frac{T_2}{T_1}\right) + R \ln\left(\frac{v_2}{v_1}\right) \tag{5}$$

Suppose some air expands over a volume ratio (v_2/v_1) of 5 and the temperature ratio (T_1/T_2) for this expansion is 1.495. You can use equation (5) to find the change in specific entropy. For air, $c_v = 0.718$ kJ/kg K and $R = 0.287$ kJ/kg K.

89

+0.1732 kJ/kg K

We have:

$$\begin{aligned} s_2 - s_1 &= 0.718 \ln(1/1.495) + 0.287 \ln(5) \\ &= 0.718 \times (-0.4021) + 0.287 \times 1.6094 \\ &= -0.2887 + 0.4619 \\ &= +0.1732 \text{ [kJ/kg K]} \end{aligned}$$

One possible process between the end states given would have been a reversible polytropic expansion of index 1.25, with heat transfer to the air and a positive work of expansion. For this process we see the entropy increase is 0.1732 kJ/kg K.

90

Air is contained behind a diaphragm in half the volume of an insulated rigid container. The other half of the volume is a vacuum. As the air is released to fill the whole volume, what is the increase in specific entropy? For air, $R = 0.287$ kJ/kg K. (To start, apply the First Law to see whether there is any change of temperature, and again use equation (5).)

91

0.1989 kJ/kg K

The process is adiabatic, but not reversible, and no work is done against the vacuum so from the First Law there is no change of internal energy, and so there is no change of temperature. In the expansion the volume doubles. From equation (5) we have:

$$s_2 - s_1 = R \ln(v_2/v_1)$$

as the temperature term disappears. Hence:

$$\begin{aligned} s_2 - s_1 &= 0.287 \times 0.6931 \\ &= 0.1989 \text{ [kJ/kg K]} \end{aligned}$$

This is the entropy increase in an isolated system as the system proceeds from a non-equilibrium to an equilibrium state.

It is possible to get two more relationships from equation (5). By substituting for v in the last term using the ideal gas rule we get:

$$s_2 - s_1 = c_V \ln\left(\frac{T_2}{T_1}\right) + R \ln\left(\frac{T_2 p_1}{T_1 p_2}\right)$$

and by combining the temperature ratio terms together this gives:

$$s_2 - s_1 = c_P \ln\left(\frac{T_2}{T_1}\right) - R \ln\left(\frac{p_2}{p_1}\right) \tag{6}$$

Finally we can substitute for T in the first term again using the ideal gas rule:

$$s_2 - s_1 = c_P \ln\left(\frac{p_2 v_2}{p_1 v_1}\right) - R \ln\left(\frac{p_2}{p_1}\right)$$

which gives, on combining the pressure ratio terms together:

$$s_2 - s_1 = c_P \ln\left(\frac{v_2}{v_1}\right) + c_V \ln\left(\frac{p_2}{p_1}\right) \tag{7}$$

The choice of equation to use really depends on which properties are known in any particular problem. It is much better to remember how these equations are derived rather than to try and remember the equations themselves. Remember they give the entropy change between two known end states; the process between these end states does not matter, it may be either reversible or irreversible.

92

We shall now see how we can use these results to construct T–s diagrams for gases. These will consist of lines at constant volume and constant pressure which may be drawn using equations (5) and (6) from Frames 88 and 91. We shall be using these diagrams later on, when we are dealing with internal combustion engines and gas turbines.

First we have to choose a datum state with a subscript of 0 at which specific entropy is zero. As long as the datum state is specified, it does not matter too much which state you choose. We can construct a constant volume line from equation (5) in Frame 88:

$$s_2 - s_1 = c_V \ln\left(\frac{T_2}{T_1}\right) + R \ln\left(\frac{v_2}{v_1}\right)$$

With $v_2 = v_1 = v_0$, the entropy of a state 2 relative to the datum state 0 is given by:

$$s_2 - s_0 = c_V \ln\left(\frac{T_2}{T_0}\right)$$

This equation is used to plot a constant volume line. Thus with the chosen v_0 and T_0, s_0 is zero, and at T_2, s_2 has the value given by this equation. By selecting a range of values of T_2, a range of values of s_2 is calculated and plotted for the line at v_0. This line is shown in the diagram in the next frame.

93

The diagram shows a single constant volume line on the T–s plane.

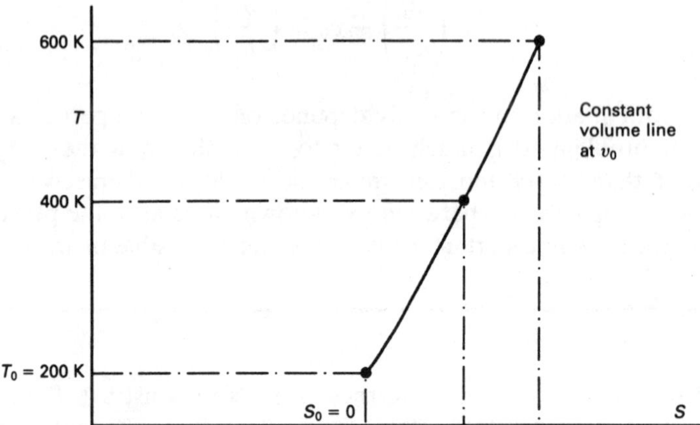

Why does the slope of the line increase with temperature?

94

Because of the logarithmic function of T_2/T_0

To illustrate this, choose a datum at 200 K, and show that s at 400 K and 600 K is 0.496 and 0.786 kJ/kg K, respectively, with c_V = 0.715 kJ/kg K. Next we can start a new constant volume line at $v = v_2$ and at $T = T_0$. From equation (5) in Frame 88 we get:

$$s_2 - s_0 = R \ln\left(\frac{v_2}{v_0}\right)$$

Again with s_0 at zero, with this value of s_2 as a starting point we can construct another line for the volume of v_2.

In a similar way, using equation (6) from Frame 91:

$$s_2 - s_1 = c_P \ln\left(\frac{T_2}{T_1}\right) - R \ln\left(\frac{p_2}{p_1}\right)$$

we can construct lines at constant pressure. At the pressure of $p_2 = p_1 = p_0$, we have:

$$s_2 - s_0 = c_P \ln\left(\frac{T_2}{T_0}\right) \quad (8)$$

and for other pressures we have:

$$s_2 - s_0 = -R \ln\left(\frac{p_2}{p_0}\right)$$

A diagram showing lines at both constant volume and constant pressure is given below.

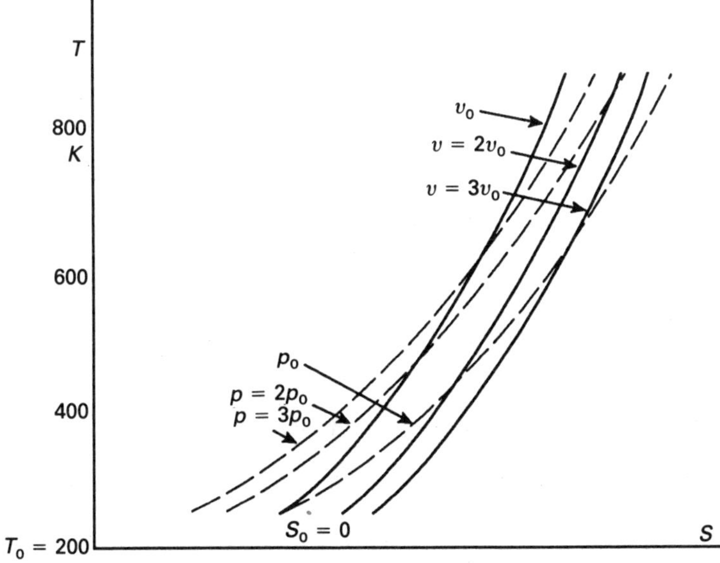

The lines of constant volume have a steeper slope than those of constant pressure. We can see why this is by finding ds/dT and hence dT/ds for both a constant volume and constant pressure line. Also we remember that c_p is greater than c_v.

From the constant volume line relationship in Frame 92:

$$s_2 - s_0 = c_V \ln\left(\frac{T_2}{T_0}\right)$$

we get:
$$\frac{ds}{dT} = \frac{c_V}{T}$$

which means that:
$$dT/ds = T/c_V$$

Now find dT/ds for the constant pressure line from equation (8) above.

95

Similarly for a constant pressure line, using:

$$s_2 - s_0 = c_P \ln\left(\frac{T_2}{T_0}\right)$$

we shall get: $$ds/dT = c_P/T$$

and hence: $$dT/ds = T/c_P$$

Since c_P has the larger value, the slope of the constant pressure line at any particular temperature is less than the slope of the constant volume line.

96

Lastly in this programme we shall see how equation (3) obtained in Frame 78:

$$T\,ds = dh - vdp$$

may be used in conjunction with the steady flow energy equation for an open system, which may be expressed in differential terms as:

$$dQ - dW = d\left(h + \frac{V^2}{2} + gz\right)$$

We stipulate that if the flow is **reversible**, then $dQ = dQ_R$ and $dQ_R = T\,ds$, so the steady flow energy equation becomes:

$$T\,ds - dW = d\left(h + \frac{V^2}{2} + gz\right)$$

In this equation $T\,ds = dh - vdp$, and for zero work dW, we get:

$$dh - vdp = d\left(h + \frac{V^2}{2} + gz\right)$$

or: $$-vdp = d\left(\frac{V^2}{2} + gz\right)$$

For the flow to be reversible it must be frictionless, so this result is in fact the *Euler equation* of fluid mechanics.

If, in the previous equation, the flow is **incompressible**, which means that the specific volume of the fluid remains constant, then we may write:

$$-\frac{dp}{\varrho} = d\left(\frac{V^2}{2} + gz\right)$$

This result is the **Bernoulli equation** of fluid mechanics. These results are examples of how the specific areas of different fields of study can merge.

97

In the following problems involving the use of tables or charts, your answers may vary slightly depending on the sources you use.

1. At 0.8 bar, in the wet steam region, the rate of change of temperature with pressure is 33.649 K/bar; using the Clausius–Clapeyron equation, calculate the absolute temperature. [366.65 K]
2. Calculate the entropy of steam at the following states, using your tables:
   (a) 1.6 MN/m², $x = 0.5$; [4.383 kJ/kg K]
   (b) 16.2 MN/m², $x = 0.8$; [4.939 kJ/kg K]
   (c) MN/m² and 325°C. [6.558 kJ/kg K]
3. From your h–s chart, read off the specific entropy of steam at the following conditions:
   (a) the critical point;
   (b) dry saturated at 0.01 bar, 0.14 bar, 18 bar and 90 bar;
   (c) 10 bar, 0.92 dry; 120 bar, 0.88 dry; 20 bar, 400°C; 100 bar, 700°C.
   [(a) 4.4, (b) 8.98, 8.04, 6.37, 5.66, (c) 6.23, 5.26, 7.13, 7.17 in kJ/kg K]
   You may care to check your answers against values from your tables.
4. In a Carnot engine using steam, isothermal heat transfer to the steam between the saturated liquid and vapour states at 200 bar gives an entropy increase of 0.914 kJ/kg K. Isothermal heat rejection is at 40 bar. Calculate (a) the heat transfer to the steam at 200 bar, check your value agrees with h_{fg} at 200 bar, and (b) the Carnot efficiency. [(a) 583.9 kJ/kg, (b) 18.064 %]
5. Calculate the entropy of Refrigerant 12 at the following states:
   (a) − 10°C, $x = 0.5$; [0.4050 kJ/kg K]
   (b) 9.607 bar, 10 K of superheat; [0.7065 kJ/kg K]
   (c) 15.26 bar, 20 K of superheat. [0.7261 kJ/kg K]
6. Calculate the entropy of air in the following examples relative to the given datum:
   (a) at 400 K relative to air at 290 K at the same volume; [0.2309 kJ/kg K]
   (b) at 500 K relative to air at 300 K at the same pressure; [0.5134 kJ/kg K]
   (c) at v of 0.45 m³/kg relative to v of 0.9 m⁹/kg at the same temperature.
   [− 0.1989 kJ/kg K]

98

Points to remember

A heat engine is a continuously operating system at the boundary of which there are continuous heat and work interactions.

Kelvin–Plank statement of the Second Law: it is impossible to construct a cyclic process whose sole result is the flow of heat from a single heat reservoir and the performance of an equivalent amount of work.

Clausius statement of the Second Law: it is impossible for any device to convey heat from one region to another at a higher temperature, except by performing work on that device.

Heat engine efficiency: $\eta = (Q_1 - Q_2)/Q_1$.

COP(refrigerator) = Q_2/W; COP(heat pump) = Q_1/W.

COP(heat pump) = 1 + COP(refrigerator).

For a reversible engine: η_R = 1/COP(heat pump).

The efficiencies of all reversible engines operating between the same two temperatures are the same.

The efficiency of the Carnot cycle, between temperatures T_H and T_C, is:

$$\eta_R = (T_H - T_C)/T_H$$

The Kelvin scale of thermodynamic temperature is defined in terms of the efficiency of reversible engines such that:

$$\eta_R \equiv (T_H - T_C)T_H$$

The Clausius–Clapeyron equation states:

$$T = \frac{h_{fg}}{v_{fg}} \frac{\delta T}{\delta p}$$

The change from a non-equilibrium state to an equilibrium state in an isolated system is accompanied by an entropy increase.

The Clausius inequality states that in a cycle: $\sum \frac{\delta Q}{T} \leqslant 0$.

A change of entropy is given by: $S_2 - S_1 = \int_1^2 \frac{\delta Q_R}{T}$.

Relations between properties:

$$\delta Q_R = T\,ds$$

$$T\,ds = du + p dv$$

$$T\,ds = dh - v dp$$

An adiabatic process in which there is no entropy change is an isentropic process. Entropy relationships for ideal gases, (5), (6) and (7), are given in Frames 88 and 91.

Programme 7

VAPOUR POWER CYCLES AND REVERSED VAPOUR CYCLES

1

In this programme we will continue to explore how work may be obtained from heat. We have seen how the Second Law places a restriction on what may be achieved, and now we will see how the property **entropy** will help us in our calculations. First we will look again at the steam cycle considered in Frames 69 and 71 in the previous programme. The *p–v* diagram of the cycle is shown again below. State 1 is saturated liquid at 1.2 MN/m², or 12 bar, and 188°C, and state 2 is saturated vapour at the same temperature and pressure. The lower pressure is atmospheric, 101.325 kN/m², or 1.013 25 bar and 100°C, and the dryness fractions at states 3 and 4 are 0.91 and 0.14 respectively.

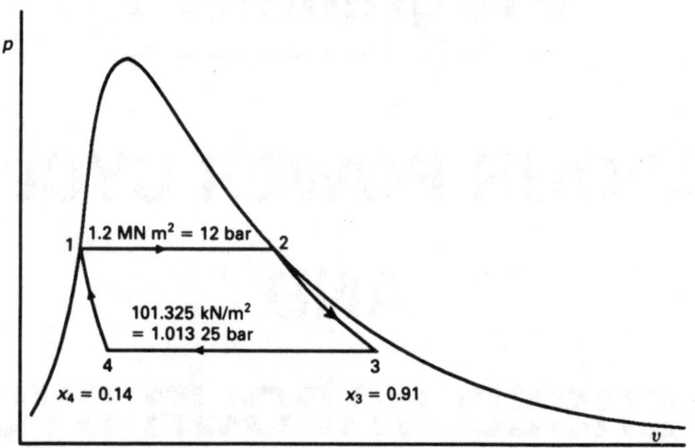

2

The same cycle is now shown again on a temperature–entropy diagram:

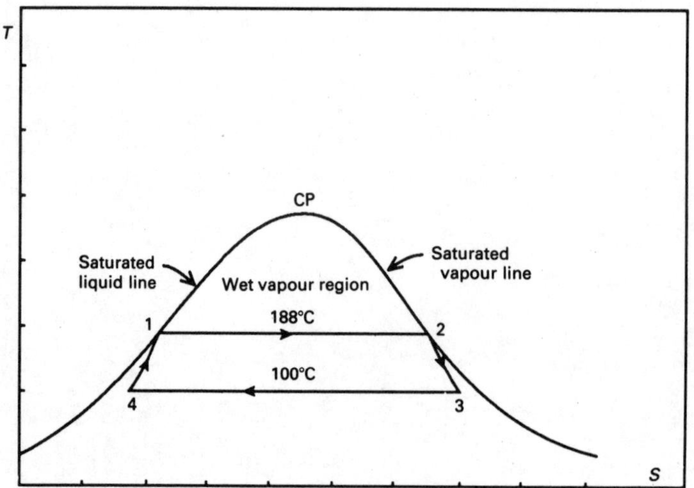

The constant pressure paths on the *p–v* diagram become *isothermal* paths at 188°C and 100°C, at which heat transfers take place, on the *T–s* diagram.

Heat transfer to the fluid causes saturated water at state 1 to become saturated steam at state 2. We will use the equation:

$$s_2 - s_1 = \int_1^2 \frac{\delta Q_R}{T}$$

from Frame 73 of the previous Programme, for unit mass of steam between states 1 and 2. State 1 is saturated water at 12 bar. The specific entropy s_1 is s_f which is 2.216 kJ/kg K from column 9 on page 4 of the Rogers and Mayhew tables. For state 2, s_2 is s_g at 12 bar which is 6.523 kJ/kg K.

Hence for heat transfer to the steam:

$$\delta Q_R/T = 6.523 - 2.216 = 4.307 \text{ [kJ/kg K]}$$

3

At state 3, the dryness fraction is 0.91 at 1.013 25 bar. We can use:

$$s = (1 - x)s_f + xs_g$$

from Frame 85 of the previous programme to find the specific entropy at state 3. From the same tables, $s_f = 1.307$ kJ/kg K and $s_g = 7.355$ kJ/kg K at 1.013 25 bar, so:

$$\begin{aligned} s_3 &= (1 - 0.91) \times 1.307 + 0.91 \times 7.355 \\ &= 6.8107 \text{ [kJ/kg K]} \end{aligned}$$

This value is slightly greater than the value for state 2, and you can see that state 3 is shown to the right of state 2 on the entropy axis. Because the adiabatic process between states 2 and 3 was not reversible, the entropy has increased. Using the same values for s_f and s_g you should now be able to calculate the specific entropy at state 4 where the dryness fraction is 0.14.

4

The answer is that $s_4 = 2.1537$ kJ/kg K.

We have that:

$$\begin{aligned} s_4 &= (1 - 0.14) \times 1.307 + 0.14 \times 7.355 \\ &= 2.1537 \text{ [kJ/kg K]} \end{aligned}$$

5

Then, using these values of s_3 and s_4, for heat transfer from the steam:

$$\begin{aligned}\delta Q_R/T &= s_4 - s_3\\ &= 2.1537 - 6.8107\\ &= -\,4.657\ [\text{kJ/kg K}]\end{aligned}$$

For the whole cycle, therefore:

$$\begin{aligned}\delta Q_R/T &= +\,4.307 - 4.657\\ &= -\,0.350\ [\text{kJ/kg K}]\end{aligned}$$

Thus by using the entropy function, our answer agrees with the result we got in Frame 71 of the previous programme. The value we obtained for s_4 is less than the value of s_1 so state 4 is to the left of state 1 on the entropy axis in the figure in Frame 2. Again, the process between states 4 and 1 was not reversible. If in this cycle $s_3 = s_2$ and $s_1 = s_4$, what could we say about the processes 2–3 and 4–1?

6

The processes are isentropic and reversible

In Frame 69 of the previous programme we saw that these two processes, though adiabatic, were not reversible because of friction in the expansion and compression respectively. If these processes were reversible they would then be vertical lines on the temperature–entropy diagram. With the heat transfers already being reversible we would then have a Carnot cycle which is a rectangle on the temperature–entropy diagram, shown on the next page.

Because they were not reversible we obtained a result in agreement with the statement of the Clausius inequality in Frame 68 in the previous programme. If we now consider the Carnot cycle, we can say that with $s_3 = s_2$ and $s_4 = s_1$, the heat transfer to the steam is:

$$\begin{aligned}Q_H &= (s_2 - s_1)\,T_H\\ &= (6.523 - 2.216) \times (188.0 + 273.15)\\ &= 1986.2\ [\text{kJ/kg}]\end{aligned}$$

The heat transfer from the steam is:

$$Q_C = (s_4 - s_3)\, T_C$$
$$= (2.216 - 6.523) \times (100.0 + 273.15)$$
$$= -\ 1607.2\ \text{[kJ/kg]}$$

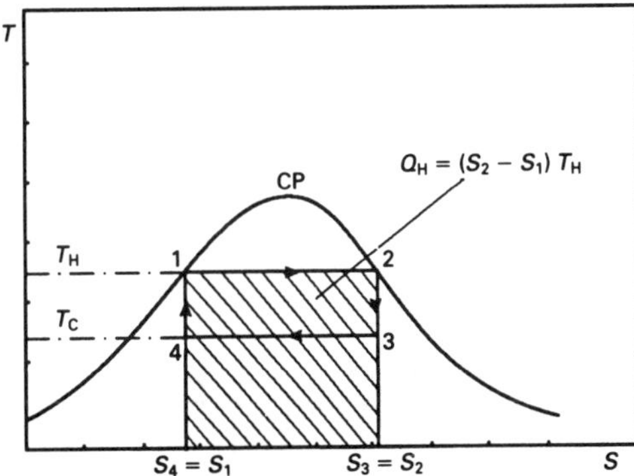

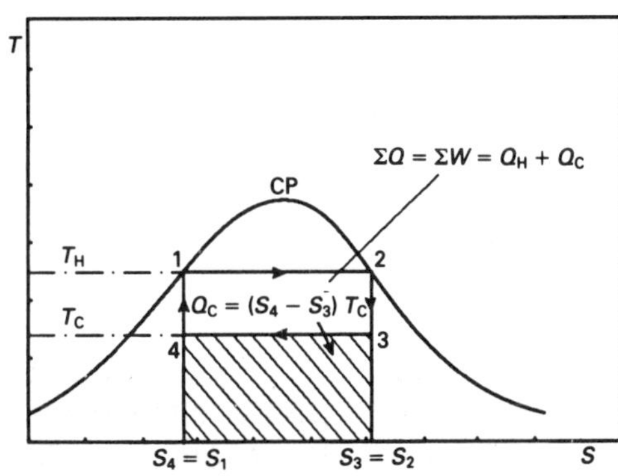

The diagrams above show the two rectangular areas which represent these two heat transfer terms. Remembering the First Law tells us that the net work transfer in a cycle is equal to the net heat transfer, what is the net work output of this cycle, and what is the cycle efficiency?

7

$$\boxed{W = 379.0\ \text{kJ/kg},\ \eta = 19.08\%}$$

The net heat transfer is: $\Sigma\, Q = 1986.2 - 1607.2 = 379.0$ [kJ/kg] $= \Sigma\, W$. This is the net work output of the cycle and is the area enclosed by the cycle on the *T–s* diagram. The cycle efficiency is:

$$\eta = \frac{\text{net work}}{\text{heat input}} = \frac{\Sigma Q}{Q_H} = (379.0/1986.2) \times 100\% = 19.08\%$$

8

Even though we are using the Carnot cycle which is the most efficient thermodynamic cycle for a heat engine, for the chosen conditions we have to reject over four-fifths of the heat supplied.

The same result must follow from the efficiency of the Carnot cycle expressed in terms of the thermodynamic temperatures between which the cycle operates. We have just seen that the cycle efficiency is given by:

$$\eta = \frac{\Sigma Q}{Q_H} = \frac{Q_H + Q_C}{Q_H}$$

where Q_c has its correct negative sign, or with a positive sign for Q_c the expression is:

$$\eta = \frac{Q_H - Q_C}{Q_H} = \frac{T_H - T_C}{T_H}$$

by the definition of thermodynamic temperature. The expressions for Q_H and Q_C in Frame 6 will also give this result. Now you can use the values of T_H and T_C to confirm the efficiency of the cycle.

9

$$\boxed{\eta = 19.08\%}$$

$$\eta = (T_H - T_C)/T_H = [(188.0 - 100.0)/(188.0 + 273.15)] \times 100\%$$
$$= 19.08\%$$

Now we will consider steam between the same conditions and with the same state points as in this example, in steady flow in the type of plant depicted in Frame 14 of the previous programme. The steam in passing through the plant executes a Carnot cycle. The plant is shown again below. Reversible isothermal heat transfers at T_H and T_C respectively occur in the boiler and condenser, and positive work of expansion is done in the turbine and negative work is done on the condensed steam in the feed pump. The state points at positions on the plant diagram are shown on the T–s diagram. The turbine and feed-pump processes are isentropic, giving us vertical lines on the T–s diagram.

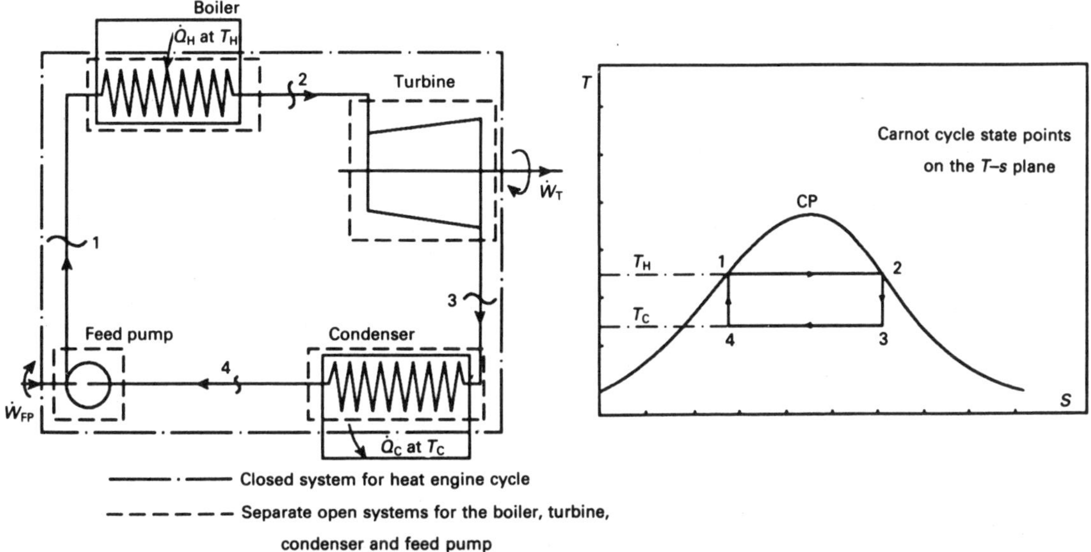

The steady flow energy equation, from Frame 40 of Programme 3, may be written for unit mass flow rate of steam with no changes in potential energy:

$$\dot{Q} - \dot{W}_S = (h_2 - h_1) + \left(\frac{V_2^2}{2} - \frac{V_1^2}{2}\right)$$

where 1 and 2 refer to the states of the fluid entering and leaving an open system, and $\dot{W}_S$ is the rate of shaft work at the boundary. This equation may be applied to each of the four open systems in turn, within the closed system of the heat engine shown in the figure. In the plant diagram, the fluid circulates continuously in the direction shown and the interactions with the closed system boundary are $\dot{Q}_H$ at the boiler, $\dot{W}_T$ at the turbine, $\dot{Q}_C$ at the condenser, and $\dot{W}_{FP}$ at the feed pump. We shall apply the steady flow energy equation to the boiler and turbine respectively, using the state points from the plant diagram, and we will neglect the kinetic energy terms because they are small in comparison with other terms. Thus, for unit mass flow:

Boiler: $$\dot{Q}_H = h_2 - h_1$$

Turbine: $$-\dot{W}_T = h_3 - h_2$$

It is assumed the fluid enters the turbine at the same state that it leaves the boiler. Making similar assumptions, you should be able to obtain corresponding energy equations for the condenser and feed pump.

10

Condener: $\dot{Q}_C = h_4 - h_3$
Feed pump: $-\dot{W}_{FP} = h_1 - h_4$

Steam enters the condenser at state 3 and leaves at state 4. The wet steam at state 4 enters the feed pump and leaves as saturated water at state 1. We will now use these energy equations for the components in the plant to find the heat supplied, the net work and the cycle efficiency. At the boiler entry at state 1 we have saturated water at 12 bar, and at state 2 at the boiler exit we have saturated steam at 12 bar. The heat transfer in the boiler is therefore given by:

$$\begin{aligned}\dot{Q}_H &= h_2 - h_1 = h_g - h_f\\ &= 2784 - 798\\ &= 1986.0 \text{ [kJ/kg], for unit mass flow of steam}\end{aligned}$$

The values of h_g and h_f came from page 4 of the Rogers and Mayhew tables.

For the turbine we have h_2 = 2784 kJ/kg, and we need to find h_3. Without the entropy function we could not find h_3, but now we know that $s_3 = s_2$. Hence we know the state of the steam at point 3 since we know the pressure is 1.013 25 bar. Using any of the equations from Frame 85 of the previous programme you should be able to find the dryness fraction after the isentropic expansion from the dry saturated state at 12 bar to the wet state at 1.013 25 bar. Then, knowing the dryness fraction at 1.013 25 bar, find the enthalpy at state 3.

11

$x_3 = 0.8624$, $h_3 = 2365.28$ kJ/kg

Again, using Rogers and Mayhew data in this example, with s_2 = 6.523 kJ/kg K, and at 1.013 25 bar s_f = 1.307, and s_{fg} = 6.048 kJ/kg K,

we have: $$6.523 = 1.307 + x_3\, 6.048$$

Hence: $$x_3 = 0.8624$$

Using the enthalpy values h_f and h_{fg} at the same state we have:

$$\begin{aligned}h_3 &= 419.1 + 0.8624 \times 2256.7\\ &= 2365.28 \text{ [kJ/kg]}\end{aligned}$$

We can now use the equation from Frame 9 to find the turbine work. Thus:

$$
\begin{aligned}
-\dot{W}_T &= h_3 - h_2 \\
&= 2365.28 - 2784.0 \\
&= -418.72 \text{ [kJ/kg]} \\
\dot{W}_T &= +418.72 \text{ [kJ/kg]}
\end{aligned}
$$

12

Turbine work can also be found from the *h–s* chart. We find the position on the chart of the initial state of the steam, in this case saturated steam at 12 bar, and from this point draw a vertical line downwards to meet the pressure line at the final state. For a final pressure of 1.013 25 bar we have to take 1.0 bar. The method is shown in the figure.

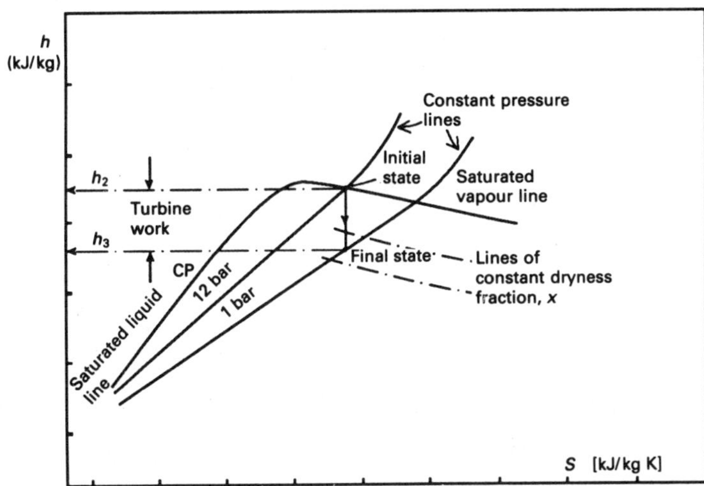

The specific enthalpy at the two end states, and the final state of the steam, can then be read from the chart. The overall process is rather quicker, but not quite as accurate as using the tables.

These steps in finding the turbine work are important. For practice, from both tables and chart, now find the turbine work with an isentropic expansion between the dry saturated condition at 40 bar, and a final pressure of 1.5 bar.

13

h_2 = 2801 kJ/kg, s_2 = 6.070 kJ/kg K, x_3 = 0.801,
h_3 = 2249.65 kJ/kg K, W_T = +551.35 kJ/kg

The full solution is not given; you should be able to follow the methods in the previous two frames to obtain x_3, h_3, and W_T.

Returning to the Carnot cycle, we now need to use the equations in Frame 10 to find the condenser heat transfer and the feed-pump work. For these we need h_1 and h_4; we know that state 1 is saturated liquid at 12 bar, so we can look up h_1, and we know that $s_4 = s_1$ and as we already know that s_1 is 2.216 kJ/kg K, we can calculate x_4 and h_4. Do that.

14

h_1 = 798 kJ/kg, x_4 = 0.1503, h_4 = 758.28 kJ/kg

State 1 is saturated liquid at 12 bar, so h_1 is 798 kJ/kg K from Rogers and Mayhew tables and using the values given for s_f and s_{fg} at 1.013 25 bar, we have:

$$2.216 = 1.307 + x_4\,6.048$$

$$\therefore \quad x_4 = 0.1503$$

Then at 1.01325 bar: $h_4 = 419.1 + 0.1503 \times 2256.7$

$$\therefore \quad h_4 = 758.28\ [\text{kJ/kg}]$$

Now for the condenser, using:

$$\dot{Q}_C = h_4 - h_3$$

from Frame 10, and h_3 of 2365.28 kJ/kg from Frame 11, we have:

$$\dot{Q}_C = 758.28 - 2365.28$$

$$= -1607\ [\text{kJ/kg}]$$

This is heat transfer from the steam, so it has to be negative. Now from the formula from Frame 10, calculate the feed-pump work.

$$\boxed{\dot{W}_{FP} = -39.72 \text{ kJ/kg}}$$

The required formula is:

$$-\dot{W}_{FP} = h_1 - h_4$$

and with h_1 of 798 kJ/kg and h_4 of 758.28 kJ/kg, we have:

$$-\dot{W}_{FP} = 798 - 758.28 = 39.72 \text{ [kJ/kg]}$$

$$\therefore \quad \dot{W}_{FP} = -39.72 \text{ [kJ/kg]}$$

The negative sign is consistent with work done on the fluid.

This work has to come from somewhere, and as the turbine work output is 418.72 kJ/kg from Frame 11, we can take it from that. This means that the net work output of the cycle, $\Sigma\, W$, is:

$$\begin{aligned} \sum W &= \dot{W}_T + \dot{W}_{FP} \\ &= 418.72 - 39.72 \\ &= 379.0 \text{ [kJ/kg]} \end{aligned}$$

You can check that this agrees with the value obtained from consideration of heat transfers in Frame 7. The efficiency of the cycle is $\Sigma\, W/Q_H$ which gives:

$$\begin{aligned} \eta &= 379.0/1986.0 \times 100\% \\ &= 19.08\% \end{aligned}$$

which must, and does, agree with the previously obtained value.

In power-producing thermodynamic cycles, some negative work is always involved. As it is provided from the positive work produced, it is important to keep it to a small value so that the net work is as large as possible. The relative amount of net work can be expressed by the **work ratio**, where:

$$\text{Work ratio} = \frac{\text{Net work output}}{\text{Positive work}}$$

From the figures above, what is the work ratio of this Carnot cycle?

16

0.905

The net work is 379.0 kJ/kg and the positive work is 418.72 kJ/kg, so:

$$\text{Work ratio} = 379.0/418.72 = 0.905$$

Work ratio is an important parameter. Clearly in any cycle its value should be as close as possible to 1. In exploring the concept of work ratio it is instructive for us today to recall how the first steam engines evolved. At the time only steam at atmospheric pressure was available, produced in cauldrons, or large kettles. One of the main needs was to pump water out of mines, and this was achieved with a beam engine. One end of the beam was attached to the piston in a cylinder, and the other end was attached to the water pump. This is illustrated below.

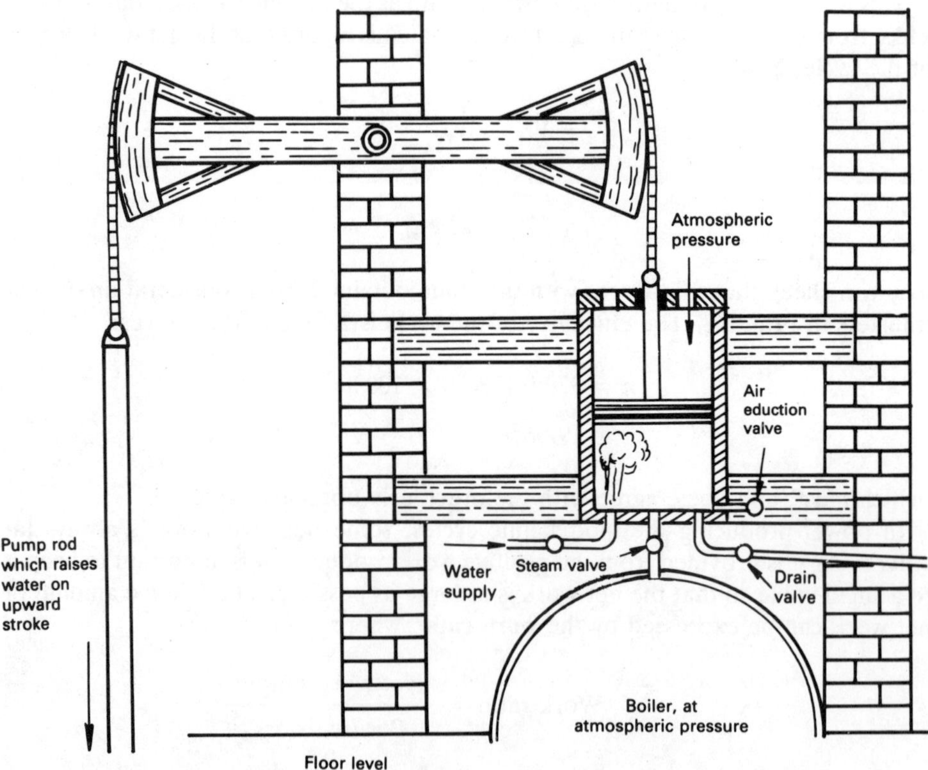

With steam at only atmospheric pressure, how could steam be used to make a piston move, bearing in mind that there would be atmospheric pressure on both sides of the piston?

17

By condensing the steam to create a vacuum

In the pumping cycle the pump rod is heavier than the piston, so the cycle starts with the piston at the top of the cylinder. Steam enters the cylinder and displaces the air through the eduction valve, which is then closed, and then water is injected into the cylinder to condense the steam. (All valve operations were performed by hand!) A vacuum is created and the pressure difference across the piston is around 1 bar. As the cylinder was often very large this created a tremendous force for the water pumping stroke. As the piston descended and the pump rod rose, water was forced upwards and at the end of the stroke the rod descended again by its own weight. In these engines the atmosphere was doing the positive work and the force of gravity was doing the negative work. Subject to the constraints of making and sealing large cylinders, it was relatively easy to design and build a successful atmospheric beam engine. Newcomen's engine of 1712 was a well known example.

However, they were very inefficient as the cylinder was repeatedly heated by the steam and cooled by the condensing water. Watt's pumping engine of 1788 had a separate condenser so the cylinder maintained its temperature and a considerable saving in fuel was achieved.

The next steps were to evolve a rotary motion from the rocking beam, to introduce automatic valve gear, and to use steam above atmospheric pressure. It was at this stage that the existence of negative work began to make itself known. So that more precise measurements could be made, the first engine indicator was made by Watt in 1803. We briefly discussed indicator diagrams in relation to petrol engines in Programme 2.

18

In Frame 42 of Programme 2 we met the term indicated mean effective pressure. This is the average pressure difference over the piston stroke between the pressure curve of the positive work and the pressure curve of the negative work. In a steam engine cylinder, negative work is done in displacing the expanded steam from the cylinder into the condenser, and in compressing the residual steam after the exhaust valve has closed. For useful net work to be done, bearing in mind the effects of friction in the engine, the boiler pressure would have to exceed the condenser pressure comfortably.

A steam engine indicator diagram is shown in the next frame.

19

The figure below shows the form of a steam engine indicator diagram.

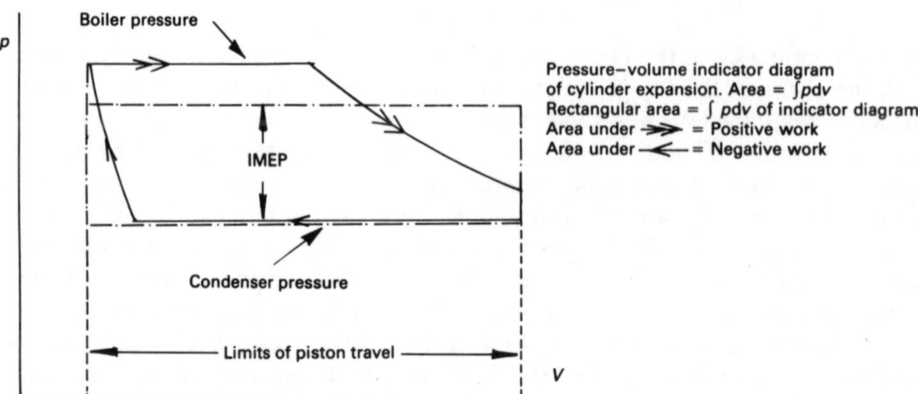

Insufficient steam pressure was often the cause of failure of early steam engines. While many accidents occurred with boiler explosions, another reason for failure was that the importance of heat transfer surfaces was not understood. A good boiler was one of the reasons for the success of Stephenson's *Rocket* in the Rainhill locomotive trials of 1829.

We will again consider the topic of work ratio when we consider I.C. engines later.

20

We have seen that the work ratio of the Carnot cycle considered in earlier frames at the particular conditions chosen was 0.905.

The **specific steam consumption** (SSC) is another useful performance indicator of a steam power cycle. This is a measure of the work produced in relation to the steam consumed, and is calculated in kilograms of steam per kilowatt-hour of net work. A kilowatt-hour [kWh] is 3600 kJ (see Frame 23 of Programme 2) and the net work is measured in kJ/kg. Thus:

$$\text{SSC} = \frac{3600\ [\text{kJ/kWh}]}{\text{Net Work}\ [\text{kJ/kg}]}$$

$$= \frac{3600}{\text{Net Work}}\ [\text{kg/kWh}]$$

What is the SSC for the steam Carnot cycle we have been considering? The net work was given in Frame 15 as 379.0 kJ/kg.

21

9.5 kg/kWh

The net work is 379.0 kJ/kg, so:

$$\text{SSC} = \frac{3600}{379.0}$$

$$= 9.5 \text{ [kg/kWh]}$$

For any particular set of conditions chosen this figure stays constant, but the power output will depend directly on the mass flow of steam. Thus for a net work figure of 379.0 kJ/kg, the power output for a steam flow of 1 kg/s would be 379.0 [kJ/kg] × 1 [kg/s], or 379.0 [kW].

In the design of any steam power plant the SSC should be as low as possible, since the amount of steam flow determines the size of plant required. The net work, therefore, should be as high as possible. This can be achieved by reducing the negative work and by using conditions to give as much enthalpy drop as possible in the turbines. We shall soon see how this can be done.

In a steam power plant, the net work is 450 kJ/kg. Calculate (a) the steam flow rate to give an output of 3 MW and (b) the SSC.

22

(a) 6.67 kg/s, (b) 8.0 kg/kWh

(a) The required output is 3 MW, or 3000 kW, so the steam flow rate is:

$$\dot{m} = \frac{3000 \text{ [kW]}}{450 \text{ [kJ/kg]}}$$

$$= 6.67 \text{ [kg/s]}$$

(b) SSC = 3600/450 = 8.0 [kg/kWh].

Although the Carnot cycle has the highest possible theoretical efficiency, it is not used for a number of practical reasons. Expansion from the saturated steam condition produces wet steam in the turbine at state 3 (see the figure below), and this causes rapid wear of the turbine blades. In addition, condensation has to be stopped at state 4 so that the wet steam which enters the feed pump is delivered to the boiler as saturated water at state 1. This is a highly impractical thing to try and do, and it also leads to a relatively large negative work term and a low work ratio. Also the SSC is not particularly good and can be improved upon.

The Rankine cycle is a logical development of the Carnot cycle, which in its various forms will overcome these objections. J. M. Rankine was Professor of Civil Engineering at Glasgow University in the second half of the last century. The essential features of the plant in a simple Rankine cycle are similar, i.e. boiler, turbine, condenser and feed pump. The feed pump is now a much simpler water pump, so it would not be possible to attempt to operate on the Carnot cycle. The *T–s* diagram of the simple Rankine cycle is also shown below. The boiler entry is at state 1 which is a compressed liquid state. In the boiler the compressed water is heated up to its saturation temperature and then to the saturated vapour state at 2 where it enters the turbine. The steam leaves the turbine after an isentropic expansion at state 3 where it enters the condenser to leave as saturated water at state 4 at entry to the feed pump. At the moment the turbine expansion is the same as in the Carnot cycle, so the wet steam problem is not yet overcome. In fact, you can see from the *T–s* diagram that as the boiler pressure increases, the wet steam problem becomes worse.

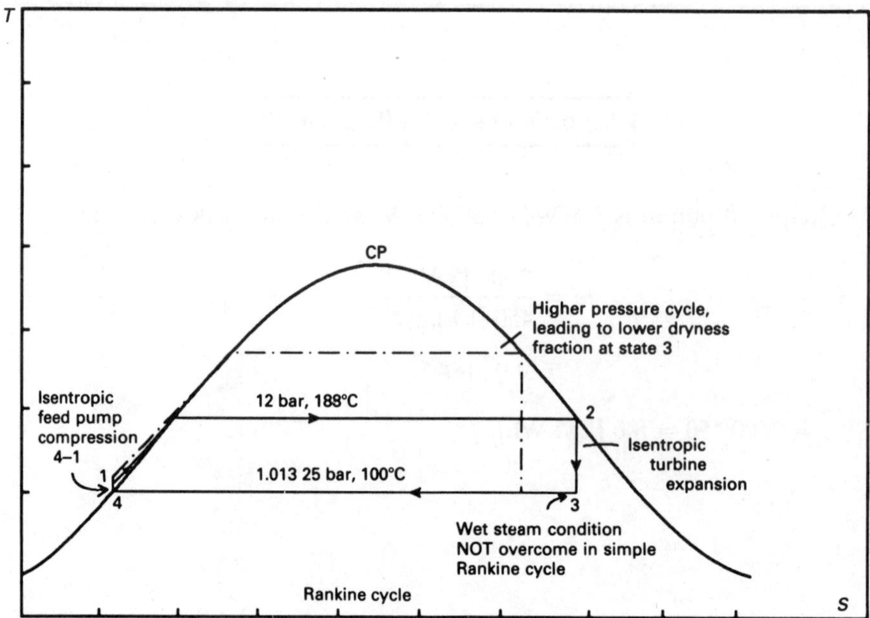

24

We could make a direct comparison with the Carnot cycle by taking the same pressure limits as before, and consider a simple Rankine cycle operating between 12 bar and 1.013 25 bar. In such a cycle the only differences are found in the state points at 4, which is saturated water, and at 1, which is compressed water. Heat transfer to the fluid now occurs over a temperature range from state 1 to the saturation value at 12 bar. Heat transfer from the fluid occurs until complete condensation is achieved at state 4. The problem of wet steam in the turbine will be overcome in the Rankine cycle with superheat which will be considered later.

We will consider a Rankine cycle at the higher pressure of 10 MN/m², or 100 bar. Saturated steam at 10 MN/m² (state 2) enters the turbine and expands isentropically to 100 kN/m², or 1 bar. What are (a) the dryness fraction of the steam at the lower pressure and (b) the specific enthalpy drop in the turbine?

25

(a) $x = 0.713$, (b) 700.3 kJ/kg

Using the Haywood tables in this example, $h_2 = h_g$ at 10 MN/m² which is 2727.7 kJ/kg. At the same state $s_g = 5.62$ kJ/kg K. At 100 kN/m², $s_f = 1.303$ kJ/kg K and $s_g = 7.36$ kJ/kg K, hence at 100 kN/m²:

$$5.62 = 1.303\,(1 - x) + 7.36\,x$$

$$\therefore \quad x = 0.713 \qquad \text{[Answer (a)]}$$

At the same pressure, $h_f = 417.5$ kJ/kg and $h_{fg} = 2257.9$ kJ/kg, so:

$$h_3 = 417.5 + 0.713 \times 2257.9$$
$$= 2027.4 \text{ [kJ/kg]}$$

For unit mass flow in the turbine:

$$-\dot{W}_T = h_3 - h_2$$
$$= 2027.4 - 2727.7$$

$$\therefore \quad \dot{W}_T = 700.3 \text{ [kJ/kg]} \qquad \text{[Answer (b)]}$$

26

Steam enters the condenser at 100 kN/m^2 and 0.713 dry. It leaves as saturated water at 100 kN/m^2, state 4, for which h_1 is 417.5 kJ/kg. At state 1 we have compressed water at 100°C and 10 MN/m^2, for which h is 426.5 kJ/kg, from page 16 of the Haywood tables. The difference between these two values is the feed-pump work for unit mass flow:

$$\dot{W}_{FP} = 417.5 - 426.5$$
$$= -9.0 \text{ [kJ/kg]}$$

The heat supplied to the boiler for unit mass flow is:

$$\dot{Q}_H = h_2 - h_1$$
$$= 2727.7 - 426.5$$
$$= 2301.2 \text{ [kJ/kg]}$$

Now calculate the SSC and the cycle efficiency.

27

SSC = 5.207 kg/kWh, η = 30.04%

For the net work:

$$\sum \dot{W} = \dot{W}_T + \dot{W}_{FP}$$
$$= 700.3 - 9.0$$
$$= 691.3 \text{ [kJ/kg]}$$

The SSC is then:

$$\text{SSC} = \frac{3600}{691.3}$$
$$= 5.207 \text{ [kg/kWh]}$$

The cycle efficiency is:

$$\eta = \frac{691.3}{2301.2} \times 100\% = 30.04\%$$

Then, the work ratio for this cycle is:

$$\text{WR} = 691.3/700.3 = 0.987$$

28

If we had obtained our results for a cycle between pressures of 12 bar and 1.013 25 bar for a direct comparison with the earlier Carnot cycle example, we would have got η = 17.67%, SSC = 8.62 kg/kWh and WR = 0.997. You could check that. We can now compare our results for the original Carnot cycle between 12 bar and 1.013 25 bar with results for Rankine cycles between the two pressure ranges.

		η	x_3	SSC	WR
Carnot cycle,	12 bar – 1.013 25 bar	19.08%	0.862	9.5 kg/kWh	0.905
Rankine cycle,	12 bar – 1.013 25 bar	17.67%	0.862	8.62 kg/kWh	0.997
Rankine cycle,	10 MN/m^2 – 0.1 MN/m^2 (100 bar – 1 bar)	30.04%	0.713	5.21 kg/kWh	0.987

Thus we see the benefits of the higher pressure and the associated higher saturation temperature of 311°C at 100 bar, compared with 188°C at 12 bar. Although in comparison with the Carnot cycle at the same pressures the SSC and WR are improved in the Rankine cycle, the efficiency is lower because some heat is now added at lower temperatures to the compressed water to bring it up to saturation temperature. We saw earlier in Frame 23 that the dryness fraction at state 3 falls as the higher pressure increases in the Rankine cycle, and the value of 0.713 is unacceptably low in the example here. To overcome the problem we need to move the vertical line of the isentropic turbine expansion to the right, so that state point 3 is closer to the dry saturated condition. What would then be the condition of the steam, leaving the boiler and entering the turbine?

29

Superheated

Superheat is used in the Rankine cycle for three reasons. The first is to achieve an acceptable dryness fraction in the turbine, and the second is to raise the average temperature at which heat is supplied. This means that the cycle efficiency is increased. The third reason is to increase the possible enthalpy drop in the turbine so that the SSC is reduced. The *h–s* diagram for the turbine expansion in a Rankine cycle with superheat is shown in the next frame. This shows at a glance how the turbine enthalpy drop has increased, and how the problem of wet steam in the turbine has been overcome.

30

The h–s diagram is given below.

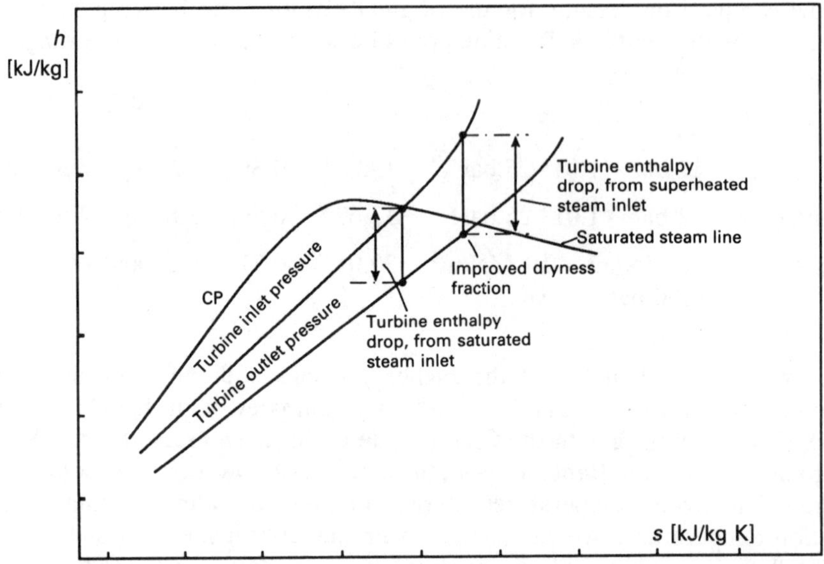

The provision of superheat requires a separate superheater within the boiler plant. This is because the heat transfer characteristics of superheated steam are different from those of wet steam. The upper limit of superheat temperature is about 600°C, well above the critical temperature.

We can consider a Rankine cycle with superheat between the same pressures as before, 10 MN/m² and 0.1 MN/m², so that states 4 and 1 will be the same. We can use the h–s chart to find the turbine enthalpy drop. Consider two cases: superheat to 500°C and to 600°C; calculate in each case the cycle efficiency and SSC. From Frame 26, h_1 = 426.5 kJ/kg and the feed-pump work is −9.0 kJ/kg.

31

500°C: η = 32.93%, SSC = 3.708 kg/kWh
600°C: η = 34.54%, SSC = 3.261 kg/kWh

Using the h–s chart supplied with the Haywood tables, at 10 MN/m² and 500°C, h_2 is 3375 kJ/kg, and dropping a vertical to the 0.1 MN/m² pressure line we find h_3 is 2395 kJ/kg. This gives a turbine work of 980 kJ/kg, and a net work of 971.0 kJ/kg. The heat supplied is the difference between h_2 (3375 kJ/kg) and h_1 (426.5 kJ/kg), which is 2948.5 kJ/kg.

Then:
$$\eta = \frac{971.0}{2948.5} \times 100\%$$
$$= 32.93\%$$

and:
$$\text{SSC} = \frac{3600}{971.0}$$
$$= 3.708 \text{ [kg/kWh]}$$

You will have noticed from the chart that x_3 is 0.875 and this is on the borderline of what is satisfactory.

The result at 600°C is obtained in the same way. For this condition x_3 is 0.925, which is acceptable.

32

In using the h–s chart in these examples you will probably have noticed that the enthalpy drop in the turbine could be increased by expanding to pressures below atmospheric. This can be done by using a vacuum condenser.

Thus we could expand from 10 MN/m^2 and 500°C down to 0.004 MN/m^2, or 0.04 bar, and obtain a lower value of h_3 of 1990 kJ/kg, with x_3 at about 0.767. This would give us an enthalpy drop of (3375 − 1990) = 1385 kJ/kg, which has increased by some 41% over the previous figure of 980 kJ/kg. However, the dryness fraction of 0.767 would be totally unacceptable, and this can be overcome by using the **reheat cycle**.

In the reheat cycle, the expansion takes place initially to an intermediate pressure, after which the steam is reheated to its top temperature and then expanded to the condenser pressure. The plant layout and the h–s diagram of the expansions are given below.

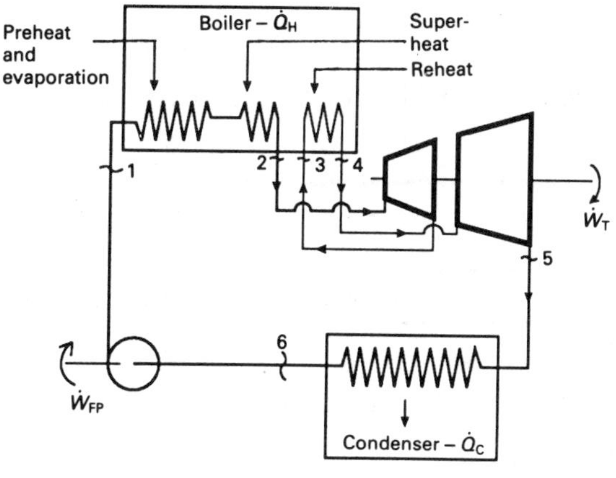

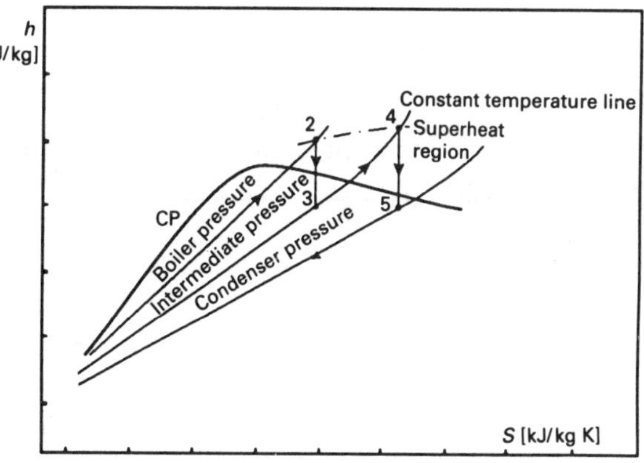

33

Suppose we expand from 10 MN/m² and 500°C down to 0.5 MN/m², and then reheat to 500°C and expand to 0.004 MN/m². The first expansion is between state points 2 and 3, and the second between points 4 and 5. The saturated liquid condition at the condenser exit, previously state 4, now becomes state 6. These particular turbine conditions are shown in the diagram below.

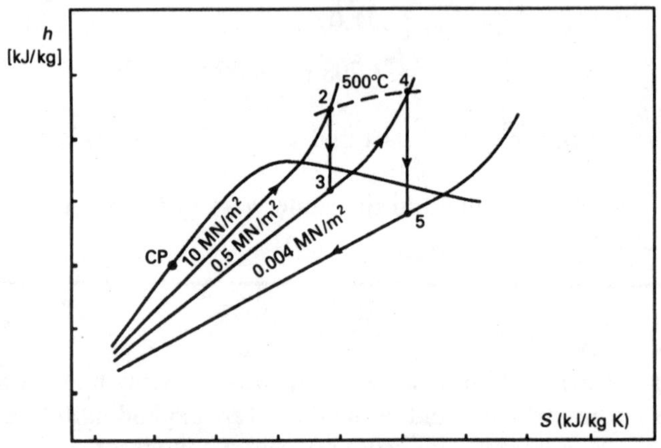

From the h–s chart we can pick out the following property values:

h_2 at 10 MN/m² and 500°C = 3375 kJ/kg
h_3 at 0.5 MN/m² = 2658 kJ/kg, with $s_3 = s_2$ and $x_3 = 0.955$
h_4 at 0.5 MN/m² and 500°C = 3485 kJ/kg
h_5 at 0.004 MN/m² = 2440 kJ/kg, with $s_5 = s_4$ and $x_5 = 0.950$

The total turbine work per unit mass flow is then:

$$\dot{W}_T = (3375 - 2658) + (3485 - 2440)$$
$$= 717 + 1045 = 1762 \text{ [kJ/kg]}$$

From the Haywood tables, what is h_f at 0.004 MN/m², and what is h in the compressed liquid state at 10 MN/m²?

34

$h_f = 121.4$ kJ/kg, $h = 130.6$ kJ/kg

From the Haywood tables, h_f at 0.004 MN/m², at 29°C, is 121.4 kJ/kg. To find h at 10 MN/m² and 29°C we have to interpolate between 25° and 50°C:

$$h = 114 + (217.8 - 114) \times \frac{(29 - 25)}{(50 - 25)}$$

$$= 130.6 \text{ [kJ/kg]}$$

This is h_1 in the cycle.

Thus the feed-pump work for unit mass flow is:

$$\dot{W}_{FP} = 121.4 - 130.6$$
$$= -9.2 \text{ [kJ/kg]}$$

Now we can find the net work:

$$\sum \dot{W} = \dot{W}_T + \dot{W}_{FP}$$
$$= 1762 - 9.2$$
$$= 1752.8 \text{ [kJ/kg]}$$

The heat supplied is:

$$\dot{Q}_H = (h_2 - h_1) + (h_4 - h_3)$$
$$= (3375 - 130.6) + (3485 - 2658)$$
$$= 4071.4 \text{ [kJ/kg]}$$

Now you can find the cycle efficiency and the SSC.

35

$\eta = 43.05\%$, SSC = 2.05 kg/kWh

$$\eta = \frac{1752.8}{4071.4} \times 100\% = 43.05\% \qquad \text{SSC} = \frac{3600}{1752.8} = 2.05 \text{ [kJ/kg]}$$

With the previous example as a model, do the following problems, using the h–s chart where possible.

1. In a Rankine cycle with reheat, the high pressure turbine entry condition is 200 bar and 500°C, and the low pressure turbine entry condition is 10 bar and 500°C. Expansions are isentropic and the condenser pressure is 0.1 bar. Calculate (a) the cycle efficiency and (b) the SSC. [(a) 42.30%, (b) 2.165 kg/kWh]
2. Repeat with superheat temperatures of 600°C. [(a) 45.06%, (b) 1.864 kg/kWh]

The Rankine cycle is an ideal heat engine cycle. This means that all processes are reversible. The actual cycle with steam as the working fluid, upon which power plant for the production of electricity is based, is a practical realisation of the Rankine cycle. Thus, in the ideal cycle, there is no pressure loss through the boiler or condenser or through any of the connecting pipe work, and we have isentropic processes in the turbine and feed pump. In practice, as you can imagine, pressure losses must occur in the system, and the flow of steam through the turbine, though considered to be adiabatic, is not isentropic.

Diagram (a) below shows a real turbine expansion on the *h–s* plane in comparison with the corresponding isentropic expansion.

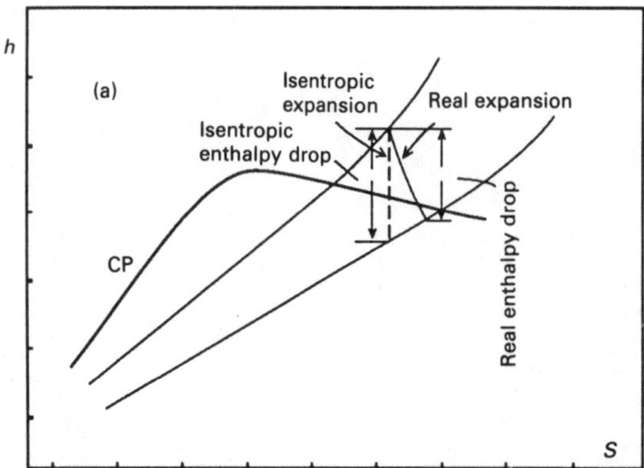

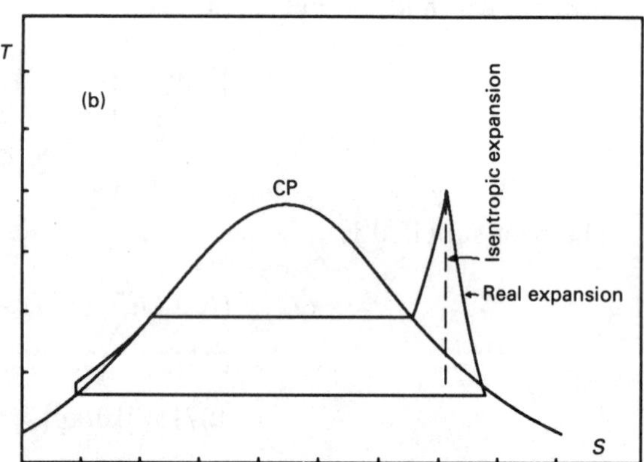

In the real expansion, there is an entropy increase, and the form of the constant pressure lines shows that the work output is less than that of the isentropic expansion. There is friction and turbulence as the expanding steam passes between the turbine blades, and this causes some of the kinetic energy of the steam to become enthalpy again, so the enthalpy drop is less than if there were no irreversibilities. The same expansion into the wet region is shown on the *T-s* diagram in (b). The extent of the irreversibilities which occur in a real expansion may be given a value: we use a term called the **isentropic efficiency** which relates the work obtained in a real expansion to the work of an isentropic expansion from the same initial state to the same final pressure. Thus:

$$\text{Isentropic efficiency} = \frac{\text{Actual work}}{\text{Isentropic work}}$$

Since from the steady flow energy equation, the work term of an adiabatic machine is directly related to the enthalpy drop through the machine, the isentropic efficiency then becomes:

$$\text{Isentropic efficiency} = \frac{\text{Actual enthalpy drop}}{\text{Isentropic enthalpy drop}}$$

37

In the test on a machine it is possible to measure the real enthalpy drop by measuring the power output of the turbine. The isentropic enthalpy drop may be calculated in the way we have seen, and so the isentropic efficiency of a turbine may be found. In the superheat cycle with reheat just considered, the isentropic enthalpy drops in the high and low pressure turbines were given in Frame 33 as 717 kJ/kg and 1045 kJ/kg, respectively. The feed-pump work was − 9.2 kJ/kg. We now assume an isentropic efficiency of 0.82 for both turbines; calculate the new net turbine work and the specific steam consumption.

38

1435.6 kJ/kg, 2.508 kg/kWh

In the first turbine, the isentropic work was found to be 717 J/kg. The isentropic efficiency is 0.82, hence:

$$\text{Actual work} = 717 \times 0.82 = 587.9 \text{ [kJ/kg]}$$

Similarly for the second turbine:

$$\text{Actual work} = 1045 \times 0.82 = 856.9 \text{ [kJ/kg]}$$

The net work is then: $\Sigma W = 587.9 + 856.9 - 9.2 = 1435.6$ [kJ/kg]
The specific steam consumption is:

$$\text{SSC} = \frac{3600}{1435.6} = 2.508 \text{ [kg/kWh]}$$

These real turbine expansions are shown on *h–s* axes below.

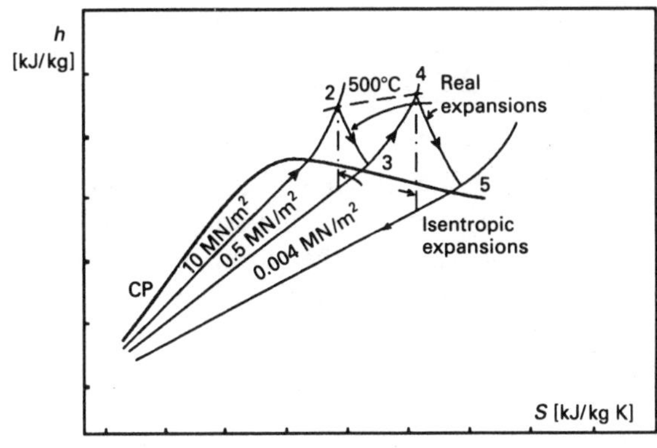

39

Since the actual enthalpy drop in the first turbine is less than the isentropic value, the enthalpy at state 3 will be slightly greater than it was before. In Frame 33, h_2 is given as 3375 kJ/kg. Thus:

$$\begin{aligned} h_3 &= h_2 - \text{(actual enthalpy drop)} \\ &= 3375 - 587.9 \\ &= 2787.1 \text{ [kJ/kg]} \end{aligned}$$

Steam with this enthalpy is now reheated to 500°C for the next turbine. As this new figure for h_3 is slightly higher than it was before, the heat supplied in the reheating process will be correspondingly less. The total heat supplied is:

$$\dot{Q}_H = (h_2 - h_1) + (h_4 - h_3)$$

With h_1 = 130.6 kJ/kg from Frame 34, and h_4 = 3485 kJ/kg from Frame 33, we have:

$$\begin{aligned} \dot{Q}_H &= (3375 - 130.6) + (3485 - 2787.1) \\ &= 3942.3 \text{ [kJ/kg]} \end{aligned}$$

With the net work of 1435.6 kJ/kg given in Frame 38, what are the new values of cycle efficiency and SSC for this cycle?

40

$$\boxed{\eta = 36.42\%,\ \text{SSC} = 2.508 \text{ kg/kWh}}$$

$$\text{Cycle efficiency} = \frac{\text{Net work}}{\text{Heat supplied}} = \frac{1436.6}{3942.3} \times 100\% = 36.42\%$$

$$\text{SSC} = \frac{3600}{1435.6} = 2.508 \text{ kg/kWh}$$

Thus we get an efficiency of 36.42% with an SSC of 2.508 kg/kWh, compared with 43.05% and 2.05 kg/kWh (from Frame 35) for the cycle with isentropic expansions.

Since the non-isentropic expansions in the turbines are the only losses considered, the reduction in performance is in line with the value of isentropic efficiency of the two turbines. The compression in the feed-pump will also be non-isentropic, but as the work ratio is so close to 1 with such a small feed-pump work term, this loss is of secondary importance and is not considered. The results that we have for cycle efficiency and SSC also assume no frictional losses in the turbine bearings, or a mechanical efficiency of 1.

41

Let us now suppose that this reheat cycle with the non-isentropic turbine expansions is to be used to generate 80 MW of electrical power. We have in mind to burn a particular fuel and we need to know at what rate it must be supplied to the boilers so that we can find an overall efficiency for the plant.

The power output of a turbine is less than that indicated by the enthalpy drop through the machine owing to bearing friction, and when a turbine drives a generator the electrical output of the generator is less than the power input from the turbine because of heating losses in the windings of the generator. These losses may be combined in a generator efficiency which may be expressed as:

$$\text{Generator efficiency} = \frac{\text{Electrical power output}}{\text{Turbine power input}}$$

If we take a generator efficiency of 94% and we have specified that the electrical power output is 80 MW, what in terms of enthalpy drop × mass flow will be the net power output from the turbines?

42

85.106 MW

For 80 MW of electrical power with a generator efficiency of 94%, the net steam turbine power output must be 80/0.94 or 85.106 MW.

From Frame 38, the net work output of the turbines is 1435.6 kJ/kg of steam, so we can find the flow rate of steam required:

$$\text{Mass flow rate of steam} = \frac{85.106\ [\text{MW}] \times 10^3\ [\text{kW/MW}]}{1435.6\ [\text{kJ/kg}]} = 59.28\ [\text{kg/s}]$$

Also from Frame 39, the heat supplied to the steam is 3942.3 kJ/kg. What is the rate of heat supply to the steam from the boilers in MW?

233.70 MW

The rate of heat supply $= 3942.3\ [\text{kJ/kg}] \times 59.28\ [\text{kg/s}] \times 10^{-3}\ [\text{MW/kW}]$

$= 233.70\ [\text{MW}]$

This rate of heat supply is obtained by burning fuel. This produces hot combustion gases which give up their energy to the steam by processes of heat transfer. The gases are cooled as a result but they must always be hotter than the fluid being heated for the heat transfer to take place. This means a considerable amount of the energy of the fuel is lost as the combustion gases leave the boiler. Obviously steps are taken to reduce these losses as much as possible, by means shown in the diagram below.

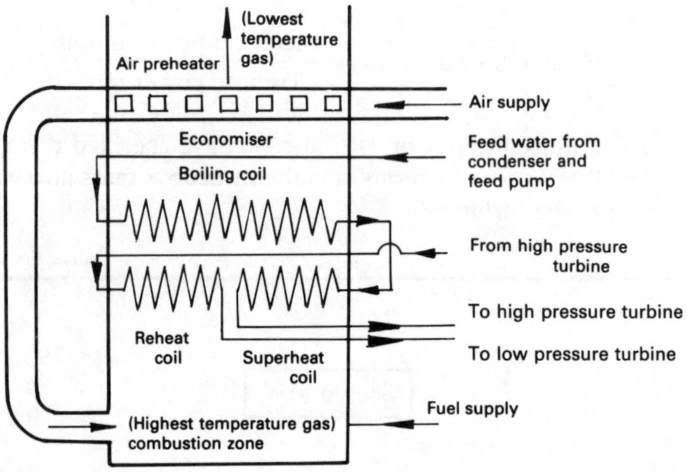

For example, the boiler feed water may be passed through an economiser consisting of tubes in the exhaust duct. Here the water is preheated before entering the boiler so that less heating is required in the boiler itself. Secondly an **air preheater** may be used in which the air supply to the combustion zone is heated by the exhaust gases, so that less fuel is used to achieve the required combustion temperature. The amount of energy in the fuel reaching the steam is quantified by a term called **combustion efficiency**, where:

$$\text{Combustion efficiency} = \frac{\text{Energy transferred to steam}}{\text{Energy released from fuel}}$$

44

If we assume a combustion efficiency of 80%, what would be the required rate of energy release from the fuel?

292.13 MW

The rate of heat transfer to the steam was 233.70 MW, representing an energy loss from the combustion gases, so the rate of energy release from the fuel is:

$$\text{Rate of energy release} = \frac{233.70}{0.80}$$

$$= 292.13 \text{ [MW]}$$

Every kg of fuel burnt releases 44 MJ. This is called the **calorific value** of the fuel, a term to be discussed in Programme 9. We can now calculate the fuel flow rate. What is it?

45

6.64 kg/s

The rate of fuel flow must be given by:

$$\dot{m}_f = \frac{292.13 \text{ [MW]}}{44 \text{ [MJ/kg]}} = 6.64 \text{ [kg/s]}$$

We can now consider the plant overall thermal efficiency. This is a measure of 'what comes out' in relation to 'what goes in'. We know the output is 80 MW of electrical power, and the input is 6.64 kg/s of fuel of calorific value 44 MJ/kg. The plant efficiency may be expressed as:

$$\text{Plant thermal efficiency} = \frac{\text{Energy flow equivalent of output}}{\text{Energy flow equivalent of input}}$$

What is this efficiency?

46

$$\boxed{27.38\%}$$

We have seen that the heat input from the fuel is 6.64 kg/s of fuel of calorific value of 44 MJ/kg, which gives a heat input of 292.13 MW, and the output as electrical power is 80 MW, hence:

$$\text{Plant thermal efficiency} = \frac{80.0\ [\text{MW}]}{6.64\ [\text{kg/s}] \times 44\ \text{MJ/kg}} \times 100\%$$

$$= 27.38\%$$

Finally, in Frame 38 we found that the SSC based on a net work of 1435.6 kJ/kg was 2.508 kg/kWh. This value of SSC is based on the turbine output. The generator output is less than the turbine output in the ratio of the generator efficiency, so we can find a value of SSC in terms of the generator output which will be greater than it is in terms of the turbine output, again in the ratio of the generator efficiency. Thus:

$$\text{SSC (Generator output)} = \frac{\text{SSC (turbine output)}}{0.94}$$

$$= \frac{2.508}{0.94} = 2.67\ [\text{kg/kWh}]$$

The plant thermal efficiency of 27.38% compares with 36.42% for the efficiency of the actual cycle with non-isentropic expansions in the turbines, calculated in Frame 40, and with 43.05% for the Rankine cycle with reheat calculated in Frame 35. The plant efficiency of 27.38% can be improved by the use of an economiser and air preheater, but it is beyond our scope here to go in to this in detail.

47

We have seen how the efficiency of the Rankine cycle has been improved by adding superheat and by expanding to lower pressures. As well as by adding superheat, the available turbine enthalpy drops can also be increased by raising the cycle peak pressure. By using vacuum condensers, heat rejection takes place at lower temperatures, but it also means that some heat has to be supplied at lower temperatures in bringing the boiler feed water up to its saturation temperature at the prevailing boiler pressure. While *rejecting* heat at lower temperatures improves thermodynamic efficiency, this gain is partially lost when some heat has to be *added* at the lower temperatures.

The addition of heat at the lower temperatures can be avoided by using what is called **regenerative feed heating**. The *T-s* diagram of a simple Rankine cycle modified in this way is shown below and is described in the next frame.

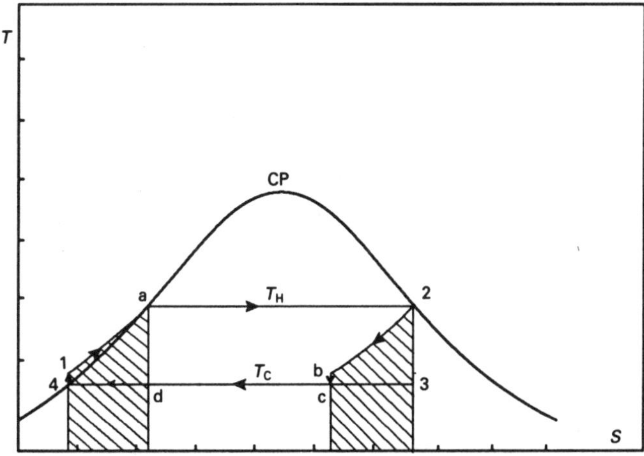

48

As we know, heat is supplied between state points 1 and 2; if heating of the feed water between points 1 and a could be done reversibly by extracting heat from the steam in the turbine expanding between points 2 and 3, then the heat supplied by the boiler would only be necessary to raise steam at the constant temperature of T_H. The reversible heat transfer required to bring the water up to T_H is the shaded area under the line 1–a. The turbine expansion is no longer adiabatic with heat extracted, and it follows the path 2–b–c. The area under 2–b to the base line is the heat extracted to heat the water from 1 to a, and for **reversible heat transfer the two shaded areas are equal**. This is an important illustration of the use of the *T–s* diagram. Heat rejection at T_c is then from point c to point 4. The length of the line 4–c equals the length a–2, because with the two shaded areas being equal, the lengths 4–d and c–3 are equal. The diagram is then equivalent to a Carnot cycle with heat addition at T_H between points a and 2, and with heat rejection at T_c between points 3 and d. The cycle efficiency must then be equal to that of the Carnot cycle between these temperatures.

The **regenerative cycle** is a practical cycle which attempts to use this idea. While it is not possible to pass the boiler feed water in counter flow through the turbine, an approach to this is achieved by using some steam bled off at one or more stages in the expansion.

49

The diagram below shows a plant layout for a cycle with superheat with a single regenerative feed water heater. In practice, several may be used over the temperature range between boiler and condenser saturation temperatures. In the diagram, steam is extracted from the turbine at a state 3 which is at a pressure between the boiler and condenser pressures, and is passed to a heater in which water from the condenser is delivered at the same pressure as compressed water. This feed water leaves the heater as saturated water at the same pressure, together with the bled steam which has now condensed. Both then pass through another feed pump to be delivered to the boiler. The temperature of water entering the boiler is now higher than the condensation temperature, and so the average temperature of heat addition in the boiler is higher and an improvement in cycle efficiency is achieved.

The *T–s* diagram for a practical regenerative cycle with a single feed water heater is also shown. For unit mass flow of steam round the cycle, *y* kg of steam is bled off at state 3 and (1 − *y*) kg passes through the remainder of the turbine and condenser.

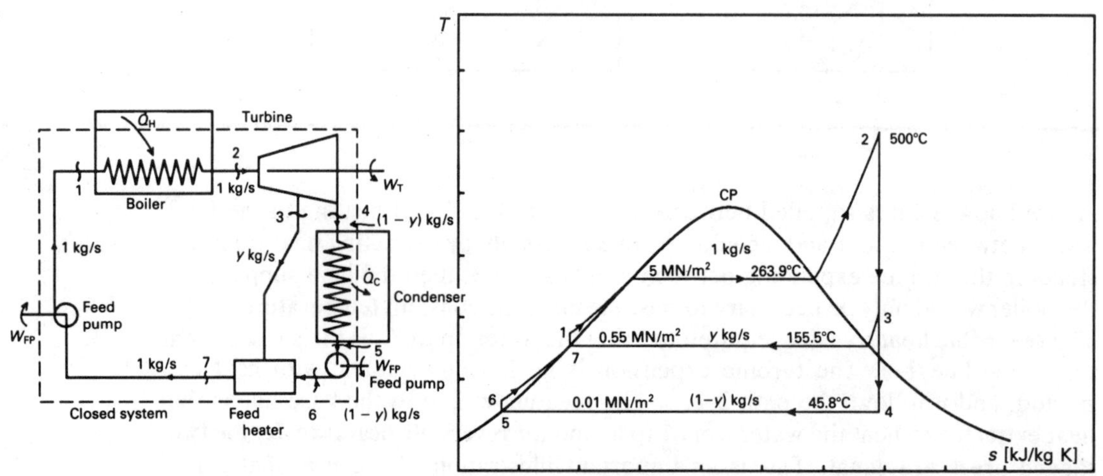

To show the benefit of the regenerative cycle we shall consider both a Rankine cycle with superheat and a regenerative cycle with superheat, operating between 5 MN/m^2 and 500°C and 0.01 MN/m^2, with a single feed water heater in the regenerative cycle. The pressure of the steam bleed can be the saturation pressure corresponding to the average of the saturation temperatures at 5 MN/m^2 and 0.01 MN/m^2. These temperatures are 263.9°C and 45.8°C, giving an average of 154.9°C. We will take a bleed pressure of 0.55 MN/m^2, having a saturation temperature of 155.5°C. Thus the minimum temperature at which heat is supplied is now 155.5°C instead of 45.8°C.

First, show that for the Rankine cycle with superheat between 5 MN/m^2 and 0.01 MN/m^2, the efficiency is 37.56% and the SSC is 2.961 kg/kWh.

50

The working is not given as the methods may be followed exactly as in Frames 29–32.

The first thing to do in the regenerative cycle is to find the amount of bled steam, y. We need to consider an energy balance on the feed heater. A control volume for this is shown in the figure below. We may take unit mass flow of steam in the cycle.

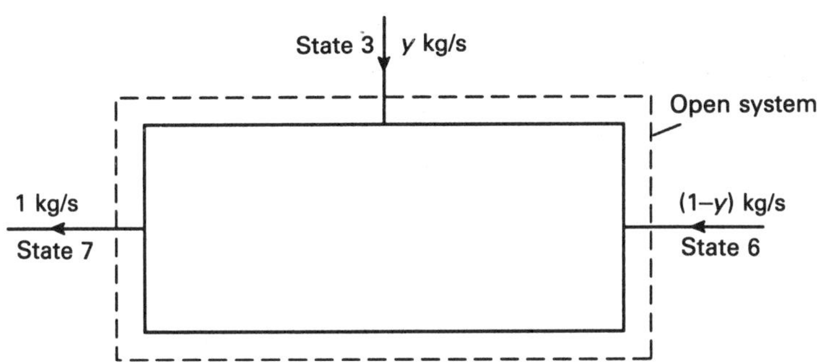

There are two flow streams into the heater: compressed water from the condenser at state 6 having mass flow of $(1 - y)$ kg/s, and y kg/s of steam from the turbine at state 3. The flow leaving the heater is 1 kg/s of saturated water at state 7. There are energy transfers between streams in the heater; there are no heat transfers across the control surface. The enthalpy flows in and out of the heater must be equal. Thus:

$$y\, h_3 + (1 - y)\, h_6 = h_7$$

We can find all the enthalpy values and so we can find y. Water enters the heater at state 6 as saturated water at 0.01 MN/m^2 compressed to 0.55 MN/m^2. The enthalpy increase across this feed pump is small, so we may take $h_6 = h_5$. This is h_f at 0.01 MN/m^2, which is 191.8 kJ/kg.

State 3 is at 0.55 MN/m^2 after an isentropic expansion from 5 MN/m^2 and 500°C. At this state, the specific entropy is 6.987 kJ/kg K. So now find h_3 and h_7 (which is h_f at 0.55 MN/m^2) and then you can solve the equation above to find y.

51

You should find that x_3 is 0.936, h_3 is 2463.2 kJ/kg, h_7 is 655.8 kJ/kg and y is 0.174 kg/s.

52

With $s_3 = 6.987$ kJ/kg K, at 0.55 MN/m^2 $s_g = 6.787$ kJ/kg K, hence state 3 is superheated. By interpolation using the Haywood tables, we find that at 0.55 MN/m^2 and 200°C, $s = 7.022$ kJ/kg K; then by interpolating between 7.022 and 6.787 kJ/kg K we find the temperature is 193.42°C. At 0.55 MN/m^2 and 193.42°C, by two further interpolations we find h is 2859.1 kJ/kg. This is the required value of h_3.

$$\therefore \qquad h_3 = 2859.1 \text{ [kJ/kg]}$$

Then, with $h_6 = h_5$: $$y\, h_3 + (1 - y)\, h_5 = h_7$$

$$\therefore \qquad 2859.1\, y + (1 - y) \times 191.8 = 655.8$$

$$\therefore \qquad y = 0.174 \text{ [kg/s]}$$

Next, knowing y you can find the turbine work:

$$\dot{W}_T = 1.0 \times (h_2 - h_3) + (1 - y) \times (h_3 - h_4)$$

You will first need to find h_4.

53

$h_4 = 2213.8$ kJ/kg, $\dot{W}_T$ 1195.6 kJ/kg

The specific entropy at state 4 is 6.987 kJ/kg K, so x_4 is given by:

$$6.987 = 0.649\,(1 - x_4) + 8.151\, x_4$$

$$\therefore \qquad x_4 = 0.845$$

$$\therefore \qquad h_4 = 191.8 + 0.845 \times 2392.9$$

$$= 2213.8 \text{ [kJ/kg]}$$

Then: $$\dot{W}_T = 1.0 \times (h_2 - h_3) + (1 - y) \times (h_3 - h_4)$$

$$= (3433.5 - 2859.1) + (1 - 0.174) \times (2859.1 - 2213.8)$$

$$= 1107.4 \text{ [kJ/kg]}$$

54

State 1 is compressed water at 155.5°C and 5 MN/m^2 (see Frame 49), which is found from page 16 of the Haywood tables by interpolating between 4 and 6 MN/m^2, and between 150°C and 175°C. The result is 658.8 kJ/kg. h_7 is 655.8 kJ/kg, so the feed-pump work is the difference, which is 3.7 kJ/kg. The net turbine work is then:

$$\sum W_T = 1107.4 - 3.0$$
$$= 1104.4 \text{ [kJ/kg]}$$

The SSC is then given by: $$\text{SSC} = \frac{3600}{1104.4} = 3.26 \text{ [kg/kWh]}$$

The heat supplied per unit mass flow is:

$$\dot{Q}_H = h_2 - h_1$$
$$= 3433.5 - 658.8 = 2774.7 \text{ [kJ/kg]}$$

Hence: $$\eta = \frac{1104.4}{2774.7} \times 100\% = 39.80\%$$

Thus compared with the Rankine cycle results in Frame 49, the cycle efficiency has improved from 37.56% to 39.80%, while the SSC has risen slightly from 2.961 kg/kWh to 3.26 kg/kWh. The steam bled in the turbine means the turbine work is reduced, giving a larger SSC, but in spite of this the cycle efficiency has improved. Small improvements in cycle efficiency are always worthwhile bearing in mind that electricity is generated on a 24 hours per day basis. Further improvement would be possible with more than one feed-water heater; 3 or 4 are normally employed.

55

There is another variation of the Rankine cycle which needs to be mentioned. This is the **combined heat and power cycle**, or CHP cycle. In this the plant is sited where locally the waste heat may be used. While electricity may be distributed over many miles, it is not feasible to do the same with the waste heat.

Usually the condenser pressure is chosen so that a useful waste heat temperature is available. This may be at a higher pressure than normal. The rate at which heat is provided from such a plant may be called **thermal power**. A typical application is in district heating.

56

We will look again at the Rankine cycle with superheat from Frame 30. The plant is shown below, modified to use the waste heat from the condenser.

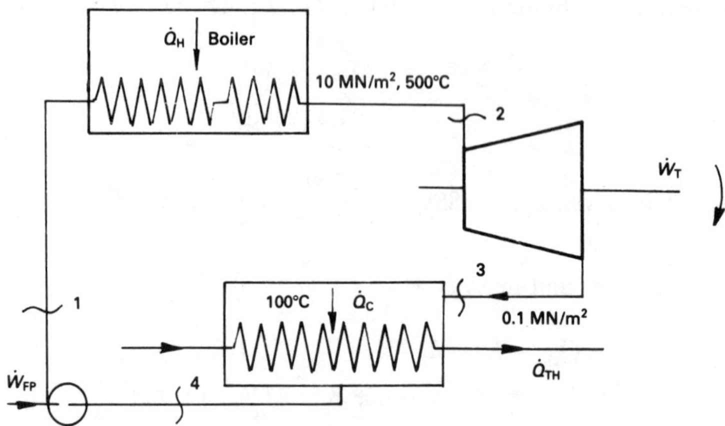

In this cycle steam at 10 MN/m² and 500°C expands to 0.1 MN/m², so that heat at a useful 100°C may be obtained from the condenser. We found the net work was 971.0 kJ/kg (Frame 31), and h_3 at the turbine exit was 2395 kJ/kg. This is also the entry condition to the condenser, from which the exit condition is saturated water at 0.1 MN/m² so h_4 is 417.5 kJ/kg. From Frame 10 we have:

$$
\begin{aligned}
Q_C &= h_4 - h_3 \\
&= 417.5 - 2395 \\
&= -1977.5 \text{ [kJ/kg]}
\end{aligned}
$$

This is negative since it is a heat transfer *from* the steam. Suppose that 70% of this will be used, and the plant is required to produce 30 MW of electrical power; what is the output of thermal power?

For 30 MW of power with a net work of 971.0 kJ/kg, the steam flow rate is:

$$\dot{m} = \frac{30 \times 1000}{971.0} = 30.90 \text{ [kg/s]}$$

The thermal power is then:

$$\dot{Q}_{TH} = 30.90 \times 1977.5 \times 0.7 \times 10^{-3} = 42.77 \text{ [MW]}$$

57

The Rankine steam cycle can also use the exhaust of a gas turbine as its source of heat in what is called a **combined cycle**. Today, gas-fired plant of this type is rapidly replacing conventional coal-fired plant. We shall be returning to this later when dealing with gas turbines in Programme 10.

In pursuit of maximum fuel economy, the combined and CHP cycles are often used together, so that waste heat recovery takes place from a combined gas turbine–steam cycle. A CHP plant at Heathrow uses gas turbines and is rated at 13 MW electrical power and 18 MW of thermal power.

Also other prime movers may be used in the CHP cycle. Thus on a much smaller scale a road vehicle engine, tuned to run on natural gas, may be used to drive an electrical generator. Waste heat from the exhaust and cooling water is recovered for domestic heating purposes.

58

While increasingly the waste heat from power cycles is being used, there is also a growing interest in using waste heat as a *source* of power. Waste heat from some processes, and from some natural sources of heat, may be at too low a temperature to be of any use in a heating role. Nevertheless, very large quantities of energy may be involved. There is thus renewed interest in the Rankine cycle for its own sake, which was appearing to be become just an auxiliary to the gas turbine cycle.

The Rankine cycle can use any vapour, and substances may be chosen to fit the temperature of the heat source. For example, experimental plant has been built to use the internal energy of the sea in the tropics. Typically the surface temperature is 30°C, so sea water at this temperature could be used to make ammonia boil at 25°C. From your tables, e.g. the Haywood tables, page 24, you can see that at this temperature the saturation pressure is 1.001 MN/m^2. Heat is rejected to the sea at lower levels, where the temperature would be around 6°C, giving a condensation temperature of say 10°C.

Thus we can run a Rankine cycle plant between 25°C and 10°C, using free 'fuel'. What is the Carnot efficiency at these temperatures?

59

5.03%

The Carnot efficiency, Frame 34, Programme 6, is: $\eta = (T_H - T_C)/T_H$. This gives 15/298.15 which is 0.0503, or 5.03%. We will look at an ammonia cycle at these temperatures in the next frame.

60

A diagram of the type of plant and the *T–s* diagram of the ammonia Rankine cycle are shown below. Also, as you can see, only small temperature increments are involved in the heating and condensation processes, so that large quantities of sea water are flowing through the plant.

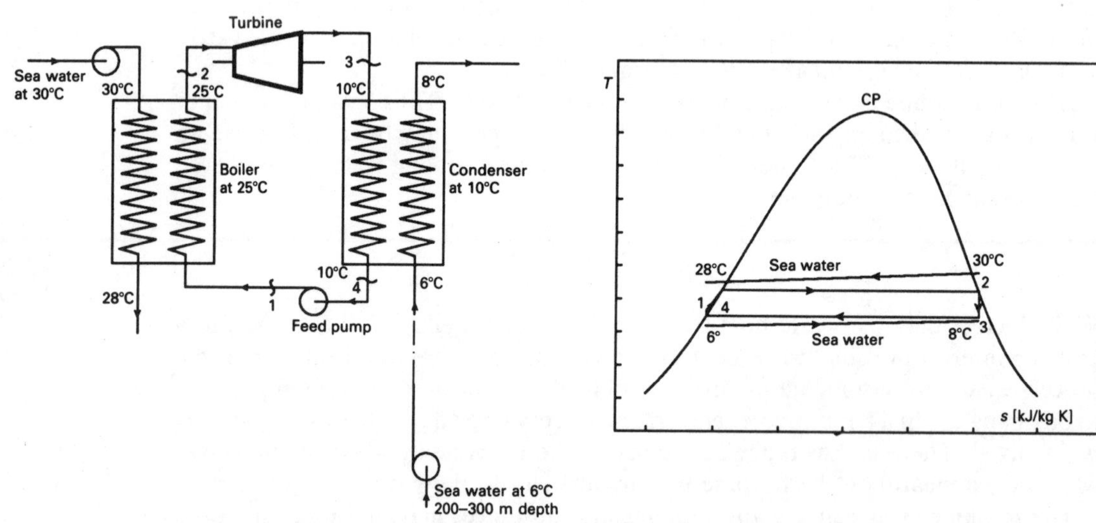

Work through the simple Rankine cycle to find the turbine isentropic work, and the heat supplied, neglecting feed-pump work. You will find that the ideal Rankine cycle efficiency is 4.97% which is not far below the Carnot efficiency, as the two temperature limits are close together.

61

We have $h_2 = 1466$ kJ/kg, $s_2 = 5.039$ kJ/kg K, at 25°C, and at 10°C s_f is 0.881 kJ/kg K, $s_g = 5.214$ kJ/kg K, $h_f = 227.7$ kJ/kg and $h_g = 1454$ kJ/kg.

Then: $$5.039 = (1 - x_3)\,0.881 + 5.214\,x_3$$

∴ $$x_3 = 0.960$$

and: $$h_3 = (1 - 0.960) \times 227.7 + 0.960 \times 1454$$

$$= 1404.9 \text{ [kJ/kg]}$$

62

The turbine work is then: 1466 − 1404.9 = 61.10 [kJ/kg]

The heat supplied is: 1466 − 227.7 = 1238.3 [kJ/kg]

The cycle efficiency is: $$\eta = \frac{61.10}{1238.3} \times 100\% = 4.93\%$$

There are two main reasons why we would not use steam in a cycle at these temperatures. The first is that the saturation pressure of steam is well below atmospheric pressure, which would lead to many practical difficulties in the design and operation of the plant. The second reason may be seen if we compare the volume flow rate of vapour into the turbine, for both ammonia and steam.

In the ammonia cycle the turbine specific work was 61.10 kJ/kg. At 25°C, v_g for ammonia is 0.128 m³/kg. If the power plant was designed to produce 1 MW, what would be the volume flow rate of dry saturated ammonia at 25°C into the turbine?

63

2.09 m³/s

The mass flow rate of ammonia is 1000 [kW]/61.10 [kJ/kg], which gives a flow rate of 16.37 [kg/s]. At 25°C, v_g is 0.128 m³/kg, so:

$$\dot{V} = 16.37\ [\text{kg/s}] \times 0.128\ [\text{m}^3/\text{kg}]$$
$$= 2.09\ [\text{m}^3/\text{s}]$$

If you repeat this whole exercise with steam between the same temperatures of 25°C and 10°C, you will find that the turbine specific work is 124.52 kJ/kg, so for the same power output the mass flow would be 8.031 kg/s, which is about half that for the ammonia cycle. However, at 25°C, v_g for steam is 43.4 m³/kg, giving a volume flow rate of 348.54 m³/kg, or 167 times as much. Thus with steam a vastly greater turbine would be required, and we see that in practical cycles the characteristics of the working fluid are important for the particular application.

64

Before continuing with reversed cycles, you should make sure you can do these problems on the various forms of the Rankine cycle.

1. In a Rankine cycle with reheat, steam enters the high pressure turbine at 30 bar and at 400°C in which it expands to a pressure of 3 bar. At this pressure the steam is reheated to 400°C and it enters the low pressure turbine in which it expands to the condenser pressure of 0.1 bar. Taking isentropic expansions in the turbines, calculate (a) the SSC and (b) the cycle efficiency.
[(a) 2.857 kg/kWh, (b) 34.86%]
2. Repeat this first problem for turbine inlet conditions of 150 bar at 500°C, and 15 bar at 500°C, with a condenser pressure of 0.5 bar. [(a) 2.599 kg/kWh, (b) 37.62%]
3. Reconsider the cycle in problem **2** with turbine isentropic efficiencies of 0.85 and 0.8 for the high and low pressure turbines respectively, and recalculate the SSC and cycle efficiency. [(a) 3.177 kg/kWh, (b) 31.51%]
4. A steam power plant operates under the conditions given in problems **2** and **3** and it has a 500 MW electrical power output with an electrical generator efficiency of 0.95, and in the boiler there is a combustion efficiency of 0.8 as fuel of calorific value of 44 MJ/kg is burnt. Calculate (a) the steam flow rate, (b) the fuel flow rate and (c) the overall plant efficiency.
[(a) 464.4 kg/s, (b) 47.46 kg/s, (c) 23.94%]
5. In a regenerative Rankine cycle, steam enters a single turbine at 50 bar and 450°C, and is condensed at 0.5 bar. Some steam is bled at a pressure of 3 bar and is passed to a single feed-water heater. Calculate (a) the SSC and the cycle efficiency for the simple cycle without feed-water heating and (b) the amount of bled steam, kg/kg, and the SSC and efficiency for the regenerative cycle.
[(a) 3.827 kg/kWh, 31.66%; (b) 0.0954 kg/kg, 3.941 kg/kWh, 33.22%]
6. The plant in problem **4** rejects heat at a pressure of 0.5 bar. In addition to the 500 MW of electrical power it is now required to give thermal power from the condenser waste heat. Assume that 65% of the available enthalpy drop in the condenser is used. What is the thermal power available? [743.7 MW]
7. Waste heat at 80°C is available, and it is proposed to use R-12 refrigerant in a cycle to produce 5 MW of power. Heat will be rejected at 15°C. The turbine has an assumed isentropic efficiency of 80%. Calculate: (a) the cycle efficiency, (b) the mass flow rate of R-12 and (c) the volume flow rate entering the turbine.
[(a) 12.59%, (b) 244.08 kg/s, (c) 6.57 m^3/kg]

65

In Programme 6 we met the idea of a reversed heat engine. In Frame 15 we saw that in contrast to a power-producing heat engine cycle, everything is proceeding in reverse: net work is done on the cycle, and via the closed system on which this work is done, heat transfer occurs from a region of low temperature to a region of high temperature. The working fluid within the closed system flows in the opposite direction to the flow in a normal heat engine. In Frame 19 of Programme 6 we saw that the efficiency of a reversed cycle is expressed in terms of a coefficient of performance. In a refrigerator we are looking at the heat transfer at the low temperature, and the coefficient of performance is given by:

$$\text{COP(refrigerator)} = Q_2/(Q_1 - Q_2)$$

The figure shows a reversed Carnot cycle for a vapour on T–s coordinates in which the heat transfers Q_1 and Q_2 become Q_H and Q_C and take place at the constant temperatures of T_H and T_C.

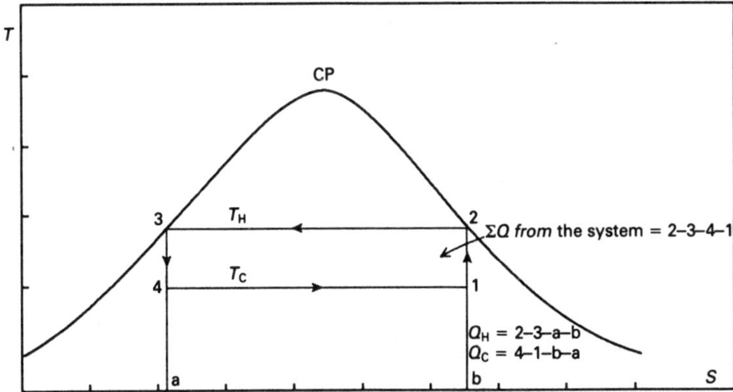

In terms of the absolute temperatures T_H and T_C, the performance of the refrigeration cycle is then given by:

$$\text{COP(refrigerator)} = T_C/(T_H - T_C)$$

In speaking of 'low' and 'high' temperatures in a refrigeration cycle, the terms are relative. Thus, they could be −25°C and 35°C respectively. What is the COP for a refrigerator working on the reversed Carnot cycle at these temperatures?

66

The answer is 4.136.

The temperatures T_H and T_C as absolute values are 308.15 K and 248.15 K respectively, so the COP is given by:

$$\text{COP(refrigerator)} = 248.15/(308.15 - 248.15)$$
$$= 248.15/60.0 = 4.136$$

67

The state points 1 to 4 in the figure in Frame 65 occur in sequence as the fluid passes in the direction shown within the closed system in the figure below.

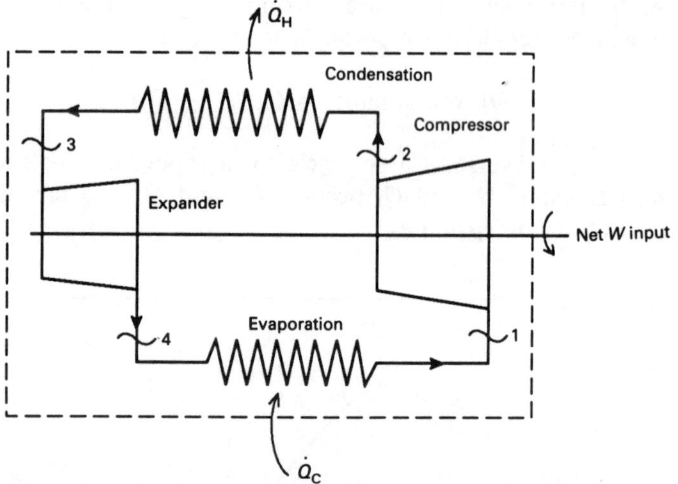

Wet vapour is compressed isentropically at state 1 to become saturated vapour at state 2. This is condensed to saturated liquid at state 3 as a heat transfer $\dot{Q}_H$ at the temperature T_H takes place. The saturated liquid is expanded isentropically from state 3 to state 4, and evaporation at T_C takes place owing to the heat transfer $\dot{Q}_C$.

In practice, this cycle is not used. It is not possible to compress a wet vapour as at state 1 without damaging the compressor, so the evaporation at T_C continues until the vapour is dry. Also the use of a work-producing expander between states 3 and 4 is unnecessary. The practical vapour compression cycle is introduced in the next frame.

If the COP of a reversed Carnot cycle refrigerator is 4.136, what is the COP of the corresponding heat pump? (See Frame 19 in the previous programme if you cannot remember.)

68

5.136

In any reversed cycle we have that:

$$\text{COP(heat pump)} = 1 + \text{COP (refrigerator)}$$

Hence: $$\text{COP(heat pump)} = 1 + 4.136 = 5.136$$

In a practical vapour compression cycle there is a reciprocating piston type compressor, a condenser, an expansion valve and an evaporator. These are shown in the figure below.

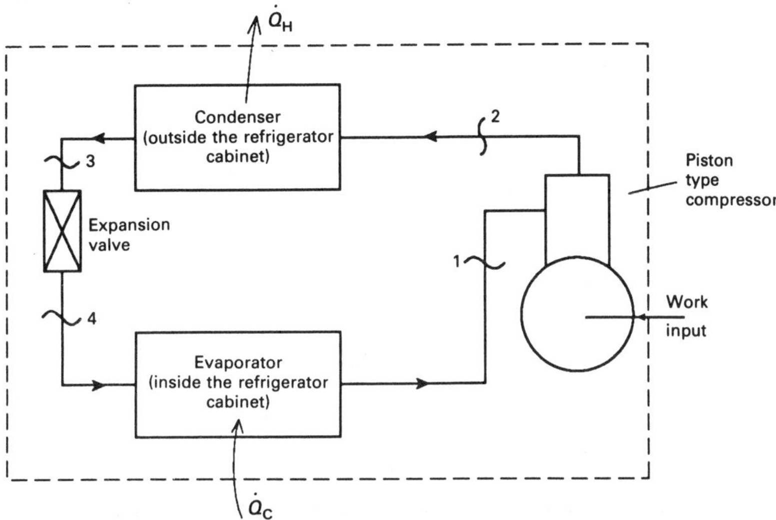

Low temperature saturated vapour enters the compressor at state 1. After isentropic compression it becomes superheated vapour at the higher pressure at state 2. It is cooled and condensed in the condenser to become saturated liquid at state 3. The liquid then passes through an expansion valve to the lower pressure again. This last process takes place at **constant enthalpy** and was first discussed in Frame 64 of Programme 3. It is an adiabatic process, but there is an entropy increase between states 3 and 4. At state 4, the fluid is a wet vapour of low dryness fraction. From this description of the cycle and before looking at the next frame, see if you can show the cycle on *T–s* coordinates for a vapour.

69

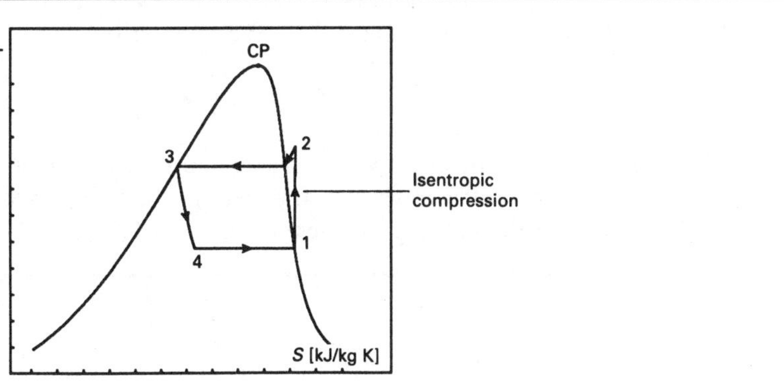

The cycle is shown above on the *T–s* plane.

70

The pressure–enthalpy diagram is also used to show the reversed cycle, and the same state points are shown on these coordinates below.

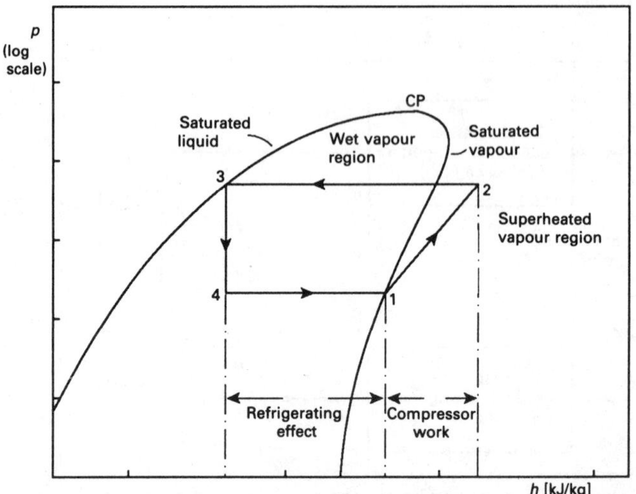

The **work of compression** and the **refrigerating effect** are two important characteristics of the refrigeration cycle, and these appear on the *p–h* diagram. Consequently the *p–h* chart is used for quick calculations on reversed cycles.

Fluid enters the compressor at state 1 and leaves at state 2, so if we assume the flow is adiabatic, from the steady flow energy equation we have:

$$-\dot{W} = h_2 - h_1$$

The work term (per unit mass flow of refrigerant) is negative for the fluid in the compressor and positive for the motor driving it. The positive value of $h_2 - h_1$ may be seen on the figure.

In the evaporator, fluid enters at state 4 and leaves at state 1 as a result of the heat transfer from the cold space, Q_C. We may write:

$$\dot{Q}_C = h_1 - h_4$$

This is the refrigerating effect which is shown on the *p–h* diagram. Bearing in mind that the COP for a refrigerator is refrigerating effect/compressor work, what is the COP in terms of the enthalpy values, h_1, h_2, and h_4?

71

The answer is: COP = $(h_1 - h_4)/(h_2 - h_1)$

Since: Refrigerating effect = $h_1 - h_4$

and: Compressor work $= h_2 - h_1$

then: $\text{COP} = (h_1 - h_4)/(h_2 - h_1)$

We will now work through a refrigeration cycle using Refrigerant 12. We shall take saturation temperatures of $-25°C$ in the evaporator and 35°C in the condenser. This means that the evaporator and condenser pressures are 1.237 and 8.477 bar respectively. Remember that heat transfer to the evaporator must take place from a cold space which is slightly warmer, at say $-20°C$, and the condenser must give out heat to the surroundings, at say 25°C. The cycle is shown on the *p–h* diagram below.

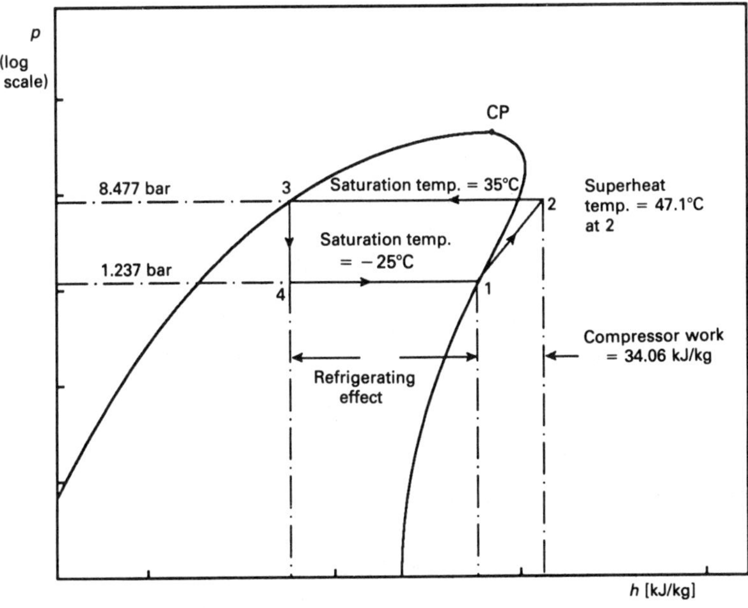

The fluid enters the compressor at state 1 which is saturated vapour at $-25°C$, at 1.237 bar. We assume in an ideal cycle that the fluid is compressed isentropically to state 2. The pressure at state 2 is the saturation pressure at 35°C, which is 8.477 bar. We shall need h_g and s_g at $-25°C$, from your R-12 tables. What are they?

72

$h_g = 176.48$ kJ/kg, $s_g = 0.7127$ kJ/kg K

These values came from page 13 of the Rogers and Mayhew tables.

73

This value of entropy, 0.7127 kJ/kg K, is also the entropy at state 2, at 8.477 bar. By inspection of the tables you can see that this value occurs in the superheat region at this pressure. At saturation s_g = 0.6839 kJ/kg K and at 15 K of superheat, it is 0.7196 kJ/kg K. The exact superheat is given by:

$$\text{Superheat} = 15 \times \frac{(0.7127 - 0.6839)}{(0.7196 - 0.6839)}$$

$$= 12.10 \text{ K}$$

This means that the fluid temperature at this state is 47.1°C. Knowing the superheat you can now interpolate to find the enthalpy at state 2.

74

$$\boxed{h_2 = 210.54 \text{ kJ/kg}}$$

At saturation, h = 201.45 kJ/kg, and at 15 K of superheat it is 212.72 kJ/kg, so at 12.10 K of superheat it is:

$$h_2 = 201.45 + (212.72 - 201.45) \times \frac{12.1}{15}$$

$$= 210.54 \text{ [kJ/kg]}$$

Next we find the compressor work, per unit mass flow:

$$-\dot{W} = h_2 - h_1$$

$$= 210.54 - 176.48$$

$$= 34.06 \text{ [kJ/kg]}$$

$$\therefore \quad \dot{W} = -34.06 \text{ [kJ/kg]}$$

This is shown on the p–h diagram and is negative for the fluid but positive for the motor driving the compressor.

Here is another calculation of isentropic compressor work:

Dry saturated ammonia (Refrigerant 717 in your tables) at 6.149 bar, 10°C, (state 1) is compressed to 11.67 bar (state 2). What are (a) the degree of superheat at state 2 and (b) the compressor work?

75

h_1 = 1454.3 kJ/kg, s_1 = 5.213 kJ/kg K, superheat at 2 = 26.44 K, h_2 = 1543.79 kJ/kg, W = −89.49 kJ/kg

You should be able to get these results using the methods of the previous two frames. For these exact figures, the Rogers and Mayhew tables were used. We now return to the refrigeration cycle problem.

In the condenser, the fluid is first cooled to the saturation temperature of 35°C and is then condensed to the saturated liquid state at 35°C, 8.477 bar (state 3). At this state, h_f is 69.55 kJ/kg. Next the fluid passes through an expansion valve to the lower pressure of 1.237 bar, at state 4. What is h_4?

76

h_4 = 69.55 kJ/kg

This is the throttling expansion referred to in Frame 68, and it takes place at constant enthalpy. We now have both h_1 and h_4 so we can find the refrigerating effect (RE). Thus:

$$\begin{aligned} \text{RE} &= h_1 - h_4 \\ &= 176.48 - 69.55 \\ &= 106.93\ [\text{kJ/kg}] \end{aligned}$$

Finally, the COP, which is RE/compressor work, is given by:

$$\begin{aligned} \text{COP} &= \frac{106.93}{34.06} \\ &= 3.14 \end{aligned}$$

77

It is possible to increase the refrigerating effect and hence the COP by subcooling the fluid at state 3. This means that in the condenser the fluid is cooled below the saturation temperature. Also some superheat is possible in the fluid before it enters the compressor. This is feasible since the main body of a refrigerator is warmer than the freezer compartment. These changes are shown in the figure below.

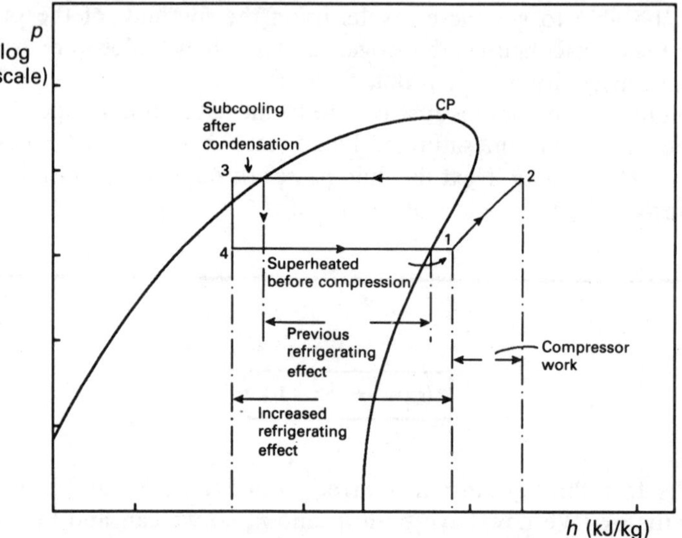

We shall repeat the calculations for this cycle with superheat to −10°C at state 1 (15 K of superheat), and 10 K of undercooling at state 3. This time we shall use the *p–h* chart.

78

State points 1 and 2 for the compressor work are shown on the *p–h* chart below.

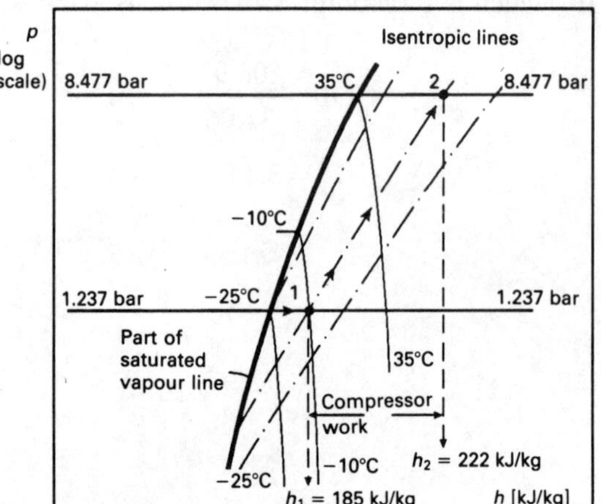

To find state point 1, find the −25°C temperature on the saturated vapour line (this point will correspond to the correct pressure), and move horizontally to the right to meet the −10°C line in the superheat region. The specific enthalpy is then read as 185 kJ/kg. Then move up the constant entropy line from this point to meet the higher pressure line of 8.447 bar. State point 2 is where these lines cross. The specific enthalpy is read as 222 kJ/kg. What then is the compressor specific work?

79

−37 kJ/kg

For unit mass flow in the compressor:

$$-\dot{W} = h_2 - h_1$$

$$\therefore \quad \dot{W} = 185 - 222 = -37 \text{ [kJ/kg]}$$

To find the specific enthalpy after undercooling at state 3, we find the 35°C point on the saturated liquid line, then move horizontally to the left to meet the 25°C line. In the subcooled liquid region the isotherms are vertical. The specific enthalpy is then read as 60 kJ/kg.

Alternatively we can use the tables and say:

$$h_3 = h_f - c_P \Delta T$$
$$= 69.55 - c_P \times 10$$

Some information for c_p of R-12 saturated liquid is given on page 15 of the Rogers and Mayhew tables. We see that it is 0.902 kJ/kg K at 250 K and 0.980 kJ/kg K at 300 K. In undercooling from 35°C to 25°C the average temperature is 303.15 K, so we may take the value of 0.980 kJ/kg K at 300 K.
Hence:

$$h_3 = 69.55 - 0.980 \times 10$$
$$= 59.75 \text{ [kJ/kg]}$$

Then h_4 is again the same as h_3, hence taking h_3 as 60 kJ/kg:

$$\text{RE} = h_1 - h_4$$
$$= 185 - 60$$
$$= 125 \text{ [kJ/kg]}$$

And finally: $\text{COP} = 125/37 = 3.38$

80

Thus we see that the refrigerating effect has increased, in round figures, from 107 to 125 kJ/kg, or by 16.8%, and the COP from 3.14 to 3.38, or by 7.6%. You can see from the slope of the isentropic lines in the superheat region why, with initial superheat before compression, the compressor work has increased so that the improvement in the COP is not as good.

So far we have assumed the vapour compression process to be isentropic; in reality this is not so. The process is still regarded as adiabatic but because of turbulence and friction in the compression there is an entropy increase and to achieve the desired pressure the work input is greater than it otherwise would be. A non-isentropic compression is shown below on both *T–s* and *p–h* coordinates.

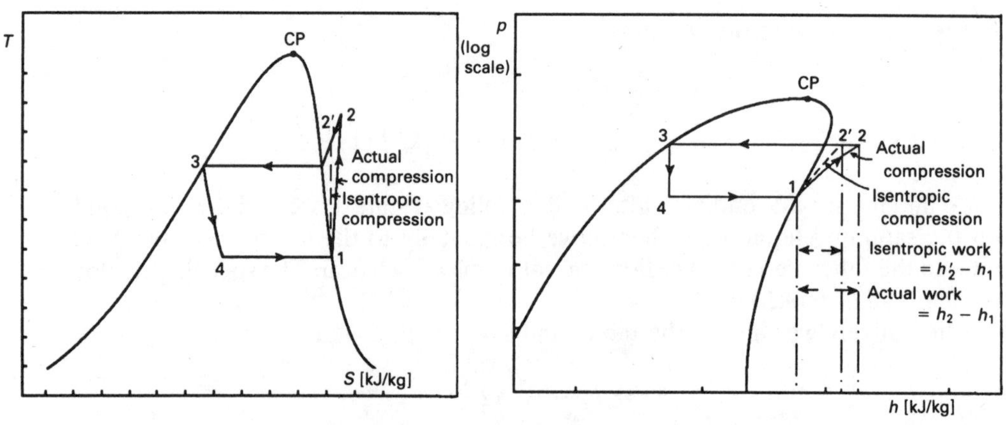

The actual work of compression may be measured and the isentropic work of compression may be calculated. A comparison of the two gives an isentropic efficiency of the compression process. Then, with the work term being the enthalpy difference across the compressor, with h_1 as the enthalpy of the fluid before compression, h_2' is the enthalpy after isentropic compression, and h_2 the actual enthalpy after compression, we have:

$$\eta_{is} = \frac{h_2' - h_1}{h_2 - h_1}$$

81

In the first example considered in Frames 71–76, we found that the isentropic work of compression was 34.06 kJ/kg, and after compression there was 12.1 K of superheat. If in this case there had been a compressor isentropic efficiency of 70%, the actual work of compression would be:

$$\text{Compressor work} = \frac{34.06}{0.7}$$

$$= 48.66 \text{ [kJ/kg]}$$

With h_1 as 176.48 kJ/kg, h_2 would then be 225.14 kJ/kg at the higher pressure of 8.477 bar. This value is higher than the range of the tables, being beyond the 30 K superheat limit, but is readily found on the *p–h* chart. State points 3 and 4 are not affected, so the refrigerating effect is unchanged. What is the new value of the COP?

82

COP = 2.2

The RE was 106.93 kJ/kg from Frame 76, and with the work at 48.66 kJ/kg, the COP is 106.93/48.66, which is 2.2.

The heat pump cycle is the same in principle as the refrigeration cycle, except that we are now more interested in the heat output from the condenser. At the same time the heat input from a cold source is important. This has to be available at all times for the heat pump to work, and it must not be subject to icing which can seriously affect the heat flow. (Refrigerators are designed for heat transfer to the evaporator to be under iced conditions.) The saturation conditions in the evaporator and condenser are set to appropriate values, which could be 15°C for the heat source and 60–70°C for the heat delivery. A *p–h* diagram for a heat pump cycle is given below.

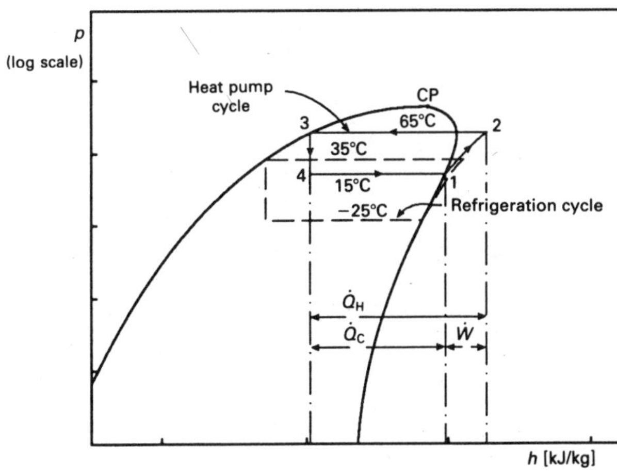

83

From the p–h diagram in the previous frame you can see that for unit mass flow:

$$\dot{Q}_H = h_2 - h_3$$
$$\dot{Q}_C = h_1 - h_4$$
$$\dot{W} = h_2 - h_1$$

and so: $$\dot{Q}_H = \dot{Q}_C + \dot{W}$$

Thus problems on both heat pump and refrigeration cycles may be readily solved using the p–h chart.

84

So far our calculations on reversed cycles have been in terms of unit mass flow of refrigerant fluid. The plant then has to be sized in terms of the required refrigeration capacity or the heat pump output. A refrigeration load is sometimes expressed in terms of the equivalent production of tons of ice per day. This load equivalent is of American origin, where the ton is 2000 lb mass. It is equivalent to a heat extraction rate of 3.517 kW.

Similarly the output of heat pumps is rated in kW. In the science of building heat transfer, the heat loss rate may be found for particular temperature conditions; this is then the rate at which heat must be supplied to the building to maintain the required temperature. For a heat pump output:

$$\dot{Q}_H = \dot{m}\,(h_2 - h_3)\ [\text{kW}]$$

If $(h_2 - h_3)$ in a heat pump cycle is 120 kJ/kg, and the required output is 10 kW, what is the mass flow rate of fluid, $\dot{m}$?

85

0.0833 kg/s

The heating requirement, $\dot{Q}_H$, is 10 kW, and $(h_2 - h_3)$ is 120 kJ/kg, so:

$$10.0 = \dot{m}\,(215.32 - 95.74)$$
$$\therefore \quad \dot{m} = 0.0833\ [\text{kg/s}]$$

We also need to know the size of the compressor. Fluid enters the compressor at the outlet of the evaporator, either of specific volume v_g at the appropriate pressure, or of specific volume v if superheated. The compressor cylinder of diameter D with a piston stroke L will have a swept volume of:

$$V_s = \frac{\pi D^2 L}{4}$$

Because of flow restrictions at the valves, the effective swept volume is often less, so we can say:

$$V_{se} = \frac{\eta_V \pi D^2 L}{4}$$

where η_V is the volumetric efficiency. The compressor has a two stroke cycle, sucking in fresh charge with each downward motion of the piston, so for a compressor speed of N rev/s, for a continuity of volume we can write:

$$N \frac{\eta_V \pi D^2 L}{4} = \dot{m} \, v_g$$

Using this formula, what is the compressor speed when $\dot{m}$ is 0.0833 kg/s of saturated R-12 vapour at 5°C, when D is 50 mm, L is 60 mm and η_V is 0.8?

86

2528 rev/min

v_g at 5°C is 0.0475 m^3/kg, so we have:

$$\frac{0.8 \pi \times 0.05^2 \times 0.06}{4} \times N = 0.0836 \times 0.0475$$

$$\therefore \quad N = 42.13 \text{ [rev/s]}$$

$$= 2528 \text{ [rev/min]}$$

87

To conclude this programme, the following reversed cycle problems may be solved either by using tables or the *p–h* chart.

1. A refrigeration cycle operates between evaporator and condenser saturation temperatures of −20°C and 30°C respectively. Fluid enters the compressor as saturated vapour. Calculate the refrigerating effect, the COP, the fluid peak temperature after compression and the mass flow rate of refrigerant for a load of 1.5 kW, using (a) R-12 and (b) R-717, ammonia.

   [(a) 114.14 kJ/kg, 4.055, 39.89°C, 0.0131 kg/s, (b) 1096.9 kJ/kg, 4.092, 110.93°C, 0.0014 kg/s]

2. A heat pump cycle using R-12 has a saturation pressure of 4.233 bar in the evaporator and a saturation temperature of 70°C in the condenser. The compressor has an isentropic efficiency of 80.0%. In the condenser there is undercooling by 10 K to 60°C. The specific heat of the liquid may be taken as 0.988 kJ/kg K. The compressor has a single cylinder of 50 mm bore with a piston stroke of 70 mm, and a volumetric efficiency of 80%. The heat pump output is 20 kW. Calculate (a) the COP, (b) mass flow of fluid, (c) compressor speed and (d) the percentage of the output above 70°C, at 70°C and below 70°C.

   [(a) 3.845, (b) 0.1566 kg/s, (c) 1360 rev/min, (d) 10.55%, 81.71%, 7.74%]

3. A heat pump using R-12 is driven by a steam engine running on the Rankine cycle. In the steam cycle, steam enters the turbine at 20 bar and 300°C and expands with an isentropic efficiency of 80% to a pressure of 0.1992 bar, so that heat is rejected at 60°C. In the heat pump cycle, the evaporator saturated temperature is 5°C and heat rejection in the condenser is also at 60°C after cooling from the superheat condition. The compressor has an isentropic efficiency of 75%. The compressor is driven by the steam turbine, so the work requirement of the compressor equals the work output of the turbine, and the steam cycle condenser heat output and heat pump output are added together. Neglecting the feed-pump work in the steam cycle, calculate (a) the mass flow of R-12 in the heat pump per kg of steam flow in the Rankine cycle and (b) the ratio of heat supplied to the steam cycle to the combined heat output from the steam cycle condenser and heat pump. [(a) 18.62 kg of R-12/kg of steam, (b) 1.63]

Programme 8

COMPRESSIBLE FLOW AND WORK TRANSFER IN TURBINE BLADING

1

In the previous programme we have seen how we may calculate the amount of work that may be obtained from steam delivered to a turbine and expanded to a given pressure. This is one example of a work interaction between a fluid in motion and a rotating wheel with blades. Other examples are a gas turbine, and a gas or air compressor. We now need to get an insight into how this work interaction is achieved in practice.

As an in-depth study is rather specialised we shall be following an elementary treatment, and we shall look at some aspects of **one dimensional steady state compressible flow** insofar as it enables us to study this work interaction. As well as the First and Second Laws of Thermodynamics, we shall need to evoke the principle of conservation of mass and Newton's Second Law of Motion.

The flow can take place in ducts, or in the spaces between fixed and moving blades of turbines or compressors.

2

To develop our ideas, we shall consider initially steady adiabatic flow of a perfect gas. We shall assume the flow to be frictionless and reversible. We shall assume also that changes in curvature of the axis of the passage and changes in cross-sectional area are gradual without any discontinuities. The diagram shows sections of (a) a flow duct and (b) a flow passage between fixed and moving blades in a turbine.

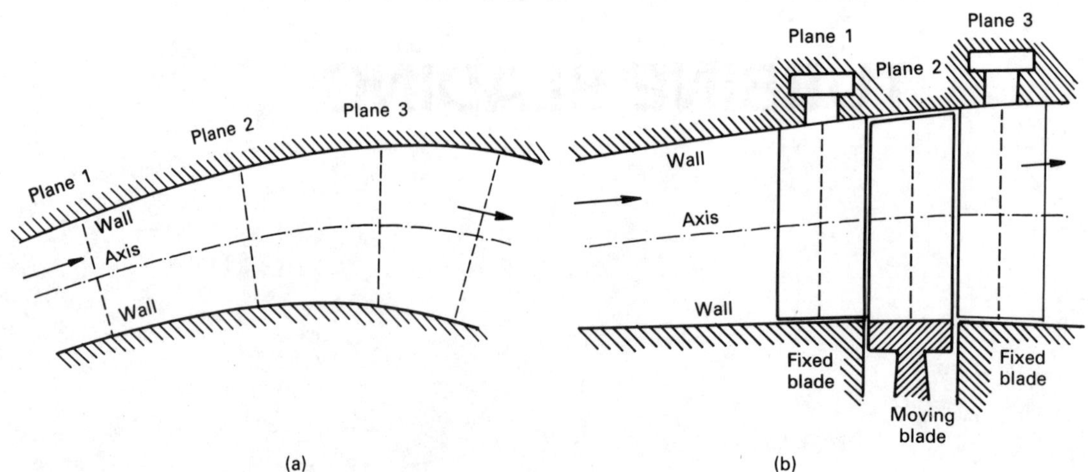

All properties are assumed to be uniform across the planes normal to the axis of flow. Thus in any plane shown in the figure, the velocity, for example, is the same at the axis and at the wall of the passage, but it will vary from one plane to another.

However, with adiabatic, reversible and frictionless flow, one property will not change from one plane to the next. Which property is that?

3

Entropy

Frictionless, reversible and adiabatic flow is isentropic flow, or flow with constant entropy.

Consider a duct through which a perfect gas is flowing. The gas enters the open system at state 1 and leaves at state 2. For a perfect gas, $pv = RT$, c_V = constant (Frame 94, Programme 4), and for reversible adiabatic flow, pv^γ = constant (Frame 67, Programme 5).

There is no work interaction and the flow is adiabatic, so in applying the First Law in the steady flow energy equation, both $\dot{Q}$ and $\dot{W}$ are zero. We also assume there is no change in potential energy:

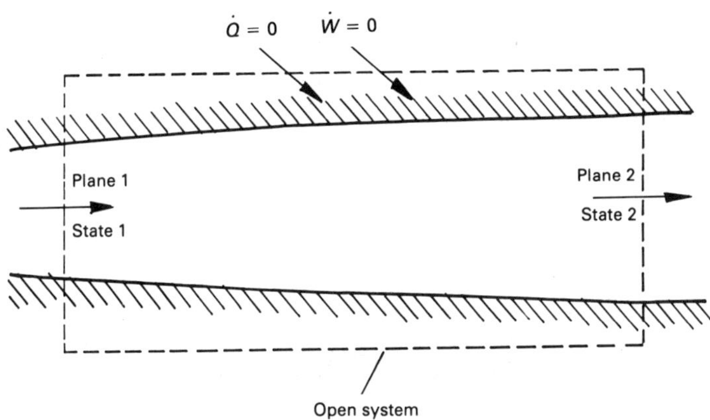

$$0 = h_2 - h_1 + \frac{V_2^2 - V_1^2}{2}$$

Hence:
$$\frac{V_2^2 - V_1^2}{2} = c_P(T_1 - T_2) = \frac{R\gamma}{\gamma - 1}(T_1 - T_2)$$

since we are dealing with a perfect gas. Now substitute pv for RT in the right-hand side.

$$\frac{V_2^2 - V_1^2}{2} = \frac{\gamma}{\gamma - 1}(p_1 v_1 - p_2 v_2)$$

Thus $p_1 v_1 = RT_1$ and $p_2 v_2 = RT_2$. This equation relates velocity, pressure and specific volume at any two planes in the flow. Suppose plane 1 is within a reservoir just before the entrance to the duct, where the pressure is p_1. The approach velocity is small and we can say that at the entrance to the duct V_1 is zero.

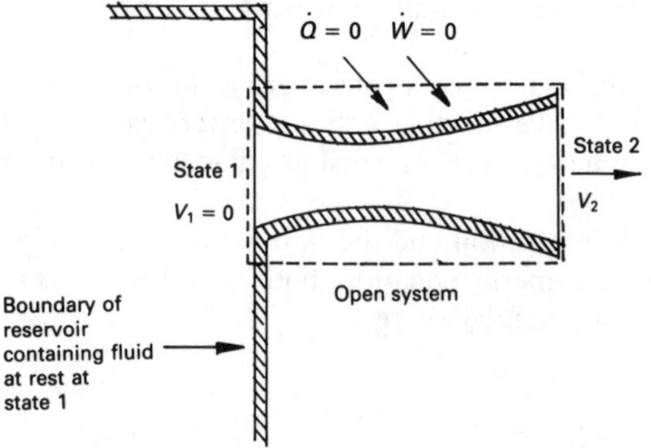

Then the previous equation gives us:

$$V_2 = \left(\frac{2\gamma}{\gamma - 1} p_1 v_1 \left(1 - \frac{p_2 v_2}{p_1 v_1}\right)\right)^{1/2}$$

Then from pv^γ = constant, we get:

$$\frac{v_2}{v_1} = \left(\frac{p_1}{p_2}\right)^{1/\gamma}$$

which when we substitute into the previous line gives:

$$V_2 = \left(\frac{2\gamma}{\gamma - 1} p_1 v_1 \left(1 - \left(\frac{p_2}{p_1}\right)^{(\gamma-1)/\gamma}\right)\right)^{1/2} \tag{1}$$

This is the velocity at plane 2 where the pressure has fallen to p_2 when plane 1 is the exit plane of a large gas reservoir where the fluid is at rest. Use this formula to calculate V_2 when p_1 is 200 kN/m², or 200 kPa, $v_1 = 0.72$ m³/kg, $p_2/p_1 = 0.5$ and $\gamma = 1.4$.

425.55 m/s

Since p_1 is in kPa units, a factor of 10^3 is needed to give the answer in m/s. Thus:

$$V_2 = \left(\frac{2 \times 1.4}{0.4} \times 200 \times 10^3 \times 0.72 \left(1 - 0.5^{0.2857}\right)\right)^{1/2}$$
$$= 425.55 \text{ [m/s]}$$

In addition to the expression for V_2 (equation (1) in the previous frame), we can find expressions for v_2 and A_2. This is done in the next frame. Then we can plot V_2, v_2 and A_2 against the pressure ratio p_2/p_1, which has a value of 1 at the origin and decreases to the right. The three lines are shown in the figure below.

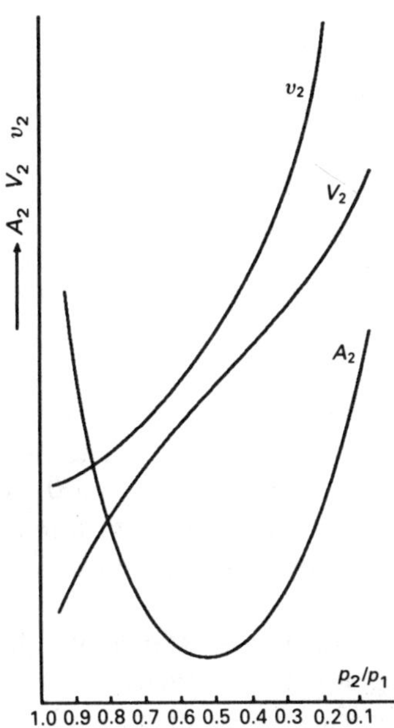

6

Thus from pv^γ = constant, we can write:

$$v_2 = v_1\left(\frac{p_1}{p_2}\right)^{1/\gamma} \tag{2}$$

The area of cross-section of the duct at the second plane is A_2, and if the mass flow is $\dot{m}$ kg/s where the velocity is V_2 and the specific volume v_2, then by the principle of continuity of mass:

$$\dot{m} = \frac{A\ V}{v} = \frac{A_2V_2}{v_2}$$

so for unit mass flow: $$A_2 = v_2/V_2 \tag{3}$$

Equations (1) to (3) were used to obtain the figure in the previous frame. We can see that if we make the pressure at some position in the nozzle relative to the inlet pressure our independent variable, then both the velocity and specific volume increase, but they do so in such a way that the area, from equation (3), reaches a minimum value at a pressure ratio between 0.6 and 0.5. At lower values the area increases again.

The area A_2 has a minimum value. What does this mean for *the mass flow per unit area* of gas passing through the duct?

7

Mass flow per unit area has a *maximum*

If a duct has a minimum area at some point along its length, then with a constant mass flow along the duct, the mass flow per unit area must have a **maximum**. Thus if you had a duct that merely reduced in area along its length, the pressure at the outlet could never fall to a value lower than about 0.5 of the inlet pressure. If the duct *had* a minimum area, then depending on how much the area increased along the duct after this minimum, the outlet pressure would fall to values well below the value at the point of the minimum.

This minimum area is called the **throat**. If in a duct the pressure is falling and the velocity is increasing, the duct is called a **nozzle**. On the other hand, if the velocity is falling and the pressure is increasing, the duct is called a **diffuser**. In both a nozzle and a diffuser it is possible for the velocity of the flow to exceed the local velocity of sound in the fluid. When this happens the flow is called **supersonic**. Flow at speeds below the velocity of sound may be called **subsonic**. Next we shall derive an equation which will tell us how we must make the area change if we want to design both subsonic and supersonic nozzles and diffusers.

8

We may write the steady flow energy equation for flow in a duct in differential form:

$$\mathrm{d}h + \mathrm{d}(V^2/2) = 0 \tag{1}$$

Then from the First and Second Laws we have seen that:

$$T\mathrm{d}s = \mathrm{d}h - v\mathrm{d}p \qquad \text{(Frame 78, Programme 6)}$$

and so for an isentropic flow in which $\mathrm{d}s = 0$ we have:

$$\mathrm{d}h = v\mathrm{d}p \tag{2}$$

From the principle of continuity of mass, AV/v = constant, so we can write:

$$\log A + \log V - \log v = 0$$

and then:

$$\frac{\mathrm{d}A}{A} + \frac{\mathrm{d}V}{V} - \frac{\mathrm{d}v}{v} = 0 \tag{3}$$

We now combine (1) and (2):

$$v\mathrm{d}p + \mathrm{d}(V^2/2) = 0$$

or:

$$v\mathrm{d}p + V\mathrm{d}V = 0 \tag{4}$$

From equation (4) we can write:

$$\mathrm{d}V = -v\mathrm{d}p/V$$

and if we combine this with equation (3) we get:

$$\frac{\mathrm{d}A}{A} = \frac{\mathrm{d}v}{v} + \frac{v\mathrm{d}p}{V^2}$$

and so:

$$\frac{\mathrm{d}A}{A} = v\mathrm{d}p\left(\frac{1}{V^2} + \frac{1}{v^2}\frac{\mathrm{d}v}{\mathrm{d}p}\right) \tag{5}$$

This is continued in the next frame.

9

From your knowledge of fluid mechanics you should know that the velocity of propagation, a, of a small pressure wave, or sound wave, is the root of the rate of change of pressure with density, with entropy constant. Thus:

$$a^2 = \left(\frac{dp}{d\rho}\right)_{s\,=\,\text{const}}$$

The velocity a is called the **sonic velocity**. The density ρ is $1/v$, so:

$$\frac{d\rho}{dv} = -\frac{1}{v^2} \quad \text{and then:} \quad \frac{1}{v^2}\frac{dv}{dp} = -\frac{1}{a^2}$$

When we substitute this into equation (5) we get:

$$\frac{dA}{A} = v dp\left(\frac{1}{V^2} - \frac{1}{a^2}\right)$$

In this equation V is the local velocity in a duct at conditions at which the local sonic velocity is a and the specific volume of the fluid at the same point is v.

10

This equation will tell us quite a lot. First, the term dA/A is the ratio of the variation in cross-sectional area, dA, to the area itself. If it is positive, it means the area is *increasing*, and obviously if it is negative the area is *decreasing*. In the group $v dp$, dp is the pressure differential, and if this is positive the pressure is increasing in the direction of flow, and if negative the pressure is decreasing. The bracketed term contains a, the sonic velocity, and V, the local gas velocity, and if V is less than a then this whole term is positive, and if V is greater than a then the term is negative.

Think about a nozzle first. The pressure is falling in the direction of flow and V is less than a. What can you deduce about the way the area of the duct is changing?

11

The duct is converging

The duct must be converging. The bracketed term is positive with V less than a, and $v dp$ is negative, so dA/A must also be negative.

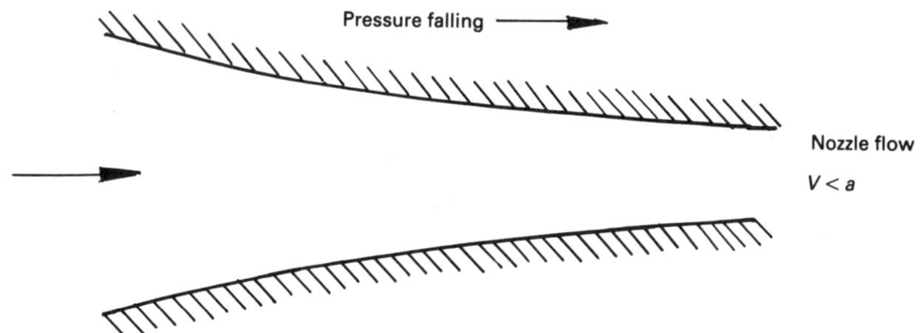

You may wonder why we have taken such trouble to prove the area of the nozzle is converging. But this was only with V less than a. A nozzle is a duct in which the pressure is falling, it is not necessarily a converging duct. Look again at the equation in Frame 9; what happens when $\mathrm{d}A/A = 0$? What is then the value of V?

12

At the minimum area $\mathrm{d}A/A = 0$; then $V = a$

When $\mathrm{d}A/A$ is not changing, the nozzle has stopped converging, and so we have reached the minimum area, or **throat**. The right-hand side of the equation in Frame 9 can only be zero when $V = a$.

Thus you have deduced that the **velocity becomes sonic at the throat of a nozzle**.

It follows that for V to exceed a, the ratio $\mathrm{d}A/A$ must be positive with the pressure falling. When $\mathrm{d}A/A$ is positive the nozzle is said to be diverging, and so to achieve velocities that are supersonic the nozzle must first converge and then diverge. This is called a **convergent–divergent nozzle**. These facts are summarised in the figure in the next frame.

13

A convergent–divergent nozzle is shown in the figure below.

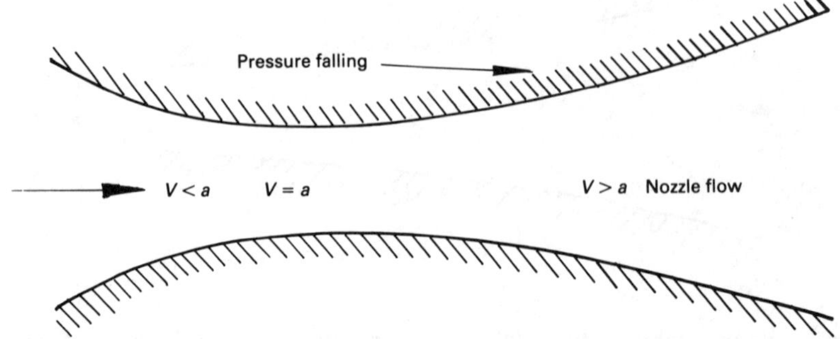

Now think about a diffuser. The velocity is now falling and the pressure is increasing. How must the area change if initially V is greater than a, and secondly if V is less than a ?

14

$V > a$, duct converges; $V < a$, duct diverges

In a diffuser dp is positive, and so if the velocity is initially supersonic the duct must converge, and if the velocity is initially subsonic the duct must diverge. A diffuser must be converging–diverging if it is to diffuse a flow from a supersonic condition down through subsonic flow.

We now need to see how we can design a nozzle to give us a required velocity of flow. We shall need to see whether this velocity is subsonic or supersonic, so that we can see what sort of nozzle we need. We will continue to work with a perfect gas and we shall consider steam later.

From equation (1) in Frame 4 we can write:

$$V_2 = \left(\frac{2\gamma}{1-\gamma} p_1 v_1 \left[\left(\frac{p_2}{p_1}\right)^{(\gamma-1)/\gamma} - 1\right]\right)^{1/2}$$

We shall use this expression for the velocity in the continuity equation to find the mass flow per unit area. Thus:

$$\dot{m} = \frac{A_2 V_2}{v_2} \quad \text{so:} \quad \frac{\dot{m}}{A_2} = \frac{V_2}{v_2}$$

and: $$v_2 = \left(\frac{p_1}{p_2}\right)^{1/\gamma} v_1$$

and then: $$\frac{\dot{m}}{A_2} = \left(\frac{2\gamma}{1-\gamma}\,\frac{p_1 v_1}{v_1^2}\left[\frac{p_2}{p_1}\right]^{2/\gamma}\left(\left[\frac{p_2}{p_1}\right]^{(\gamma-1)/\gamma} - 1\right)\right)^{1/2}$$

which gives: $$\frac{\dot{m}}{A_2} = \left(\frac{2\gamma}{1-\gamma}\,\frac{p_1}{v_1}\left[\left(\frac{p_2}{p_1}\right)^{(\gamma+1)/\gamma} - \left(\frac{p_2}{p_1}\right)^{2/\gamma}\right]\right)^{1/2}$$

From this expression we can find the pressure ratio (p_2/p_1) at which $(\dot{m}/A_2)$ becomes a maximum. How do we do that?

15

By putting $\mathrm{d}(\dot{m}/A_2)/\mathrm{d}(p_2/p_1) = 0$

We differentiate $(\dot{m}/A_2)$ with respect to (p_2/p_1) to find the pressure ratio at which $(\dot{m}/A_2)$ becomes a maximum. **This will give us the pressure at the throat of the nozzle**. By putting $\mathrm{d}(\dot{m}/A_2)/\mathrm{d}(p_2/p_1) = 0$, we obtain:

$$\frac{\gamma+1}{\gamma}\left(\frac{p_2}{p_1}\right)^{1/\gamma} - \frac{2}{\gamma}\left(\frac{p_2}{p_1}\right)^{(2-\gamma)/\gamma} = 0$$

Next we can divide this expression by $(p_2/p_1)^{1/\gamma}$ which gives:

$$\frac{\gamma+1}{\gamma} - \frac{2}{\gamma}\left(\frac{p_2}{p_1}\right)^{(1-\gamma)/\gamma} = 0$$

From this, what is then p_2/p_1 at the throat?

16

$$\frac{p_2}{p_1} = \left(\frac{2}{\gamma + 1}\right)^{\gamma/(\gamma-1)}$$

Here, p_2 is the pressure at the throat, and this expression is referred to as the **critical pressure ratio**.

This means that knowing γ for the gas and the inlet pressure p_1 we can find the throat pressure p_2. This gives us the pressure ratio at which the mass flow per unit area becomes a maximum, or at which the nozzle is said to be **choked**.

Since we are dealing with a perfect gas we can also write:

$$\frac{T_2}{T_1} = \left(\frac{p_2}{p_1}\right)^{(\gamma-1)/\gamma} \qquad \text{(Frame 69, Programme 5)}$$

and so:

$$\frac{T_2}{T_1} = \left[\left(\frac{2}{\gamma + 1}\right)^{\gamma/(\gamma-1)}\right]^{(\gamma-1)/\gamma}$$

$$= \frac{2}{\gamma + 1}$$

This is the temperature ratio corresponding to the critical pressure ratio, so we can also find the temperature of the gas at the throat, T_2, if we know T_1 at inlet. If air, for which γ is 1.4, enters a nozzle at 400 kPa and at 400 K, what are the pressure and temperature at the throat?

17

211.31 kPa, 333.33 K

The pressure ratio is $(2/2.4)^{1.4/0.4}$ which gives 0.5283, so that p_2 is 0.5283 of p_1, which is 211.31 kPa. The temperature ratio is (2/2.4), giving a throat temperature of 333.33 K.

In Frame 12 we saw that the sonic velocity of the fluid occurs at the throat. Now we need an expression for it in terms of the temperature of the fluid. For isentropic flow of a perfect gas, for which $\rho = 1/v$, we can write:

$$p/\rho^\gamma = \text{constant}$$

or:

$$\log p - \gamma \log \rho = \text{constant}$$

and then:
$$\frac{\mathrm{d}p}{p} - \gamma\frac{\mathrm{d}\rho}{\rho} = 0$$

$$\therefore \quad \frac{\mathrm{d}p}{\mathrm{d}\rho} = \frac{\gamma p}{\rho}$$

We then use the expression for a, the velocity of sound, from Frame 9:

$$a^2 = \left(\frac{\mathrm{d}p}{\mathrm{d}\rho}\right)_{s=\text{const}}$$

$$\therefore \quad a^2 = \frac{\gamma p}{\rho} = \gamma RT \quad \text{for a perfect gas}$$

$$\therefore \quad a = (\gamma RT)^{1/2}$$

Thus, for air at the throat at 333.33 K, for which $\gamma = 1.4$ and $R = 0.287$ kJ/kg K, the sonic velocity is:

$$a = (1.4 \times 0.287 \times 10^3 \times 333.33)^{1/2}$$

$$= 365.97 \text{ [m/s]}$$

This throat velocity can also be calculated from the steady flow energy equation:

$$\frac{V_2^2 - V_1^2}{2} = c_P\,(T_1 - T_2)$$

(see the last equation in Frame 3), taking state 1 to be reservoir conditions with V_1 zero, with T_1 = 400 K and T_2 = 333.33 K, and with the specific heat at constant pressure = 1.005 kJ/kg K. You should get the same answer to three significant figures.

18

366.07 m/s

With V_1 zero we get:

$$V_2 = [2 \times c_P \times (T_1 - T_2)]^{1/2}$$

$$= [2 \times 1.005 \times 1000 \times (400 - 333.33)]^{1/2}$$

$$= 366.07 \text{ [m/s]}$$

Both answers agree to three significant figures, the difference being due to rounding off errors in the values of γ, R and c_P.

In Frame 54 of Programme 3 we saw that in adiabatic flow, the enthalpy and kinetic energy term can be combined to give the **stagnation enthalpy**, from which we can obtain the total head or stagnation temperature, T_T, where:

$$T_T = T + \frac{V^2}{2c_P}$$

In the adiabatic nozzle or diffuser with no heat and work transfer terms, it follows that the total head temperature remains constant throughout. In the examples so far we have taken the velocity at entry to the nozzle to be zero, so consequently the total head and static temperatures at entry are equal. When a flow is slowed down so that its static temperature rises, there is also a rise in pressure. The pressure at inlet to a nozzle where the velocity is zero is a **total pressure**. In any nozzle calculation the throat pressure is always found from the total head inlet pressure, so if an inlet velocity is given, the inlet total head conditions of temperature and pressure must first be found.

This will be illustrated with an example in the next frame.

19

At the inlet of a convergent air nozzle the velocity is 200 m/s, the temperature is 550 K and the pressure is 350 kPa. The mass flow is 10 kg/s. Nothing is said about the exit conditions from the nozzle, but if we assume the nozzle discharges to atmosphere at say 100 kPa, the pressure ratio p_2/p_1 across the nozzle is 100/350 or 0.286. In Frame 17 we saw that for air the pressure ratio at the throat for air with $\gamma = 1.4$ was 0.5283. Thus a convergent nozzle could easily operate under choked conditions with the given inlet pressure. This implies that the velocity at the throat is sonic. In a convergent nozzle the throat is at the exit plane. For the conditions given we will calculate the pressure, temperature and velocity at the throat, and the diameter of the nozzle at inlet and at the throat for the mass flow of 10 kg/s. For air, take $\gamma = 1.4$, $c_P = 1.005$ kJ/kg K.

First of all, with an inlet velocity of 200 m/s we must find the inlet total head conditions. The total head temperature at inlet will be:

$$T_{T_1} = 550 + \frac{200^2}{2 \times 1005}$$

$$= 569.90 \text{ K}$$

Notice that we used c_P in J/kg K units. Total pressure is defined assuming reversible as well as adiabatic conditions, so we can use the equation from Frame 69 of Programme 5 to give us:

$$\frac{p_T}{p} = \left(\frac{T_T}{T}\right)^{\gamma/(\gamma-1)}$$

where p_T is the total pressure corresponding to the total temperature T_T.

If T_T is the inlet total temperature of 569.9 K, $T = 550.0$ K and $p = 350$ kPa, what is the inlet total pressure, p_T ?

20

396.39 kPa

p_T is the total head pressure at inlet p_{T_1}, and so:

$$p_{T_1} = 350 \times (569.90/550)^{1.4/0.4} = 396.39 \text{ [kPa]}$$

For the inlet velocity of 200 m/s, the total head or stagnation inlet temperature is 569.90 K, and the total pressure is 396.39 kPa. Later we shall use these values to find the pressure and temperature at the throat. First, we will find the inlet area of the nozzle for a mass flow of 10 kg/s.

21

To find the inlet area of the nozzle we need first to find the specific volume of air in the state at which it is moving, which is at a static temperature of 550 K and a static pressure of 350 kPa. For air with $\gamma = 1.4$ and $c_P = 1.005$ kJ/kg K, R will be 0.2871 kJ/kg K (you can check that using $R = c_P^{(\gamma-1)/\gamma}$), so using $pv = RT$ we get:

$$v_1 = \frac{0.2871 \times 550}{350}$$

$$= 0.4512 \text{ [m}^3\text{/kg]}$$

Then, for continuity:

$$\dot{m} = \frac{A\,V}{v}$$

so:

$$10 = \frac{A \times 200}{0.4512}$$

$$\therefore \quad A = 0.0226 \text{ [m}^3\text{]}$$

The diameter of the nozzle at inlet is therefore 0.1696 m.

Now, what are the values of the pressure and temperature at the throat? Use the total head inlet conditions we have just found and, if you need to, refer back to Frame 16 for the formulae to use.

22

209.41 kPa, 474.92 K

The pressure at the throat is given by:

$$p_2 = 396.39 \times \left(\frac{2}{2.4}\right)^{1.4/(1.4-1)}$$

$$= 209.41 \text{ [kPa]}$$

and the temperature is:

$$T_2 = \frac{2}{2.4} \times 569.9$$

$$= 474.92 \text{ K}$$

We can find the velocity at the throat by calculating the sonic velocity at this temperature. Thus:

$$a = (\gamma RT)^{1/2}$$

$$= (1.4 \times 0.2871 \times 10^3 \times 474.92)^{1/2}$$

$$= 436.91 \text{ [m/s]}$$

Finally you can calculate the specific volume at the throat, and then find the area of cross-section and the diameter.

23

$v_2 = 0.6511$ m³/kg, $A_2 = 0.0149$ m³, diameter $= 0.1377$ m

$$v_2 = \frac{0.2871 \times 474.92}{209.41} = 0.6511 \text{ [m}^3\text{/kg]}$$

From continuity:

$$10 = \frac{A_2 \times V_2}{v_2}$$

$$\therefore \quad A_2 = 10 \times 0.6511/436.91$$

$$= 0.0149 \text{ [m}^3\text{]}$$

Hence the diameter at the throat is 0.1377 m.

Now here is another similar problem. Air enters a convergent nozzle at 250 kPa and 400 K, at a velocity of 150 m/s. The inlet diameter is 0.15 m. Calculate the mass flow, the inlet total head conditions, and the pressure and temperature at the throat, then find the nozzle diameter at the throat. Take, for air, $\gamma = 1.4$, $c_P = 1.005$ kJ/kg K and $R = 0.2871$ kJ/kg K.

24

$v = 0.4594\ \text{m}^3/\text{kg}$, $\dot{m} = 5.77\ \text{kg/s}$, $T_T = 411.19\ \text{K}$, $p_T = 275.36\ \text{kPa}$,
Throat: $p = 145.47\ \text{kPa}$, $T = 342.66\ \text{K}$, $v = 0.6763\ \text{m}^3/\text{kg}$,
$V = 371.12\ \text{m/s}$, $d = 0.1157\ \text{m}$

By following the previous example in Frames 19–23, you should be able to obtain these answers.

We have seen how in a convergent nozzle the conditions at the throat may be found from the inlet total head temperature and pressure. These are called the **design conditions** but the nozzle may not necessarily operate at these conditions. Suppose a nozzle discharges into a larger duct which has a valve in it, as shown below. Operation of this valve will control the pressure in the duct beyond the nozzle exit; this pressure is called the nozzle back pressure.

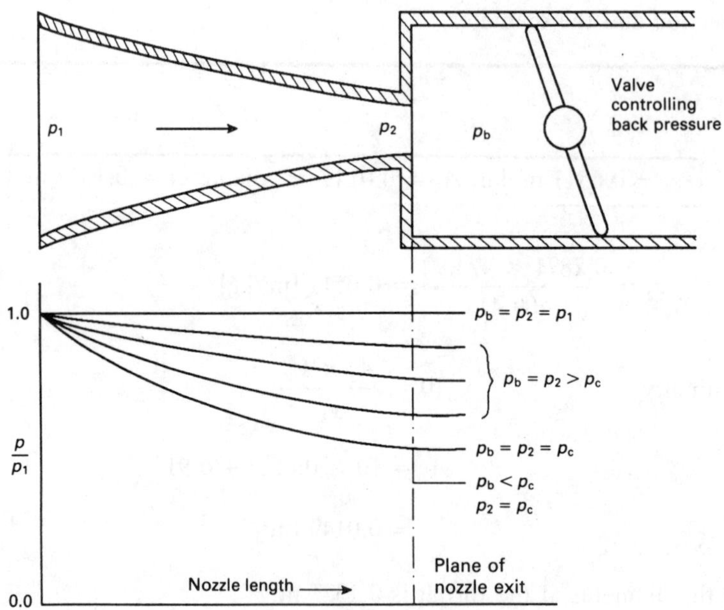

25

In the figure, p is the pressure at any point in the nozzle, while p_1 is the inlet total pressure, p_2 is the pressure at the exit (the throat) and p_C is the critical or design throat pressure for choking conditions. p_b is the back pressure after discharge from the nozzle. As the back pressure falls from the inlet pressure to the critical pressure a series of curves of pressure distribution is obtained, and as the back pressure falls the mass flow increases to its maximum value. When the back pressure falls below the critical pressure there is a pressure drop after the nozzle exit; the throat pressure remains at p_C and mass flow remains at its maximum. The curve of mass flow against the ratio of back pressure to inlet pressure is shown below.

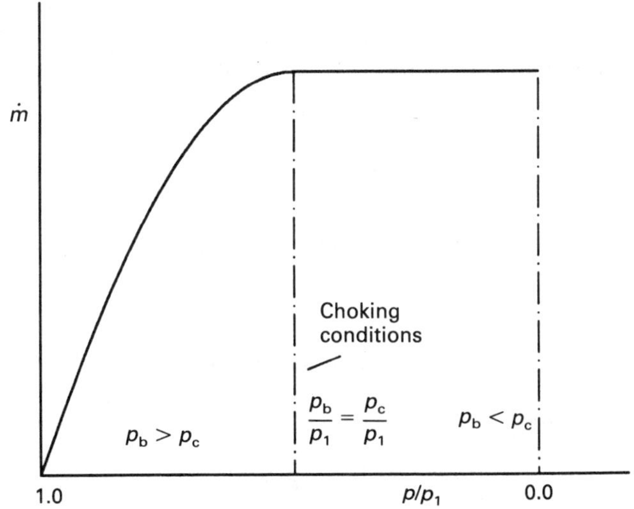

Once the throat pressure reaches the critical value, the mass flow remains constant however much the back pressure continues to drop. At these conditions, the velocity in the plane of the throat is sonic, and as the flow expands into the lower pressure region a shock wave occurs to bring the velocity back to a subsonic value. A shock will always occur in an unsustainable sonic or supersonic flow to restore conditions to subsonic values. In the example in Frames 19–23 the pressure at the throat was 209.41 kPa, so we can imagine a pressure drop to atmospheric pressure at the plane of the nozzle exit.

We can achieve a supersonic flow in a nozzle by adding a further section to the nozzle in which sonic flow occurs at the throat. Does this extra section converge or diverge?

26

Diverge

27

From Frame 12 we saw that the duct or nozzle must diverge from the throat to achieve supersonic flow. If we know the velocity we want to achieve, we can calculate the temperature and pressure at outlet and the area of cross-section for a given mass flow. Now we will continue with the example we worked through in Frames 19–23: we had total head inlet conditions of 569.9 K temperature and 396.39 kPa pressure, and a throat temperature and pressure of 474.92 K and 209.41 kPa, giving a throat sonic velocity of 436.91 m/s. We will suppose we want a supersonic velocity of 600 m/s at the nozzle exit. From the total head inlet temperature of 569.9 K we can find the outlet temperature, T_3, using the steady flow energy equation:

$$0 = c_P T_3 + \frac{V_3^2}{2} - c_P T_{T1}$$

$$0 = 1.005 \times 10^3 \times T_3 + 600^2/2 - 1.005 \times 10^3 \times 569.9$$

$$\therefore \quad T_3 = 390.8 \text{ K}$$

At this temperature the sonic velocity is given by:

$$\begin{aligned} a &= (\gamma R T)^{1/2} \\ &= (1.4 \times 0.2871 \times 10^3 \times 390.8)^{1/2} \\ &= 396.33 \text{ [m/s]} \end{aligned}$$

The ratio of actual velocity to the sonic velocity is called the **Mach number**, so for the velocity of 600 m/s it is 1.514. Using the relationship between pressure ratio and temperature ratio for an isentropic expansion (see Frame 16 if you need to), what is the pressure at the nozzle exit?

28

105.83 kPa

Using:

$$\frac{p_3}{p_1} = \left(\frac{T_3}{T_1}\right)^{\gamma/(\gamma-1)}$$

we get:

$$\frac{p_3}{396.39} = \left(\frac{390.8}{569.9}\right)^{3.5}$$

hence $$p_3 = 105.83 \text{ [kPa]}$$

Up to now we have assumed that the flow in nozzles is isentropic. This is reasonably true for the subsonic converging portion of a nozzle, but is less accurate in the diverging portion of a supersonic nozzle. Frictional losses occur, which means the kinetic energy of the flow is partially degraded into enthalpy. Because of these irreversibilities the entropy will increase from plane to plane (as depicted in Frame 2), in the direction of flow. Nevertheless, the steady flow energy equation must still hold, and for the velocity of 600 m/s the outlet temperature is still 390.8 K. We would find, however, that to achieve this velocity with friction in the flow, the expansion would have to be to an outlet pressure lower than the previous value of 105.83 kPa. By measuring the actual outlet pressure we can, after some calculation, determine the **isentropic efficiency** for the nozzle expansion, which is defined:

$$\text{Isentropic efficiency} = \frac{h_{T_1} - h_3}{h_{T_1} - h_3'}$$

where $(h_{T_1} - h_3)$ is the actual enthalpy drop, and $(h_{T_1} - h_3')$ is the isentropic enthalpy drop. This efficiency is related to the expansion over the whole nozzle, even though the losses occur in the diverging portion only.

29

Suppose we are told that the isentropic efficiency of this nozzle is 90%. This means we can calculate the isentropic outlet temperature since we can assume a constant c_P throughout which cancels:

$$0.9 = \frac{569.9 - 390.8}{569.9 - T_3'}$$

Hence: $$T_3' = 370.9 \text{ K}$$

Thus, in this instance, if the flow had been isentropic, the oulet temperature would be lower than it actually was and the velocity would have been higher than 600 m/s. Because of frictional losses, the velocity was reduced to 600 m/s and the outlet temperature was correspondingly higher at 390.8 K. To achieve these results with friction present, the pressure at outlet will be lower than before, and is calculated in the next frame.

30

With γ for air at 1.4, the nozzle outlet pressure is given by:

$$\frac{p_3}{396.39} = \left(\frac{370.9}{569.9}\right)^{\gamma/(\gamma-1)}$$

$$\therefore \qquad p_3 = 88.07 \text{ [kPa]}$$

We see then that because of the losses, expansion to a lower pressure has to take place to achieve the required velocity. By measuring the outlet velocity and pressure, the isentropic efficiency, which was stated in this exercise, can be deduced by calculation. The expansion, between a total head pressure of 396.39 kPa, to a pressure of 88.07 kPa, is shown on the *T–s* plane in the figure below. In Frame 94 of Programme 6 we saw how to construct lines of constant pressure on a *T–s* diagram for air.

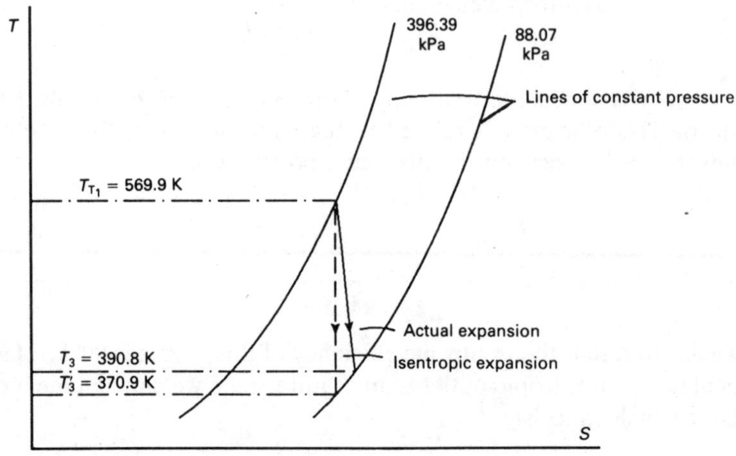

Calculate the specific volume of the flow at 600 m/s, and the area of nozzle at its exit for a mass flow of 10 kg/s. The pressure and temperature are 88.07 kPa and 390.8 K, and *R* is 0.2871 kJ/kg K.

31

$$\boxed{v_3 = 1.274 \text{ m}^3\text{/kg}, \ A_3 = 0.0212 \text{ m}^2}$$

At the supersonic outlet, $v_3 = \dfrac{0.2871 \times 390.82}{88.07} = 1.274 \text{ [m}^3\text{/kg]}$

From continuity: $$10 = \frac{A_3 \times V_3}{v_3}$$

$$\therefore \quad A_3 = 10 \times 1.274/600.0$$
$$= 0.0212\ [\text{m}^3]$$

Hence the diameter at the nozzle exit is 0.1643 m.

Now, using the example in Frames 19–23, with the methods of Frame 26 onwards as your guide, here is another problem on a convergent–divergent nozzle. From inlet total head conditions of 1000 kPa and 800 K, it is required to produce a supersonic flow of 5 kg/s at 800 m/s. The isentropic efficiency of the nozzle over its whole length is 89%. Calculate the throat and exit areas, and the Mach number at exit. As before, for air take $\gamma = 1.4$, $c_p = 1.005$ kJ/kg K and $R = 0.2871$ kJ/kg K.

32

Throat: p = 528.28 kPa, T = 666.67 K, v = 0.3623 m^3/kg, V = 517.65 m/s, A = 0.0035 m^2, Exit: p = 125.6 kPa, T = 481.59 K, v = 1.101 m^3/kg, a = 439.97 m/s, M = 1.8183, A = 0.0069 m^2

The solution is not given, but you should be able to obtain these results. The Mach number at the exit is found by calculating the sonic velocity at the temperature of the exit condition.

It is possible that a nozzle designed to produce supersonic flow is not operating at its intended conditions. The back pressure may be higher or lower than the design value. The pressure distribution in a convergent–divergent nozzle is shown below for a range of nozzle back pressures.

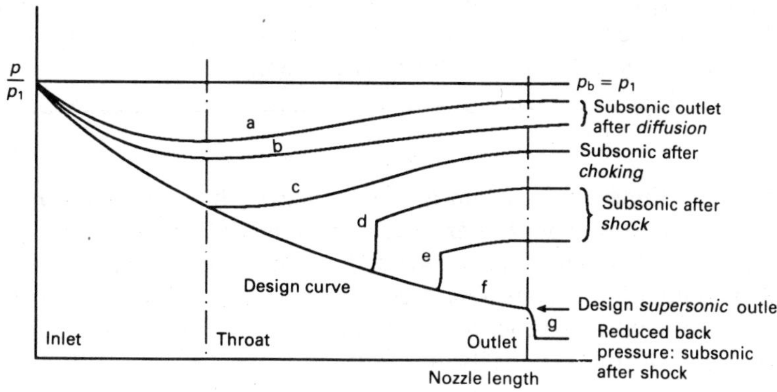

33

Initially the back pressure equals the inlet pressure, so there is no flow. The back pressure is reduced, in curves a and b, and the pressure at the throat is seen to be lower than it is at outlet. The velocity is highest at the throat, but is subsonic, and the diverging portion of the nozzle is acting as a subsonic diffuser. In curve c, sonic conditions occur at the throat, and the nozzle becomes choked. The diverging portion again acts as a subsonic diffuser. The next two curves, d and e, show a shock wave in the diverging portion. The back pressure is low enough for a supersonic expansion to start in the diverging portion, but the flow cannot now diffuse back to subsonic conditions to meet the prevailing back pressure, and so a shock occurs. Curve f represents the design condition in which a supersonic expansion is complete to the design outlet velocity. The final curve g represents an expansion after the supersonic outlet, so again shocks occur to restore the velocity to a subsonic value.

For the same set of curves we can also plot the velocity against the length of the nozzle, as in the figure below.

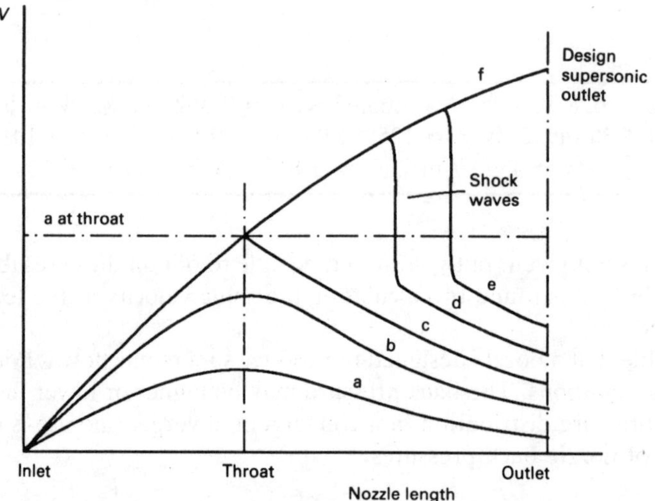

Only the design curve will give the required supersonic velocity at outlet. For all the other curves, either as a result of diffusion, or as the result of a shock wave followed by diffusion, the final velocity is subsonic.

34

Next we shall think about steam flow in a nozzle, but first here are some more problems on gas flow in nozzles and subsonic diffusers. In the diffuser problem you only need to think about using the steady flow energy equation, or the fact that the total head enthalpy is constant, together with the pressure–temperature relationship for an isentropic process. All the problems involve air, for which $R = 0.287$ kJ/kg K, $\gamma = 1.4$ and $c_p = 1.0045$ kJ/kg K.

1. Flow in a subsonic diffuser is isentropic. The inlet velocity is 200 m/s, the pressure is 110 kPa and the temperature is 300 K. Calculate the temperature and pressure when the velocity has dropped to 100 m/s. [314.93 K, 130.39 kPa]
2. In a convergent nozzle, total head inlet conditions are a pressure of 500 kPa and a temperature of 700 K. The outlet, or throat, area is 0.0015 m². Calculate the mass flow for choking conditions. [1.146 kg/s]
3. For the nozzle in problem **2**, calculate the mass flow if the outlet pressure is 400 kPa, with isentropic flow. [0.957 kg/s]
4. An air nozzle has total head inlet conditions of 1000 kPa pressure and a temperature of 800 K, and a mass flow of 2 kg/s. Flow is isentropic only to the throat, overall the isentropic efficiency is 0.9. Calculate (1) the throat temperature, pressure and diameter and (2) the exit pressure, temperature and diameter for a velocity of 900 m/s.
[(1) 666.67 K, 528.3 kPa, 42.2 mm; (2) 56.51 kPa, 396.81 K, 75.5 mm]
5. An air nozzle has an inlet area of 0.003 m², with a velocity of 100 m/s at a pressure of 600 kPa and a temperature of 850 K. Calculate the inlet total head conditions and the mass flow. Flow is isentropic; calculate the pressure, temperature and area at the throat. [612.39 kPa, 854.98 K, 0.738 kg/s; 323.51 kPa, 712.48 K, 0.0009 m²]
6. A nozzle has an outlet velocity of 800 m/s at a pressure of 100 kPa and a temperature of 300 K. The isentropic efficiency is 0.9. Calculate the inlet total head conditions and the pressure and temperature at the throat.
[1953.3 kPa, 618.57 K; 1031.9 kPa, 515.47 K]

35

In Frame 4, equation (1), the velocity V_2 at the outlet of a duct in which the inlet velocity was zero was derived with reference to a perfect gas.
The equation is:

$$V_2 = \left[\frac{2\gamma}{\gamma - 1} p_1 v_1 \left(1 - \left(\frac{p_2}{p_1}\right)^{(\gamma-1)/\gamma}\right)\right]^{1/2}$$

which may be written as:

$$V_2 = \left(\frac{2\gamma}{1 - \gamma} p_1 v_1 \left(\left(\frac{p_2}{p_1}\right)^{(\gamma-1)/\gamma} - 1\right)\right)^{1/2}$$

The same equation can be derived with an index n for the isentropic expansion of any fluid such as steam replacing γ for a perfect gas, so the result is completely general. We shall do this in the next frame.

36

Equations (1) and (2) in Frame 8 give us the steady flow energy equation in differential form states:

$$dh + d(V^2/2) = 0 \tag{1}$$

and for an isentropic process:

$$dh = v dp \tag{2}$$

If we combine these equations and integrate between inlet at state 1 and outlet at state 2, we get:

$$\frac{V_2^2 - V_1^2}{2} = -\int_1^2 v dp$$

Then we take the equation $pv^n = k$ for the isentropic expansion of steam, where as we have seen earlier n is determined for the p–v relationship which expresses the actual isentropic process. We again consider a duct in which the inlet velocity V_1 at pressure p_1 is zero, and so:

$$\frac{V_2^2}{2} = -k^{1/n}\int_1^2 \frac{dp}{p^{1/n}}$$

$$= -\frac{n}{n-1}k^{1/n}\left(p^{(n-1)/n}\right)_1^2$$

$$= \frac{n}{1-n}\left(p_2 v_2 - p_1 v_1\right)$$

Then using $pv^n = k$, we can see that:

$$V_2 = \left(\frac{2n}{1-n}p_1 v_1\left(\left(\frac{p_2}{p_1}\right)^{(n-1)/n} - 1\right)\right)^{1/2}$$

This is the same equation as that in Frame 35, except that n replaces γ. In the isentropic expansion of steam, the appropriate value of n is equivalent to γ for a perfect gas. As far as the mathematics are concerned, there is no difference, so we can use the same derivation as previously for p_2/p_1. With the same steps as in Frames 15 and 16 we shall get:

$$\frac{p_2}{p_1} = \left(\frac{2}{n+1}\right)^{n/(n-1)}$$

This gives the pressure at the throat of a steam nozzle in terms of the inlet pressure and the index of isentropic expansion for steam. Other properties at the throat, e.g. temperature, dryness fraction, or superheated state, specific volume and velocity, must be found from the use of the steam tables, or where possible from the *h–s* chart, and from the steady flow energy equation.

We cannot use:
$$\frac{T_2}{T_1} = \frac{2}{n+1}$$

to find the temperature at the throat of a steam nozzle. Why not?

37

This equation depends on the use of the ideal gas rule

If we refer back to Frame 69 in Programme 5 we can see that the relationships between pressure and temperature ratios in reversible adiabatic or isentropic processes with ideal gases involve the use of the ideal gas rule. Consequently there is no corresponding throat to inlet temperature ratio term for a steam nozzle.

We will now work through a problem on a steam nozzle to see what is involved. Total head conditions at inlet are a pressure of 50 bar, or 5 MN/m^2, and a temperature of 400°C. The steam expands to a pressure of 20 bar, or 2 MN/m^2, with an isentropic efficiency of 84%. Find (a) the state of the steam at the throat and (b) the velocity of the steam and the diameter of the nozzle at both the throat and outlet for unit mass flow.

The steam is superheated at inlet, and you will recall from Frame 29 in Programme 5 that if a process is wholly superheated then n for an isentropic expansion is 1.3. We will assume this to be so. Using $n = 1.3$, what is the pressure at the throat?

38

The answer is 27.285 bar, or 2.729 MN/m^2.

Using: $\dfrac{p_2}{p_1} = \left(\dfrac{2}{n+1}\right)^{n/(n-1)}$ with $n = 1.3$, we get:

$$p_2 = 50 \times (2/2.3)^{1.3/0.3} = 50 \times 0.5457 = 27.285 \text{ bar} = 2.729 \text{ [MN/m}^2\text{]}$$

39

At the inlet conditions of 50 bar and 400°C from page 7 of the Rogers and Mayhew steam tables we see that h_1 is 3196 kJ/kg and s_1 is 6.646 kJ/kg K. The entropy at the throat has the same value since the expansion to the throat is isentropic. Knowing this we can find the superheat at the throat, by using the same methods that we used in an isentropic steam turbine expansion in the previous programme. By interpolating between 20 bar and 30 bar on page 7, we find s at 27.285 bar and 300°C is 6.603 kJ/kg K and at 27.285 bar and 350°C it is 6.802 kJ/kg K. You can check these figures and then find the degree of superheat with s = 6.646 kJ/kg K.

40

310.8°C

Since we have s = 6.603 kJ/kg K at 27.285 bar and 300°C and 6.802 kJ/kg K at 350°C, with the actual s = 6.646 kJ/kg K, the state is given by:

$$\text{temperature} = 300 + 50 \times \frac{6.646 - 6.603}{6.802 - 6.603}$$

$$= 310.8°\text{C}$$

You should be fully conversant with interpolation by this time, so we will not give all the workings that have to be done. We now have to find h and v at the throat conditions of 27.285 bar and 310.8°C. To do this we interpolate between 20 bar and 30 bar at both 300°C and 350°C to find the values at 27.285 bar, and then at this pressure we interpolate between 300°C and 350°C to find the values at 310.8°C. The results are given below.

41

At 27.285 bar	300°C	350°C	310.8°C	
h =	3003.14	3122.70	3028.97	[kJ/kg]
v =	0.0932	0.1036	0.0954	[m³/kg]

42

Then between the inlet and the throat the enthalpy drop is:

$$h_1 - h_2 = 3196 - 3028.97$$
$$= 167.03 \text{ [kJ/kg]}$$

The velocity at the throat is then given by:

$$V_2 = (2 \times 10^3 \times 167.03)^{1/2}$$
$$= 577.98 \text{ [m/s]}$$

For continuity at the throat:

$$\dot{m} = \frac{A \times V}{v}$$

With $\dot{m} = 1$ kg/s, what is the diameter of the nozzle at the throat?

43

14.5 mm

For unit mass flow, $A = v/V$ and $v = 0.0954$ m^3/kg from frame 41, so:

$$A = 0.0954/577.98 = 0.000165 \text{ [m}^2\text{]}$$

The diameter to give this area is 14.5 mm.

The expansion continues to 20 bar, or 2 MN/m^2, with an isentropic efficiency over the whole nozzle of 84%. In dealing with the diverging portion, we start by taking an isentropic expansion, so we know the entropy is 6.646 kJ/kg K at 20 bar. From the tables we can see that the steam is still superheated at 20 bar, with a temperature between 250°C and 300°C. By interpolation we find the temperature is 272.4°C. At this temperature we find the enthalpy, h_3', is 2958.21 kJ/kg K, so over the whole nozzle the isentropic enthalpy drop is:

$$h_1 - h_3' = 3196 - 2958.21$$
$$= 237.79 \text{ [kJ/kg]}$$

44

The actual enthalpy drop is 84% of this, which is 199.74 kJ/kg. This means the actual h_3 is given by:

$$h_3 = 3196 - 199.74$$
$$= 2996.26 \text{ [kJ/kg]}$$

Using this value of h, recalculate the actual superheat at the nozzle outlet, and then find v_3 at the nozzle outlet.

45

288.12°C, 0.1222 m³/kg

The effect of the non-isentropic expansion has been to raise the superheat temperature at outlet from 272.4°C to 288.12°C. Next we can find the outlet velocity:

$$V_3 = (2 \times 10^3 \times 199.74)^{1/2}$$
$$= 632.04 \text{ [m/s]}$$

Then using $v_3 = 0.1222$ m³/kg, we can find the area of cross-section A_3:

For unit mass flow: $A_3 = 0.1222/632.04$

$$= 0.0001933 \text{ [m}^2\text{]}$$

Finally, the diameter is 15.69 mm. This example has shown how the solution of a steam nozzle problem can be fairly lengthy. We could have found the specific enthalpy figures needed rather quicker using the *h–s* chart, but we would still have had to use the steam tables to find values of specific volume.

You should now do the steam nozzle problems in the next frame, and then we shall consider the work interaction between a fluid and a moving set of blades.

46

1. A steam nozzle with total head inlet conditions of 60 bar and 350° expands isentropically to a pressure of 30 bar. Is this nozzle choked? Calculate the velocity of steam at the exit plane and the exit diameter for a mass flow of 1.5 kg/s. [Yes, critical pressure is 32.74 bar, 568.24 m/s, 15.62 mm]

2. Reconsider problem **1** for flow with an isentropic efficiency of 0.88. Calculate the state of the steam at the exit plane, and the new values of exit velocity and exit diameter. [30 bar 266.4°C, 533.07 m/s, 16.29 mm]
3. Steam enters a nozzle with total head conditions of 20 bar and 300°C, and expands with an isentropic efficiency of 0.9 to a pressure of 5 bar. Find the conditions at the throat and the throat velocity and diameter for a mass flow of 2 kg/s; and find the state of the steam at the exit and the exit velocity and diameter.
[10.915 bar 224.79°C, 536.19 m/s, 31.2 mm; 5 bar 154.9°C, 733.57 m/s, 36.2 mm]

47

We will now think about the work interaction between a flowing fluid and a rotating turbine or compressor wheel. Turbine and compressor wheels are of two types called *axial* flow and *radial* flow, which indicates the predominant flow direction of fluid as it passes through the wheel. To consider the principles we may think about an axial flow wheel. This has blades, pointing radially outwards, at the outer diameter of the wheel, shown in the diagram in the next frame, and *axial* flow occurs in the spaces between the blades. The blades have a *leading edge*, in the plane of the wheel where the fluid enters the space between the blades, and a *trailing edge*, in the plane of the wheel where the fluid leaves the space between the blades. We have to see what happens to the fluid between these two planes.

The fluid has inlet and outlet velocities at the two planes of the wheel which have absolute values of V_1 and V_2 respectively. These absolute velocities have three components of velocity mutually at right angles, which are the axial, radial and whirl, or tangential, components, V_a, V_r and V_w, respectively. We have to consider *changes* in these components between the inlet and outlet planes of the wheel. Thus, changes in the axial and radial components cause forces in the wheel and its rotor in these directions. A change in the axial component will cause an end thrust on the shaft of the wheel, and a change in the radial component can cause a bending force on the rotor shaft. Both of these factors must be considered in any design study, but it is only changes in the whirl component which will affect the rotation of the wheel and which leads to a work interaction.

48

The diagram below is of part of an axial flow wheel, and it shows the leading and trailing edges of the blades, together with the inlet absolute velocity and its axial, radial and whirl components.

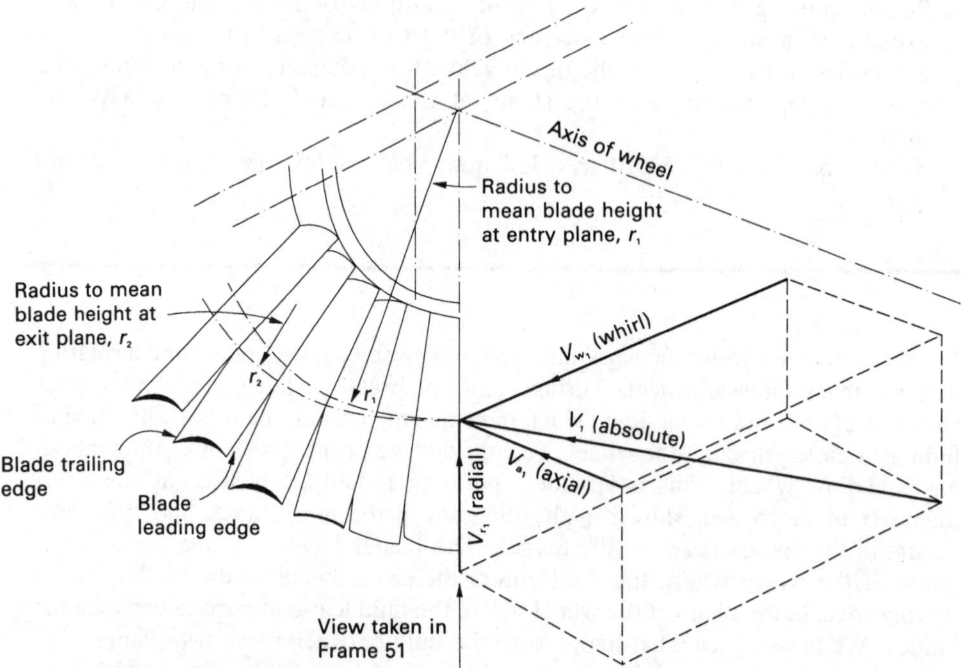

We may assume that we have only axial and whirl components to consider and the figure below shows the fluid with absolute velocity V_1, and with the components V_{a1} and V_{w1} about to enter the plane of the wheel.

49

We can assume the fluid enters and leaves the space between the blades at the mean height of the blade, which is at a radius of r_1 at entry and r_2 at exit. We also assume conditions are steady so there is a constant mass flow rate of $\dot{m}$ through the blades. A mass of fluid Δm_1 in the entry stream at radius r_1 would have angular momentum of:

$$\Delta m_1 \, r_1^2 \, \omega_{F1}$$

where ω_{F1} is the angular rate of rotation, which is V_{w1}/r_1. Similarly a mass of fluid Δm_2 in the exit stream would have an angular momentum of:

$$\Delta m_2 \, r_2^2 \, V_{w2}/r_2$$

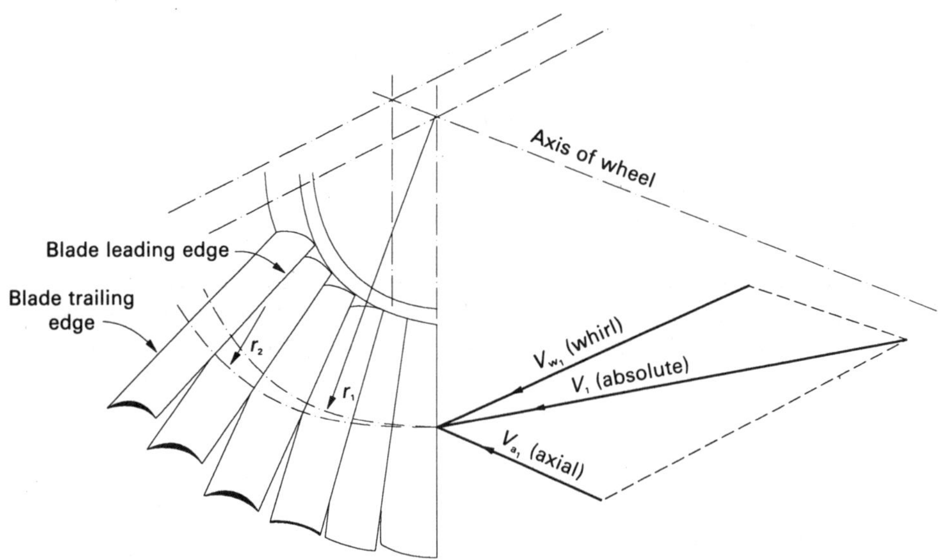

As the mass flow is steady at the rate of $\dot{m}$, the *rate* of change of angular momentum is:

$$\dot{m}\ (r_2\ V_{w2} - r_1\ V_{w1})$$

By Newton's Second Law of Motion, the rate of work done by the rotor on the fluid is the product of the angular velocity and the rate of change of angular momentum of the fluid. The angular velocity is U/r, where U is the peripheral speed at the particular radius, so if we now assume the entry and exit radii are the same, the rate of work done *by the rotor on the fluid* is:

$$\dot{W} = \dot{m}\ U\ (V_{w2} - V_{w1})$$

The rate of work done *by the fluid on the rotor* is:

$$\dot{W} = \dot{m}\ U\ (V_{w1} - V_{w2})$$

By Newton's third law of motion these expressions result from equal and opposite forces acting between the fluid and the rotor.

Examine the units of these results to satisfy yourself that they are correct.

50

The units are: [kg/s][m/s][m/s]
= [kg m/s^2][m/s] = [N][m/s] = [J/s] = rate of work

51

In Frame 81 of Programme 2 we saw that rate of shaft work, or rotor work, is torque × rate of angular rotation of the shaft, so in the expression in Frame 49 the change in $\dot{m}\ V_w r$ represents the torque, and U/r is the rate of angular rotation.

We can now look at velocity diagrams of steam turbine blading, and from them calculate the work interaction. There are two types of steam turbine blading, called **impulse** and **reaction** blading.

The impulse turbine is historically older, and it consists of a set of steam nozzles and a turbine wheel designed to receive high velocity steam. The set of nozzles plus the turbine wheel is called a **stage**. In an impulse stage, all the expansion of steam takes place in the nozzles, and the blades are relatively short and thick to withstand the forces that arise. This means that impulse stages are used in the initial high pressure stages in a turbine where the specific volume of the steam and the flow area between the blades are low. In the diagram below we are looking radially inwards at the end of the blading (see the diagrams in Frame 48), and a single nozzle is shown in section adjacent to the blading. The nozzle will direct steam to the mid-height of the blade, and there will be perhaps three or four nozzles spaced equally circumferentially.

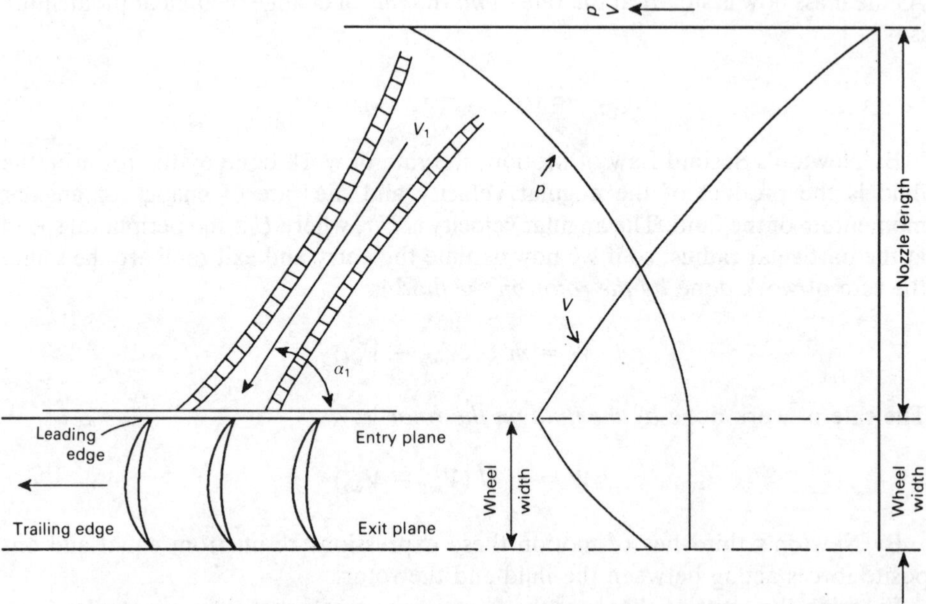

The diagram also shows how the pressure will fall along the length of the nozzle and how correspondingly the velocity will increase. This absolute velocity of the steam will then fall as the steam passes between the blades.

We can now consider the velocity triangles for the steam entering and leaving the blades. The velocity triangle at entry is shown below.

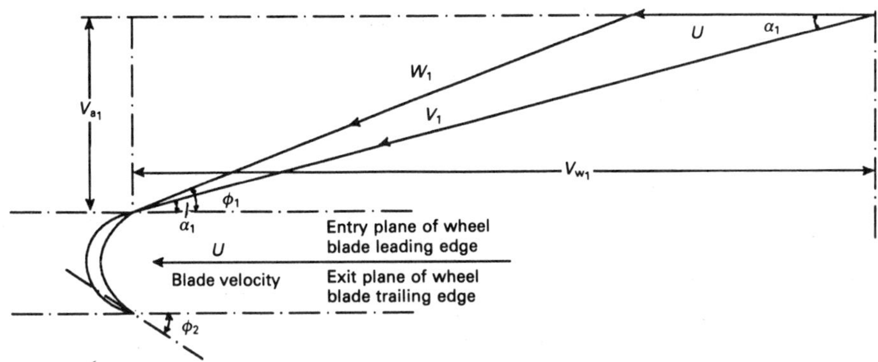

In the diagram, V is used for steam velocity, either as absolute velocity or as components, W is used for relative velocity and U is used for the peripheral velocity of the blading at its mid-height. Relative to the plane of the wheel the angle of the nozzle is α_1, and ϕ_1 is the angle between a tangent to the curve of the blade at the leading edge of the blade and the plane of the wheel. It is called the blade inlet angle. The steam is therefore directed onto the wheel at the angle α_1, and as this angle is fixed there is only one blade velocity U that will produce a velocity triangle in which W_1 is at the correct angle of ϕ_1. At this design speed the turbine will operate at its greatest efficiency. Starting from rest, the angle of W will be much in error and losses due to turbulence and friction will occur; these losses will diminish as the blade velocity increases and the angle of W approaches its correct value of ϕ_1. Then the losses will increase again if U is allowed to exceed its design value.

Velocity, as you should know, is a vector quantity with both magnitude and direction, and you will have learned the rules for adding velocity vectors from G. E. Drabble's *Dynamics* in this series. The vector addition in the diagram may be written as:

Velocity of steam relative to blade (W) + Velocity of blade relative to wheel axis (U) = Velocity of steam relative to wheel axis (V)

The angle ϕ_2 is the blade outlet angle, between the tangent to the curve of the blade at its trailing edge and the plane of the wheel. See if you can sketch the triangle for the steam leaving the blades using the velocities W_2 and V_2 and the same blade velocity U. Begin by relating the three velocities by an equation in words.

Velocity of steam relative to blade (W) = Velocity of blade relative to wheel axis (U) + Velocity of steam relative to wheel axis (V)

The velocity diagram at the blade outlet is shown below.

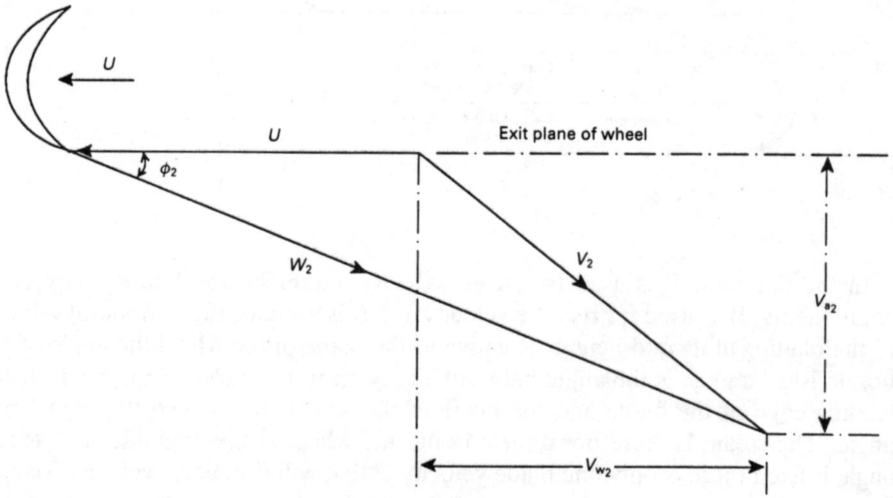

We can now combine both velocity diagrams into a single diagram based on a common value of U.

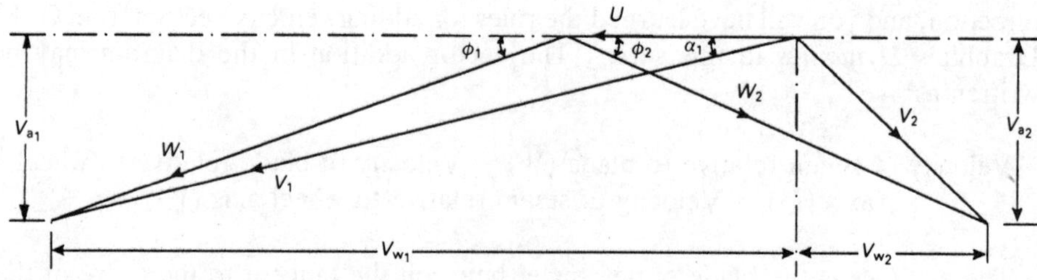

In this diagram we can show the axial and whirl components at both inlet and outlet. We are using subscripts 1 and 2 for inlet and outlet respectively.

From the expression for the work term in Frame 49, which velocities do we need to calculate the rate of work?

54

U, V_{w1} and V_{w2}

To calculate U we need to know the rate of rotation of the turbine wheel. The peripheral velocity of the blade will increase from the blade root, (where it joins the wheel), to the blade tip, so we take U to be the blade velocity at the mean blade height. Thus:

$$U = \frac{\pi D N}{60} \text{ m/s}$$

where D is the diameter of the turbine wheel at the mean blade height and N is the rate of rotation in rev/min. We obtain V_{w1} and V_{w2} from the combined velocity diagram, so we need to know V_1 and the nozzle and blade angles. The steam velocity V_1 is found using the methods developed earlier in this programme.

If we assume there is no friction between the steam and the blading then the relative velocity W_2 will equal W_1. In practice, W_2 is less than W_1 owing to friction, which is accounted for in a velocity coefficient, K, where:

$$K = W_2/W_1$$

A typical value for K is 0.9.

The whirl components of the steam velocity V_{w1} and V_{w2} act in opposite directions, so V_{w2} is negative with respect to V_{w1}. The difference between these two velocities in the equation:

$$\dot{W} = \dot{m}\, U\, (V_{w1} - V_{w2})$$

becomes a sum of their actual values from the velocity diagram, so the rate of work becomes:

$$\dot{W} = \dot{m}\, U\, (V_{w1} + V_{w2}) \text{ [J/s]} = \text{[W]}$$

The units of this equation are [W] with velocities in [m/s] and $\dot{m}$ in [kg/s].

55

If in an impulse turbine stage the blade velocity, U, is 65 m/s, V_{w1} is 464 m/s, V_{w2} is 200 m/s and $\dot{m}$ is 2 kg/s, what is the power output in kW?

56

86.32 kW

From the previous frame:

$$\begin{aligned}\dot{W} &= \dot{m}\, U\,(V_{w1} + V_{w2}) \\ &= 2.0 \times 65 \times (464 + 200) \times 10^{-3} \\ &= 86.32\ [\text{kW}]\end{aligned}$$

The **blade efficiency** is a measure of the efficiency of the conversion of the kinetic energy of the steam leaving the nozzles into useful work. Thus in an isentropic expansion of steam in the nozzles, the enthalpy drop in the steam condition is converted entirely into kinetic energy:

$$\Delta h = V_1^2/2$$

and the work done on the blading is $U(V_{w1} + V_{w2})$ per unit mass flow, so the blade efficiency is given by:

$$\eta_B = \frac{U\,(V_{w1} + V_{w2})}{V_1^2/2}$$

$$= \frac{2\,U\,(V_{w1} + V_{w2})}{V_1^2}$$

If, in the example in the previous frame, V_1 is 500 m/s, what is the blade efficiency?

57

34.53%

Using the formula for blade efficiency developed in the previous frame:

$$\eta_B = [2 \times 65 \times (464 + 200)/500^2] \times 100\%$$
$$= 34.53\%$$

This is a relatively low value of blade efficiency which is due to a high value of V_1, and the steam leaving the stage would have a relatively high velocity V_2, which represents wasted energy.

With an isentropic nozzle expansion, the blade efficiency would also be equal to the *stage efficiency*, η_S; if the nozzles have an isentropic efficiency of η_{is}, then:

$$\eta_S = \eta_B \, \eta_{is}$$

We will now work through a complete example of an impulse turbine stage. Steam enters the nozzles at 10 bar and at 200°C and expands isentropically to 8 bar. (Remember this is only a single stage in a turbine so the pressure drop is relatively small.) The diameter of the turbine wheel to the mean blade height is 400 mm, and the wheel rotates at 3000 rev/m. The nozzle angle is 20° and the blade entry and exit angles are equal at 25.08°. Assume flow is frictionless through the turbine so that there is no change in relative velocity and $K = 1$ (Frame 54). Calculate the absolute velocities of the steam entering and leaving the turbine wheel, the power of the stage per unit mass flow, and the blade efficiency. Draw the blade velocity diagram.

First you can calculate the isentropic enthalpy drop in the nozzle, and the inlet velocity, V_1.

58

45.25 kJ/kg, 300.83 m/s

59

At 10 bar and 200°C, h = 2829 kJ/kg and s = 6.695 kJ/kg K, from page 7 of the Rogers and Mayhew tables. At 8 bar, s is also 6.695 kJ/kg K, so we see that the steam is still superheated. By interpolation we find there is 6.15 K of superheat, so we find h is 2783.75 kJ/kg. Then:

$$\Delta h = 2829 - 2783.75$$
$$= 45.25 \text{ [kJ/kg]}$$

The velocity V_1 is then given by:

$$V_1^2/2 = 45.25 \times 10^3$$
$$\therefore \quad V_1 = 300.83 \text{ [m/s]}$$

Next you can find U. The diameter at the mean blade height was 400 mm, and the rate of rotation was 3000 rev/min.

60

62.83 m/s

$$U = \pi \times 0.4 \times 3000/60$$
$$= 62.83 \text{ [m/s]}$$

We can now draw the velocity diagram, which is shown below.

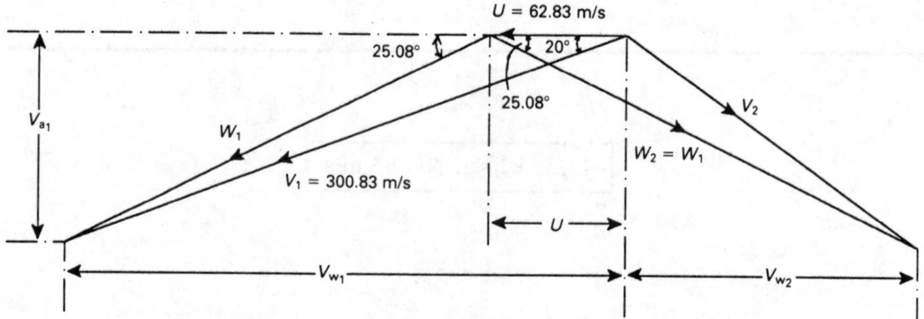

Measurements may be taken from a drawing made to scale, but it is more accurate to calculate the required lengths. From the diagram we see that:

$$\begin{aligned} V_{w1} &= 300.83 \cos 20° \\ &= 282.69 \text{ [m/s]} \end{aligned}$$

To find V_{w2} we first need to find W_1 from V_{a1}:

$$\begin{aligned} V_{a1} &= V_1 \sin 20° \\ &= 300.83 \times 0.3420 \\ &= 102.89 \text{ [m/s]} \end{aligned}$$

Then: $$V_{a1}/W_1 = \sin \phi_1$$

$\therefore$ $$\begin{aligned} W_1 &= 102.89/\sin 25.08° \\ &= 242.73 \text{ [m/s]} \end{aligned}$$

W_2 will be the same, as $K = 1$. You should now be able to find V_{w2}.

61

157.02 m/s

To find V_{w2}:

$$(U + V_{w2})/W_2 = \cos \phi_2$$

$\therefore$ $$\begin{aligned} V_{w2} &= W_2 \cos \phi_2 - U \\ &= 242.73 \cos 25.08° - 62.83 \\ &= 157.02 \text{ [m/s]} \end{aligned}$$

The power output for unit rate of mass flow is then:

$$\begin{aligned} \dot{W} &= 62.83 \times (282.69 + 157.02) \times 10^{-3} \\ &= 27.63 \text{ [kW/kg]} \end{aligned}$$

Now calculate the blade efficiency.

62

61.06%

$$\eta_B = 2 \times 27.63 \times 10^3/300.83^2 = 0.6106, \text{ or } 61.06\%$$

What has happened to the remaining energy that was available in the enthalpy drop in the nozzle? This would amount to (100 – 61.06)% of 45.25 kW which is 17.62 kW. We need to find the kinetic energy of the steam leaving the turbine which is $V_2^2/2$. To find V_2 we need V_{a2} which is the same as V_{a1}, which is 102.89 m/s. Then:

$$V_2 = (V_{w2}^2 + V_{a2}^2)^{1/2}$$

$$= (157.02^2 + 102.89^2)^{1/2}$$

$$= 187.73 \text{ [m/s]}$$

The kinetic energy is then:

$$V_2^2/2 = 187.73^2(2 \times 10^3)$$

$$= 17.62 \text{ [kW]}$$

which is the missing amount. We can thus apply the steady flow energy equation to the turbine to give:

$$-\dot{W} = \Delta h + V_2^2/2$$

The turbine is adiabatic with no losses and no entry velocity to the stage, but the exit velocity is significant and has to be accounted for. In the previous programme we assumed in our calculations that the enthalpy drop across a turbine could be converted entirely into work. In Frame 57 we came across a low value of blade efficiency owing to a high velocity of steam leaving the turbine. Shortly we shall see how some of this available energy can be used in a second turbine wheel.

63

In this example we assumed there was no friction as the steam passed over the moving blade. As in normal practice, we also had equal blade inlet and outlet angles. From the velocity diagram in Frame 60 you can see that this means that V_{a1} and V_{a2} are equal. With no change in axial velocity through the turbine there will be no end thrust on the shaft of the rotor. With friction in the blades so that W_2 is less than W_1 and with equal blade angles, V_{a2} will then be less than V'_{a1} and so end thrust will then occur. We can see what this will be if $K = 0.9$.

We have that $W_1 = 242.73$ m/s and $V_{a1} = 102.89$ m/s, from Frame 60, so W_2 is 0.9×242.73 m/s, which is 218.46 m/s. Then:

$$V_{a2} = W_2 \times \sin 25.08°$$

$$= 218.46 \times 0.4239$$

$$= 92.60 \text{ [m/s]}$$

For mass flow $\dot{m}$, the end thrust is $\dot{m} \times$ change in velocity, in [N]. Hence for unit mass flow the thrust is $(102.89 - 92.60)$, or 10.29 N.

Now as an exercise go through the previous example again, and calculate the power output when there is friction with $K = 0.9$. The nozzle angle is 20° and the moving blade inlet and outlet angles are 25.08°. We found that U is 62.83 m/s, V_1 is 300.83 m/s and V_{w1} is 282.69 m/s.

64

$W_2 = 218.46$ m/s, $V_{w2} = 135.03$ m/s, $\dot{W} = 26.25$ kW/kg, $\eta_B = 58.01\%$

Following the example in Frames 60–62, you should be able to get these results. We can see that both the power output and the blade efficiency are reduced owing to blade friction, as we should expect.

65

We now return to the question of how to reduce the kinetic energy of the steam leaving the turbine. The turbine wheel casing is extended, and after the existing turbine wheel, a ring of fixed blades and a second turbine wheel are added. This is shown in the diagram in the next frame.

66

The figure below is from a similar viewpoint to the figure in Frame 51.

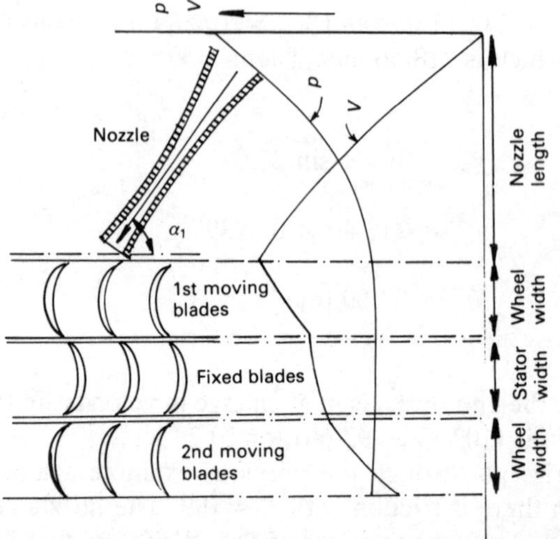

Steam leaves the first wheel with an absolute velocity V_2 and enters the passage between the fixed blades. In the absence of friction it leaves at the same velocity (at $V_3 = V_2$), and at angle α_3 which for the second wheel is equivalent to the nozzle angle α_1. The blade velocity at the mean blade height is the same in both wheels, and a second pair of velocity diagrams may be drawn for the second wheel. This is similar to the pair of diagrams for the first wheel and is shown below.

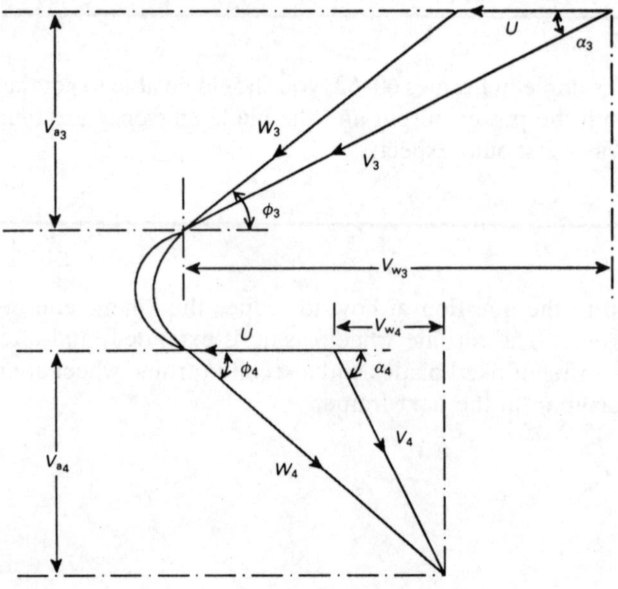

67

We can continue with the previous worked example (Frames 59–62), and see how much more useful work may be extracted. We know that U is 62.83 m/s and V_3 is 187.73 m/s. We may take the nozzle angle again to be 20° and work through to find the work output of the second wheel. First sketch out the combined velocity diagrams as we did in Frames 53 and 60, and calculate V_{a3} and V_{w3}.

68

$V_{a3} = 64.21$ m/s, $V_{w3} = 176.41$ m/s

The velocity diagram is given below:

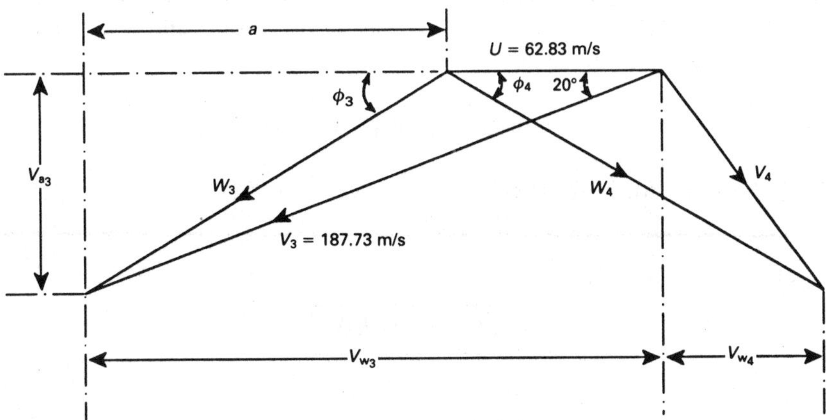

We can see that: $V_{a3}/V_3 = \sin 20°$

$$\therefore \quad V_{a3} = 187.73 \sin 20° = 64.21 \text{ [m/s]}$$

Also: $V_{w3}/V_3 = \cos 20°$

$$\therefore \quad V_{w3} = 187.73 \cos 20° = 176.41 \text{ [m/s]}$$

69

Next, from the geometry of the figure, we can find the length a, and then calculate ϕ_3 and W_3.

The length a is (176.41 − 62.83) which is 113.58 m/s.

Then: $$\tan \phi = 64.21/113.58$$

∴ $$\phi_3 = 29.48°$$

and then: $$W_3 = (64.21^2 + 113.58^2)^{1/2}$$

$$= 130.47 \text{ [m/s]}$$

In the absence of friction, W_4 is the same.

You should now be able to show that V_{w4} is 50.75 m/s, and then you can calculate the rate of work output for the wheel. The formula will be similar to that for the first wheel given in Frame 54, i.e.

$$\dot{W} = \dot{m}\, U\, (V_{w3} + V_{w4}) \qquad \text{[W]}$$

U will be the same for both wheels.

70

$$\boxed{\dot{W} = 14.27 \text{ kW}}$$

To find V_{w4} we have: $$(62.83 + V_{w4})/W_4 = \cos \phi_4$$

With $\phi_4 = \phi_3$, this gives: $$V_{w4} = 50.75 \text{ [m/s]}$$

The rate of work output for unit mass flow is $U\,(V_{w3} + V_{w4})$

$$= 62.83 \times (176.41 + 50.75) \times 10^{-3}$$

$$= 14.27 \text{ [kW]}$$

Finally you can calculate V_4. This will enable us to find the amount of kinetic energy leaving the second turbine wheel.

71

$$\boxed{V_4 = 81.84 \text{ m/s}}$$

We have: $V_4 = (50.75^2 + 64.21^2)^{1/2}$

$= 81.84$ [m/s]

We can then check the kinetic energy of the flow with outlet velocity V_4. This is $81.84^2/2 \times 10^{-3}$, which is 3.35 kW. Thus:

$$\Delta h = \dot{W} \text{ (first wheel)} + \dot{W} \text{ (second wheel)} + V_4^2/2$$

$$= 27.63 + 14.27 + 3.35$$

$$= 45.25 \text{ [kW]}$$

This represents a 92.6% conversion of the available enthalpy drop into work. We will leave impulse turbines at this point, but you must realise that in practice the effects of friction extend from a single wheel to a double wheel turbine. There will be a loss in the *absolute* velocity of the steam passing over the fixed blades between the turbine wheels, and there will be additional axial thrust arising from loss of relative velocity in the second wheel.

72

In Frame 51, the reaction turbine was mentioned. This differs from the impulse turbine in a number of ways. First, the nozzles are replaced by a complete ring of fixed or stator blades. Steam enters these blades with residual velocity from a previous reaction or impulse stage, and with a further expansion a velocity increase occurs. The steam now enters the moving blades at a much lower velocity than in the impulse stage. As the steam passes between the moving blades further expansion occurs, so the blades move partly as a result of the impulse of the steam received and partly as a result of the reaction to the expanding steam. Because of the work interaction between the steam and the moving blades, the velocity of the steam leaving the blades is less than at entry. To meet these requirements, the blade inlet and outlet angles cannot be equal, and the blade section becomes similar to that of an aerofoil, rather like an aircraft wing.

73

In a reaction turbine the velocities remain relatively low, so when a large enthalpy drop is available there is a considerable number of stages. A single stage is shown below.

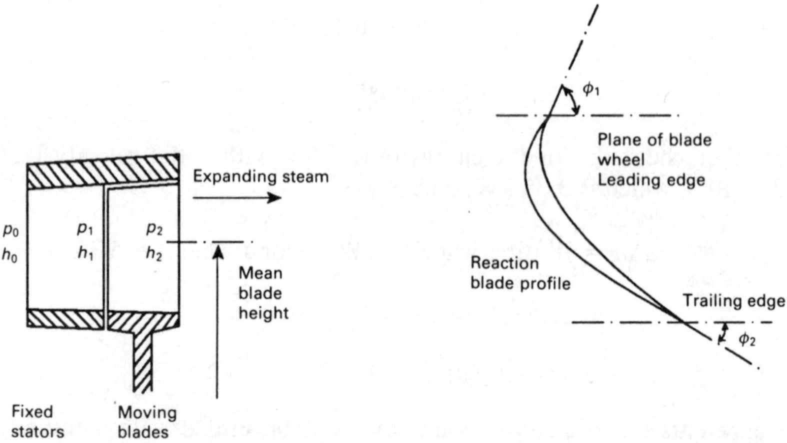

Although it is pressure drop which leads to reactive forces, the **degree of reaction** is expressed in terms of enthalpy differences. This is because the velocity changes which lead to the work interaction are directly related to the enthalpy changes, through the steady flow energy equation. The symbol Λ (Greek lambda) is used for the degree of reaction:

$$\Lambda = \frac{h_1 - h_2}{h_0 - h_2}$$

The enthalpy drop across the moving blades is $h_1 - h_2$, and across the whole stage is $h_0 - h_2$. The impulse turbine is a special case in which the degree of reaction is zero, because all the enthalpy drop occurs in the nozzles, equivalent to the fixed blades, and there is no enthalpy drop across the moving blades. Another special case is when there is an equal enthalpy drop across both fixed and moving blades. What is then the degree of reaction?

74

The answer is 50%. With an equal enthalpy drop in both fixed and moving blades, the enthalpy drop in the moving blades must be half that of the stage. This is widely used in practice.

Since in a reaction stage the relative velocity at the outlet of the moving blades is greater than it is at inlet, the velocity diagram is rather different from that of the impulse turbine. As a large number of stages are involved in a steam turbine it seems reasonable to use a common blade form in each stage, so that the axial component of the steam velocity remains constant throughout any one stage and is the same in all stages. As the steam expands through the stages, the blade size gradually increases from one stage to the next. With no change in the axial velocity component, there is no end thrust to consider. The combined velocity diagram is shown below. In this diagram $V_{a1} = V_{a2}$, and we shall assume there is 50% reaction.

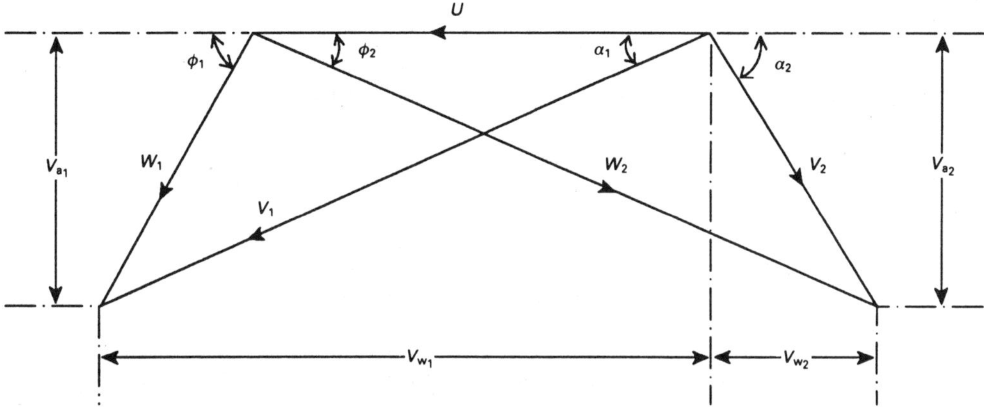

75

As in the impulse turbine, the work term is given by:

$$\dot{W} = \dot{m}\, U\, (V_{w1} + V_{w2})$$
$$= \dot{m}\, U\, V_{a1}\, (\cot \phi_1 + \cot \phi_2)$$

Consider the steady flow energy equation applied to a single stage. Since the axial velocity is constant in all stages and common blade forms are used, the exit velocity V_2 from any one stage is also the entry velocity to the fixed blades of any one stage, and so the kinetic energy terms at entry and exit for any one stage are equal. Thus:

$$\dot{W} = \dot{m}\, U\, V_{a1}\, (\cot \phi_1 + \cot \phi_2) = \dot{m}\, (h_o - h_2)$$

We now consider a control volume fixed to the rotor blades, and rotating with them. For this open system which rotates with the shaft there is no work transfer across the boundary, and entry and exit velocities relative to the control volume are W_1 and W_2. So what will be the steady flow energy equation for this open system?

76

$$h_1 - h_2 = \frac{W_2^2 - W_1^2}{2}$$

From the velocity diagram in Frame 74 this gives us:

$$h_1 - h_2 = V_{a1}^2\,(\text{cosec}^2\,\phi_2 - \text{cosec}^2\,\phi_1)/2$$
$$= V_{a1}^2\,(\cot^2\phi_2 - \cot^2\phi_1)/2$$

since $\text{cosec}^2 = 1 + \cot^2$.

The degree of reaction then gives us:

$$\Lambda = \frac{h_1 - h_2}{h_0 - h_2} = \frac{V_{a1}^2\,(\cot^2\phi_2 - \cot^2\phi_1)}{2\,U\,V_{a1}\,(\cot\phi_1 + \cot\phi_2)}$$

$$= \frac{V_{a1}}{2\,U}(\cot\phi_2 - \cot\phi_1)$$

$$= 0.5$$

for the assumed 50% degree of reaction.

$$\therefore \qquad U/V_{a1} = \cot\phi_2 - \cot\phi_1$$

Also from the velocity diagram in Frame 74, we have:

$$U/V_{a1} = \cot\alpha_1 - \cot\phi_1 = \cot\phi_2 - \cot\alpha_2$$

$$\therefore \qquad \alpha_1 = \phi_2 \text{ and } \alpha_2 = \phi_1$$

Therefore the velocity triangles are symmetrical, and $W_2 = V_1$ and $V_2 = W_1$, and all velocity diagrams are the same for all stages.

77

In the following example we shall calculate the rate of work in one stage in a reaction steam turbine where the diameter of the blade wheel at the mean blade height is 1050 mm. The wheel rotates at 3000 rev/m, and the blade outlet and inlet angles are 75° and 20° respectively. In the velocity diagram shown on the next page we need to calculate U and the total width of the diagram, which is $(V_{w1} + V_{w2})$.

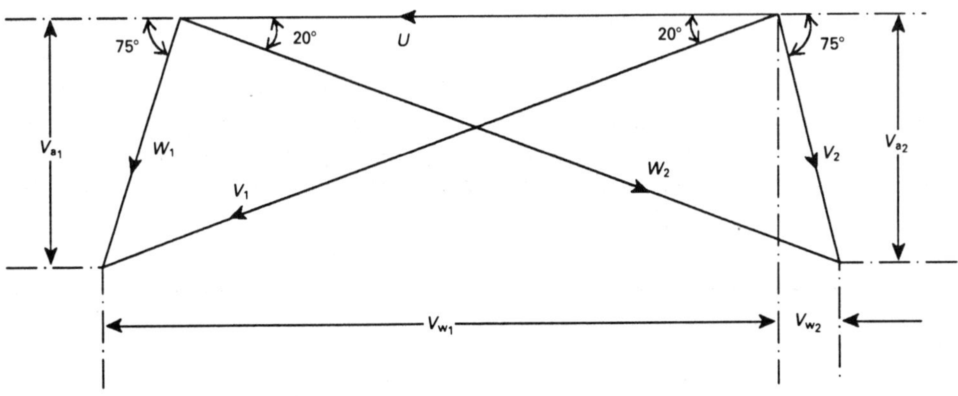

$$U = \pi \times (3000/60) \times 1050/10^3 = 164.93 \text{ [m/s]}$$

Then: $U/V_{a1} = \cot 20° - \cot 75° = 2.7475 - 0.2679 = 2.4796$

$$\therefore \qquad V_{a1} = 164.93/2.4796 = 66.516 \text{ [m/s]}$$

From symmetry: $U + V_{w2} = V_{w1}$ and $V_{a1} = V_{a2}$

Hence: $V_{a1}/(U + V_{w2}) = \tan 20°$

$$\therefore \qquad U + V_{w2} = V_{a2}/\tan 20° = 66.516/0.3640$$

$$= 182.75 \text{ [m/s]} = V_{w1}$$

$$V_{w2} = 182.75 - 164.93 = 17.82 \text{ [m/s]}$$

We now have all we need to find the rate of work per stage, using the formula given in Frame 75. What is it?

78

33.08 kW/kg

$$\dot{W} = U \times (V_{w1} + V_{w2})$$

$$\therefore \qquad \dot{W} = 164.93 \times (182.75 + 17.82) \times 10^{-3}$$

$$= 33.08 \text{ [kW/kg]}$$

79

Here is another example which follows these steps closely:

In a reaction turbine the wheel diameter at the mean blade height is 1200 mm, and the shaft rotates at 3000 rev/min. The blade inlet and outlet angles are 80° and 20° respectively, and there are 8 stages. Calculate U, V_{w1} and V_{w2}, and hence the power per unit mass flow per stage. If steam enters the first stage at 100 bar, or 10 MN/m^2, and 400°C, what is the state of the steam as it leaves the last stage if it expands to 80 bar, or 8 MN/m^2, pressure?

80

$U = 188.5$ m/s, $V_{w1} = 201.41$ m/s, $V_{w2} = 12.91$ m/s,
$\dot{W} = 40.399$ kW/kg, total $\Delta h = 323.19$ kJ/kg, 80 bar 298.75°C

Following Frame 77 you should be able to get U, V_{w1} and V_{w2}, and hence the power per stage following Frame 77. The power per stage for unit mass flow is also numerically the enthalpy drop per stage, so the total enthalpy drop is 8 times this. The specific enthalpy at 100 bar and 400°C is 3097 kJ/kg, and with an enthalpy drop of 323.19 kJ/kg we then have 2773.81 kJ/kg, which at 80 bar is just in the superheat region, at a temperature of 298.75°C.

81

As we have seen, in a reaction steam turbine there are a large number of stages with a relatively small enthalpy drop in each stage. In the previous programme we considered an isentropic efficiency for a steam turbine; we can now relate this to what happens in each stage. Because of friction, velocities in a stage are slightly less than they would otherwise be, and there is a small amount of reheating of the steam, so that the enthalpy of the steam leaving a stage is slightly more than it would be for an isentropic expansion. Remember that we assume the flow is adiabatic; in practice, this is not strictly true and some of the frictional effect causes heating of the blades. On the h–s diagram, the condition of the steam moves slightly to the right at an increased entropy, but at the same pressure for that particular stage. This is shown below.

The state of the steam as it passes through the turbine follows what is called the **condition line**. In each stage the isentropic efficiency is:

$$\eta_{\text{stage}} = \frac{\Delta h_s}{\Delta h_s'}$$

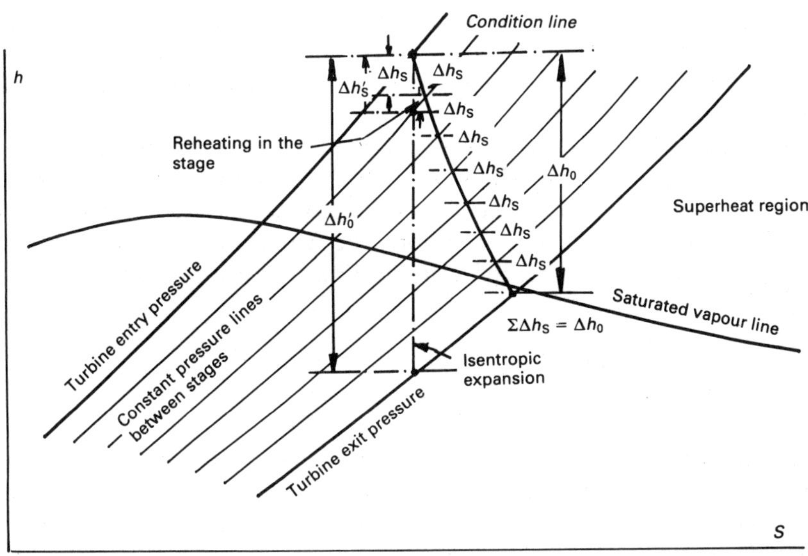

For the whole turbine, the isentropic enthalpy drop is $\Delta h_0'$ and the actual enthalpy drop is Δh_0 so the isentropic efficiency, η_S, is:

$$\eta_S = \frac{\Delta h_0}{\Delta h_0'}$$

This is the isentropic efficiency we used in the previous programme. The actual enthalpy drop Δh_0 is the sum of the values of Δh_s for all the stages, so:

$$\Delta h_0 = \sum \Delta h_s = \eta_{\text{stage}} \sum \Delta h_s'$$

$$= \eta_s \Delta h_0'$$

$$\therefore \qquad \eta_s = \eta_{\text{stage}} \frac{\sum \Delta h_s'}{\Delta h_0'}$$

$$= \eta_{\text{stage}} R$$

where R is the **reheat factor**.

82

From the diagram in the previous frame we can see that $\Sigma\ \Delta h'_s$ is larger than $\Delta h'_0$, so the reheat factor is greater than 1. This is because frictional losses in all stages, except the last one, are recovered as an increase in enthalpy which may be used in the next stage.

Suppose that in the reaction turbine considered in Frame 77 there are 10 stages, and the stage efficiency is 0.8. The steam enters the turbine at 80 bar, or 8 MN/m^2, and 400°C, and expands to 15 bar. We shall calculate the isentropic efficiency for the whole turbine and the reheat factor. The actual total enthalpy drop in the turbine is Δh_0, where $\Delta h_0 = \Sigma\ \Delta h_S$, and in Frame 77 we found that Δh_S is 33.08 kW/kg.

$$\therefore \qquad \Delta h_0 = \sum \Delta h_S = 10 \times 33.08$$
$$= 330.8\ [\text{kW/kg}]$$

We can find the isentropic enthalpy drop in the usual way. At entry the steam is at 80 bar and 400°C, at which state $h = 3139$ kJ/kg, and $s = 6.364$ kJ/kg K (from the Rogers and Mayhew tables). After expanding to 15 bar, what is the isentropic enthalpy drop, $\Delta h_0'$?

83

385.19 kJ/kg

Equating entropies at 80 bar and 15 bar, we have at 15 bar:

$$6.364 = 2.315 + 4.130\,x$$
$$\therefore \qquad x = 0.9804$$

The enthalpy at 15 bar is then:

$$h = 845 + 0.9804 \times 1947$$
$$= 2753.81\ [\text{kJ/kg}]$$
$$\therefore \qquad \Delta h_0' = 3139 - 2753.81$$
$$= 385.19\ [\text{kJ/kg}]$$

Then:
$$\eta_S = 330.8/385.19$$
$$= 0.8588$$

and since: $\eta_o = \eta_s R$

$$R = 0.8588/0.8$$
$$= 1.0735$$

This reheat factor is a measure of effect of friction in the expansion on the overall enthalpy drop.

Now calculate the turbine isentropic efficiency and the reheat factor for this case: steam enters a reaction turbine at 12 MN/m^2 and 500°C and expands through 4 stages to 7 MN/m^2, with a stage efficiency of 0.82. The actual power output of each stage is 35 kW/kg of steam flow. Follow the steps in the previous two frames.

84

At 7 MN/m^2, isentropic expansion, superheat T = 400.598°C, isentropic Δh = 162.66 kJ/kg, η_s = 0.8607, R = 1.0496

Following the previous example, you should be able to get these results.

To conclude this programme we give some problems on turbine stages.

1. In an impulse turbine stage with a single wheel that rotates at 3000 rev/m, the absolute velocity of the steam at inlet is 400 m/s with a nozzle angle of 22°. The wheel diameter at the mean blade height is 450 mm. The velocity coefficient is 0.9. There is no change in axial velocity through the wheel. Calculate the moving blade inlet and outlet angles, the work of the stage per kg of steam flow, the blade efficiency, and the angle and velocity of the steam at outlet. [26.53°, 29.75°, 39.75 kJ/kg, 49.69%, 38.05°, 243.12 m/s]

2. A second wheel is added to the stage in problem **1**. The stator blades outlet angle is 25° and the velocity coefficient applies through the stator blades. The axial velocity is constant (at a different value) through the second wheel. Calculate the blade inlet and outlet angles for the second wheel, the work per kg for the wheel, and the blade efficiency for the two wheels. [35.89°, 40.65°, 16.63 kJ/kg, 70.47%]

3. The work output for an impulse wheel is 35 kJ/kg. The wheel rotates at 3500 rev/m and the nozzle angle is 20°. The blade inlet and outlet angles are equal at 24°. There is no friction, K = 1. Calculate the wheel diameter at the mean blade height, the steam velocity at entry and the blade efficiency. [341.2 mm, 364.4 m/s, 52.72%]

4. In a reaction turbine stage with 50% reaction, the stator blade outlet angle is 20°, and the moving blade inlet angle is 70°. The blade velocity is 200 m/s. Calculate the work of the stage per kg of steam flow. [52.22 kJ/kg]

5. A reaction turbine with 10 stages operates between 40 bar and 400°C down to 10 bar pressure with an isentropic efficiency of 0.8617. The turbine shaft rotates at 3000 rev/m and the diameter to the mean blade height is 955 mm. The stator blade outlet angle is 20°. Calculate the work per stage per kg of steam flow and the moving blade inlet angle. [30 kJ/kg, 68.57°]
6. Three stages in a reaction turbine with equal work have an inlet condition of 50 bar and 450°C, and the steam expands to 30 bar. The stage efficiency is 0.82 and the reheat factor is 1.07. Calculate the overall isentropic efficiency of the three stages, the state of the steam as it leaves the third stage and the specific work of the steam per stage. [0.8774, 30 bar and 379.04°C, 44.26 kJ/kg]

Programme 9

COMBUSTION

1

Our foremost concern in studying Thermodynamics is to find out how we may produce work, and what is the efficiency of the process. In Programme 6 we met the idea of a heat engine, a device within a closed system with heat interactions at high and low temperatures at the boundary, and from which work is obtained. In the previous two programmes we have studied the steam power plant at some length. This is the classical example of the heat engine concept, which is now finding new applications with alternative vapours. A heat source is produced by the combustion of fuel *external* to the closed system in which the working fluid circulates, and similarly an *external* heat sink is provided via the condenser. Work is obtained where a rotating shaft passes through the closed system boundary. These are the characteristic features of a heat engine. We now need to think about other prime movers which use fuel and produce work. How many different types can you name?

2

Spark-ignition engines
Compression-ignition engines
Gas turbine engines
Fuel cells

The spark-ignition engine includes two and four stroke engines, in both reciprocating and rotary form, which may burn petrol, specialised liquid fuels for high performance, or gas. The four stroke cycle was described in Frames 2 and 41 of Programme 2. The four events, suction of fresh fuel–air mixture, compression, ignition by a spark and expansion, and exhaust of the burnt fuel, are accomplished in only two strokes of the piston in the two stroke cycle. In the compression-ignition, or diesel engine, there may again be either two or four strokes, but now only air is sucked into the cylinder, the compression ratio is much greater, and fuel is injected at the end of compression where the high temperature of the air causes ignition. Both types of engine may be supercharged which means that the fresh charge of air–fuel mixture or just air is forced in under pressure. In the early days such engines were said to be 'blown'.

The gas turbine engine is a machine through which the working fluid flows continuously, rather than intermittently as in a petrol or diesel engine. It often produces its power in the form of jet thrust rather than in a rotating shaft. The fuel cell is another device which burns fuel to produce work. All these inventions will be considered in the final programme.

They all may be represented symbolically as an open system containing some device into which flows air and fuel, and from which flows combustion products and an output of work, which is electrical work for the fuel cell. In addition, all these devices give out heat. Fuel is burnt and work is done, but these machines do not

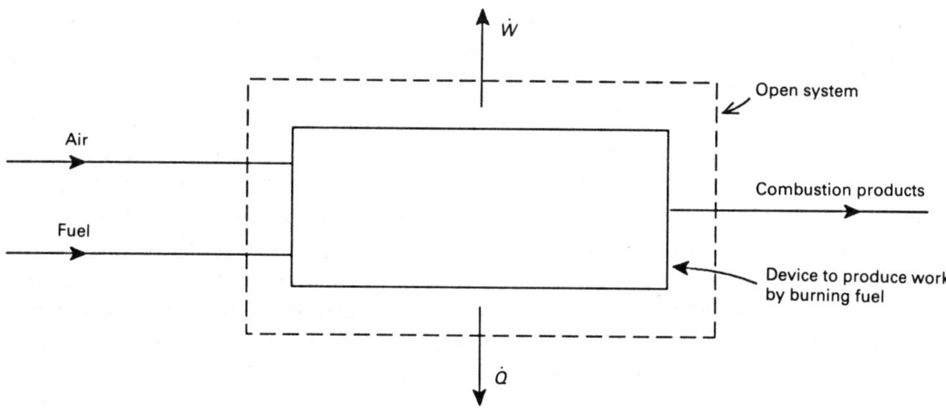

have the characteristic features of heat engines mentioned in the previous frame. Only the gas turbine engine in its closed cycle form is a true heat engine.

3

The gas turbine engine may be built in both open and closed cycle form. Aircraft gas turbine engines are open cycle engines through which air and fuel flow continuously. The closed cycle form is shown below. Examples do exist in industrial applications, though they are rare.

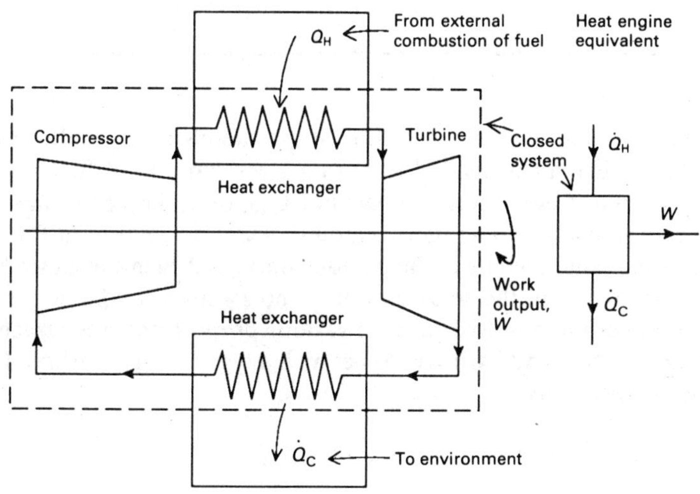

4

Air, or possibly some other gas, circulates within the closed system in the direction shown. It leaves the compressor at a high pressure, and enters a heat exchanger where it receives a heat transfer Q_H at a high temperature. It goes on to expand through the turbine back to the original pressure. Finally it passes through a second heat exchanger where it gives out a heat transfer Q_C at a low temperature, from which it enters the compressor again. The heat source is from the external combustion of fuel, and heat is rejected to a heat sink such as the atmosphere. The contents of the closed system are equivalent to the heat engine shown on the right.

Compared with the open cycle machine, which consists of a compressor, combustion system and turbine only, the closed cycle gas turbine is bulky and expensive. But since the products of combustion do not pass through the turbine, it can burn cheap and abrasive fuels such as powdered coal.

Though not rare like the closed cycle gas turbine, the steam power plant is similarly bulky and expensive. Thus both types of *heat engine* involve large and costly heat transfer equipment. Increasingly nowadays we find that electrical power is being obtained from much cheaper open cycle gas turbine engines, and similarly, air and road transport rely entirely on *internal* combustion turbines or engines. So we see that power for many applications is obtained very extensively from plant that may not be categorised as a heat engine.

It is obviously important, therefore, that we need to examine these power-producing machines in which fuel is burnt *internally* and which do not conform to the heat engine concept. From the Second Law we shall see how much work may be produced theoretically, but this will follow after we have looked at the First Law aspects of combustion.

5

So that we may do that, we must first look at the chemistry of combustion in order to find out how much air is needed to burn a given amount of fuel.

In writing chemical equations, the symbol O_2, for example, means 1 mole of oxygen molecules, each containing two atoms. As we have seen in Programme 4, all moles have the same volume in the gaseous state, and each substance has its own molecular mass. Moles in the solid state have no significant volume.

Fuels contain carbon and hydrogen in various proportions, plus traces of other substances, and carbon and hydrogen react with oxygen to give carbon dioxide and water vapour respectively:

$$C + O_2 = CO_2$$

$$H_2 + \tfrac{1}{2} O_2 = H_2O$$

The symbol C represents one mole of carbon molecules, which each contains one atom of carbon, and H_2 represents one mole of hydrogen molecules each containing two atoms. Thus one mole of carbon requires one mole of oxygen to give one mole of carbon dioxide (there is no volume change in this reaction as carbon by itself is a solid), and one mole of hydrogen requires half a mole of oxygen to give one mole of water. Methane is a gaseous fuel (natural gas) and a mole of methane molecules is written CH_4, which means that in one molecule there is one atom of carbon and four atoms of hydrogen. How many molecules of carbon and hydrogen are there in the methane molecule?

6

Carbon – 1, Hydrogen – 2

The single carbon atom is one molecule of carbon, and the four atoms of hydrogen are two molecules. We now need to find out how many moles of oxygen are needed to burn one mole of methane. We can do this by seeing what the products are and then we can count up the oxygen needed. From the previous frame we saw that **one atom or molecule of carbon gives one molecule of carbon dioxide, and two atoms or one molecule of hydrogen give one molecule of water**. We can express this:

$$C \rightarrow CO_2$$

$$H_2 \rightarrow H_2O$$

How many moles of carbon dioxide and water will we get from one mole of methane?

7

$$CH_4 \rightarrow CO_2 + 2\ H_2O$$

8

We get one mole of carbon dioxide and two of water. The rule in Frame 6 is important since it enables you to work out what the products will be from any fuel molecule consisting of carbon and hydrogen. Methane is the first in a series of substances called **paraffins**; the next two substances in the series are called ethane and propane:

Ethane: C_2H_6

Propane: C_3H_8

How many moles of carbon dioxide and water vapour will each of these give?

9

$$C_2H_6 \rightarrow 2\ CO_2 + 3\ H_2O$$
$$C_3H_8 \rightarrow 3\ CO_2 + 4\ H_2O$$

Provided we know the molecular composition of the fuel substance, we can write down in this way the moles of carbon dioxide and water that we shall get in the products of combustion. Now we can start to work out the amount of oxygen needed. We simply add up the number of oxygen molecules we can see in the products. Thus for the combustion of methane we had:

$$CH_4 \rightarrow CO_2 + 2\ H_2O$$

which means that we have one molecule in the carbon dioxide and one atom in each of the water molecules, or one molecule in the two, which gives us a total of two molecules altogether. Then we can write down the complete equation for the combustion of methane in oxygen:

$$CH_4 + 2\ O_2 = CO_2 + 2\ H_2O$$

We know the equation is correct because we have one atom of carbon, four atoms of hydrogen and four atoms of oxygen on each side. Now see if you can write the corresponding equations for the combustion of ethane and propane, C_2H_6 and C_3H_8, in oxygen.

10

$$C_2H_6 + 3\tfrac{1}{2}\ O_2 = 2\ CO_2 + 3\ H_2O$$
$$C_3H_8 + 5\ O_2 = 3\ CO_2 + 4\ H_2O$$

The three atoms of oxygen in the water in the first example gives one and a half molecules, and a total of three and a half molecules. You can check that we have the same numbers of atoms of all three chemical elements on both sides of each of these equations.

When a fuel burns in some device it is important that the oxygen in the air can get to all the fuel molecules present. We saw in Programme 5 that a gas mixture can be a pure substance only when the composition is uniform throughout. In practice this is often not so, and in combustion there can be too much fuel in parts of the combustion zone for the air present, while elsewhere there could be very little fuel for the available oxygen. Thus some fuel remains unburnt in some zones, but is burnt completely in others. This is the reason for smoke and soot appearing in combustion products. We can consider the overall effect of what happens. A mole of fuel contains a vast number of molecules, so for example, it can be that say 95% of the molecules burn completely, and 5% do not.

11

Poor distribution of fuel can lead to unburnt fuel passing through even though overall there is sufficient oxygen in the air available, but where overall there is insufficient oxygen, we assume all the hydrogen in the fuel burns completely and some of the carbon burns to carbon monoxide. This reaction is written:

$$C + \tfrac{1}{2}\ O_2 = CO$$

Suppose 95% of the carbon in a mole of methane burns to carbon dioxide and the remainder burns to carbon monoxide. How many moles of oxygen are needed? The combustion of the carbon is given in these two equations:

$$0.95\ C + 0.95\ O_2 = 0.95\ CO_2$$
$$0.05\ C + \tfrac{1}{2} \times 0.05\ O_2 = 0.05\ CO$$

12

The oxygen required for the carbon is 0.95 + 0.025 moles, which is 0.975 moles. The oxygen required for the hydrogen is one mole as before, so the total is 1.975 moles, instead of 2 moles for complete combustion. Thus:

$$CH_4 + 1.975\ O_2 = 0.95\ CO_2 + 0.05\ CO + 2\ H_2O$$

The propane molecule is C_3H_8. Suppose 90% of the carbon burns to carbon dioxide and the remainder to carbon monoxide, how many moles of each will there be in the products?

13

$$\boxed{2.7\ CO_2 + 0.3\ CO}$$

The carbon in a mole of C_3H_8 could produce 3 moles of carbon dioxide; but only 90% of the carbon burns to carbon dioxide, so 2.7 moles are produced and the remaining 0.3 moles of carbon becomes carbon monoxide.

The hydrogen in the propane mole is burnt completely to water as before, so we can write:

$$C_3H_8 \rightarrow 2.7\ CO_2 + 0.3\ CO + 4\ H_2O$$

To find the number of moles of oxygen needed, we simply have to add up the number of atoms of oxygen on the right-hand side and divide by two. This oxygen is then included on the left-hand side to complete the combustion equation, which is given in the next Frame.

14

$$\boxed{C_3H_8 + 4.85\ O_2 = 2.7\ CO_2 + 0.3\ CO + 4\ H_2O}$$

The number of oxygen atoms is 2.7 × 2, plus 0.3 in the carbon monoxide, and 4 in the water, so the number of moles of oxygen is given by:

$$X = \frac{(2.7 \times 2 + 0.3 + 4)}{2} = 4.85$$

Alternatively we can say that if X is the number of moles of oxygen, we can see that there are 2.7 moles in the carbon dioxide, 0.15 in the carbon monoxide and 2.0 in the water, so X must be 4.85.

In combustion systems, oxygen normally occurs as a constituent of air, so we need to find the amount of air needed to give the required oxygen. Air contains 20.95% O_2 and 78.09% N_2 by volume plus traces of other gases, but for practical purposes we take the composition to be 21% O_2 and 79% N_2. Thus for every mole of oxygen required there would be:

$$0.79/0.21 = 3.76 \text{ moles of nitrogen}$$

Thus we can say:

4.76 moles of air may be written: $O_2 + 3.76\ N_2$

1 mole of air may be written: $0.21\ O_2 + 0.79\ N_2$

Looking back to Frame 5, how much air is needed to burn one mole of carbon and one mole of hydrogen?

15

Carbon: 4.76 moles of air
Hydrogen: 2.38 moles of air

The carbon requires one mole of oxygen, which means 4.76 moles of air, and the hydrogen requires half a mole of oxygen, or 2.38 moles of air. The combustion equations in air would then be:

$$C + O_2 + 3.76\ N_2 = CO_2 + 3.76\ N_2$$

and:

$$H_2 + \tfrac{1}{2}\ O_2 + \tfrac{1}{2} \times 3.76\ N_2 = H_2O + \tfrac{1}{2} \times 3.76\ N_2$$

or:

$$H_2 + 0.5\ O_2 + 1.88\ N_2 = H_2O + 1.88\ N_2$$

In the last line you can see that there are 2.38 moles of air on the left. Now write the equation for the combustion of methane, CH_4, in air.

16

$$CH_4 + 2\ O_2 + 7.52\ N_2 = CO_2 + 2\ H_2O + 7.52\ N_2$$

For every mole of methane we need 2 moles of oxygen, which means we have 2 × 3.76 which is 7.52 moles of nitrogen. Thus we have 9.52 moles of air. As before, the moles of nitrogen do not enter the reaction.

This equation is for the complete combustion of methane without any oxygen left over, so it is said to be **chemically correct**. In the equation there is one mole of methane and 9.52 moles of air, which gives an **air/fuel ratio by volume** of 9.52 to 1. The air/fuel ratio for chemically correct combustion is called the **stoichiometric** air/fuel ratio. When the fuel is a gas, the air required for combustion is expressed in terms of the air/fuel ratio by volume.

In Frame 10 we saw that ethane, C_2H_6, requires 3.5 moles of oxygen, and propane, C_3H_8, requires 5 moles for chemically correct combustion. Find the air needed and the stoichiometric air/fuel ratio by volume in these two cases.

17

$$C_2H_6\text{: 16.67 to 1; } C_3H_8\text{: 23.8 to 1}$$

The chemically correct equation for ethane is:

$$C_2H_6 + 3.5\ O_2 + 3.5 \times 3.76\ N_2 = 2\ CO_2 + 3\ H_2O + 3.5 \times 3.76\ N_2$$

$$\therefore\ C_2H_6 + 3.5\ O_2 + 13.17\ N_2 = 2\ CO_2 + 3\ H_2O + 13.17\ N_2$$

$$\therefore\ \text{Moles of air} = 3.5 + 13.17$$

$$= 16.67$$

and the air/fuel ratio is 16.67 to 1.

For propane:

$$C_3H_8 + 5\ O_2 + 5 \times 3.76\ N_2 = 3\ CO_2 + 4\ H_2O + 5 \times 3.76\ N_2$$

$$\therefore\ C_3H_8 + 5\ O_2 + 18.8\ N_2 = 3\ CO_2 + 4\ H_2O + 18.8\ N_2$$

$$\therefore\ \text{Moles of air} = 5 + 18.8$$

$$= 23.8$$

and the air/fuel ratio is 23.8 to 1. We can see that as the number of atoms in the fuel molecule increases, the air/fuel ratio for chemically correct combustion also increases.

Usually fuel is burnt with an excess of air present, so that there is oxygen left over after combustion. The main reason for doing this is that with the chemically correct air, the temperature after combustion is often too high for the intended purpose. As we shall see later, if more air is used then the combustion temperature will be lower. Another advantage of using more air is that combustion is more likely to be complete. If the stoichiometric air/fuel ratio is $(A/F)_{st}$ and the actual air/fuel ratio is (A/F), then the **excess air** is given by:

$$\text{Excess air} = \frac{(A/F) - (A/F)_{st}}{(A/F)_{st}} \times 100\%$$

If methane is burnt with an air/fuel ratio of 15 to 1, what is the percentage excess air? In Frame 16 we found that $(A/F)_{st}$ for methane was 9.52 to 1.

18

57.56%

The excess air is (15 − 9.52)/9.52 which is 0.5756 or 57.56%. Fuels will only burn within a range of certain limits of air/fuel ratio. When there is excess air, the mixture is said to be **weak**, and when there is insufficient air, the mixture is said to be **rich**, and fuels have weak and rich flammability limits. These limits can depend on the temperature of the environment; thus the petrol engine of a car needs a lower air/fuel ratio, or a richer mixture, for starting when the engine is cold. This is provided by the use of the choke. Quite rapidly the cylinder walls reach their normal running temperature and then the weaker normal mixture is used.

We will return to the question of what are the actual flammability limits of fuels a little later on.

19

As we shall see later, the temperature of the combustion products depends on the air/fuel ratio, so it is possible that a specified combustion temperature (as in a gas turbine engine, for example) requires the air/fuel ratio to be outside the weak limit. The air is then split into two streams; combustion takes place at more or less stoichiometric conditions, and the combustion products are then diluted and cooled by the excess air.

Suppose methane is burnt at an air/fuel ratio of 15 to 1. There will be 15 moles of air and only 9.52 are needed, so there are 5.48 moles of air in which the oxygen is not used. What is the chemical equation for the combustion of methane at this air/fuel ratio? The products will contain O_2 as well as the usual CO_2, H_2O and N_2.

20

$$CH_4 + 3.15\ O_2 + 11.85\ N_2 = CO_2 + 2\ H_2O + 1.15\ O_2 + 11.85\ N_2$$

With 21% oxygen by volume, 15 moles of air will contain 3.15 moles of oxygen and 11.85 moles of nitrogen. Only 2 moles of oxygen will be used, so 1.15 moles will be left over. The 11.85 moles of nitrogen appear on both sides of the equation.

The right-hand side of this equation lists the **combustion products** and we can find the percentage composition by volume of these products. The total number of moles is 1 + 2 + 1.15 + 11.85, which is 16.0. The volume composition is then:

$$CO_2 = 1/16 \times 100\% = 6.25\%$$

$$H_2O = 2/16 \times 100\% = 12.5\%$$

$$O_2 = 1.15/16 \times 100\% = 7.19\%$$

$$N_2 = 11.85/16 \times 100\% = 74.06\%$$

Sometimes the water content of the combustion products is condensed out, and the remainder is called the **dry exhaust products**. What would be the percentage composition of the dry exhaust products in this case?

21

$CO_2 = 7.143\%$, $O_2 = 8.214\%$, $N_2 = 84.643\%$

When the 2 moles of water are removed there are 14 moles of products remaining, so the percentage composition is:

$$CO_2 = 1/14 \times 100\% = 7.143\%$$

$$O_2 = 1.15/14 \times 100\% = 8.214\%$$

$$N_2 = 11.85/14 \times 100\% = 84.643\%$$

In Frame 8 we saw that methane, CH_4, ethane, C_2H_6, and propane, C_3H_8, are the first three substances in the series of hydrocarbons called paraffins. A general formula for these substances is C_nH_{2n+2}. As the number of atoms in the molecule gets larger, the boiling point of the substance rises so that the fuel becomes a liquid at normal temperatures. Propane is a gas at atmospheric pressure and temperature, but may be liquefied under pressure and stored in bottles. (You should know the reason for this – it is because at atmospheric temperature propane is below its critical temperature – see Frame 69 in Programme 4.)

Octane is a liquid fuel in the paraffin series with 8 carbon atoms. What is its chemical formula?

22

C_8H_{18}

The number of hydrogen atoms must be:

$$8 \times 2 + 2 = 18$$

In dealing with liquid fuels we use air/fuel ratios by mass, rather than by volume. To find an air/fuel ratio by mass we need first of all to write the chemical equation in terms of moles. See if you can do that for octane burnt at its stoichiometric air/fuel ratio.

23

$$C_8H_{18} + 12.5\ O_2 + 47\ N_2 = 8\ CO_2 + 9\ H_2O + 47\ N_2$$

Because the octane molecule contains 8 atoms of carbon, we know that this will give us 8 moles of carbon dioxide in the products; similarly we know that the 18 atoms of hydrogen will give us 9 moles of water. The 8 moles of carbon dioxide and 9 moles of water will require 12.5 moles of oxygen.

The 12.5 moles of oxygen mean that there will be 12.5 × 3.76 moles, or 47 moles, or nitrogen. Now we can find the masses represented by the terms in this equation. The molecular masses of individual atoms or molecules are:

$$C - 12;\ H - 1,\ O_2 - 32,\ N_2 - 28$$

The molecular mass of octane is therefore:

$$12 \times 8 + 1 \times 18 = 114$$

As we work in kg, a kmol of octane has a mass of 114 kg. The mass of air is:

$$12.5 \times 32 + 47 \times 28 = 1716\ [\text{kg}]$$

The mass of fuel is 114 kg and the mass of air is 1716 kg. Hence:

$$(A/F)_{st} = \frac{1716}{114}$$

$$= 15.053 \text{ to } 1$$

24

The paraffins are straight chain hydrocarbons, so called because of their molecular structure. Benzene is a liquid and is an example of another type of hydrocarbon, in which the structure is a closed ring. The structure of the molecule looks like this:

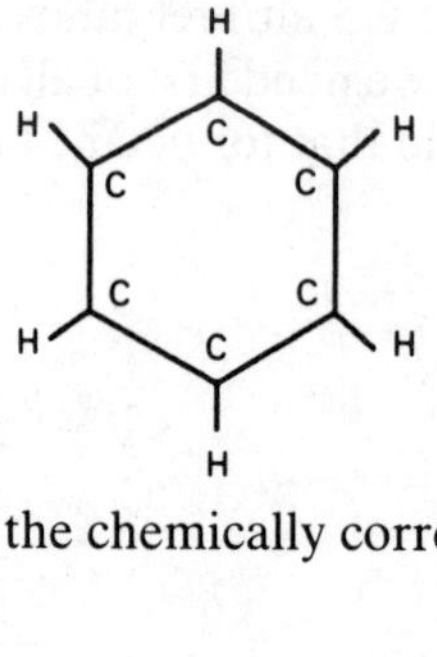

H
H C H
C C
C C
H C H
H

The formula is C_6H_6. What is the chemically correct equation for the combustion of benzene in air?

25

$$C_6H_6 + 7.5\ O_2 + 28.2\ N_2 = 6\ CO_2 + 3\ H_2O + 28.2\ N_2$$

With 6 atoms each of carbon and hydrogen we know that we will get 6 moles of carbon dioxide and 3 moles of water in the products. This will require 7.5 moles of oxygen, and 7.5 × 3.76 moles of nitrogen. What is now the stoichiometric air/fuel ratio by mass?

26

13.2 to 1

The masses involved are:

$$C_6H_6 = 6 \times 12 + 6 \times 1 = 78\ [kg]$$
$$7.5\ O_2 = 7.5 \times 32 = 240\ [kg]$$
$$28.2\ N_2 = 28.2 \times 28 = 789.6\ [kg]$$
$$\text{or: } 35.7 \text{ moles of air} = 1029.6\ [kg]$$

The mass of fuel is 78 kg and the mass of air is 1029.6 kg, so the air/fuel ratio by mass is 1029.6/78, which is 13.2 to 1.

Air/fuel ratios by mass other than stoichiometric can be used in a term called the **mixture strength**. This is defined as:

$$\text{Mixture strength} = \frac{(A/F)_{st}}{(A/F)} \times 100\%$$

Both weak and rich mixtures can be expressed in terms of mixture strength:

(A/F) is greater than $(A/F)_{st}$; mixture is weak, and mixture strength $< 100\%$

(A/F) is less than $(A/F)_{st}$; mixture is rich, and mixture strength $> 100\%$

27

The terms **weak** and **rich mixture** indicate there is a deficiency or excess of fuel in relation to the air present for chemically correct combustion.

A term called the **fuel air equivalence ratio**, ϕ, is also used. This uses the reciprocal of air fuel ratio, thus:

$$\phi = \frac{(F/A)}{(F/A)_{st}}$$

Effectively it is the same as mixture strength; when, for example, the mixture strength is 120%, the fuel air equivalence ratio is 1.2. The ratio ϕ is useful in expressing various combustion characteristics of fuels; thus for most fuels the flammability limits for normal combustion lie between the weak to rich limits of $0.6 < \phi < 1.6$.

One mole of benzene is reacting with 30 moles of air. What is the mixture strength?

28

119%

Since a mole of air contains 21% oxygen and 79% nitrogen by volume, we can work out the mass of air:

$$\text{Mass of oxygen} = 30 \times 0.21 \times 32 = 6.3 \text{ moles} \times 32 \text{ kg/mole} = 201.6 \text{ [kg]}$$

$$\text{Mass of nitrogen} = 30 \times 0.79 \times 28 = 23.7 \text{ moles} \times 28 \text{ kg/mole} = 663.6 \text{ [kg]}$$

$$\therefore \quad \text{Mass of air:} \qquad = 865.2 \text{ [kg]}$$

The mole of benzene has a mass of 78 kg, so the air/fuel ratio by mass is 865.2/78 which is 11.09 to 1. $(A/F)_{st}$ is 13.2 to 1 from Frame 26.

Then the mixture strength = 13.2/11.09 × 100% = 119%.

The equivalence ratio, ϕ, is thus 1.19.

When the mixture is rich, and there is not enough air present, it is reasonable to assume that all the hydrogen in the fuel burns completely and as much carbon as possible burns to carbon dioxide. The remainder of the carbon burns to carbon monoxide.

29

To find how much carbon dioxide and how much carbon monoxide we shall get in this example, we can say let X be the number of moles of CO_2 and then $(6 - X)$ must be the number of moles of CO, since there are 6 atoms of carbon in the benzene molecule.

We can then write the chemical equation as follows:

$$C_6H_6 + 6.3\ O_2 + 23.7\ N_2 = 3\ H_2O + X\ CO_2 + (6 - X)\ CO + 23.7\ N_2$$

The number of moles of O_2 and N_2 come from the previous frame. We know that there must be 6.3 moles of oxygen on the right hand-side, so:

$$1.5 + X + (6 - X)/2 = 6.3$$

$$\therefore \quad X = 3.6$$

The complete equation becomes:

$$C_6H_6 + 6.3\ O_2 + 23.7\ N_2 = 3\ H_2O + 3.6\ CO_2 + 2.4\ CO + 23.7\ N_2$$

The right-hand side represents the combustion products of this reaction.

Knowing the molecular masses, we can find the **gravimetric analysis** of the products. This is the composition by mass of the gases present. Individual atomic or molecular masses were given in Frame 23; from them we can find that the molecular masses of the products are:

$$H_2O = 18;\ CO_2 = 44;\ CO = 28;\ \text{and } N_2 = 28$$

Then we can find the actual masses of products present:

$$\text{Mass of water} = 3 \times 18 = 54\ [\text{kg}]$$

$$\text{Mass of carbon dioxide} = 3.6 \times 44 = 158.4\ [\text{kg}]$$

Now find the masses of carbon monoxide and nitrogen, and hence the percentage composition by mass of the exhaust products.

30

$CO = 67.2$ kg, $N_2 = 663.6$ kg

$H_2O = 5.73\%$, $CO_2 = 16.79\%$, $CO = 7.12\%$, $N_2 = 70.36\%$

The calculations follow in the next frame.

31

Mass of carbon monoxide = 2.4 × 28 = 67.2 [kg]

Mass of nitrogen = 23.7 × 28 = 663.6 [kg]

The total mass is: 54 + 158.4 + 67.2 + 663.6 = 943.2 [kg]

Hence the percentages by mass of gases present may be found:

$$H_2O = 54/943.2 \times 100\% = 5.73\%$$

$$CO_2 = 158.4/943.2 \times 100\% = 16.79\%$$

$$CO = 67.2/943.2 \times 100\% = 7.12\%$$

$$N_2 = 663.6/943.2 \times 100\% = 70.36\%$$

Now here is another problem that brings together a number of things we have learned in the previous few frames. Octane, C_8H_{18}, burns in air at a mixture strength of 115%. Deduce the chemical equation for the reaction, and calculate the gravimetric analysis of the *dry* exhaust products.

32

$C_8H_{18} + 10.87\ O_2 + 40.87\ N_2 = 4.74\ CO_2 + 3.26\ CO + 9\ H_2O + 40.87\ N_2$
CO_2: 14.44%, CO: 6.32%, N_2: 79.24%

We found in Frame 23 that the stoichiometric air/fuel ratio for this reaction is 15.053 to 1. Hence with a mixture strength of 115% the actual air/fuel ratio is 15.053/1.15, which is 13.09 to 1. This is an air/fuel ratio by mass. From this we have to work round to find the proportions by mole of reactants. We can take 1 mole of octane, which has a mass of 114 kg. So we have 114 × 13.09 kg of air. We know that 1 mole of air consists of 0.21 moles of oxygen and 0.79 moles of nitrogen, and the corresponding mass is:

$$\text{Mass of 1 mole of air} = 0.21 \times 32 + 0.79 \times 28 \text{ kg}$$
$$= 28.84 \text{ [kg]}$$

Hence the number of moles of air present must be:

$$\frac{114 \times 13.09}{28.84} = 51.74 \text{ moles}$$

The oxygen present will be: $0.21 \times 51.74 = 10.87\ O_2$
and the nitrogen present will be: $0.79 \times 51.74 = 40.87\ N_2$.

The left-hand side of the chemical equation can then be written:

$$C_8H_{18} + 10.87\ O_2 + 40.87\ N_2$$

and using the technique of Frame 29 we can write the right-hand side as:

$$9\ H_2O + X\ CO_2 + (8 - X)\ CO + 40.87\ N_2$$

From the balance of oxygen molecules we have:

$$10.87 = 4.5 + X + (8 - X)/2$$

$$\therefore \quad X = 4.74$$

$$\text{and} \quad (8 - X) = 3.26$$

The complete chemical equation is therefore:

$$C_8H_{18} + 10.87\ O_2 + 40.87\ N_2 = 4.74\ CO_2 + 3.26\ CO + 9\ H_2O + 40.87\ N_2$$

The masses of the dry exhaust products are:

$$CO_2 = 4.74 \times 44 = 208.56\ [\text{kg}]$$

$$CO = 3.26 \times 28 = 91.28\ [\text{kg}]$$

$$N_2 = 40.87 \times 28 = 1144.36\ [\text{kg}]$$

The total mass is 1444.20 kg, hence the gravimetric analysis is:

$$CO_2 = 208.56/1444.2 \times 100\% = 14.44\%$$

$$CO = 91.28/1444.2 \times 100\% = 6.32\%$$

$$N_2 = 1144.36/1444.2 \times 100\% = 79.24\%$$

33

As the number of atoms in the liquid fuel molecule increases, the boiling point continues to rise and at normal conditions the liquid becomes increasingly viscous. For these fuels and for solid fuels such as the various types of coal, the chemical formula for the substance becomes complex and it is more usual to express the carbon and hydrogen content as a percentage by mass. Thus we could say that a heavy fuel oil contains 87.6% carbon by mass and 12.4% hydrogen by mass.

Next we will consider such a fuel, and find the stoichiometric air/fuel ratio by mass, and the volumetric composition of the *dry* exhaust products.

34

We start by taking 100 kg of a particular fuel, which has a composition of 87.6 kg of carbon and 12.4 kg of hydrogen. How many moles of carbon and hydrogen will there be in the 100 kg of this fuel?

35

C: 7.3 moles, H_2: 6.2 moles

The moles of carbon and hydrogen present in 100 kg of fuel can be found by dividing the actual masses by their molecular masses:

$$\text{Moles of carbon} = 87.6/12 = 7.3 \text{ moles}$$

$$\text{Moles of hydrogen} = 12.4/2 = 6.2 \text{ moles}$$

These are the proportions by mole of carbon and hydrogen present in the fuel, so we could write the chemical formula as $C_{7.3}H_{12.4}$. Notice that there are 6.2 moles of H_2 giving $H_{12.4}$ in the formula.

As we know, every mole of oxygen in air is accompanied by 3.76 moles of nitrogen, giving a total of 4.76 moles. Let's say there are $X \times 4.76$ moles of air needed to burn 100 kg of the fuel. We can now construct the chemical equation:

$$7.3\ C + 6.2\ H_2 + X\ (O_2 + 3.76\ N_2) = 7.3\ CO_2 + 6.2\ H_2O + 3.76\ X\ N_2$$

Now add up all the oxygen molecules on the right-hand side and find X.

36

$X = 10.4$

There are 7.3 moles of oxygen in the carbon dioxide and 6.2/2 moles in the water, so we can say:

$$X = 7.3 + 6.2/2 = 10.4$$

Then, nitrogen present $= 10.4 \times 3.76$
$= 39.1$ moles

The complete chemical equation is then:

$$7.3\ C + 6.2\ H_2 + 10.4\ O_2 + 39.1\ N_2 = 7.3\ CO_2 + 6.2\ H_2O + 39.1\ N_2$$

To find the air/fuel ratio by mass, we have 100 kg of fuel and the mass of air is:

$$10.4 \times 32 + 39.1 \times 28 = 1427.6\ [\text{kg}]$$

The air/fuel ratio is then 1427.6/100 = 14.28 to 1.

The dry exhaust products will consist of carbon dioxide and nitrogen only, as there is no excess air. The total number of moles is 7.3 + 39.1 which is 46.4, so:

$$CO_2 = 7.3/46.4 \times 100\% = 15.73\%$$

$$N_2 = 39.1/46.4 \times 100\% = 84.27\%$$

Now here is a similar problem. A sample of coal consists of 93.2% carbon and 6.8% hydrogen by mass. Calculate the stoichiometric air/fuel ratio by mass and the volumetric composition of the dry exhaust gases with 100% excess air present. First deduce the chemical equation for the correct amount of air as we did in the previous example, and then double the quantity of air for the second part of the problem.

37

13.0 to 1, CO_2 = 8.78%, O_2 = 10.71%, N_2 = 80.51%

Following the method of the previous two frames, then doubling the quantity of air, we get:

$$7.77\ C + 3.4\ H_2 + 18.94\ O_2 + 71.21\ N_2 =$$

$$7.77\ CO_2 + 3.4\ H_2O + 9.47\ O_2 + 71.21\ N_2$$

Notice that the amount of oxygen on the right is half the amount on the left, as it should be with 100% excess air. The number of moles of dry products is 88.45 which gives the above percentages of CO_2, O_2 and N_2.

38

In these previous two problems, we started with the known composition of the fuel and found the composition of the combustion products. We can use a reverse procedure to find the composition of the fuel by measuring the composition of the products.

Consider the following problem. A solid fuel of unknown chemical composition is burnt in air and the dry products of combustion have the following composition by volume: 11.37% carbon dioxide, 5.6% oxygen and 83.03% nitrogen. Calculate the chemical composition of the fuel, assuming it to be carbon and hydrogen only, and calculate the air/fuel ratio.

We can say that the fuel consists of X kg carbon per 100 kg, and $100 - X$ kg hydrogen. This 100 kg of fuel reacts with Y moles of air to produce Z moles of dry exhaust gas. A mole of air is $(0.21\ O_2 + 0.79\ N_2)$. Each mole of dry exhaust gas contains CO_2, O_2 and N_2 in the proportions given. The chemical equation is then:

$$\frac{X}{12}C + \frac{100 - X}{2}H_2 + Y(0.21\ O_2 + 0.79\ N_2) =$$

$$Z(0.1137\ CO_2 + 0.056\ O_2 + 0.8303\ N_2) + (100 - X)/2\ H_2O$$

We need to find the unknowns X, Y and Z, and we can get the necessary equations since for each substance we can equate the number of molecules of that substance on each side of the chemical equation above. Thus for carbon we have $X/12$ molecules on the left and $Z \times 0.1137$ molecules on the right, so we have:

$$\text{Carbon:} \qquad X/12 = Z \times 0.1137$$

With three unknowns we need three simultaneous equations to solve. Two more may be written for the molecules of oxygen and nitrogen. What are they?

39

Oxygen: $0.21\ Y = 0.1137\ Z + 0.056\ Z + (100 - X)/4$
Nitrogen: $0.79\ Y = 0.8303\ Z$

These three equations may now be solved. From the equation for carbon we have: $X = 1.3644\ Z$, and from the equation for nitrogen: $Y = 1.051\ Z$. These results are substituted into the equation for oxygen to give:

$$0.2207\ Z = 0.1697\ Z + 25.0 - 0.3411\ Z$$

$$\therefore \qquad Z = 63.76$$

This is the number of moles of dry exhaust products. From the equation for nitrogen we have:

$$0.79\ Y = 0.8303 \times 63.76$$

$$\therefore \qquad Y = 67.0$$

This is the number of moles of air. The percentage of carbon X is 1.3644 Z, which is 87.0%, so the percentage of hydrogen is 13.0%.

With 67.0 moles of air, the mass of air is:

$$67.0 \times (0.21 \times 32 + 0.79 \times 28) = 1932.28\ [\text{kg}]$$

For 100 kg of fuel the air/fuel ratio is 19.32 to 1.

40

Now try these problems. If you get stuck go through the above frames again. It is important to be familiar with the chemistry of combustion problems before moving on to First Law considerations.

1. A fuel gas consists of 40% H_2, 55% CH_4 and 5% CO (carbon monoxide). It burns completely to CO_2 and H_2O. Calculate the stoichiometric air/fuel ratio by volume, and the percentage excess air for an air/fuel ratio of 7 to 1. Also calculate the percentage dry exhaust gas composition by volume for this air/fuel ratio. [5.45 to 1, 28.38%, 9.295% CO_2, 5.035% O_2, 85.67% N_2]
2. Ethane, C_2H_6, burns to completion in air and gives a dry exhaust gas analysis by volume of 8.1% CO_2, 8.1% O_2 and 83.8% N_2. Calculate the air/fuel ratio by volume and the percentage excess air. [26.19 to 1, 57.11%]
3. Octane, C_8H_{18}, burns at an air/fuel ratio by mass of 17 to 1. Calculate the composition of the exhaust products by volume. [11.16% CO_2, 2.25% O_2, 12.55% H_2O, 74.04% N_2]
4. Octane, C_8H_{18}, burns as a rich mixture with a mixture strength of 105%. All the hydrogen burns to water and as much carbon as possible burns to CO_2, the remainder to CO. Calculate the composition of the dry exhaust products by volume. [13.03% CO_2, 2.3% CO, 84.67% N_2]
5. A fuel consists of equal proportions by mass of octane, C_8H_{18}, and heptane, C_7H_{16}. Calculate the stoichiometric air/fuel ratio by mass. [15.083 to 1]
6. A solid fuel consists of 84% carbon and 11.0% hydrogen by mass, and the remainder is incombustible material, or ash. Calculate the stoichiometric air/fuel ratio by mass, and the composition of the gaseous combustion products by mass. [13.39 to 1, 21.48% CO_2, 6.9% H_2O, 71.62% N_2]
7. Dry exhaust gases of the combustion of a liquid fuel contain 11.15% CO_2, 5.72% O_2 and 83.13% N_2 by volume. Calculate the proportions by mass of carbon and hydrogen in the fuel, and the air/fuel ratio by mass. [86.5% carbon, 13.5% hydrogen, 19.61 to 1]

41

We now move on to think about the First Law in relation to combustion. You will recall (e.g. from Frame 91 of Programme 4) that the First Law energy equation for a closed system of a gas or a mixture of gases is written:

$$Q - W = U_2 - U_1$$
$$= m\, c_V\, (T_2 - T_1)$$

where m is the mass of the single gas or the mass of the gas mixture, and c_V is the specific heat at constant volume for the single gas, or gas mixture.

Suppose the gas mixture was hydrogen and oxygen, or methane and air, for example, and as such was capable of a chemical reaction as we have described. We would express the First Law equation:

$$Q - W = U_2' - U_1'$$

where the superscript ′ indicates a chemically reacting system.

42

The figure on the next page shows a rigid reaction vessel containing, say, a small known mass of liquid fuel and oxygen (the reactants), and a means of obtaining ignition. The reactants are at room temperature, at say 25°C, as their initial state. The reaction vessel is surrounded by a jacket containing a known mass of water, and it is possible to measure any temperature change very accurately.

A spark is passed, ignition takes place, and the vessel then contains the products of combustion, carbon dioxide, water vapour and any unused oxygen, at a high temperature. Then cooling takes place, and the reaction vessel, its contents and the water jacket reach an equilibrium temperature just a few degrees warmer than the initial temperature. Knowing the mass of the reaction vessel and the mass of water and their respective specific heats, we can calculate the heat transfer from the hot combustion gases to the reaction vessel and the water. Since the temperature rise is very small the reaction is almost isothermal, and it is possible to make a correction to the measured heat transfer at the equilibrium temperature to obtain a corresponding figure for an exactly isothermal reaction. This is a standard laboratory technique in calorimetry experiments. We can then apply the First Law to the process:

$$Q = U_P' - U_R'$$

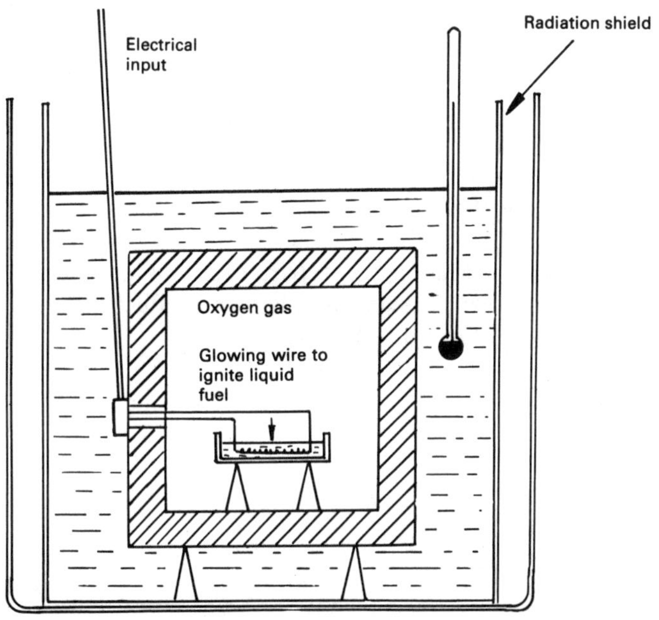

Subscripts P and R refer to products and reactants respectively, and Q is the measured heat transfer for combustion of the small known mass of fuel. We can calculate what Q would be for a kmol of the known fuel. W is zero since the vessel is rigid and no work is done. If we choose the temperature to be 25°C, which is the accepted datum state for chemically reacting systems, the new calculated Q will then be the increase in internal energy per kmol of the chemically reacting system at datum state 0:

$$Q = U'_P - U'_R$$
$$= [\Delta U']_0$$

What is the sign of this heat transfer?

43

Negative

Being a heat transfer *from* the system, it is negative. This means that $[\Delta U']_0$ is also a negative quantity. If we burn a known quantity of fuel, say hydrogen, methane, octane or whatever, and measure the heat transfer we can evaluate $[\Delta U']_0$ per kmol of that reactant at 25°C. This internal energy quantity will give us the difference between two points at 25°C on the separate graphs of internal energy against temperature for the known reactants and known products. The two points U_R' and U_P' at 25°C are shown in the figure below.

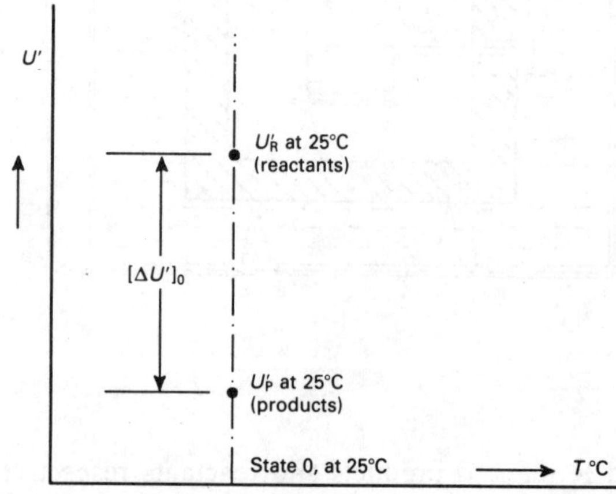

44

The products of the chemical reaction will consist of carbon dioxide, water vapour and unused oxygen. Knowing the content of the reactants we started with, we can calculate precisely the mass of each constituent in the products, which makes up the gas mixture. Each constituent will have its own specific heat at constant volume, and so we can calculate c_V for the gas mixture, using the method given in Frame 90 of Programme 5.

Since product temperatures can be very high, we must consider the effects of real gas specific heats which vary with temperature. Specific heats at constant pressure over a range of temperature for these product gases are given in the Rogers and Mayhew tables. From these we can calculate values of c_V over the same range of temperature, since we know that $R = c_P - c_V$, where R is the gas constant. Then we can calculate c_V for the gas mixture at each temperature over the required range, which would be from 275 K to about 3500 K. Each value of c_V for the mixture multiplied by the temperature would give a value of U_P' for that temperature which is then plotted as a graph of U'_P against temperature. The graph passes through the point of U'_P at 25°C we had in the previous figure. This graph is the lower of the two curves in the figure below.

In the same way we can go on to calculate c_V for the reactants and obtain another curve of U'_R against temperature. To the values of U'_R obtained we must add the value of $[\Delta U']_0$ at 25°C and plot the results as the upper curve in the figure below.

The vertical distance between the two curves at any temperature is the value of $[\Delta U']$ at that temperature. It is not necessary to measure $[\Delta U']$ for all temperatures since it may be calculated at any temperature from $[\Delta U']_0$.

We shall see how this can be done in the next frame.

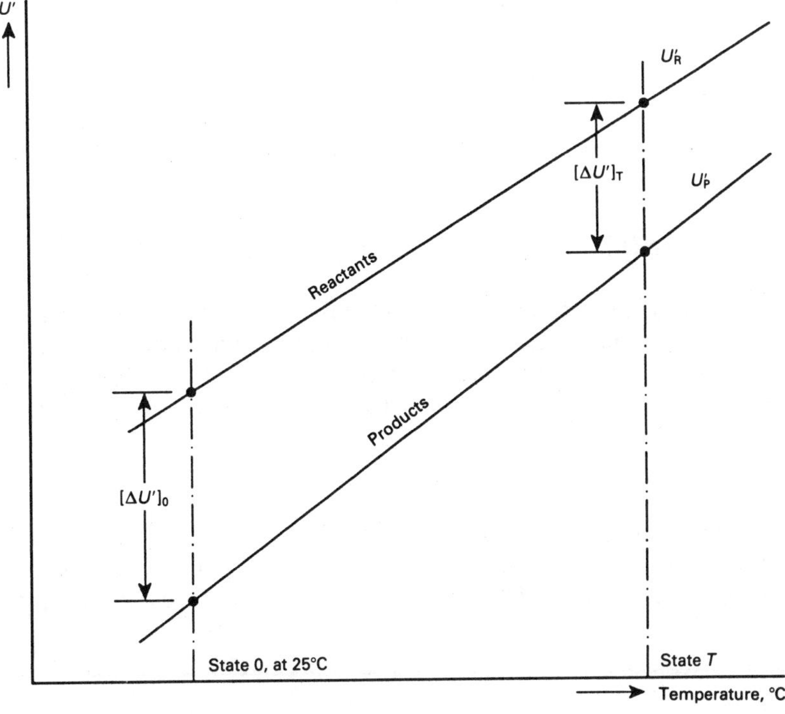

45

We will now use two subscripts for chemically reacting systems to denote first reactant or product, and then temperature. Thus $U'_{R,1}$ is the internal energy of reactants at temperature 1. At the datum temperature, T_0, of 25°C which carries the subscript 0 we therefore already have that:

$$U'_{P,0} - U'_{R,0} = [\Delta U']_0$$

Similarly, we can write for another temperature T_1, with subscript 1:

$$U'_{P,1} - U'_{R,1} = [\Delta U']_1$$

We wish to see how we can find $[\Delta U']_1$ from $[\Delta U']_0$. We start by taking the left-hand side of the previous equation and by subtracting and adding like terms we can say:

$$U'_{P,1} - U'_{R,1} = (U'_{P,1} - U'_{P,0}) - (U'_{R,1} - U'_{R,0}) + (U'_{P,0} - U'_{R,0})$$

The last term in this equation is $[\Delta U']_0$ and the left-hand side is $[\Delta U']_1$, so we get:

$$[\Delta U']_1 = (U'_{P,1} - U'_{P,0}) - (U'_{R,1} - U'_{R,0}) + [\Delta U']_0$$

The first two bracketed terms on the right may be calculated from the masses and specific heats of the individual constituents in the products and reactants respectively and the temperature difference between state 1 and state 0. With the last term $[\Delta U']_0$ brought forward, we can rewrite this equation as follows:

$$[\Delta U']_1 = [\Delta U']_0 + \sum (mc_V)_P(T_1 - T_0) - \sum (mc_V)_R(T_1 - T_0)$$

Consider this example: methane burns stoichiometrically in oxygen only, with the products in the vapour phase. For methane $[\Delta U']_0 = -802\ 310$ kJ/kmol at 25°C. We wish to calculate $[\Delta U']_1$ at 125°C. First, what is the chemical equation for this reaction?

46

$$\boxed{CH_4 + 2\,O_2 = CO_2 + 2\,H_2O}$$

This equation was first given in Frame 9. The masses of the constituents involved and mean values of c_V between 25°C and 125°C are as follows:

Methane, CH_4	mass =	16 [kg]	c_V = 1.856 [kJ/kg K]
Oxygen, O_2		2 × 32 [kg]	0.670 [kJ/kg K]
Carbon dioxide, CO_2		44 [kg]	0.704 [kJ/kg K]
Water vapour, H_2O		2 × 18 [kg]	1.421 [kJ/kg K]

Therefore for the products we can write:

$$\Sigma\,(mc_V)_P(T_1 - T_0) = (44 \times 0.704 + 36 \times 1.421) \times (125 - 25)$$
$$= 8213.2\ [\text{kJ/kmol}]$$

Using this as a guide, what is $\Sigma\,(mc_V)_R(T_1 - T_0)$ for the reactants?

47

7257.6 kJ/kmol

For the reactants of 16 kg of CH_4 and 64 kg of O_2 we have:

$$\Sigma\,(mc_V)(T_1 - T_0) = (16 \times 1.856 + 64 \times 0.670) \times (125 - 25)$$
$$= 7257.6\ [\text{kJ/kmol}]$$

Now, from the previous frame we can now write:

$$[\Delta U']_1 = [\Delta U']_0 + \Sigma\,(mc_V)_P(T_1 - T_0) - \Sigma\,(mc_V)_R(T_1 - T_0)$$
$$= -802\,310 + 8213.2 - 7257.6 = -801\,354.4\ [\text{kJ/kmol}]$$

If we know the mean specific heat values over the required temperature range, we can find $[\Delta U']_1$ at any temperature from the value of $[\Delta U']_0$ at 25°C. In the example, only oxygen was present in the reaction. If the reaction had been in air, even with excess air, the result must be the same. The extra terms involved will be the same for both products and reactants, and consequently cancel out.

Here is a similar exercise. Calculate $[\Delta U']_1$ for the combustion of ethane, C_2H_6, in oxygen given that $[\Delta U']_0$ at 25°C is −1 429 100 kJ/kmol. Use the same values of c_V given in Frame 46; c_V for ethane is 1.710 kJ/kg K. You worked out the chemical equation for this case in Frame 10. $[\Delta U']_1$ is again at 125 °C.

$$\boxed{-1\ 427\ 865.4 \text{ kJ/kmol}}$$

The chemical equation is:

$$C_2H_6 + 3\tfrac{1}{2}\, O_2 = 2\ CO_2 + 3\ H_2O$$

with masses of 30 kg of C_2H_6, 112 kg of O_2, 88 kg of CO_2 and 54 kg of H_2O. The energy equation is:

$$\begin{aligned}[\Delta U']_1 &= [\Delta U']_0 + \sum (mc_V)_P(T_1 - T_0) - \sum (mc_V)_R(T_1 - T_0) \\ &= -1\ 429\ 100 + (88 \times 0.704 + 54 \times 1.421) \times (125 - 25) \\ &\qquad - (30 \times 1.710 + 112 \times 0.670) \times (125 - 25) \\ &= -1\ 427\ 865.4\ [\text{kJ/kmol}]\end{aligned}$$

In these two reactions, the water in the products was in the vapour phase. If the vapour is condensed, the heat transfer Q and consequently the value of $[\Delta U']$ will be greater. This will lead to a second internal energy curve for the products, as shown below.

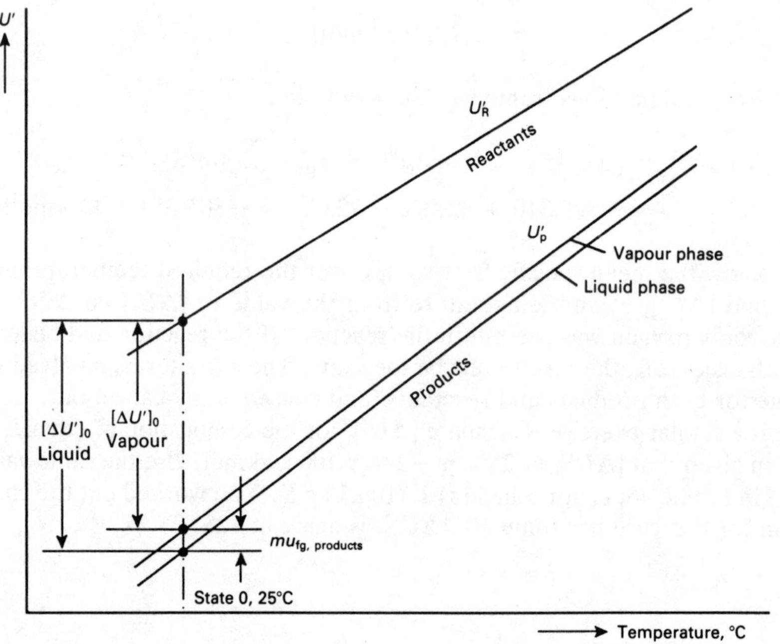

The vertical distance between the two internal energy curves for the products will be $u_{fg} \times$ mass of water involved, where u_{fg} is taken at the temperature of the reaction. From our steam tables we can calculate that u_{fg} at 25°C is 2304.4 kJ/kg. For the methane reaction the water produced is kg/kmol of methane, so $u_{fg} \times$ mass = 82 958.4 kJ/kmol. What is $[\Delta U']_0$ for methane with products in the liquid phase? From the figure you can see that it will be a larger negative value than $[\Delta U']_0$ with the products in the vapour phase, which is −802 310 kJ/kmol.

49

−885 268.4 kJ/kmol

By inspection of the figure in the previous frame we have that:

$$[\Delta U']_{0,\ \text{liquid}} = [\Delta U']_{0,\ \text{vapour}} - m\, u_{\text{fg, products}}$$

$$= -802\ 310 - 82\ 958.4$$

$$= -885\ 268.4\ [\text{kJ/kmol}]$$

Now repeat this exercise for propane, C_3H_8, for which $[\Delta U']_{0,\ \text{vapour}}$ is −2 039 224 kJ/kmol. What is $[\Delta U']_{0,\ \text{liquid}}$?

50

−2 205 140.8 kJ/kmol

There will be 4 moles of water produced, so we may write:

$$[\Delta U']_{0,\ \text{liquid}} = [\Delta U']_{0,\ \text{vapour}} - m\, u_{\text{fg, products}}$$

$$= -2\ 039\ 224 - 4 \times 18 \times 2304.4$$

$$= -2\ 205\ 140.8\ [\text{kJ/kmol}]$$

51

In considering liquid fuels, the phase of the reactants is also important. Thus petrol is supplied to a car as a liquid but is burnt in the engine as a vapour. A second internal energy curve for liquid reactants can be plotted, as shown below.

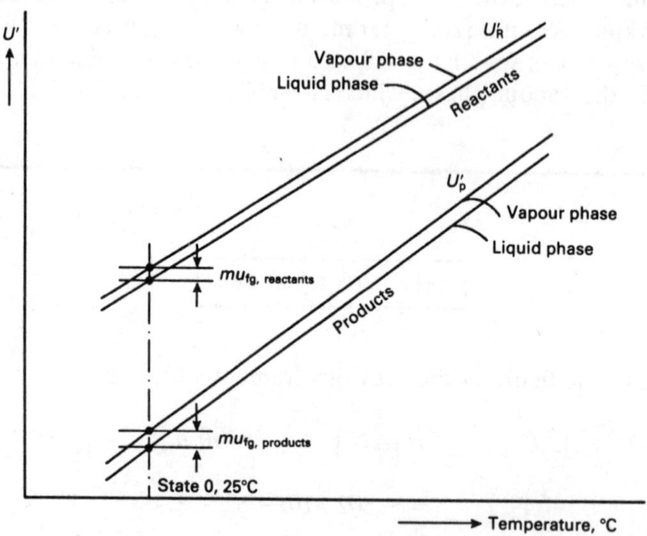

We can see that with a liquid fuel, $[\Delta U']$ is reduced. This is because energy has to be used to evaporate the liquid. Thus for octane, C_8H_{18}, the internal energy of evaporation is appoximately 364 kJ/kg, so:

$$u_{fg,\ reactants} = 364\ [kJ/kg] \times 114\ [kg/kmol]$$
$$= 41\ 496\ [kJ/kmol]$$

52

The internal energy of reaction of octane, $[\Delta U']_0$, is $-5\ 124\ 856$ kJ/kmol with the octane and the water in the products both in the vapour phase. Extending the nomenclature of Frame 49 we can express this as:

$$_{vapour}[\Delta U']_{0,\ vapour} = -5\ 124\ 856\ [kJ/kmol]$$

where the first subscript refers to the reactants and the second refers to the products. By inspection of the figure in Frame 51 we can see that:

$$_{liquid}[\Delta U']_{0,\ vapour} = {}_{vapour}[\Delta U']_{0,\ vapour} + m\ u_{fg,\ reactants}$$

Thus, in this case for octane:

$$_{\text{liquid}}[\Delta U']_{0,\ \text{vapour}} = -5\ 124\ 856 + 41\ 496$$
$$= -5\ 083\ 360\ [\text{kJ/kmol}]$$

53

We have seen that values of $[\Delta U']_0$ (with added subscripts to denote phase of reactants and products) can be obtained by measuring the heat transfer after the reaction which brings the temperature of the products back to the temperature of the reactants. Thus we move from state 1 to state 2 in the figure below.

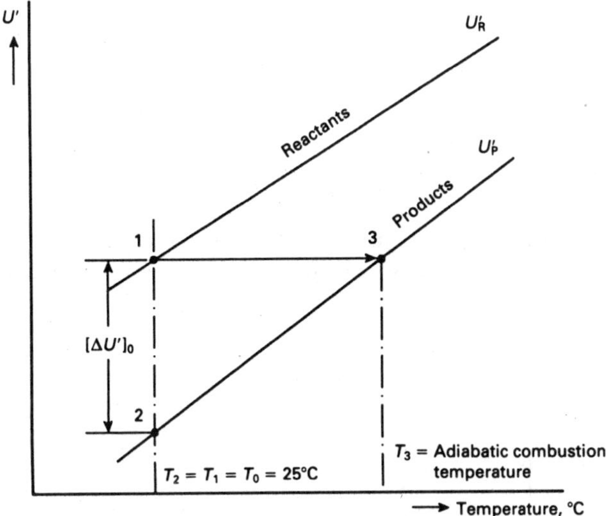

Since there is combustion we have to move from the internal energy curve of the reactants to that of the products, and if there is no heat transfer the internal energy is constant and we go to state 3. The temperature at state 3 is called the **adiabatic combustion temperature**. From the First Law equation:

$$Q - W = U_3 - U_1$$

with both Q and W zero. For the chemical reaction we have:

$$0 = U'_{P,3} - U'_{R,1}$$

Thus, with no boundary interaction of heat or work, the internal energy of reactants and products is the same as we can see in the figure.

Points 2 and 3 on the product curve are the extremes of what can happen in practice. Suppose there was a heat transfer equal to, say, half the value of $[\Delta U']_0$, how could we show the final state in the figure?

54

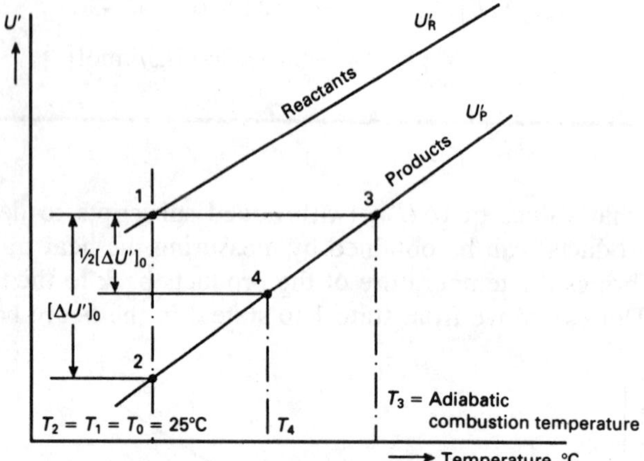

In the figure above, we have taken half the distance between points 1 and 2, and then drawn a horizontal line to the product curve, to give a final state 4. State points 2, 3 and 4 are achieved in three different situations; normally we call the final condition state 2. In the figure below we show just states 1 and 2 for this last case where there has been a heat transfer of half $[\Delta U']_0$. Next we must see how we can calculate T_2.

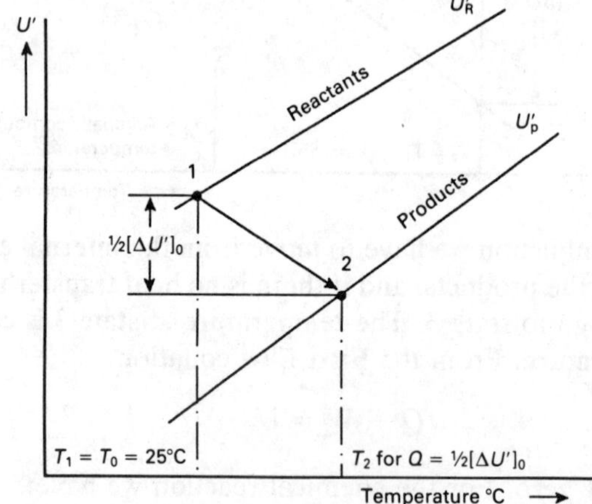

55

We take the First Law equation for a closed system, which we apply to the chemically reacting system, thus:

$$Q - W = U'_{P,2} - U'_{R,1}$$

Then we *subtract* and *add* $U'_{P,0}$ to the right-hand side, and *add* and *subtract* $U'_{R,0}$ to the right hand-side, as we did in Frame 45, to give:

$$Q - W = (U'_{P,2} - U'_{P,0}) - (U'_{R,1} - U'_{R,0}) + (U_{P,0} - U'_{R,0})$$

The last term on the right is equal to $[\Delta U']_0$, so we get:

$$Q - W = (U'_{P,2} - U'_{P,0}) - (U'_{R,1} - U'_{R,0}) + [\Delta U']_0$$

We then write the first two terms on the right-hand side in the form we used in Frame 45. We express the difference in internal energy between the two temperatures, 2 and 0, or 1 and 0, as the product of mass, specific heat and temperature difference, summed for all the different gases present. Thus:

$$Q - W = \Sigma\,(mc_V)_P(T_2 - T_0) - \Sigma\,(mc_V)_R(T_1 - T_0) + [\Delta U']_0$$

This is an important result that can be applied to a reacting closed system with any boundary interaction of Q and/or W. In our particular example we had a rigid boundary so W was zero, and state 1 was the datum state 0, so the equation simplifies to give:

$$Q = \Sigma\,(mc_V)_P(T_2 - T_0) + [\Delta U']_0$$

Study this last equation carefully. If $Q = [\Delta U']_0$, what will be the value of T_2? Secondly, if Q is zero, what can we say about the value of T_2? (Look at the figure in Frame 53.)

56

$T_2 = T_0$; T_2 is at a maximum

If you study this equation you should be able to see that if $Q = [\Delta U']_0$ then the product energy term must be zero, and it must follow that $T_2 = T_0$. Then if Q is zero, T_2 will have its highest possible value, governed by the contents of the product energy term. The final temperature T_2 will depend on the mass of product gases; if there is a lot of excess oxygen or air so that $\Sigma\,(mc_V)_P$ is large, then T_2 must be low. Thus the product temperature in combustion can be carefully controlled by varying the air/fuel ratio.

57

Now we will work through this example to find T_2 in a reaction in a closed system where there is a heat transfer. The gas ethylene, C_2H_4, burns in 50% excess air in a closed system, initially at 25°C. There is a heat transfer of −660 000 kJ/kmol, and for ethylene $[\Delta U']_0 = -1\ 323\ 170$ kJ/kmol, with the products in the vapour phase. What is the final temperature?

We may use the following values of c_V of the product gases as being mean values over the temperature range: CO_2 = 0.913, H_2O = 1.585, O_2 = 0.757, N_2 = 0.789, all in kJ/kg K.

As the reaction starts at 25°C we do not have to consider the reactants in the energy equation, so we can use the second equation in Frame 55.

First, what is the chemical equation for the stoichiometric combustion of ethylene, and then what is the equation with 50% excess air?

58

$$C_2H_4 + 3\ O_2 + 3 \times 3.76\ N_2 = 2\ CO_2 + 2\ H_2O + 11.28\ N_2$$
$$C_2H_4 + 4.5\ O_2 + 4.5 \times 3.76\ N_2 = 2\ CO_2 + 2\ H_2O + 1.5\ O_2 + 16.92\ N_2$$

One mole of C_2H_4 will give 2 moles each of CO_2 and H_2O requiring 3 moles of oxygen, so we shall need 4.5 moles of oxygen with 50% excess air.

Now we can use the energy equation from Frame 55:

$$Q = \sum (mc_V)_P(T_2 - T_0) + [\Delta U']_0$$

With $(mc_V)_P$ for the product gases in the order in which they occur in the chemical equation, we get:

$$\therefore\ -660\ 000 = (2 \times 44 \times 0.913 + 2 \times 18 \times 1.585 + 1.5 \times 32 \times 0.757$$
$$+ 16.92 \times 28 \times 0.789) \times (T_2 - T_0) + (-1\ 323\ 170)$$
$$\therefore\ +1\ 323\ 170 - 660\ 000 = 547.54 \times (T_2 - T_0)$$
$$\therefore\ (T_2 - T_0) = 1211.2\ \text{K}$$
$$\therefore\ T = 1236.2°\text{C}$$

Now use this as a model to solve a similar problem with ethylene, using the same specific heats. Ethylene initially at 25°C burns in a closed system in 150% excess air. During combustion there is a heat transfer of −350 000 kJ/kmol. Calculate the final temperature.

59

$$\boxed{1123.3°C}$$

Here is a summary of the solution. The chemical equation is:

$$C_2H_4 + 7.5\ O_2 + 28.2\ N_2 = 2\ CO_2 + 2\ H_2O + 4.5\ O_2 + 28.2\ N_2$$

Using the same energy equation:

$$\therefore\quad -350\ 000 = (2 \times 44 \times 0.913 + 2 \times 18 \times 1.585 + 4.5 \times 32 \times 0.757 + 28.2 \times 28 \times 0.789) \times (T_2 - T_0) + (-1\ 323\ 170)$$

$$\therefore\quad +1\ 323\ 170 - 350\ 000 = 869.41 \times (T_2 - T_0)$$

$$\therefore\quad (T_2 - T_0) = 1119.4\ \text{K}$$

$$\therefore\quad T_2 = 1144.4°\text{C}$$

In this example, the excess air is greater, tending to lower the final temperature, but the heat transfer is less, which by itself would raise the final temperature, so the answer is not far removed from the previous one.

60

The general equation for a closed reacting system in Frame 55 included a work term, so we may imagine the equation applied to a fuel/air mixture in an engine cylinder with displacement work being done as the piston recedes down the cylinder. In the figure below, the closed system is shown in its initial and final states, and these states are also shown on the energy–temperature diagram in the next frame.

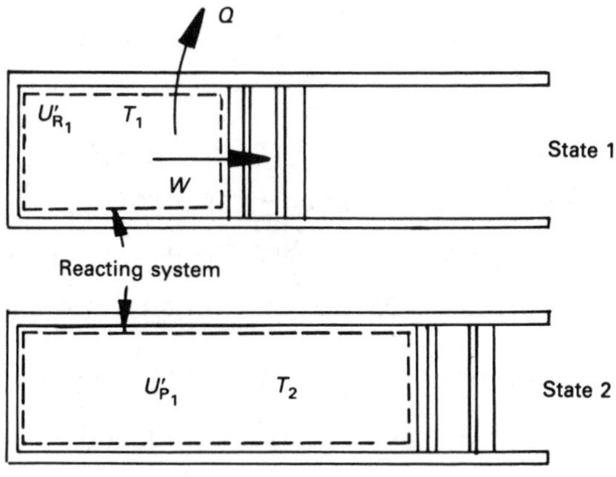

61

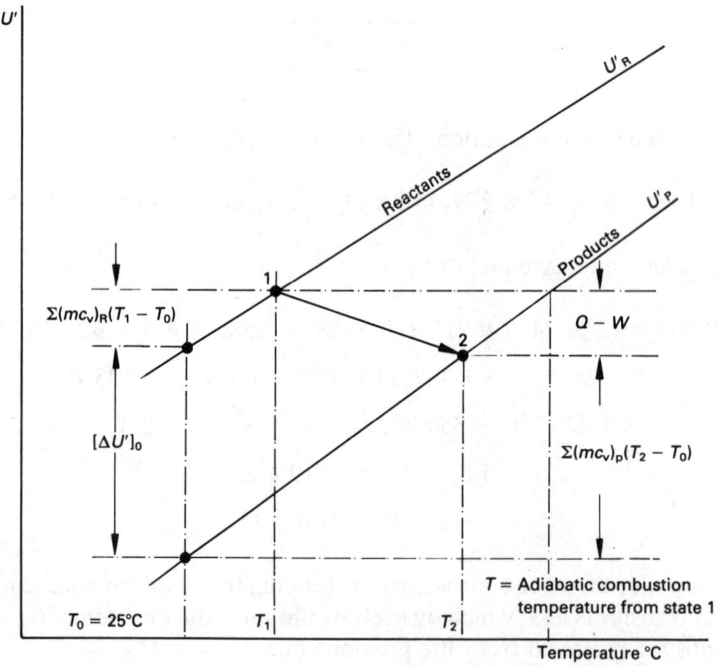

As Q from a system is negative and W by or from the system is positive, a work of expansion (as we have already seen in earlier programmes) will lower the final temperature in the same way that Q does. In the examples in Frames 57–59, the same final temperature will result in each case if conditions are adiabatic and W is equal to the previous value of Q.

In the example with 150% excess air, if T_2 is 1100°C, and Q is $-180\ 000$ kJ, let's see what W will be.

The chemical equation is:

$$C_2H_4 + 7.5\ O_2 + 7.5 \times 3.76\ N_2 = 2\ CO_2 + 2\ H_2O + 4.5\ O_2 + 28.2\ N_2$$

The energy equation is:

$$Q - W = \Sigma\ (mc_V)_P(T_2 - T_0) + [\Delta U']_0$$

$$\therefore\ Q - W = (2 \times 44 \times 0.913 + 2 \times 18 \times 1.585 + 4.5 \times 32 \times 0.757 + 28.2 \times 28 \times 0.789) \times 1075 + (-1\ 323\ 170)$$

$$\therefore\ -180\ 000 - W = 869.41 \times 1075 - 1\ 323\ 170$$

$$\therefore\ W = +208\ 554\ [\text{kJ/kmol}]$$

62

Now consider the same reaction, but with 100% excess air, with $Q = -150\ 000$ kJ/kmol and $W = +530\ 000$ kJ/kmol. What is T_2?

The answer is 932.83°C.

The chemical equation is:

$$C_2H_4 + 6\ O_2 + 6 \times 3.76\ N_2 = 2\ CO_2 + 2\ H_2O + 3\ O_2 + 22.56\ N_2$$

The energy equation, as before, is:

$$Q - W = \sum (mc_V)_P\ (T_2 - T_0) + [\Delta U']_0$$

$$\therefore\ Q - W = (2 \times 44 \times 0.913 + 2 \times 18 \times 1.585 + 3 \times 32 \times 0.757 + 22.56 \times 28 \times 0.789) \times (T_2 - T_0) + (-1\ 323\ 170)$$

$$\therefore\quad -150\ 000 - 530\ 000 = 708.47 \times (T_2 - T_0) - 1\ 323\ 170$$

$$\therefore\quad (T_2 - T_0) = 907.83\ \text{K}$$

$$\therefore\quad T_2 = 932.83\text{°C}$$

So far we have considered closed systems involving heat and work transfers and the internal energies of reacting gases or vapours. The figure in Frame 2 depicted an open system through which flowed streams of air and fuel. Which energy equation would we need for that?

63

The steady flow energy equation

For this we would use the First Law equation for open systems, or the steady flow energy equation, which, as you know, involves enthalpy rather than internal energy terms. We first met this equation in Programme 3. A reacting system with steady flow in a control volume is shown below.

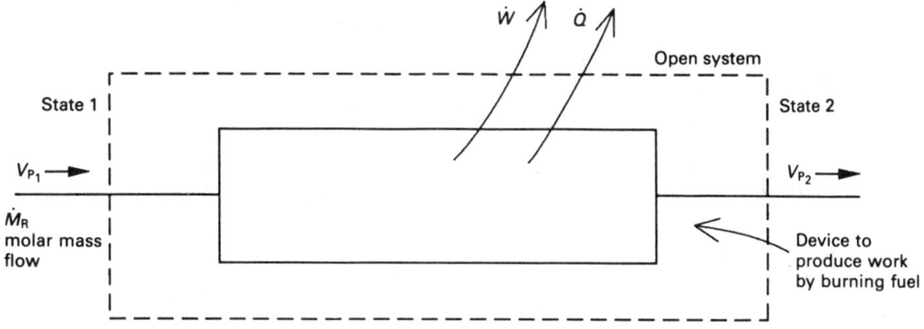

64

Assuming no change in potential energy, the energy equation would be:

$$\dot{Q} - \dot{W} = H'_{P,2} - H'_{R,1} + \dot{M}_R\left(\frac{V^2_{P,2} - V^2_{R,1}}{2}\right)$$

which has units of W or kW. This is written for 1 mole of reactants of mass $\dot{M}_R$ flowing per second. In many cases it would also be possible to neglect the kinetic energy terms.

All that we have done so far involving the First Law equation for closed systems (described in Frames 42–44) may be repeated using the steady flow energy equation for open systems. Thus with combustion in steady flow with a heat transfer and no work so that the products are at the same datum temperature of 25°C as the reactants, then the heat transfer measured is $[\Delta H']_0$ for the combustion of 1 kmol of reactant. Kinetic energy terms are assumed negligible.

The energy equation is then:

$$\begin{aligned} Q &= H'_{P,0} - H'_{R,0} \\ &= [\Delta H']_0 \end{aligned}$$

A diagram with enthalpy terms, similar to the figure in Frame 44 with internal energy terms, is given below. This shows the term $[\Delta H']_0$, and the lines showing the variation of enthalpy of the products and reactants with temperature.

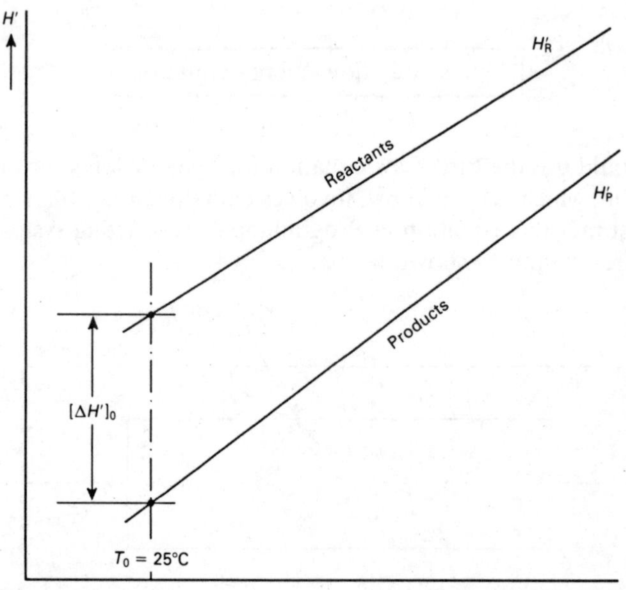

65

The term $[\Delta H']_0$ is the enthalpy of reaction at 25°C. It has a negative value and units of kJ/kmol. Values for common reactants are given on page 21 of the Rogers and Mayhew steam tables. Values of **calorific value** for similar reactions are given in the Haywood tables. This is defined later in Frame 84.

Values of $[\Delta H']$ at other temperatures may be found from $[\Delta H']_0$, using a method similar to that in Frame 45. Thus we may write:

$$[\Delta H']_1 = [\Delta H']_0 + (H'_{P,1} - H'_{P,0}) - (H'_{R,1} - H'_{R,0})$$

You can check that we can get this equation by adding and subtracting like terms as we did before. The last two terms on the right again may be calculated from the masses and specific heats of the constituents and the temperature difference between state 1 and state 0. Thus:

$$[\Delta H']_1 = [\Delta H']_0 + \sum (mc_P)_P(T_1 - T_0) - \sum (mc_P)_R(T_1 - T_0)$$

An example involving this equation would be exactly similar to the example in Frames 46 and 47. Try this one. $[\Delta H']_0$ for the combustion of ethane, C_2H_6, is −1 427 860 kJ/kmol, at 25°C. What is $[\Delta H']_1$ at 325°C? Mean values of c_P for the gases involved are: C_2H_6: 2.402, O_2: 0.956, N_2: 1.049, CO_2: 0.978, H_2O: 1.926, all in kJ/kg K units. First write the chemical equation, and then work out the enthalpy terms to find the answer. If you need them, molecular masses are given in Frame 23.

66

$C_2H_6 + 3.5\ O_2 + 13.17\ N_2 = 2\ CO_2 + 3\ H_2O + 13.17\ N_2$
$[\Delta H']_1 = -1\ 424\ 579$ kJ/kmol

We met the chemical equation for the combustion of ethane in Frame 17. The second part of the solution is given in the next frame.

67

The enthalpy terms are:
For products: $\Sigma\ (mc_P)_P(T_1 - T_0) = (2 \times 44 \times 0.978 + 3 \times 18 \times 1.926 + 13.17 \times 28 \times 1.049) \times (325 - 25)$

$$= 576.9 \times 300$$

$$= 173\ 069\ [\text{kJ/kmol}]$$

For reactants: $\Sigma\ (mc_P)_R(T_1 - T_0) = (30 \times 2.402 + 3.5 \times 32 \times 0.956 + 13.17 \times 28 \times 1.049) \times (325 - 25)$

$$= 565.96 \times 300$$

$$= 169\ 788\ [\text{kJ/kmol}]$$

Then: $[\Delta H']_1 = -1\ 427\ 860 + 173\ 069 - 169\ 788$

$$= -1\ 424\ 579\ [\text{kJ/kmol}]$$

68

We have seen that values of $[\Delta H']_0$ are tabulated in tables; what about values of $[\Delta U']_0$ that we have used previously? The answer is that we can find these from the measured values of $[\Delta H']_0$, so it is not in fact necessary to measure $[\Delta U']_0$ in separate experiments. You will recall the relationship between specific enthalpy and specific internal energy:

$$h = u + pv$$

Or, for some given mass:

$$H = U + pV$$

For a reacting system we may write:

$$[\Delta H']_0 = [\Delta U']_0 + \Delta[pV']_0$$

The term $\Delta[pV']_0$ is the change in pV between reactant and product states. For the reactants:

$$p_R V'_R = n_R\, R\, T_0$$

and for the products:

$$p_P V'_P = n_P\, R\, T_0$$

Hence $\Delta[pV']_0$ will be given by:

$$\Delta[pV']_0 = [p_P V'_P - p_R V'_R]_0$$
$$= (n_P - n_R)\, R\, T_0$$
$$\therefore \qquad [\Delta H']_0 = [\Delta U']_0 + (n_P - n_R)\, R\, T_0$$

From this result we can see that when the number of moles of reactants and products are the same, $[\Delta H']_0$ and $[\Delta U']_0$ are also the same. Is this true for the combustion of methane? (See Frame 9.)

69

Yes

The chemical equation is:

$$CH_4 + 2\ O_2 = CO_2 + 2\ H_2O$$

so there are 3 moles of both reactants and products. If we had included the 7.52 moles of nitrogen on both sides of the equation, the result would be the same.

Consider the combustion of octane vapour with reactants and products in the vapour phase:

$$C_8H_{18} + 12\tfrac{1}{2}\ O_2 = 8\ CO_2 + 9\ H_2O$$

For this reaction. $[\Delta H']_0 = -5\ 116\ 180$ kJ/kmol, and there are $13\frac{1}{2}$ moles of reactants and 17 moles of products. To find $[\Delta U']_0$ we may write:

$$[\Delta U']_0 = [\Delta H']_0 - (n_P - n_R)\, R\, T_0$$
$$= -5\ 116\ 180 - (17 - 13.5) \times 8.3145 \times 298.15$$
$$= -5\ 124\ 856\ [\text{kJ/kmol}]$$

$[\Delta H']_0$ for benzene vapour, C_6H_6, is $-3\ 169\ 540$ kJ/kmol. What is $[\Delta U']_0$? You need only consider the oxygen required to find n_P and n_R.

3 170 779 kJ/kmol

The chemical equation is:

$$C_6H_6 + 7\tfrac{1}{2}\,O_2 = 6\,CO_2 + 3\,H_2O$$

so there are 9 moles of products and $8\frac{1}{2}$ moles of reactants. We then have:

$$[\Delta U']_0 = -3\,169\,540 - (9 - 8.5) \times 8.3145 \times 298.15$$
$$= -3\,170\,779\ [\text{kJ/kmol}]$$

The value of $[\Delta H']_0$, like the value of $[\Delta U']_0$, is affected by phase changes, so for products in the liquid or vapour phase, with reactants in the liquid phase, we have:

$$_{\text{liquid}}[\Delta H']_{0,\ \text{liquid}} = {}_{\text{liquid}}[\Delta H']_{0,\ \text{vapour}} - m\,h_{\text{fg, products}}$$

Similarly, when the phase of the reactants may change we have:

$$_{\text{liquid}}[\Delta H']_{0,\ \text{vapour}} = {}_{\text{vapour}}[\Delta H']_{0,\ \text{vapour}} + m\,h_{\text{fg, reactants}}$$

The subscript before $[\Delta H']_0$ indicates the phase of the reactants; the subscript after indicates the phase of the products.

These differences in the values of $[\Delta H']_0$ are shown in the figure below.

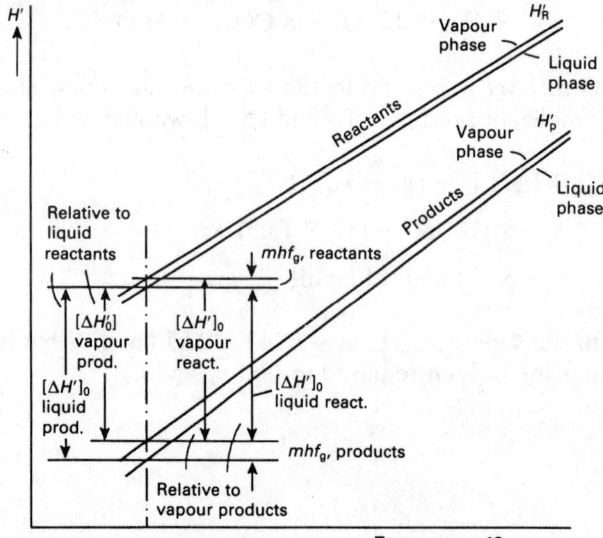

In the combustion of octane, considered in Frame 69, $[\Delta H']_0$ with products (water) in the vapour phase is $-5\ 116\ 180$ kJ/kmol. What is the corresponding value with the water condensed? First you need to find how much water is produced by the combustion of one mole of octane, C_8H_{18}. You will also need to look up h_{fg} for steam at 25°C.

71

$-5\ 511\ 752$ kJ/kmol

The reaction, given in Frame 69, is:

$$C_8H_{18} + 12\tfrac{1}{2}\ O_2 = 8\ CO_2 + 9\ H_2O$$

Thus we have 9 moles of water produced, and h_{fg} at 25°C is 2441.8 kJ/kg. We shall use the equation:

$$[\Delta H']_{0,\ \text{liquid}} = [\Delta H']_0,\ \text{vapour} - m\ h_{fg,\ \text{products}}$$

$$\therefore \quad [\Delta H']_{0,\ \text{liquid}} = -5\ 116\ 180 - 9 \times 18 \times 2441.8$$

$$= -5\ 511\ 752\ [\text{kJ/kmol}]$$

72

We shall now consider combustion in steady flow, with possible heat and work interactions. The energy equation will have similarities with that for a closed system which we studied in Frames 55–62. Thus if we neglect potential and kinetic energy terms, for 1 kmole of reactants we have:

$$\dot{Q} - \dot{W} = H'_{P,2} - H'_{R,1}$$

By adding and subtracting like terms at the datum temperature for the reacting system:

$$\dot{Q} - \dot{W} = (H'_{P,2} - H'_{P,0}) - (H'_{R,1} - H'_{R,0}) + (H'_{P,0} - H'_{R,0})$$

$$= (H'_{P,2} - H'_{P,0}) - (H'_{R,1} - H'_{R,0}) + [\Delta H']_0$$

What do we do next?

73

We then write the first two terms on the right-hand side in the form we have used previously, thus:

$$\dot{Q} - \dot{W} = \sum (mc_P)_P(T_2 - T_0) - \sum (mc_P)_R(T_1 - T_0) + [\Delta H']_0$$

The terms in this important relationship appear in the figure below.

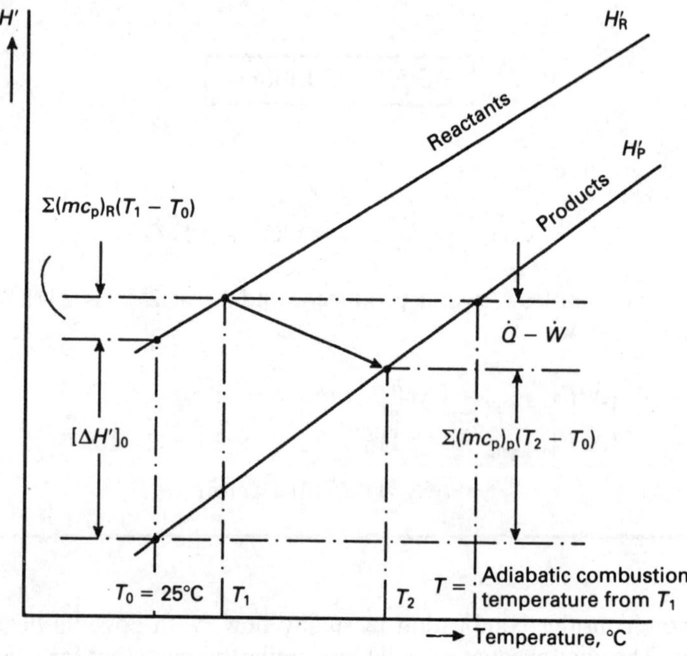

74

The use of this equation is illustrated in this example. A gas turbine engine burns methane with 350% excess air, and the air flow is 60 kg/s. There is a heat transfer from the machine of 296 kW, and 10 MW of shaft power is produced. Assume both air and fuel enter 320 K; what is the exhaust gas temperature? $[\Delta H']_0$ with reactants and products in the vapour phase is −802 310 kJ/kmol. The following values of c_P may be used, all in kJ/kg K units. Reactants: CH_4 = 2.226, O_2 = 0.918, N_2 = 1.040. Products: CO_2 = 1.102, O_2 = 1.017, N_2 = 1.086, H_2O = 2.047. Oxygen and nitrogen appear in both reactants and products because of the two different temperature ranges.

First we have to write the chemical equation. One mole of methane requires 2 moles of oxygen. How much oxygen and nitrogen will there be in the reactants with 350% excess air?

75

$$\boxed{9\ O_2 + 33.84\ N_2}$$

With 2 oxygen moles needed, 350% excess air means another 7 oxygen moles will be present. This must be accompanied by 9 × 3.76 = 33.84 moles of nitrogen. The complete chemical equation is:

$$CH_4 + 9\ O_2 + 33.84\ N_2 = CO_2 + 2\ H_2O + 7\ O_2 + 33.84\ N_2$$

Next we need to establish the energy equation, written for 1 mole of methane, so we have to find the values of $\dot{Q}$ and $\dot{W}$. We know that the air flow is 60 kg/s. In the chemical equation the mass of air is:

$$9 \times 32 + 33.84 \times 28 = 1235.52\ [\text{kg}]$$

The mass of methane is 16 kg, so for the mass flow rate of air of 60 kg/s, the mass flow rate of methane is then:

$$\frac{16}{1235.52} \times 60\ [\text{kg/s}] = 0.777\ [\text{kg/s}]$$

For these flow rates the heat transfer is 296 kW, and the power is 10 MW. So what would Q and W be, in units of kJ/kmol, for a flow rate of 16 kg of methane per second?

76

$$\boxed{Q = -6\ 095.24\ \text{kJ/kmol},\ W = 205\ 920.2\ \text{kJ/kmol}}$$

These results are given by:

$$Q = \frac{-296\ [\text{kJ/s}]}{0.777\ [\text{kg/s}]} \times 16\ [\text{kg/kmol}] = -6095.24\ [\text{kJ/kmol}]$$

and:

$$W = \frac{10\ 000\ [\text{kJ/s}]}{0.777\ [\text{kg/s}]} \times 16\ [\text{kg/kmol}] = 205\ 920.2\ [\text{kJ/kmol}]$$

77

We can now consider the energy equation with T_2 unknown, $T_1 = 320$ K, or 46.85°C, and $T_0 = 25$°C. First we have to work out the terms $\Sigma\,(mc_P)_P(T_2 - T_0)$ and $\Sigma\,(mc_P)_R(T_1 - T_0)$. From the chemical equation in Frame 75, for the products we have:

$$\sum (mc_P)_P(T_2 - T_0)$$

$$= (44 \times 1.102 + 36 \times 2.047 + 7 \times 32 \times 1.017 + 33.84 \times 28 \times 1.086) \times (T_2 - T_0)$$

$$= 1378.99 \times (T_2 - 25)$$

T_2 is the unknown we are trying to find. What is the corresponding term for the reactants?

78

28 086.43 kJ/kmol

For the reactants:

$$\sum (mc_P)_R\,(T_1 - T_0)$$

$$= (16 \times 2.226 + 9 \times 32 \times 0.918 + 33.84 \times 28 \times 1.040) \times (T_1 - T_0)$$

$$= 1285.42 \times (46.85 - 25)$$

$$= 28\,086.43\ [\text{kJ/kmol}]$$

We can now complete the energy equation:

$$\dot{Q} - \dot{W} = 1378.99 \times (T_2 - 25) - 28\,086.43 + [\Delta H']_0$$

$$\therefore\ -6095.24 - (+205\,920.2) = 1378.99 \times (T_2 - 25) - 28\,086.43 - 802\,310$$

where the units are kJ/kmol.

$$\therefore \qquad T_2 = 473.43\text{°C}$$

79

This example may be used as a model for similar problems, such as part (c) of problem **8** in Frame 89.

We must now move on to see what the Second Law has to say about combustion processes. From the First Law we can see how to calculate the temperature of combustion products given the magnitude of any heat and work interactions at the boundary of the closed or open system, but the First Law will not tell us how much work we might expect to get by burning fuel in an engine. Thus, the First Law energy equation of Frame 73 does not tell us how much of $[\Delta H']_0$ may be converted into work, so we have to evoke the Second Law to find out.

In Frame 76 of Programme 6 we saw that in general in irreversible changes:

$$dQ \leqslant T\,dS$$

Combustion is a good example of an irreversible change, so for a chemically reacting system we may write:

$$dQ \leqslant T\,dS'$$

Applying the First Law to chemically reacting systems, for a closed system (a) in the absence of work, we have:

$$dQ = dU'$$

and (b) at constant pressure, or for an open system per unit mass flow in the absence of work, we have:

$$dQ = dH'$$

Substituting for dQ in these two cases we get:

$$dU' - T\,dS' \leqslant 0 \qquad \text{(a)}$$

and

$$dH' - T\,dS' \leqslant 0 \qquad \text{(b)}$$

for closed (a) and open (b) systems respectively, in the absence of work.

If we now think about chemical reactions in which the reactants and products are at the same temperature, the previous two equations for infinitesimal changes may be integrated to give:

$$[\Delta U'] - T\,[\Delta S'] \leqslant 0 \qquad \text{for (a)}$$

and

$$[\Delta H'] - \mathrm{T}\,[\Delta S'] \leqslant 0 \qquad \text{for (b)}$$

which in terms of reactants and products may be expressed as:

$$(U' - T\,S')_P - (U' - T\,S')_R \leqslant 0 \qquad \text{for (a)}$$

and

$$(H' - T\,S')_P - (H' - T\,S')_R \leqslant 0 \qquad \text{for (b)}$$

80

These statements tell us that isothermal chemical reactions are possible for cases (a) and (b) only if the term for the products is less than the term for the reactants. If in a proposed reaction the term for the supposed products is greater than the term for the reactants, then there is no reaction. Thus, the Second Law is telling us what is possible, and what is not.

For the simple chemical equation:

$$H_2 + \frac{1}{2} O_2 = H_2O$$

with water in the vapour phase, $[\Delta H'] - T[\Delta S']$ is $-227\ 590$ kJ/kmol at 25°C. The calculation involves finding the difference between entropy values of products and reactants, but you need not worry about how this is done, since at this level you only need an understanding in general terms of what the Second Law will tell us. The important result is that the change is negative and, as we know, the reaction takes place. Also at 25°C, $[\Delta H']_0$ is $-241\ 830$ kJ/kmol, so that the effect of the entropy change is relatively small. The numerical value of $[\Delta H'] - T[\Delta S']$ in this and in all reactions depends on the temperature of the reaction, and in fact at high temperatures the negative value starts to get smaller, so at these temperatures the greatest negative value occurs *before* the reaction is complete, and the reaction stops at this point.

This is called **dissociation**, as if the reaction had been completed and had then partially reversed. The amount of dissociation increases with the temperature. So if we have a reaction vessel full of steam, which is the product of combustion of hydrogen in oxygen, and we heat it to a high temperature, as the temperature rises the vessel will contain increasing amounts of hydrogen and oxygen, and so the reaction is going in the reverse direction. The amount of hydrogen and oxygen produced depends on the greatest negative value of $[\Delta H'] - T[\Delta S']$ at the particular temperature in the vessel.

It is also possible to produce hydrogen and oxygen from water by using electricity. What is that process called?

81

Electrolysis

The value of $[\Delta H'] - T[\Delta S']$ for the spontaneous decomposition of water into hydrogen and oxygen at 25°C is positive, so, as we know, this does not happen. We have seen however that at high temperature the value does become negative so decomposition occurs. The equations in Frame 79 were derived for spontaneous processes in the absence of any work terms. When external work is done on a system (which is *negative* work for the system), positive changes in $[\Delta H'] - T[\Delta S']$ are then possible, so hydrogen and oxygen may be obtained by passing an electric current through water containing a small amount of sulphuric acid between copper electrodes.

Now we shall think about the *positive* work done by a chemically reacting system.

82

All types of internal combustion engine, reciprocating and rotary, can be treated as an open system, case (b) in Frame 79, as shown below.

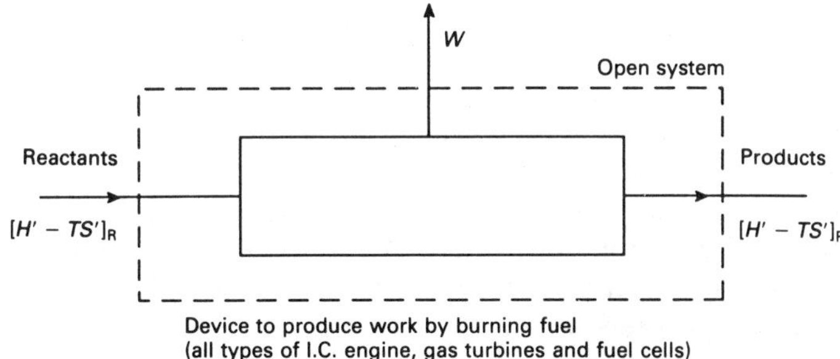

Device to produce work by burning fuel
(all types of I.C. engine, gas turbines and fuel cells)

Neglecting changes in kinetic and potential energy, the steady flow energy equation gives, for unit mole flowing per second:

$$dQ - dW = dH'$$

which is combined with $dQ \leqslant T\,dS'$ to give:

$$dW \leqslant -\,dH' + T\,dS'$$

For an isothermal change this is integrated as before, to give:

$$W \leqslant -([\Delta H'] - T[\Delta S'])$$

Here, the first term is $-[\Delta H']_0$ if the reaction is at 25°C, which for the hydrogen–oxygen reaction is $-(-241\,830)$, or $+241\,830$ kJ/kmol. This is given on page 21 of the Mayhew and Rogers steam tables. The whole of the right-hand side, $-([\Delta H]' - T[\Delta S'])$, for the same reaction at 25°C is $-(-227\,590)$, or $+227\,590$ kJ/kmol. This value came from Frame 80. This equation is telling us that the positive work possible from an isothermal reaction is equal to, or less than, this amount. The maximum work possible is therefore equal to this amount.

83

By the Second Law, the maximum work possible in an isothermal reaction is:

$$W_{max} = -([\Delta H'] - T[\Delta S'])$$

The term $-([\Delta H') - T[\Delta S'])$ represents the difference between $(H' - TS')$ for the products and reactants of the chemically reacting system. The group of properties $(H - TS)$ is the Gibbs function, G, or the Gibbs free energy.

Thus, for the chemically reacting system: $W_{max} = -[\Delta G']_0$

84

In Frame 82 we have seen that in the combustion of hydrogen, as an example of a combustion process, the maximum possible work by the Second Law in an isothermal reaction at 25°C is +227 590 kJ/kmol. The term $-[\Delta H']_0$ for the same reaction is 241 830 kJ/kmol. For all fuels, the difference between $-[\Delta H']_0$ and $-[\Delta G']_0$ is small. In practical engineering terms, in order to deduce an efficiency by which a device produces work from the combustion of fuel, a term called the **calorific value** or **heating value**, CV, is used, where:

$$\text{CV} \equiv -[\Delta H']_0$$

While the units of $[\Delta H']_0$ are kJ/kmol, the units of CV are usually kJ/kg.

We have taken the hydrogen–oxygen reaction as an example and we have seen that the figures for $-[\Delta H']_0$ and $-[\Delta G']_0$ are not very different. Consequently by the Second Law the maximum work obtainable in an isothermal reaction is close to $-[\Delta H']_0$, and therefore close to the calorific value of the fuel. This is shown on the opposite page.

In practice, much less work is obtained because the reaction is not isothermal. The figure also shows the work for adiabatic combustion with the products at T_P.

When using calorific values, we have to know whether water in the products is in the vapour or liquid phase. Thus we have a **higher** or a **lower** calorific value:

$$\text{HCV} = \text{LCV} + m\, h_{fg} \quad \text{[kJ/kg]}$$

where m is the mass of water involved and h_{fg} is the latent heat of evaporation, usually taken as 2442 kJ/kg at 25°C.

If the LCV of octane, C_8H_{18}, is 44.878 MJ/kg (with water in the vapour phase), what is the HCV, with the water condensed? 1 mole, or 114 kg of octane will produce 9 moles of water. Take h_{fg} as 2442 kJ/kg.

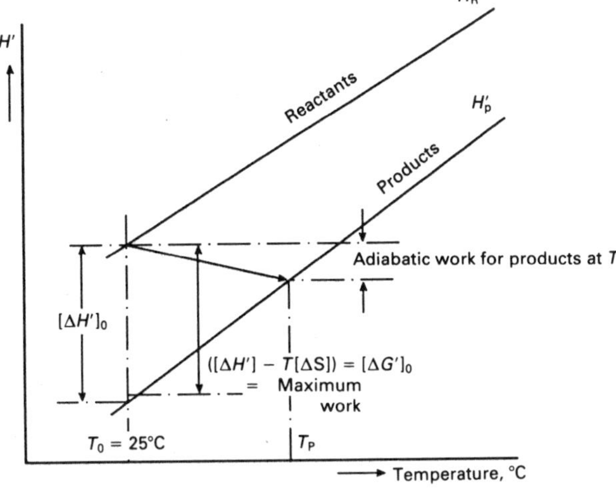

85

48.348 MJ/kg

114 kg of octane will produce 9 × 18 or 162 kg of water. So, per kg of octane there will be 162/114 or 1.421 kg of water. Thus:

$$\begin{aligned} HCV &= LCV + m\,h_{fg} \\ &= 44.878 + 1.421 \times 2442 \times 10^{-3} \text{ MJ/kg} \\ &= 48.348 \text{ [MJ/kg]} \end{aligned}$$

We can say that the **thermal efficiency** of a fuel-burning device, from our use of the Second Law, concluded in Frames 83 and 84, is:

$$\eta_{th} = \frac{\text{Work produced}}{\text{Maximum work possible}}$$

$$= \frac{W}{W_{max}} = -\frac{W}{[\Delta G']_0} \simeq \frac{W}{CV}$$

since we have seen that the maximum work obtainable in an isothermal reaction is close to the calorific value of the fuel.

86

We can say, therefore, that the **thermal efficiency** of an internal combustion device is:

$$\eta_{th} = \frac{\text{Work produced}}{\text{Energy equivalent of fuel used}}$$

This equation must be dimensionless, so we may use specific work [kJ/kg of fuel], and the energy equivalent of unit mass of fuel [kJ/kg]. As the maximum possible work is close to the energy equivalent of the fuel, the maximum possible thermal efficiency in an isothermal reaction is close to 100%. The figure for the hydrogen–oxygen reaction, using the values of $[\Delta H']_0$ and $[\Delta G']_0$ from Frame 84, is:

$$\eta_{th,\,max} = \frac{227\,590}{241\,830} \times 100\% = 94.11\%$$

An engine burns 13.71 kg of fuel per hour of calorific value 45 MJ/kg and gives an output of 60 kW. What is the thermal efficiency of the engine?

87

35%

The work output is 60 kJ/s, and the fuel used is 13.71/3600 kg/s, which is 0.003 808 kg/s, so the specific work is:

$$\frac{60}{0.003\,808} = 15\,754.9 \text{ [kJ/kg]}$$

The thermal efficiency is then:

$$\eta_{th} = \frac{15\,754.9}{45\,000} = 35\%$$

In the next programme we shall examine internal combustion engine devices to see how they perform, so that we can decide whether this result is good or not. The fuel cell, which has the same practical objectives as the internal combustion engine, operates at rather better efficiencies.

88

Here is a summary of the relationships introduced in this programme. As in all programmes, it is important to grasp the principles involved, rather than to try and memorise with little understanding.

Chemical equations may be built up from two basic reactions:

$$C + O_2 = CO_2 \quad \text{and} \quad H_2 + \frac{1}{2} O_2 = H_2O$$

The content of air may be expressed:

$$4.76 \text{ moles of air} = O_2 + 3.76\, N_2 \qquad 1 \text{ mole of air} = 0.21\, O_2 + 0.79\, N_2$$

When there is more air than required:

$$\text{Excess air} = \frac{(A/F) - (A/F)_{st}}{(A/F)_{st}} \times 100\%$$

When the mixture is either rich or weak:

$$\text{Mixture strength} = \frac{(A/F)_{st}}{(A/F)} \times 100\%$$

$$\text{Air fuel equivalence ratio, } \phi = \frac{(F/A)}{(F/A)_{st}}$$

The increase in internal energy of reaction is:

$$U'_{P,0} - U'_{R,0} = [\Delta U']_0$$

Values of $[\Delta U']$ at other temperatures may be found from:

$$[\Delta U']_1 = [\Delta U']_0 + \sum (mc_V)_P(T_1 - T_0) - \sum (mc_V)_R(T_1 - T_0)$$

The difference between values of $[\Delta U']_0$ with products in the vapour or liquid phase is:

$$[\Delta U']_{0,\text{ liquid}} = [\Delta U']_{0,\text{ vapour}} - m\, u_{fg,\text{ products}}$$

The difference with reactant in the vapour or liquid phase and products in the vapour phase is:

$$_{\text{liquid}}[\Delta U']_{0,\text{ vapour}} = {}_{\text{vapour}}[\Delta U']_{0,\text{ vapour}} + m\, u_{fg,\text{ reactants}}$$

The energy equation for a reaction in a closed system is:

$$Q - W = \sum (mc_V)_P (T_2 - T_0) - \sum (mc_V)_R (T_1 - T_0) + [\Delta U']_0$$

The increase of enthalpy of reaction is:

$$[\Delta H']_0 = H'_{P,0} - H'_{R,0}$$

Values at other temperature are found from:

$$[\Delta H']_1 = [\Delta H']_0 + \sum (mc_P)_P (T_1 - T_0) - \sum (mc_P)_R (T_1 - T_0)$$

The relation between $[\Delta H']$ and $[\Delta U']$ is:

$$[\Delta H']_0 = [\Delta U']_0 + (n_P - n_R) R T_0$$

Changes in value with products in the liquid or vapour phase:

$$[\Delta H']_{0,\text{ liquid}} = [\Delta H']_{0,\text{ vapour}} - m h_{fg,\text{ products}}$$

With products in the vapour phase and reactants in the liquid or vapour phase:

$$_{\text{liquid}}[\Delta H']_{0,\text{ vapour}} = {}_{\text{vapour}}[\Delta H']_{0,\text{ vapour}} + m h_{fg,\text{ reactants}}$$

The energy equation for reaction in steady flow is:

$$\dot{Q} - \dot{W} = \sum (mc_P)_P (T_2 - T_0) - \sum (mc_P)_R (T_1 - T_0) + [\Delta H']_0$$

For spontaneous reactions in closed (a) or open (b) systems:

$$(U' - T S')_P - (U' - T S')_R \leqslant 0 \qquad \text{for (a)}$$

and

$$(H' - T S')_P - (H' - T S')_R \leqslant 0 \qquad \text{for (b)}$$

By the Second Law, the maximum work possible in an isothermal reaction is:

$$W_{max} = -([\Delta H'] - T[\Delta S']) = -[\Delta G']_0$$

The calorific value of fuel is: $CV \equiv -[\Delta H']_0$
The efficiency of a work producing device is:

$$\eta_{th} = \frac{W}{W_{max}} = -\frac{W}{[\Delta G']_0} \simeq \frac{W}{CV}$$

To conclude this programme make sure you can do the following problems.

1. In the combustion of ethylene, C_2H_4, $[\Delta U']_0$ is $-1\ 323\ 170$ kJ/kmol at 25°C with products in the vapour phase. What is $[\Delta U']_1$ at 226.85°C (500 K)? Use the following values of c_V: $C_2H_4 = 1.595$; $O_2 = 0.681$; $CO_2 = 0.750$; $H_2O = 1.439$, all in kJ/kg K units. [−1 321 598 kJ/kmol]
2. For the above reaction, calculate $[\Delta U']_1$ with the products in the liquid phase. Take $u_{fg} = 1634$ kJ/kg at 226.85°C. [−1 380 422 kJ/kmol]
3. In the combustion of octane, C_8H_{18}, $[\Delta H']_0$ is $-5\ 116\ 180$ kJ/kmol, with both reactants and products in the vapour phase. Calculate $[\Delta H']_0$ with both reactants and products in the liquid phase. h_{fg} for C_8H_{18} is 364 kJ/kg. [−5 470 288 kJ/kmol]
4. For the combustion of ethane, C_2H_6, $[\Delta H']_0$ is − 1 427 860 kJ/kmol, with reactants and products in the vapour phase. Calculate $[\Delta H']_1$ at 195°C, (a) with products in the vapour phase and (b) with products in the liquid phase. Use the following values of c_P: $C_2H_6 = 2.199$, $O_2 = 0.941$, $CO_2 = 0.939$, $H_2O = 1.091$, all in kJ/kg K. [(a) −1 425 492 kJ/kmol, (b) −1 531 332 kJ/kmol]
5. For the given values of $[\Delta H']_0$ for the combustion of the following substances, calculate $[\Delta U']_0$. (a) H_2: −241 830; (b) C_2H_6: −1 427 860; (c) C_3H_8: −2 039 224; (d) C_6H_{14}: −3 878 170, all in kJ/kmol. [(a) −240 591, (b) −1 429 099, (c) −2 041 703, (d) −3 886 846 kJ/kmol]
6. Propane, C_3H_8, burns adiabatically in a closed system with 150% excess air, and with no work transfer. Calculate the temperature achieved with the reactants at 25°C. For the reaction, $[\Delta U']_0 = -2\ 041\ 703$ kJ/kmol. Use the following values of c_V: $CO_2 = 1.045$, $O_2 = 0.830$, $N_2 = 0.870$, $H_2O = 1.776$, all in kJ/kg K units. [1293.2°C]
7. Hexane, C_6H_{14}, burns with 100% excess air in a closed system. There is a work transfer of $+1.5 \times 10^6$ kJ/kmol, and no heat transfer. With the reactants at 25°C, calculate the final temperature. $[\Delta U']_0$ is −3 886 846 kJ/kmol. Use the following values of c_V: $CO_2 = 1.015$, $O_2 = 0.814$, $N_2 = 0.849$, $H_2O = 1.755$, all in kJ/kg K units. [1005.3°C]
8. Butane, C_4H_{10}, burns with 200% excess air in a gas turbine engine, which produces 6 MW of shaft power, and gives off 49.5 kW in heat. The air flow through the machine is 20 kg/s. $[\Delta H']_0 = -2\ 631\ 054$ kJ/kmol, and assume the reactants enter at 25°C. Use the following values of c_P: $CO_2 = 1.075$, $O_2 = 1.003$, $N_2 = 1.075$, $H_2O = 2.015$, all in kJ/kg K units. Calculate (a) the dry exhaust gas analysis, (b) the air/fuel ratio by mass, and (c) the temperature of the gases leaving the turbine. [(a) 4.43% CO_2, 14.39% O_2, 81.18% N_2, (b) 46.15/1, (c) 634.2°C]
9. In the previous example, it is required to reduce the temperature of the exhaust products to 500°C by weakening the overall air/fuel ratio. Calculate the new air flow, assuming the fuel flow, work output and heat transfer remain the same as before. Also calculate the thermal efficiency of the engine. [24.88 kg/s, 30.41%]

Programme 10

INTERNAL COMBUSTION ENGINES, GAS TURBINES AND FUEL CELLS

1

We shall start this programme by thinking about the reciprocating internal combustion engine.

The smallest spark-ignition engines are of about 0.75 kW output and are used for such things as power tools. The largest are of about 5000 kW output, are often tuned to run on natural gas, and are used for pumping in gas pipelines and for small heat and power generation schemes. In between these sizes, petrol engines are used for motor cycles, cars and light commercial vehicles, small marine craft, light agricultural vehicles and light aircraft. Diesel engines range in size up to about 20 000 kW, and are used in small marine craft, cars, light and heavy commercial vehicles, heavy off-road vehicles, railway locomotives and ships. They are also used in electric power generation.

2

As we have already seen, these engines run on either the four stroke or the two stroke cycle. The figure from Frame 2 of Programme 2 is given again below. It shows in section the cylinder, crankcase and piston of a single cylinder four stroke engine, in each of the four strokes.

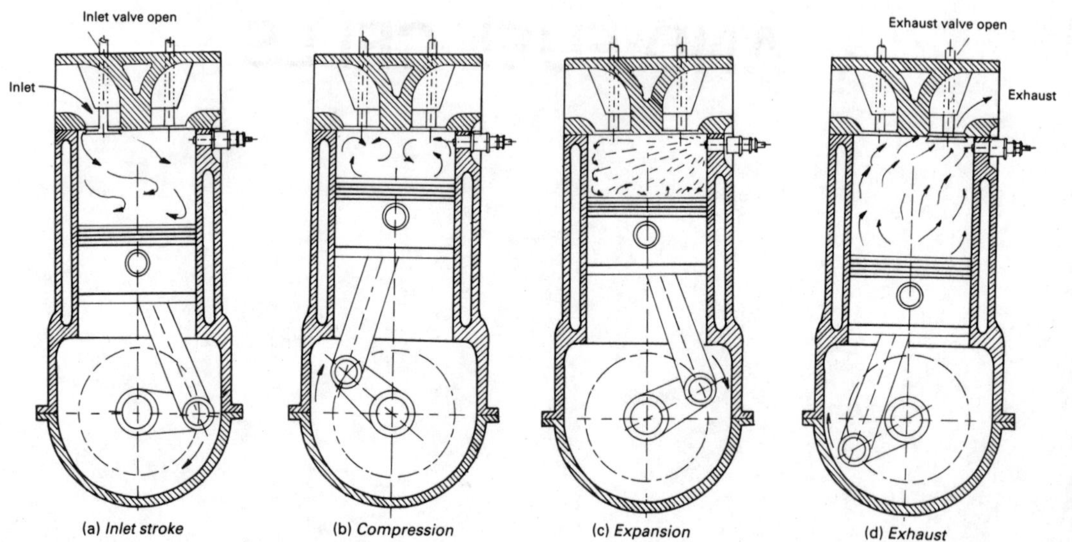

Inlet valve closes at the end of (a) and exhaust valve opens at the end of (c)

The figure shows, in (a) to (d), the four strokes of the piston which make up the cycle of events. Between the inlet valve closing at the end of the suction stroke (a) and the exhaust valve opening at the end of the power or expansion stroke (c), the flow through any one cylinder is trapped for a few milliseconds, so that while the flow is seen to be continuous for the whole multi-cylinder engine when running at its normal speed, it is nevertheless intermittent for each cylinder in each four stroke cycle. Later, we shall compare actual events in the cylinder with an idealisation based on air only as a perfect gas working substance.

3

When thinking of an idealised engine cycle, we assume air as a perfect gas is sealed within the cylinder and the same fixed mass of air undergoes a repeated cycle of events. In Frames 27 and 28 of Programme 6 we did this when analysing the Carnot cycle, in which the heat transfers were isothermal and reversible. In other cycles, as we shall see, they can also be under constant volume or constant pressure conditions. In any ideal cycle all processes are reversible, and compression and expansion processes are usually, though not always, isentropic. Positive work is done by the air in expansion and negative work is done on the air in compression, leading to a term called work ratio. We first met this in Frame 15 of Programme 7. Can you recall the definition of work ratio?

4

$$\boxed{\text{Work ratio} = \frac{\text{Net work output}}{\text{Positive work}}}$$

In a work-producing cycle there is positive work of expansion and negative work of compression of the air or working substance, so the net work produced is the algebraic sum of the two, which must be positive, or the engine will not run. This net work relative to the expansion work is an important ratio, and should be as close to unity as possible. In Frame 35 of Programme 2 we saw that in a real expansion the work obtained is less than the ideal work, and in a real compression the work input is greater than the ideal work. From these results we may define two efficiency terms, an expansion efficiency, and a compression efficiency. The expansion efficiency is defined:

$$\text{Expansion efficiency} = \eta_{\text{exp}} = \frac{\text{Actual work done}}{\text{Maximum work in the ideal process}}$$

The compression efficiency is defined in the next frame.

5

$$\text{Compression efficiency} = \eta_{\text{comp}} = \frac{\text{Work absorbed during ideal process}}{\text{Actual work to attain required pressure}}$$

In using these two efficiency terms, would the work ratio of a real cycle be less or greater than the work ratio of the corresponding ideal cycle?

6

Real work ratio is less than that of the ideal cycle

In a real cycle, though the expansion work is reduced, the net work as a difference is reduced much more, and consequently the work ratio will always be less than that of the corresponding ideal cycle.

The figure below is a pressure-volume diagram for an arbitrary ideal cycle, with shaded areas for expansion or positive work, compression or negative work, and net work. The work ratio for this cycle is:

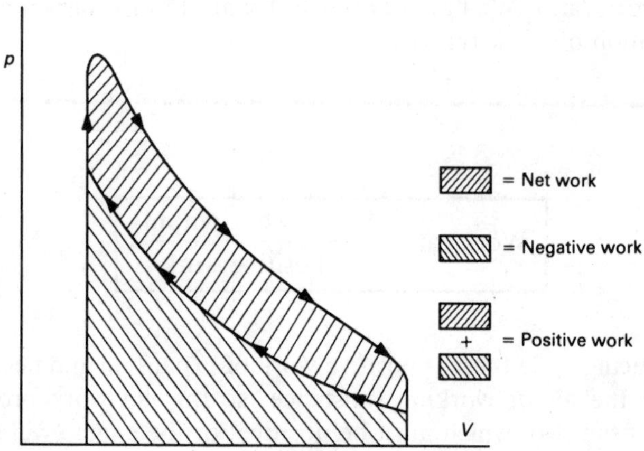

$$r_{\text{w}} = \frac{\text{Net work}}{\text{Positive work}}$$

In the ideal cycle we may define:

W_{exp} = Positive work of expansion

W_{comp} = Negative work of compression

Then: $W_{\text{net}} = W_{\text{exp}} + W_{\text{comp}}$

If, however, we write W_{comp} as a positive quantity, we can say:

$$W_{net} = W_{exp} - W_{comp}$$

It follows that: $$W_{exp} = \frac{W_{net}}{r_w}, \quad \text{and: } W_{comp} = W_{net}\left(\frac{1}{r_w} - 1\right)$$

If we now think about the real cycle with expansion and compression efficiencies of η_{exp} and η_{comp} defined in Frames 4 and 5, we can write:

$$W_{exp,R} = \eta_{exp} W_{exp}$$

and: $$W_{comp,R} = \frac{W_{comp}}{\eta_{comp}}$$

where $W_{exp,R}$ and $W_{comp,R}$ relate to the real cycle.

For the real cycle we can say:

$$W_{net,R} = \eta_{exp} \times W_{exp} - \frac{W_{comp}}{\eta_{comp}}$$

$$= \eta_{exp} \times \frac{W_{net}}{r_w} - \frac{W_{net}}{\eta_{comp}}\left(\frac{1}{r_w} - 1\right)$$

$$= W_{net}\left(\frac{\eta_{exp}}{r_w} - \frac{1 - r_w}{\eta_{comp}\, r_w}\right)$$

where r_w and W_{net} refer to the ideal cycle. If we assume the heat transfer Q to each cycle is the same (in practice, for the same top temperature in a cycle, Q in the real cycle is less because of a higher temperature after irreversible compression), and if the efficiencies of the real and ideal cycles are η_R and η_{ideal}, then:

$$\frac{\eta_R}{\eta_{Ideal}} = \frac{W_{net,R}}{W_{net}} = \left(\frac{\eta_{exp}}{r_w} - \frac{1 - r_w}{\eta_{comp}\, r_w}\right)$$

$$\therefore \quad \eta_R = \eta_{Ideal} \times \left(\frac{\eta_{exp}}{r_w} - \frac{1 - r_w}{\eta_{comp}\, r_w}\right)$$

This equation expresses the efficiency of the real cycle in terms of the ideal cycle efficiency, the work ratio of the ideal cycle, and the efficiencies of the expansion and compression processes in the real cycle. If in this equation η_{exp} and η_{comp} are both 0.85, what is the ratio of η_R/η_{Ideal} for $r_w = 0.6$?

$$\boxed{\eta_R/\eta_{Ideal} = 0.6324}$$

Using the equation in Frame 6:

$$\eta_R/\eta_{Ideal} = [(0.85/0.6) - (1 - 0.6)/(0.85 \times 0.6)]$$
$$= 0.6324$$

Thus if η_{Ideal} is 50%, the real cycle efficiency will only be 31.62%. This equation is important because, if we know the work ratio of an ideal cycle and can estimate the real expansion and compression efficiencies, it will tell us whether a real engine based on that ideal cycle is going to be worthwhile. In the figure below, using the equation in the previous frame, we have plotted the ratio of real cycle efficiency to ideal cycle efficiency against the work ratio of the ideal cycle for different values of expansion and compression efficiency, assuming Q in the real and ideal cycles is the same.

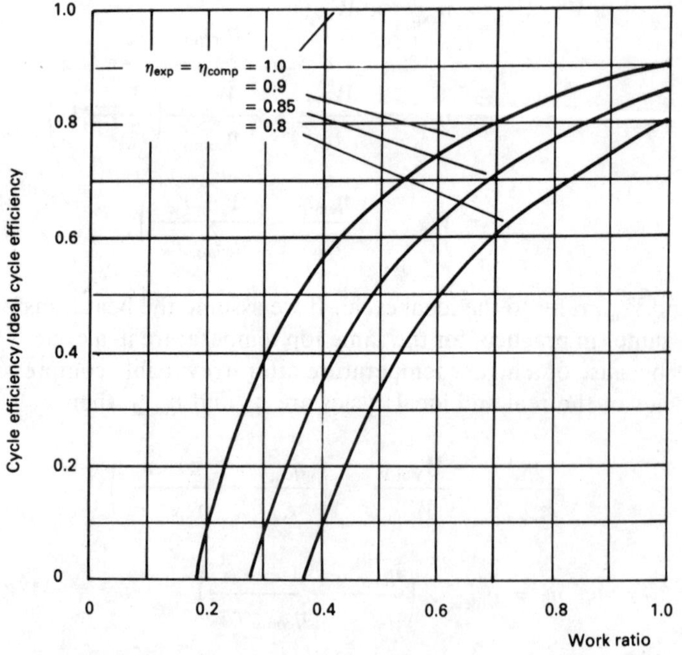

Thus in place of the calculation we did at the beginning of this frame we can find η_R/η_{Ideal} from the figure, for $r_w = 0.6$, with η_{exp} and η_{comp} both equal to 0.85. From the figure we can see that the efficiency ratio is 0.63.

As we would expect, when η_{exp} and η_{comp} are both 1.0, the ratio η_R/η_{Ideal} is also 1.0 at all values of r_w.

If we assume for a real engine $\eta_{exp} = \eta_{comp} = 0.8$, and the work ratio of the ideal cycle is 0.4, would a real engine based on that cycle be of any use?

8

No

We can see from the figure in the previous frame that the ratio of real to ideal cycle efficiency is going to be so low that the real engine will be of no great value. As actual compression and expansion efficiencies are around 0.8, it is essential, as we can see from the figure, for a real engine to be successful so that the work ratio of the ideal cycle should be as high as possible.

We know that the ideal Carnot cycle has the highest possible thermal efficiency, so now let's find the work ratio to see whether a practical Carnot cycle is going to be of any use. We first looked at the Carnot cycle with a perfect gas such as air as the working fluid in Programme 6. The cycle is shown below on both p–V and T–s coordinates.

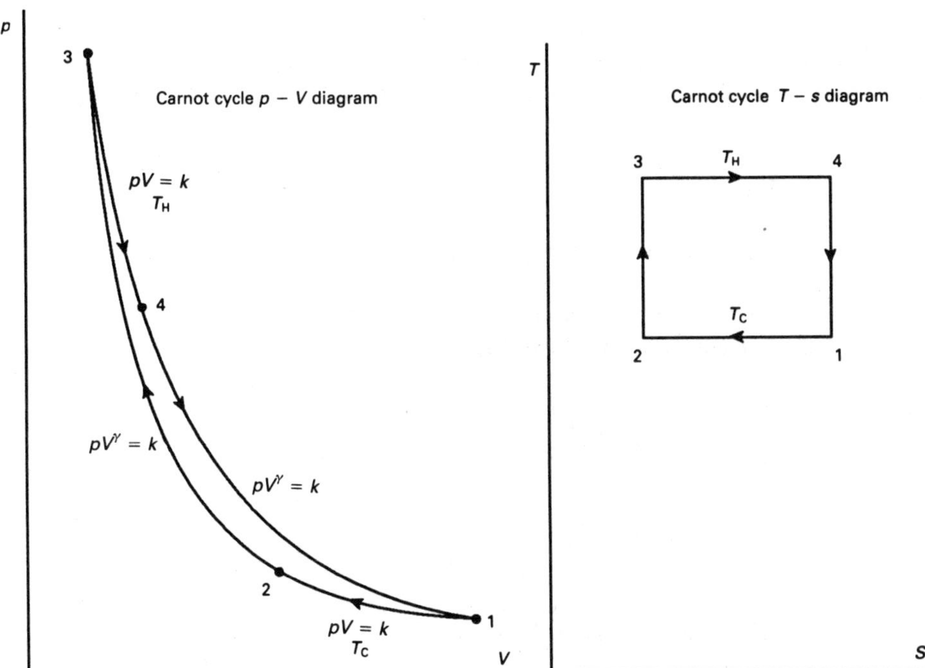

9

Isothermal compression at T_C with heat rejection takes place from state 1 to state 2, followed by isentropic compression to the higher temperature T_H at state 3. This is followed by isothermal expansion to state 4 and isentropic expansion to the original state 1.

We can write the work of compression as the sum of the isothermal work (Frame 75 in Programme 2) and the isentropic work (Frame 67 in Programme 5). The result is:

$$W_{comp} = p_1V_1 \ln V_2/V_1 + \frac{p_3V_3 - p_2V_2}{1-\gamma} = R\,T_C \ln V_2/V_1 + \frac{R(T_H - T_C)}{1-\gamma}$$

Remember that this will calculate out as a negative quantity. Now what is the work of expansion?

10

$$\boxed{W_{exp} = R_H T \ln V_4/V_3 + \frac{R\,(T_C - T_H)}{1-\gamma}}$$

This is again the isothermal work followed by the isentropic work and the result will be positive. The net work, as we saw in Frame 6, will be the sum of the two:

$$\begin{aligned} W_{net} &= R\,T_H \ln V_4/V_3 + \frac{R\,(T_C - T_H)}{1-\gamma} + R\,T_C \ln V_2/V_1 + \frac{R\,(T_H - T_C)}{1-\gamma} \\ &= R\,T_H \ln V_4/V_3 + R\,T_C \ln V_2/V_1 \end{aligned}$$

We can prove that the two volume ratios, V_4/V_3 and V_1/V_2, are the same. Since 1–4 and 2–3 are isentropics, we have that:

$$\frac{T_H}{T_C} = \left(\frac{V_2}{V_3}\right)^{\gamma-1} = \left(\frac{V_1}{V_4}\right)^{\gamma-1}$$

$$\therefore \quad \frac{V_2}{V_3} = \frac{V_1}{V_4}$$

and by cross-multiplying we have the result:

$$\frac{V_4}{V_3} = \frac{V_1}{V_2}$$

Next with $\ln V_2/V_1 = -\ln V_1/V_2$ we have:

$$W_{net} = R \ln(V_1/V_2)\,(T_H - T_C)$$

The work ratio is W_{net}/W_{exp}, which is:

$$r_w = \frac{R \ln(V_1/V_2)(T_H - T_C)}{R\,T_H \ln V_4/V_3 + \dfrac{R\,(T_C - T_H)}{1 - \gamma}}$$

Now with $V_4/V_3 = V_1/V_2$, and by dividing top and bottom by T_H we get:

$$r_w = \frac{(\gamma - 1) \ln V_1/V_2 \times \left(\dfrac{T_H - T_C}{T_H}\right)}{(\gamma - 1) \ln V_1/V_2 + \left(\dfrac{T_H - T_C}{T_H}\right)}$$

The term $(T_H - T_C)/T_H$ has a special significance. What is it? (See Frame 29 of Programme 6 if you cannot remember.)

11

It is the efficiency of the Carnot cycle

$\eta_{Carnot} = (T_H - T_C)/T_H$, so the expression for work ratio becomes:

$$r_w = \frac{(\gamma - 1) \ln V_1/V_2 \times \eta_{Carnot}}{(\gamma - 1) \ln V_1/V_2 + \eta_{Carnot}}$$

We can now establish what the work ratio would be for an ideal Carnot cycle. With air, $\gamma = 1.4$, and let T_H be 1000 K and T_C be 300 K. Also let the isothermal volume ratio V_1/V_2 be 3. First calculate the Carnot efficiency and then using the formula above, calculate r_w.

12

$$\boxed{\eta_{\text{Carnot}} = 0.7,\ r_w = 0.270}$$

The Carnot efficiency is:

$$\eta_{\text{Carnot}} = \frac{(1000 - 300)}{1000}$$
$$= 0.700$$

Then we have:

$$r_w = \frac{(1.4 - 1) \times \ln 3 \times 0.7}{(1.4 - 1) \times \ln 3 + 0.7}$$
$$= 0.270$$

From the figure in Frame 7 we can see that, with this very low work ratio for the ideal cycle, the efficiency of a real cycle with expansion and compression efficiences of 0.8 is going to be zero, and so a practical Carnot cycle with air is not going to produce any useful work.

Also from Frame 10 we have:

$$\frac{T_H}{T_C} = \left(\frac{V_2}{V_3}\right)^{\gamma-1}$$

This means the volume compression ratio between T_H of 1000 K and T_C of 300 K is going to be:

$$V_2/V_3 = (1000/300)^{1/(\gamma-1)}$$
$$= 20.29$$

We also had an isothermal volume ratio of 3 giving a total compression ratio of 60.86 to 1, which is very unrealistic, particularly as the engine will not work!

13

The Stirling cycle is another ideal cycle which has the same efficiency as the Carnot cycle, and which *can* be realised as a practical engine.

14

The Stirling engine falls in a class of its own. It is in fact a true heat engine, and *not* an internal combustion engine. The heat engine, you remember, has heat transfers at high and low temperatures at the boundary of a closed system which lead to a work output. Thus, the Stirling engine relies on *external* combustion of fuel to provide the heat source, and the working fluid is an enclosed mass of air or other gas.

The Stirling engine is remarkable because it was the first practical air engine that actually worked, and it has the same ideal efficiency as the Carnot cycle. The insight of its inventor, Rev. Dr Robert Stirling, is all the more remarkable, because at the age of 26 he had grasped both the theory and practice of heat engine design, and patented his closed cycle external combustion engine in 1816, some 9 years *before* Carnot published his work. His main occupation was that of a Scottish clergyman.

15

The ideal Stirling cycle is shown on p–V and T–s coordinates in the figure in the next frame.

Starting at state point 1, there is isothermal compression and heat rejection to an external heat sink at T_C to state point 2, from which there is constant volume heating to the top temperature T_H at state point 3. This heating is provided internally in the cycle and is discussed later. From state point 3 there is isothermal expansion at T_H with heat reception at T_H from the external combustion of fuel to the state point 4, which is at the same volume as state point 1. Constant volume cooling from point 4 to point 1 completes the cycle. The heat transfer in this last process is again within the cycle.

Although there is heat transfer in each of the four processes making up the cycle, **only the two isothermal transfers are across the closed system boundary** containing the circulating fluid.

Also in the next frame is a figure which shows in diagrammatic form how the four processes making up the cycle described above can be achieved in a practical Stirling engine.

Here is the ideal Stirling cycle shown on p–V and T–s coordinates.

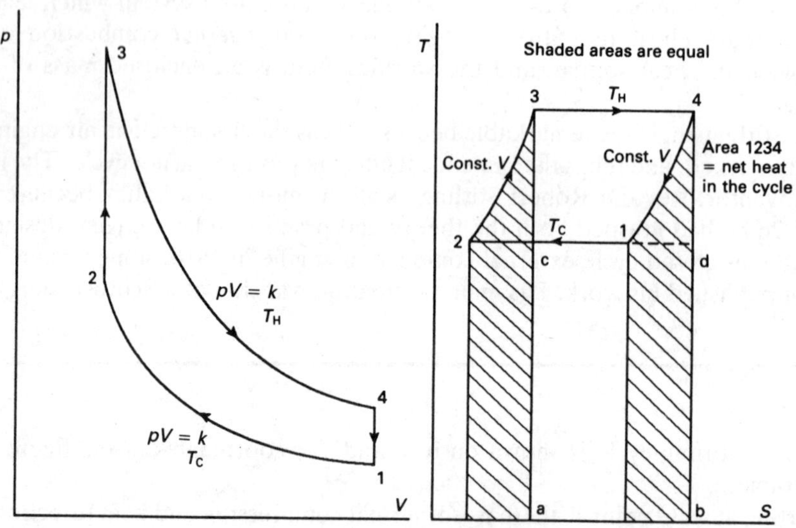

The figure below shows in diagrammatic form how a practical Stirling engine can be achieved.

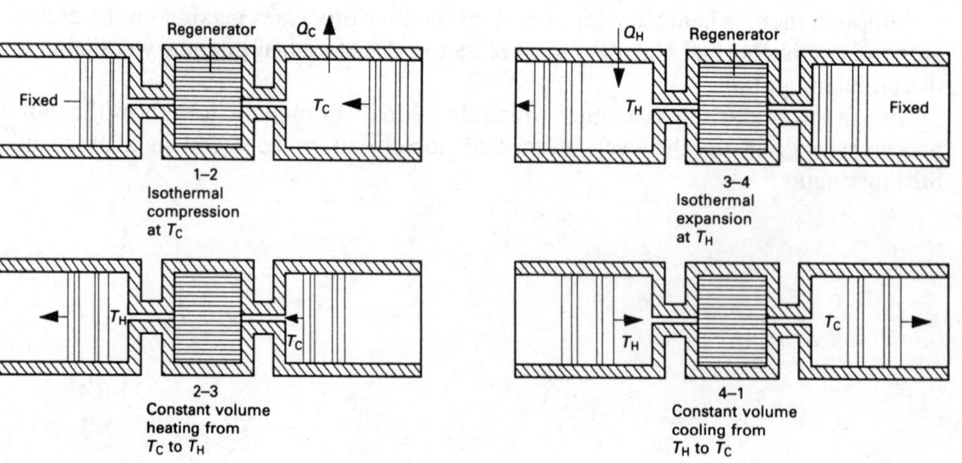

The engine has two cylinders connected by a regenerator which contains a porous matrix capable of receiving and giving heat to air passing through it. The left-hand cylinder receives heat at temperature T_H and the right-hand cylinder rejects heat at T_C. The top diagram illustrates process 1–2. The left-hand piston is fixed while the right-hand piston compresses the air isothermally at T_C. In the second diagram for process 2–3, both pistons move together and the air passes through the regenerator at constant volume and is heated from T_C to T_H. For unit mass, the quantity of heat must be $c_V (T_H - T_C)$. The regenerator loses this amount of energy in the process. This is a reversible heat transfer and the quantity of heat is shown as the left-hand shaded area on the *T–s* diagram on the previous page.

Next, in process 3–4 in the third diagram, the air expands isothermally at T_H in the left-hand cylinder while the right-hand piston is fixed. Finally, in process 4–1 shown in the last diagram, the air passes back through the regenerator at constant volume as both pistons move and is cooled from T_H back to T_C. The heat transfer is thus $c_V(T_C - T_H)$, which is shown as the right-hand shaded area on the *T–s* diagram. In this process the regenerator regains the same amount of energy it lost earlier.

A complicated mechanism is involved in creating a continuous rotary motion of a shaft while the pistons move or stay still in the sequences described.

Remembering that in isothermal processes, the heat and work transfers are equal, that in cycles the net work equals the net heat, and that we have these two definitions:

$$\text{Work ratio} = \frac{\text{Net work}}{\text{Positive work}}, \text{ and Cycle efficiency} = \frac{\text{Net work}}{\text{Positive heat}}$$

what can we say about the Cycle efficiency and work ratio for the Stirling cycle?

17

They are equal

From the *T–s* diagram we can see that heat supplied to the cycle is the area 3–4–b–a at T_H, and with the shaded areas equal line c–d equals line 2–1, and the heat rejected is area c–d–b–a at T_C. The cycle is therefore equivalent to the Carnot cycle and has the same thermal efficiency:

$$\eta_{\text{Stirling}} = \eta_{\text{Carnot}} = \frac{T_H - T_C}{T_H}$$

18

We can prove this, and that this is also the work ratio. The positive work of expansion and the heat supplied are the area under the isothermal expansion from point 3 to point 4 on the p–V diagram in Frame 16:

$$Q_H = W_{exp} = p_3V_3 \ln V_4/V_3 = R\,T_H \ln V_4/V_3$$

Similarly the work of compression, or negative work, and heat rejected are:

$$Q_C = W_{Comp} = p_1V_1 \ln V_2/V_1 = R\,T_C \ln V_2/V_1$$

The net work, and net heat, are the sum of the two:

$$Q_{net} = W_{net} = R\,T_H \ln V_4/V_3 + R\,T_C \ln V_2/V_1$$

You should now be able to prove the result given above for the cycle efficiency and work ratio for the Stirling cycle. Notice from the p–V diagram that $V_4/V_3 = V_1/V_2$.

19

From the definitions in Frame 16, we can say:

$$\eta_{cycle} = r_W = \frac{R\,T_H \ln V_4/V_3 + R\,T_C \ln V_2/V_1}{R\,T_H \ln V_4/V_3}$$

and since: $V_4/V_3 = V_1/V_2$

we can write this as:

$$\eta_{cycle} = r_W = \frac{R\,T_H \ln V_4/V_3 - R\,T_C \ln V_4/V_3}{R\,T_H \ln V_4/V_3}$$

$$= \frac{T_H - T_C}{T_H}$$

In the Carnot cycle in Frame 12, we found the work ratio was 0.27, and we chose T_H and T_C to be 1000 K and 300 K respectively, giving a cycle efficiency of 0.7. Thus for the same temperatures, both the work ratio and cycle efficiency for the ideal Stirling cycle will also be 0.7. From the diagram in Frame 7, with expansion and compression efficiencies of 0.8, what is the ratio of real to ideal thermal efficiency for a practical Stirling engine?

20

$$\eta_R/\eta_{Ideal} \simeq 0.6$$

From the figure we get an efficiency ratio of about 0.6. This means the actual thermal efficiency would be about 0.6 × 0.7, which is 0.42.

Although Stirling engines have been built and operated, they are complicated machines with an internal heat regenerator, and they also have the external heat exchangers that are characteristic of heat engines. However, there has been renewed interest in them in recent years because of their potentially high thermal efficiency.

21

Now that we know that a high work ratio in an ideal cycle is essential to enable the ideal cycle to be the basis for a practical engine, we can go on to consider the development of the modern internal combustion engine.

When engineers of the early nineteenth century were trying to build fuel-burning engines, they had only the experience of the steam engine builders for guidance. Thus, as the first steam engines were atmospheric engines (relying on the vacuum created by condensing the steam in a cylinder to give a pressure difference across the piston), early attempts were based on combustion taking place in the cylinder at atmospheric pressure. The Lenoir engine of around 1860 was the first successful internal combustion engine of this type.

We can examine the Lenoir engine and consider an ideal cycle which would correspond to what the engine actually did. The ideal cycle is shown on a *p–V* diagram below.

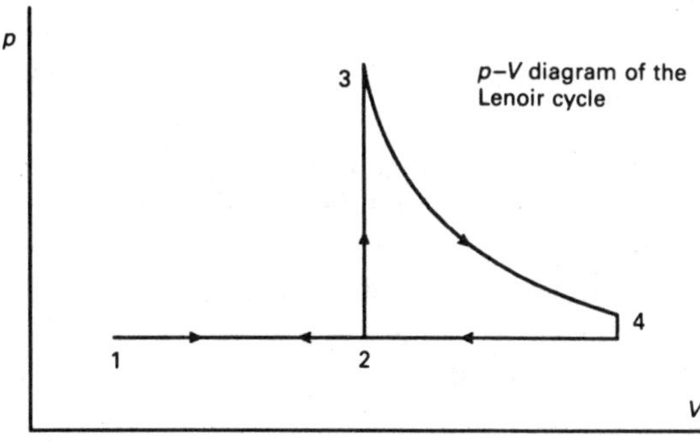

p–V diagram of the Lenoir cycle

22

Starting from point 1, a mixture of coal-gas and air was sucked in at atmospheric pressure, until the piston was about half way down the cylinder. The inlet valve then closed and ignition exploded the mixture at point 2 to raise the pressure to point 3. Work was then done on the piston face for the remainder of the stroke as the gases expanded to point 4. This was followed by an exhaust stroke from point 4 to point 1.

In an ideal cycle, we can assume the base line through points 1 and 2 on the p–V diagram is a constant pressure line at atmospheric pressure, with heat rejection from point 4 to point 2; line 2 to 3 is a constant volume line with heat addition at constant volume *replacing* the combustion process, and line 3 to 4 is an isentropic expansion to the end of the stroke.

23

Consider the area under the constant pressure line 1–2–4. As the piston moves from position 1 to position 2 there is atmospheric pressure on both sides of the piston, so no net work is done. Similarly as the piston moves from 4 to 2 there is again atmospheric pressure on both sides of the piston, so no net work is done. Positive work is done after ignition, represented by the area 2–3–4, and with no negative work, the work ratio for the Lenoir cycle is therefore 1.

The Lenoir engine was successful but not very efficient, and it suffered from mechanical problems due to the shock of the ignition process on the piston. Nevertheless a significant number were built between 1860 and 1865.

Although of historical interest in the development of reciprocating engines, the Lenoir cycle was also used in 1944–45 as the basis of the pulse-jet engine which powered the German V1 flying bomb.

24

Another atmospheric engine was introduced by Otto and Langen in 1867, but this was rapidly superseded by the Otto four stroke 'silent' gas engine of 1876. This was the basis of the now much developed four stroke petrol engine. It was discovered later that Beau de Rochas of France had described the principles of the four stroke cycle in 1862, and Otto's own patent in Germany was declared invalid.

Although it was first thought that an engine with only one working stroke in four could not possibly run, because of negative work and work done in overcoming friction, the benefits that followed from compressing the air–fuel mixture before ignition were immediate. The positive work done during the expansion stroke was much greater than anything previously achieved in the atmospheric engines. The volume compression ratio of early engines was about 3 to 1; this has steadily increased over the years and typical values are now between 8 and 12 to 1.

Also from early experience it was realised that heat losses from the cylinder charge should be kept to a minimum by increasing the speed of the engine. Since, as we saw in Programme 2, the amount of heat transfer depends on time, the duration of a cycle and *the heat transfer per cycle* are reduced by increasing the speed. Thus the amount of energy available per cycle to do useful work is increased.

The sequence of events that happen during the four strokes of the piston have already been described in the discussion of work in Programme 2, and again were referred to at the beginning of this programme. Starting with the suction stroke, what are they?

25

Suction of fresh air–fuel mixture
Compression of the mixture
Ignition and expansion of products of combustion
Exhaust of the products of combustion

These four events produce a pressure–volume diagram for the contents of the cylinder which is shown below.

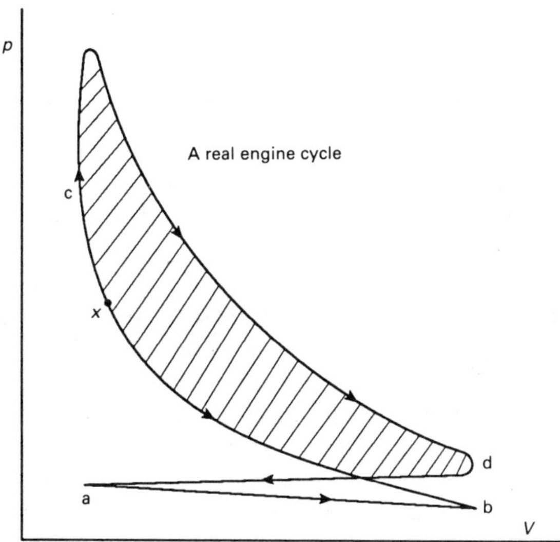

26

Line a–b is the suction stroke; the compression stroke is b–c, and ignition takes place just before the end of the stroke at around point x. Expansion is from c to d and exhaust is from d to a. The shaded enclosed area represents the net work done on the piston face, from which the power developed may be calculated.

Notice that in the first part of the expansion stroke from c the pressure is still rising. Why is this?

27

The temperature is rising rapidly with combustion of the fuel at close to constant volume conditions.

We know from Programme 9 how the temperature of a mixture rises as the chemical reactions of combustion take place, and as in this case there is very little change of volume immediately after ignition, the pressure must rise rapidly.

In the two stroke cycle, invented by Dugald Clerk in 1880, the exhaust gases escape through ports in the cylinder wall uncovered by the piston towards the end of the expansion stroke, and fresh mixture is then delivered to the cylinder by a displacer at the beginning of the compression stroke. Often the crankcase is used to do this. On the compression stroke, fresh mixture is sucked by the piston into the crankcase, and on the expansion stroke this is compressed slightly for delivery to the cylinder. In 1891, Clerk went on to develop the idea of supercharging, by using extra means of forcing fresh charge in under pressure.

28

The shaded area in the figure in Frame 25 represents the net work per cycle done against the piston of the engine. In Programme 2 (Frames 41–46), we saw how this work can be measured. An engine indicator is used to show how the pressure *actually* varies with volume in the cylinder, and from this we can calculate the indicated mean effective pressure and the work per cycle.

In order to design engines it is useful to be able to plot theoretical diagrams of cylinder pressure against volume so that we can estimate a work output. In the simplest way of doing this (which we shall only consider), we assume the working fluid is just air, and that it behaves like a perfect gas. We approximate the actual p–V lines we get from an engine indicator to a diagram made up from the well known processes we studied in Programme 5. Such cycles are known as **ideal air standard cycles**. In this programme we have already considered the Carnot, Stirling and Lenoir cycles as examples of **ideal cycles**, and used the analysis of Frames 4–7 to predict how a real cycle would roughly compare. We shall continue with this way of analysis: by looking at internal combustion engine air standard cycles, by finding the work ratio of the ideal cycle and by predicting the ratio of real to ideal cycle efficiency.

At a much more detailed level of analysis it is possible to predict theoretical p–V diagrams which conform quite closely to the real thing, but that is beyond what we can attempt here. Which of the familiar processes were used in the ideal Carnot and Stirling cycles?

29

Carnot: isothermal and isentropic
Stirling: isothermal and constant volume

The Carnot cycle in Frame 8 consists of two isothermal and two isentropic processes, and the Stirling cycle in Frame 15 consists of two isothermal and two constant volume processes. Although the term is not often used, constant volume processes are said to be **isochoric**.

We shall now look at the **Otto ideal air standard cycle**, upon which the modern petrol engine is based. This consists of two isentropic and two constant volume processes, so it is also known as the constant volume cycle.

From state 1, there is isentropic compression to state 2 over the volumetric compression ratio r_C, where $r_C = V_1/V_2$. At state 2, there is constant volume heating to state 3 which is followed by isentropic expansion to the original volume at state 4. Finally there is constant volume rejection of heat to the original state 1. These processes are shown in the figure in the next frame.

30

The p–V diagram for the ideal Otto cycle is shown below.

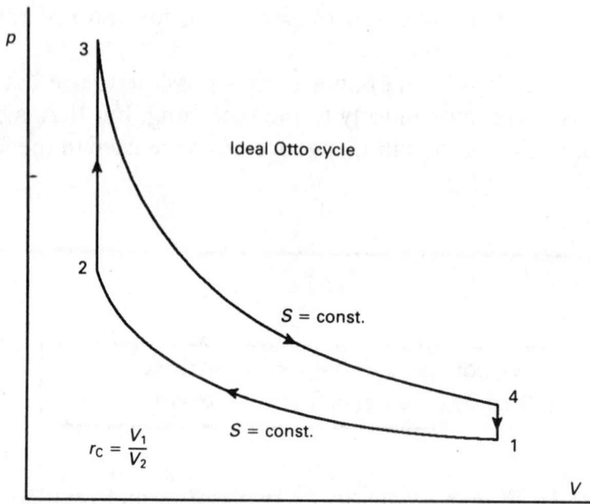

The idealised constant volume heat transfers replace combustion at the end of compression and the discharge of products into the atmosphere at the end of the expansion stroke. Now see if you can draw the corresponding T–s diagram; you can refer back to Programme 6 if necessary.

31

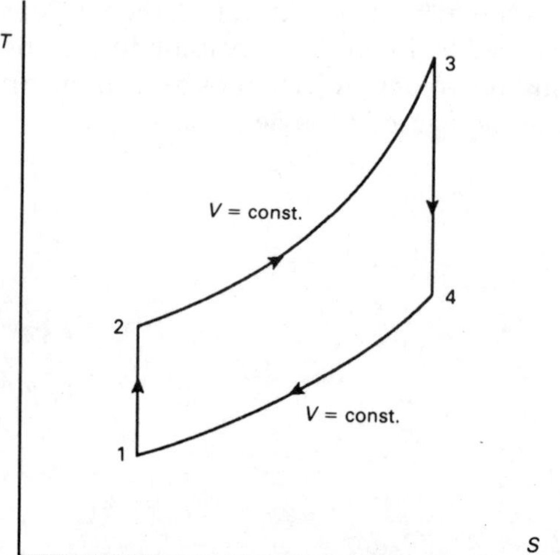

The isentropic processes 1–2 and 3–4 are vertical lines which are connected by the constant volume lines 2–3 and 4–1.

Now let's find an expression for the thermal efficiency of the Otto cycle. We have the two constant volume processes connected by two isentropic changes, so using what we learned earlier in Programme 5 we can write:

$$\frac{T_3}{T_4} = \left(\frac{V_4}{V_3}\right)^{\gamma-1} = r_C^{\gamma-1}$$

We shall also need the ratio T_2/T_1. What is it?

32

$$\boxed{T_2/T_1 = r_C^{\gamma-1}}$$

Since states 1 and 2 are also joined by an isentropic, we can write:

$$\frac{T_2}{T_1} = \left(\frac{V_2}{V_1}\right)^{\gamma-1} = r_C^{\gamma-1}$$

From this it follows that:

$$\frac{T_3}{T_4} = \frac{T_2}{T_1}, \text{ and so } \frac{T_3}{T_2} = \frac{T_4}{T_1}$$

In the cycle, the heat transfer to the air, or the heat supplied per unit mass, is:

$$Q_{\text{supplied}} = c_V(T_3 - T_2)$$

and the heat transfer from the air is:

$$Q_{\text{rejected}} = c_V(T_1 - T_4)$$

Summed for the whole cycle the heat transfer for unit mass is:

$$\sum Q = c_V(T_3 - T_2) + c_V(T_1 - T_4)$$

From Programme 3 we know that ΣQ in a cycle is equal to ΣW, the sum of the work, or net work, and we also know that the cycle efficiency is:

$$\eta_{\text{cycle}} = \frac{\text{Net work}}{\text{Heat supplied}}$$

What is the cycle efficiency in terms of these expressions for heat transfer?

33

$$\eta_{cycle} = \frac{c_V(T_3 - T_2) + c_V(T_1 - T_4)}{c_V(T_3 - T_2)}$$

You can see that this will simplify straight away to:

$$\eta_{cycle} = 1 - \frac{(T_4 - T_1)}{(T_3 - T_2)}$$

by cancelling out $c_V(T_3 - T_2)$ and c_V, and by changing the order of T_1 and T_4.
We can now rewrite this as:

$$\eta_{cycle} = 1 - \frac{T_4(1 - T_1/T_4)}{T_3(1 - T_2/T_3)}$$

$$= 1 - \frac{T_4}{T_3}$$

since from the previous frame we know that $T_1/T_4 = T_2/T_3$.
We also know from Frame 31 that:

$$\frac{T_3}{T_4} = r_C^{\gamma-1}$$

$$\therefore \qquad \eta_{cycle} = 1 - \frac{1}{r_C^{\gamma-1}}$$

This important result is highlighted by a box.
Thus, the thermal efficiency of the ideal Otto cycle depends only on the volumetric compression ratio, r_C, which is V_1/V_2, and also V_4/V_3. What is the cycle efficiency for a compression ratio of 6 to 1? For air, $\gamma = 1.4$.

34

The answer is $\eta_{cycle} = 0.512$.
We have:

$$\eta_{cycle} = 1 - 1/6^{0.4}$$

$$= 0.512$$

Notice that we have found the net work in the cycle from the net heat and so far we have not used any expressions for work involving pressure and volume. So that we can get an idea of how efficient a real Otto engine might be, we now need to find an expression for the work ratio, as we did for the Carnot and Stirling cycles.

The expansion work is the area under the isentropic process 3–4 on the p–V diagram in Frame 30, and the compression work is the area under the isentropic process 1–2. Thus:

$$W_{\text{exp}} = \frac{p_4V_4 - p_3V_3}{1 - \gamma} = \frac{R(T_4 - T_3)}{1 - \gamma}$$

and, similarly:

$$W_{\text{comp}} = \frac{R(T_2 - T_1)}{1 - \gamma}$$

What is the work ratio in terms of W_{exp} and W_{comp}? (Remember that W_{exp} and W_{comp} have positive and negative values respectively.)

35

$$r_{\text{w}} = \frac{R(T_4 - T_3)/(1 - \gamma) + R(T_2 - T_1)/(1 - \gamma)}{R(T_4 - T_3)/(1 - \gamma)}$$

Changing the order of T_3 and T_4, this simplifies to give:

$$r_{\text{w}} = 1 - \frac{(T_2 - T_1)}{(T_3 - T_4)}$$

$$= 1 - \frac{T_1(T_2/T_1 - 1)}{T_3(1 - T_4/T_3)}$$

The derivation of the expression for work ratio is continued in the next frame.

36

From the temperatures ratios in Frame 32, we can see that we can get:

$$r_W = 1 - \frac{T_1(r_C^{\gamma-1} - 1)}{T_3(1 - 1/r_C^{\gamma-1})}$$

and by dividing the numerator by $r_C^{\gamma-1}$ we can see that:

$$\boxed{r_W = 1 - \frac{T_1}{T_3} r_C^{\gamma-1}}$$

and from Frame 33: $$\boxed{\eta_{\text{cycle}} = 1 - \frac{1}{r_C^{\gamma-1}}}$$

If we look at these results for the Otto cycle, we see that increasing the *compression ratio*, r_C, raises the ideal cycle efficiency. However, lowering T_1 and raising T_3 does *not* increase the ideal cycle efficiency but raises the work ratio and, as we now know, raises the *practical* cycle efficiency.

If in the Otto cycle, r_C is 12 to 1, (a) first see how the ideal cycle efficiency has changed by doubling the value of r_C, (compared with a value of 6 in Frame 34), then with $T_1 = 300$ K, $T_3 = 2300$ K and $\gamma = 1.4$, (b) find the work ratio and the practical cycle efficiency, using the figure in Frame 7, and assuming compression and expansion efficiencies of 85%.

37

$$\boxed{\text{(a) } \eta_{\text{cycle}} = 0.630, \text{ (b) } r_W = 0.648, \ \eta = 0.422}$$

For (a), $\eta_{\text{cycle}} = 1 - 1/12^{0.4} = 0.630$
This compares with 0.512 when r_C is 6 to 1.

In (b), $$r_W = 1 - \frac{300}{2300} 12^{0.4} = 0.648$$

From the figure in Frame 7, with compression and expansion efficiencies of 0.85, the cycle efficiency ratio is about 0.67 which gives a practical cycle efficiency of 0.67 × 0.630, which is 0.422 or 42.2%.

This is a measure of the efficiency of producing indicated work against the piston face, or **indicated thermal efficiency**. As we saw in Frames 41 and 42 of Programme 2, this work may be measured by a device called an indicator. Useful work at the output shaft, which is measured by a dynamometer or brake, is called brake work, and is less since work is lost in overcoming engine friction and in driving things like the oil pump and valve gear. From this a **mechanical efficiency** can be defined:

$$\text{Mechanical efficiency} = \frac{\text{Work at output shaft}}{\text{Indicated work}}$$

The value of this will depend on the load, since, for example, at a very light load most of the indicated work is used in keeping the engine turning. At full load, the mechanical efficiency may be 80–90% so that the **brake thermal efficiency** would be in the range 33.8–38.0%.

If the Otto ideal cycle efficiency is 0.6, the work ratio is 0.7 and the mechanical efficiency of the corresponding real engine is 65% at a particular load, using the figure in Frame 7 what is the brake thermal efficiency and the power output for an indicated power of 35 kW?

38

23.8%, 22.75 kW

From the figure in Frame 7, the efficiency ratio is 0.61, so the brake thermal efficiency is:

$$\begin{aligned} \eta_{\text{brake}} &= 0.6 \times 0.61 \times 0.65 \times 100\% \\ &= 23.8\% \end{aligned}$$

The brake power is 0.65 × 35 kW which is 22.75 kW.

It is often necessary to calculate the state of the air at individual state points around the Otto ideal cycle, so in the next few frames we shall work through an example as an illustration.

39

In an Otto ideal cycle, at the beginning of compression the pressure is 100 kPa and the temperature is 300 K. The compression ratio is 9.5 to 1, and the temperature after constant volume heating is 2150 K. (a) Calculate: the pressure and temperature at the end of compression, the highest pressure in the cycle, and the pressure and temperature after expansion. Also calculate the specific work output of the cycle and the ideal cycle efficiency, without using the expression in Frame 33. (b) Show that the cycle efficiency does not change if the top temperature is increased to 2400 K. For air, γ is 1.4, R = 0.2871 kJ/kg K and c_V is 0.718 kJ/kg K.

The figure below shows the p–v diagram, with the state points labelled 1 to 4.

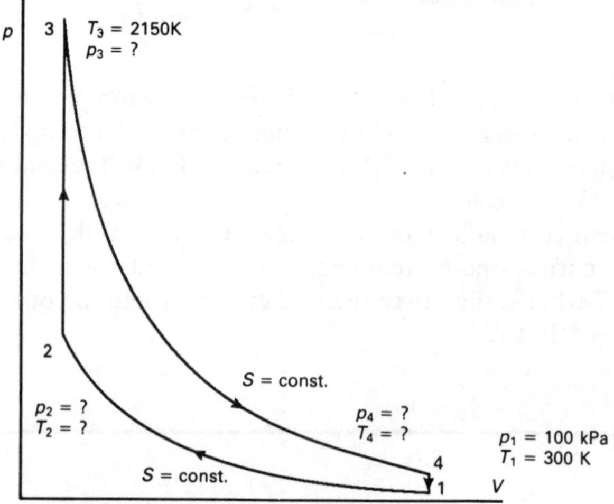

This example makes use of much that we did with perfect gases in Programme 5. For a reversible adiabatic compression, which is isentropic, we know that pv^γ = constant, so what is the pressure at the end of compression?

40

2338 kPa

We have $p_2 v_2^\gamma = p_1 v_1^\gamma$, and the volume ratio is 9.5 to 1, so:

$$p_2 = 100 \times 9.5^{1.4} = 2338 \text{ [kPa]}$$

The temperature at the end of compression, T_2 is then given by:

$$\frac{T_2}{T_1} = \left(\frac{V_1}{V_2}\right)^{\gamma-1}$$

$$= 9.5^{0.4} = 2.461$$

Hence: $$T_2 = 2.461 \times 300$$

$$= 738.3 \text{ K}$$

Next heat transfer at constant volume takes place to raise the temperature to 2150 K. With c_V given in Frame 39, what is the magnitude of Q, per unit mass, and what is the pressure at state 3?

41

1013.6 kJ/kg, 6808.5 kPa

The process 2–3 is one of heat transfer at constant volume, so the First Law equation for a closed system of unit mass says:

$$Q = u_3 - u_2$$

$$= c_V(T_3 - T_2)$$

$$= 0.718 \times (2150 - 738.3) = 1013.6 \text{ [kJ/kg]}$$

For the same constant volume process, we have:

$$\frac{p_3}{p_2} = \frac{T_3}{T_2}$$

∴ $$p_3 = p_2 \times \frac{T_3}{T_2}$$

$$= 2338 \times 2150/738.3$$

$$= 6808.5 \text{ [kPa]}$$

This will be the highest pressure in the cycle. Now again using pv^{γ} = constant, find the pressure at state 4.

42

291.2 kPa

For the expansion from 3 to 4 we have:

$$p_4 = p_3\left(\frac{V_3}{V_4}\right)^{\gamma}$$

$$= 6808.5 \times \left(\frac{1}{9.5}\right)^{1.4}$$

$$= 291.2\ [\text{kPa}]$$

From Frame 31 we also have:

$$\frac{T_3}{T_4} = \left(\frac{V_4}{V_3}\right)^{\gamma-1}$$

Hence:

$$T_4 = 2150 \times \left(\frac{1}{9.5}\right)^{0.4}$$

$$= 873.7\ \text{K}$$

Now what is the magnitude and direction of the heat transfer per unit mass at constant volume between states 4 and 1?

43

−411.9 kJ/kg, *from* the system

For process 4–1:

$$Q = c_V(T_1 - T_4)$$

$$= 0.718 \times (300 - 873.7)$$

$$= -411.9\ [\text{kJ/kg}]$$

Being negative, it is heat transfer from the system.

The net heat transfer per unit mass for the cycle is:

$$\begin{aligned}\sum Q &= 1013.6 - 411.9 \\ &= 601.7 \text{ [kJ/kg]}\end{aligned}$$

For unit mass, this must also be the specific work, since for a cycle we remember that:

$$\sum Q = \sum W$$

Hence the efficiency of the cycle must be:

$$\begin{aligned}\eta_{\text{cycle}} &= (601.7/1013.6) \times 100\% \\ &= 59.36\%\end{aligned}$$

Now confirm that this result can also be obtained using the expression in Frame 33 with $r_C = 9.5$.

44

$$\boxed{\eta_{\text{cycle}} = 59.36\%}$$

For the Otto cycle we have

$$\begin{aligned}\eta &= 1 - \frac{1}{r_C^{\gamma-1}} \\ &= 1 - \frac{1}{9.5^{0.4}} \\ &= 0.5936 = 59.36\%\end{aligned}$$

The second part of the exercise is to repeat the heat addition, expansion and heat rejection with $T_3 = 2400$ K, instead of 2150 K. Calculate the new heat addition (as we did in Frame 41), then find T_4 (see Frame 42), the new heat rejection with T_1 the same as before at 300 K, and the cycle efficiency (which as you should now expect, will not have changed).

45

$$1193.1 \text{ kJ/kg}, T_4 = 975.3 \text{ K}, -484.8 \text{ kJ/kg}, 59.36\%$$

For the heat addition:

$$Q = 0.718 \times (2400 - 738.3)$$
$$= 1193.1 \text{ [kJ/kg]}$$

The temperature after expansion is:

$$T_4 = 2400 \times (1/9.5)^{0.4} = 975.3 \text{ K}$$

The heat rejected is:

$$Q = 0.718 \times (300 - 975.3)$$
$$= -484.8 \text{ [kJ/kg]}$$

The specific work is:

$$\sum W = \sum Q = 1193.1 - 484.8$$
$$= 708.3 \text{ [kJ/kg]}$$

The cycle efficiency is:

$$\eta_{\text{cycle}} = (708.3/1\,193.1) \times 100\%$$
$$= 59.36\%$$

We did not recalculate p_3 and p_4 since we were not asked to find them, but you can confirm that they are 7762.7 and 332.0 kPa, respectively. You can get more practice at working round cycles in this way in the examples in Frame 74.

46

You may be a little worried that in this example we have increased the top temperature in the ideal cycle and yet the cycle efficiency is the same. But we found in Frame 33 that the Otto cycle efficiency depends only on the volumetric compression ratio and not on the heat transfer afterwards. We found in Frame 36 that the work ratio *does* depend on the highest temperature achieved, so now we need to find the work ratio using the expression in Frame 36 for the two values of T_3 of

2150 K and 2400 K. Then using the figure in Frame 7, find the actual cycle efficiency in the two cases with compression and expansion efficiencies of 0.8.

47

$r_w = 0.657$ and 0.692; $\eta_{real} = 33.24\%$ and 35.62%

The work ratio in each case is:

$$r_w = 1 - \frac{300}{2150} \times 9.5^{0.4} = 0.657$$

and

$$r_w = 1 - \frac{300}{2400} \times 9.5^{0.4} = 0.692$$

These work ratios give cycle efficiency ratios of 0.56 and 0.6 respectively from Frame 7, which give actual cycle efficiencies of 33.24% and 35.62%. Thus we can see that by increasing the top temperature in the cycle, the *practical* cycle efficiency has increased, which in the end is all that matters.

The actual thermal efficiencies of early Otto engines were limited by the characteristics of fuels then available. This meant that the compression ratio had to be kept low to avoid 'knocking' in the combustion process. In normal combustion a flame front travels out from the source of ignition, the spark, through the entire mixture until it is all consumed. The flame front tends to compress and heat the unburnt mixture ahead of it, which in abnormal combustion then ignites spontaneously in a detonation. This detonation produces the 'knocking' or 'pinking' sound, and if allowed to continue will damage the engine. Fuels of today have a high 'knock rating' which allows them to be used in engines having a much higher compression ratio.

48

In 1892, Rudolf Diesel introduced the idea of an engine with a much higher compression ratio than that of the Otto engines of the day, in which fuel injected at the end of the compression stroke ignited spontaneously.

This spontaneous ignition was due to the high temperature of the air after compression. After ignition at the start of the injection period, the fuel then burnt as it entered the cylinder and combustion was approximately at constant pressure. Therefore the ideal Diesel cycle with air is taken to be a cycle involving heat transfer at constant pressure as the heat addition process.

Owing to the high compression ratio, Diesel engines had to be much stronger and heavier, and it took five years for a practical engine to be developed. However, it was then soon evident that efficiencies were substantially greater than those of contemporary Otto engines.

The pressure–volume and temperature–entropy diagrams for the ideal Diesel cycle are shown below.

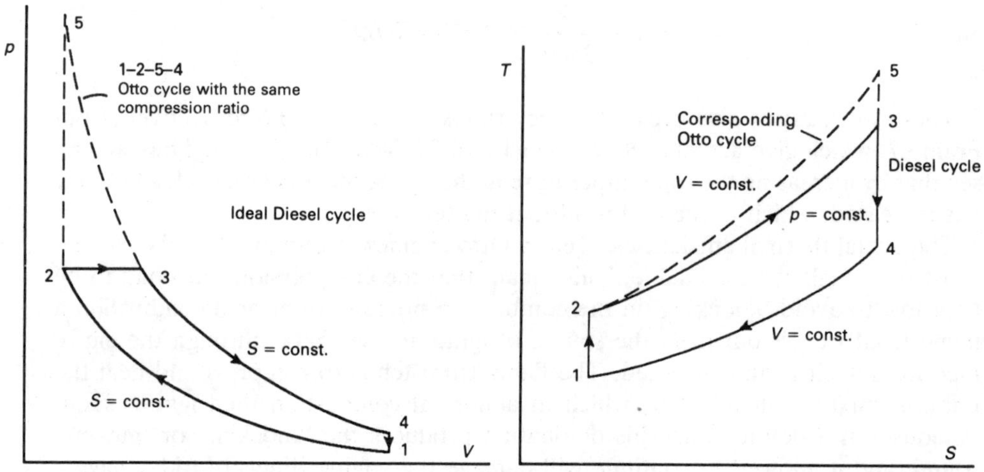

We have isentropic compression from state 1 to state 2, followed by heat transfer at constant pressure to state 3. Then there is isentropic expansion to state 4 and to conclude the cycle, as in the Otto cycle, there is heat rejection at constant volume to state 1. The diagram also shows, in dotted outline, a constant volume process to a fictitious state 5 and an isentropic expansion to state 3 so that the line 5–3–4 represents a continuous isentropic process. The cycle 1–2–5–4 would be an Otto cycle with the same compression ratio. The state point 5 is introduced because it makes it much easier to find the expression for the Diesel cycle efficiency. The Diesel and corresponding Otto cycles are also shown for comparison on the *T–s* diagram above.

50

In the Diesel air cycle there are two important ratios, the compression ratio r_C, and the volume ratio for constant pressure heat addition, r_P. We have:

$$r_C = \frac{V_1}{V_2}; \qquad r_P = \frac{V_3}{V_2}$$

We can find the thermal efficiency of the air cycle in terms of r_C and r_P. Heat supplied at constant pressure is:

$$Q_{\text{supplied}} = c_P(T_3 - T_2)$$

and heat rejected at constant volume is:

$$Q_{\text{rejected}} = c_V(T_1 - T_4)$$

Then $\Sigma\ Q$ for the cycle is also $\Sigma\ W$, and:

$$\sum W = \sum Q = c_P(T_3 - T_2) + c_V(T_1 - T_4)$$

The cycle efficiency is then:

$$\eta_{\text{cycle}} = \frac{\sum W}{\text{Heat supplied at constant pressure}}$$

$$= \frac{c_P(T_3 - T_2) + c_V(T_1 - T_4)}{c_P(T_3 - T_2)}$$

For the next step you can simplify this in terms of $(T_4 - T_1)$, (notice the change of order), $(T_3 - T_2)$ and γ. Finally, we shall see how we can express this in terms of the two volume ratios r_C and r_P.

51

$$\boxed{\eta_{\text{cycle}} = 1 - \frac{(T_4 - T_1)}{\gamma(T_3 - T_2)}}$$

This is true because γ for the gas is c_P/c_V.

52

Next we can divide the top by T_1 and the bottom by T_2:

$$\eta_{cycle} = 1 - \frac{T_1(T_4/T_1 - 1)}{\gamma\, T_2(T_3/T_2 - 1)}$$

From Frame 32 we know that:

$$\frac{T_2}{T_1} = r_C^{\gamma-1}$$

and for the constant pressure process we know that:

$$\frac{T_3}{T_2} = \frac{V_3}{V_2} = r_P$$

Hence we can say:

$$\eta_{cycle} = 1 - \frac{(T_4/T_1 - 1)}{r_C^{\gamma-1}\,\gamma(r_P - 1)}$$

Finally we have to eliminate T_4/T_1.

53

We now use the corresponding Otto cycle in the figure in Frame 49 and we know that $T_5/T_2 = T_4/T_1$, so we can say:

$$\frac{T_4}{T_1} = \frac{T_5}{T_2} = \frac{T_5}{T_3} \times \frac{T_3}{T_2}$$

and since states 3 and 5 are on the same isentropic, with $V_2 = V_5$, we have:

$$\frac{T_5}{T_3} = \left(\frac{V_3}{V_2}\right)^{\gamma-1} = r_P^{\gamma-1}$$

Also for the constant pressure process 2–3:

$$\frac{T_3}{T_2} = \frac{V_3}{V_2} = r_P$$

Hence: $$\frac{T_4}{T_1} = \frac{T_5}{T_3} \times \frac{T_3}{T_2} = r_P^{\gamma-1} \times r_P = r_P^{\gamma}$$

The ideal Diesel cycle efficiency is then:

$$\eta_{cycle} = 1 - \frac{1}{r_C^{\gamma-1}} \frac{(r_P^{\gamma} - 1)}{\gamma(r_P - 1)}$$

and for the Otto cycle:

$$\eta_{cycle} = 1 - \frac{1}{r_C^{\gamma-1}}$$

In comparing this expression with the efficiency of the Otto cycle in Frame 33, we can identify the extra terms involving r_P which occur in the Diesel cycle. What is η_{cycle} for the Diesel cycle if r_C is 15, r_P is 2.5 and γ is 1.4?

54

$$\eta_{cycle} = 58.0\%$$

We have:

$$\eta_{cycle} = 1 - \frac{(2.5^{1.4} - 1)}{15.0^{0.4} \times 1.4 \times 1.5}$$

$$= 1 - 0.420$$

$$= 0.580 = 58.0\%$$

Now we shall find the work ratio of the Diesel cycle by following the same principles we used for the Otto cycle. For the work of expansion from state 2 to state 4:

$$W_{exp} = (V_3 - V_2)p_2 + \frac{p_4V_4 - p_3V_3}{1 - \gamma}$$

Then by substituting RT for pV and by putting both terms over the denominator $(1 - \gamma)$ we get:

$$W_{exp} = \frac{(1 - \gamma)R(T_3 - T_2) + R(T_4 - T_3)}{1 - \gamma}$$

This is next simplified a little more and then we shall find the work of compression.

55

The previous expression is rearranged to give:

$$W_{\text{exp}} = \frac{\gamma R(T_3 - T_2) + R(T_2 - T_4)}{\gamma - 1}$$

Then the work of compression from state 1 to state 2 is:

$$W_{\text{comp}} = \frac{p_2V_2 - p_1V_1}{1 - \gamma} = \frac{R(T_2 - T_1)}{1 - \gamma}$$

With W_{comp} negative, the work ratio for the Diesel cycle is going to be:

$$r_{\text{W}} = (W_{\text{exp}} + W_{\text{comp}})/W_{\text{exp}}$$

As a next step, express this in terms of $(T_2 - T_1)$, $(T_3 - T_2)$, $(T_2 - T_4)$ and γ.

56

$$\boxed{r_{\text{W}} = 1 - \frac{(T_2 - T_1)}{\gamma(T_3 - T_2) + (T_2 - T_4)}}$$

From the expression $r_{\text{W}} = (W_{\text{exp}} + W_{\text{comp}})/W_{\text{exp}}$ we get:

$$r_{\text{W}} = 1 + \frac{\dfrac{R(T_2 - T_1)}{1 - \gamma}}{\dfrac{\gamma R(T_3 - T_2) + R(T_2 - T_4)}{\gamma - 1}}$$

$$= 1 - \frac{(T_2 - T_1)}{\gamma (T_3 - T_2) + (T_2 - T_4)}$$

Next we can divide top and bottom by T_2 and get:

$$r_{\text{W}} = 1 - \frac{(1 - T_1/T_2)}{\gamma (T_3/T_2 - 1) + (1 - T_4/T_2)}$$

We also know that:

$$\frac{T_1}{T_2} = \frac{1}{r_C^{\gamma-1}} \quad \text{and} \quad \frac{T_3}{T_2} = r_P$$

and using the relationship for T_4/T_1 from Frame 53 we can see that:

$$\frac{T_4}{T_2} = \frac{T_4}{T_1} \times \frac{T_1}{T_2} = r_P^{\gamma}/r_C^{\gamma-1}$$

When these expressions are used, for the work ratio of the Diesel cycle we get:

$$r_W = 1 - \frac{(1 - 1/r_C^{\gamma-1})}{\gamma(r_P - 1) + (1 - r_P^{\gamma}/r_C^{\gamma-1})}$$

In Frame 54 we found that with $r_C = 15$ and $r_P = 2.5$, η_{cycle} was 58.0% for the ideal Diesel cycle. What is the work ratio for this case?

57

$$r_W = 0.648$$

We have: $r_C^{\gamma-1} = 2.9542$; $1/r_C^{\gamma-1} = 0.3385$; $r_P^{\gamma} = 3.6067$

Then: $$r_W = 1 - \frac{(1 - 0.3385)}{1.4 \times 1.5 + (1 - 3.6067/2.9542)} = 0.648$$

With this work ratio, the cycle efficiency ratio is about 0.54 from the figure in Frame 7 so the real engine efficiency is 0.54 × 0.580 which is 0.313 or 31.3%. This would be the indicated efficiency, and with a mechanical efficiency of about 80%, the brake efficiency would be about 25%. Notice that this work ratio is exactly the same as that of the Otto cycle at a compression ratio of 12, which we found in Frames 36 and 37. A real engine efficiency of 25% would have been a high value in Diesel's day, because the compression ratios used in Otto engines were then very low.

58

The constant pressure ideal Diesel cycle is not a very good cycle as such, because of its lower work ratio. However, it was a good basis for the early low speed oil engines in which the fuel injection period was relatively long. The air cycle which corresponds more closely to the modern high speed Diesel engine is the Dual cycle. The fuel injection period is now very short so most of the combustion is taken to be at constant volume with the remainder at constant pressure. The p–V and T–s diagrams for this cycle are shown below.

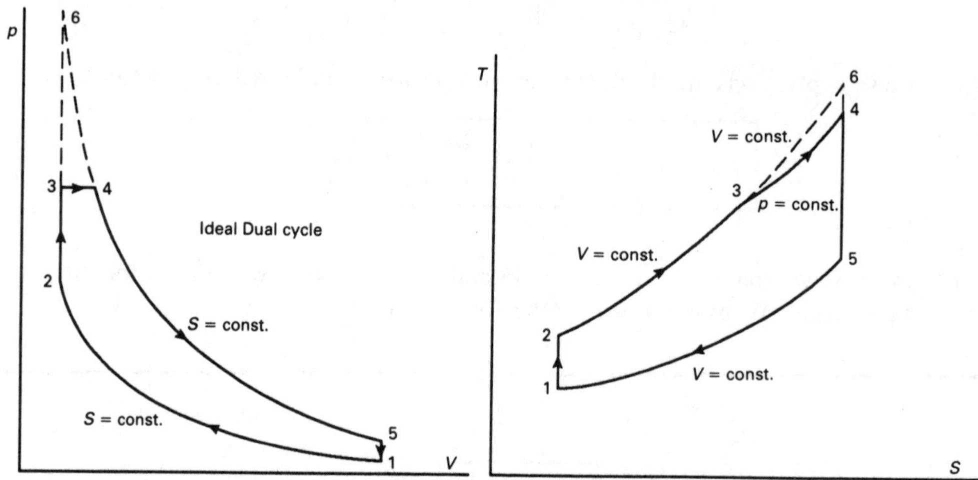

Again an equivalent Otto cycle is shown; the fictitious point 6 is useful in finding the cycle efficiency and work ratio.

There is heat addition at constant volume followed by heat addition at constant pressure, and we have the following three ratios:

$$r_C = V_1/V_2; \qquad r_H = p_3/p_2 = T_3/T_2; \qquad r_P = V_4/V_3 = T_4/T_3$$

The heat transfers per unit mass in the cycle are:

$$Q_{\text{supplied}} = c_V(T_3 - T_2) + c_P(T_4 - T_3)$$
$$Q_{\text{rejected}} = c_V(T_1 - T_5)$$

The cycle efficiency is, as before, the net work which is the net heat, divided by the heat supplied, which is:

$$\eta_{\text{cycle}} = \frac{c_V(T_3 - T_2) + c_P(T_4 - T_3) + c_V(T_1 - T_5)}{c_V(T_3 - T_2) + c_P(T_4 - T_3)}$$

The next step is to cancel out the heat supplied, divide out by c_V and reverse the order of T_1 and T_5.

59

$$\eta_{\text{cycle}} = 1 - \frac{(T_5 - T_1)}{(T_3 - T_2) + \gamma(T_4 - T_3)}$$

As we know that:

$$r_H = T_3/T_2 \text{ and } r_P = T_4/T_3$$

we shall divide the first and second terms in the denominator by T_2 and T_3 respectively, and we shall also divide the numerator by T_1, so we get:

$$\eta_{\text{cycle}} = 1 - \frac{T_1(T_5/T_1 - 1)}{T_2(r_H - 1) + \gamma\, T_3(r_P - 1)}$$

Next we separate T_2 from the denominator to give:

$$\eta_{\text{cycle}} = 1 - \frac{T_1(T_5/T_1 - 1)}{T_2[(r_H - 1) + \gamma(T_3/T_2)(r_P - 1)]}$$

Then using: $r_H = T_3/T_2$ and $T_1/T_2 = 1/r_C^{\gamma-1}$ we get:

$$\eta_{\text{cycle}} = 1 - \frac{1}{r_C^{\gamma-1}} \frac{(T_5/T_1 - 1)}{(r_H - 1) + \gamma\, r_H(r_P - 1)}$$

The last step, which we shall do in the next two frames, is to show that:

$$T_5/T_1 = r_P^{\gamma}\, r_H$$

Now we use the fictitious point 6, and with $T_5/T_1 = T_6/T_2$ we can say: 60

$$\frac{T_5}{T_1} = \frac{T_6}{T_2} = \frac{T_6}{T_4} \times \frac{T_4}{T_2} = \frac{T_6}{T_4} \times \frac{T_4}{T_3} \times \frac{T_3}{T_2} = \frac{T_6}{T_4} \times (r_P\, r_H)$$

Finally in the next Frame we use the fact that state points 4 and 6 are joined by an isentropic over a volume ratio of r_P.

61

Consequently, we have then:

$$T_6/T_4 = r_P^{\gamma-1}$$

Hence from the result in the previous frame:

$$T_5/T_1 = r_P^{\gamma-1}\, r_P\, r_H = r_P^{\gamma}\, r_H$$

The ideal Dual cycle efficiency is then:

$$\boxed{\eta_{cycle} = 1 - \frac{1}{r_C^{\gamma-1}} \frac{(r_P^{\gamma}\, r_H - 1)}{(r_H - 1) + \gamma\, r_H(r_P - 1)}}$$

What is the efficiency when r_C is 15, r_P is 1.5, and r_H is 2.5?

62

$$\boxed{\eta_{cycle} = 0.6448}$$

We have:

$$\eta_{cycle} = 1 - \frac{1}{15^{0.4}} \frac{[(1.5^{1.4} \times 2.5) - 1]}{(2.5 - 1) + 1.4 \times 2.5 \times (1.5 - 1)}$$

$$= 0.6448$$

Small high speed Diesel engines have compression ratios rather higher than 15 to 1. Now let's work through this example on an air standard Dual cycle. The compression ratio r_C is 23 to 1, and T_1 before compression is 300 K. Heat transfer to the air per unit mass is 1400 kJ/kg in total and T_3 after the constant volume process is 2600 K. With $c_V = 0.718$ kJ/kg K, $c_P = 1.005$ kJ/kg K, find r_H and r_P, and hence the ideal cycle efficiency.

For compression we have:

$$\frac{T_2}{T_1} = r_C^{\gamma-1} = 23^{0.4}$$

$$= 3.505$$

$$\therefore \quad T_2 = 3.505 \times 300$$
$$= 1051.5 \text{ K}$$

What is the constant volume heat addition per unit mass between states 2 and 3?

63

1111.8 kJ/kg

The temperature rise is:

$$2600 - 1051.5 = 1548.5 \text{ K}$$

and the heat transfer per unit mass at constant volume is then:

$$c_V(T_3 - T_2) = 0.718 \times 1548.5$$
$$= 1111.8 \text{ [kJ/kg]}$$

The heat transfer at constant pressure is then:

$$1400 - 1111.8 = 288.2 \text{ [kJ/kg]}$$

The temperature rise at constant pressure is then given by:

$$c_P(T_4 - T_3) = 288.2 \text{ [kJ/kg]}$$
$$\therefore \quad T_4 - T_3 = 288.2/1.005$$
$$= 286.7 \text{ K}$$

Then:

$$T_4 = 2600 + 286.7$$
$$= 2886.7 \text{ K}$$

Now that we know T_2, T_3 and T_4, what are r_H and r_P? (See Frame 58.)

64

$$r_H = 2.473, \ r_P = 1.110$$

From Frame 58 and with the calculated temperatures we have:

$$r_H = \frac{T_3}{T_2} = \frac{2600}{1051.5} = 2.473$$

and

$$r_P = \frac{T_4}{T_3} = \frac{2886.7}{2600} = 1.110$$

We can now find the ideal cycle efficiency:

$$\eta_{cycle} = 1 - \frac{1}{23^{0.4}} \frac{[(1.110^{1.4} \times 2.473) - 1]}{(2.473 - 1) + 1.4 \times 2.473(1.110 - 1)}$$

$$= 0.7134$$

$$= 71.34\%$$

To roughly evaluate the real engine cycle efficiency corresponding to the Dual cycle, we need, as before, to find the work ratio which is obtained from W_{exp} and W_{comp}. The expansion work, W_{exp}, is the constant pressure work from state 3 to state 4 followed by the work of the isentropic between states 4 and 5. The *p–V* diagram is given in Frame 58. See if you can get the expression for W_{exp} in terms of the temperatures of the state points, γ and R.

65

$$W_{exp} = \frac{\gamma R(T_4 - T_3) + R(T_3 - T_5)}{\gamma - 1}$$

From the *p–V* diagram in Frame 58 we can see that:

$$W_{exp} = (p_4V_4 - p_3V_3) + \frac{p_5V_5 - p_4V_4}{1 - \gamma}$$

$$= R(T_4 - T_3) + \frac{R(T_5 - T_4)}{1 - \gamma}$$

$$= \frac{\gamma R(T_4 - T_3) + R(T_3 - T_5)}{\gamma - 1}$$

Also we have:

$$W_{\text{comp}} = \frac{p_2V_2 - p_1V_1}{1 - \gamma}$$

$$= \frac{R(T_2 - T_1)}{1 - \gamma}$$

Remembering that r_W is $(W_{\text{exp}} + W_{\text{comp}})/W_{\text{exp}}$ (with W_{comp} negative), using these results you should be able to see that:

$$r_W = 1 - \frac{(T_2 - T_1)}{\gamma(T_4 - T_3) + (T_3 - T_5)}$$

To get this, we made W_{comp} positive by putting $(\gamma - 1)$ for the denominator so that it would cancel with the denominator in W_{exp}. In this result we need to divide the top by T_2 and the bottom by T_3, so that we can then use the ratios:

$$\frac{T_2}{T_1} = r_C^{\gamma-1},$$

$$\frac{T_4}{T_3} = r_P \quad \text{and}$$

$$\frac{T_3}{T_2} = r_H$$

The result of doing this is given in the next frame.

66

$$\boxed{r_W = 1 - \frac{1}{r_H} \frac{(1 - 1/r_C^{\gamma-1})}{[\gamma (r_P - 1) + (1 - T_5/T_3)]}}$$

The algebra involved follows in the next frame.

67

We have from Frame 65:

$$r_W = 1 - \frac{(T_2 - T_1)}{\gamma\,(T_4 - T_3) + (T_3 - T_5)}$$

$$= 1 - \frac{T_2(1 - T_1/T_2)}{\gamma\, T_3(T_4/T_3 - 1) + T_3(1 - T_5/T_3)}$$

$$= 1 - \frac{1}{r_H}\,\frac{(1 - 1/r_C^{\gamma-1})}{[\gamma\,(r_P - 1) + (1 - T_5/T_3)]}$$

Finally we have to get rid of T_5/T_3. We can say:

$$\frac{T_5}{T_3} = \frac{T_5}{T_1}\,\frac{T_1}{T_3} = \frac{T_6}{T_2}\,\frac{T_1}{T_3} = \frac{T_6}{T_4}\,\frac{T_4}{T_3}\,\frac{T_1}{T_2}$$

From the p–V diagram (in Frame 58), we can see that:

$$\frac{T_6}{T_4} = r_P^{\gamma-1}$$

$$\frac{T_4}{T_3} = r_P \quad \text{and}$$

$$\frac{T_1}{T_2} = \frac{1}{r_C^{\gamma-1}}$$

$$\therefore \qquad \frac{T_5}{T_3} = \frac{r_P^{\gamma}}{r_C^{\gamma-1}}$$

When substituted into r_W we get finally:

$$\boxed{r_W = 1 - \frac{1}{r_H}\,\frac{(1 - 1/r_C^{\gamma-1})}{[\gamma(r_P - 1) + (1 - r_P^{\gamma}/r_C^{\gamma-1})]}}$$

where the three ratios r_C, r_H and r_P are as previously defined. Notice the similarity with r_W for the Diesel cycle in Frame 56, which is:

$$r_W = 1 - \frac{(1 - 1/r_C^{\gamma-1})}{\gamma(r_P - 1) + (1 - r_P^{\gamma}/r_C^{\gamma-1})}$$

The work ratio of the Dual cycle improves relative to the Diesel cycle as the value of r_H increases.

Now calculate the work ratio for the previous example in Frame 64, for which $r_C = 23$, $r_H = 2.473$ and $r_P = 1.110$.

68

$$r_W = 0.6492$$

The working is:

$$r_W = 1 - \frac{1}{2.473}\,\frac{(1 - 1/23^{0.4})}{(1.4 \times 0.110 + 1 - 1.110^{1.4}/23^{0.4})}$$

$$= 0.6492$$

If we take the compression and expansion efficiencies for the modern engine as between 85 and 90%, from the figure in Frame 7 we find the cycle efficiency ratio is about 0.73 at this value of work ratio. This gives a real engine efficiency of about 0.73 × 71.34%, which is 52.08%. Remember that this is the indicated thermal efficiency, so the brake efficiency will be rather less, say about 45%.

In Frames 45 and 46 of Programme 2 we showed how the power developed by displacement work in an engine cylinder could be calculated from the cylinder volume and the indicated mean effective pressure, or IMEP. The value of IMEP produced is a direct function of the compression ratio, and a figure of 1000 kPa would be typical for a compression ratio of 9 to 1. Suppose we have to meet a brake load of 85 kW at full speed at 5000 rev/m, and we wish to know the size of engine required, and the thermal efficiency. We may assume a mechanical efficiency of 88% and a volumetric efficiency of 90%. Also, the engine will run at an air/fuel ratio of 16 to 1 by mass, and the calorific value of the fuel is 42 MJ/kg.

If the Brake power is 85 kW, what is the indicated power for the stated mechanical efficiency?

69

96.59 kW

70

Since we have:

$$\eta_{mech} = \frac{\text{Brake power}}{\text{Indicated power}}$$

the indicated power is:

$$85.0/0.88 = 96.59 \text{ [kW]}$$

Next we shall see how to find the necessary volumetric displacement of the engine to produce this power, assuming the value of 1000 kPa for the IMEP.

From Frame 45 of Programme 2 we have the power developed by an engine cylinder operating on the four stroke cycle at speed N rev/m is:

$$\text{Power} = \frac{A_P \times \text{IMEP} \times L_P \times \text{N}}{60 \times 2} \text{ [kW]}$$

where the product $A_P \times L_P$ is the cylinder displacement of the one cylinder. So for the whole engine we can say:

$$\begin{aligned}\text{Cylinder displacement [m}^3\text{]} &= 96.59 \text{ [kW]} \times \frac{60 \times 2}{\text{IMEP} \times \text{N}} \\ &= 96.59 \times \frac{60 \times 2}{1000 \times 5000} \\ &= 0.002\,32 \text{ [m}^3\text{]} \\ &= 2.32 \text{ [litres]}\end{aligned}$$

This is the displacement of the whole engine, so we could then choose a configuration of either 4 or 6 cylinders.

The volumetric efficiency is a measure of the volume of air passing through an engine relative to the displacement of the engine. Every two revolutions, the volume of air displaced would be:

$$\begin{aligned}2.32 \times 0.9 &= 2.088 \text{ litres} \\ &= 0.002\,088 \text{ [m}^3\text{]}\end{aligned}$$

With $R = 0.2871$ kJ/kg K, and taking the air at 100 kPa and 300 K, what is the mass flow of air when the engine is running at 5000 rev/m ?

71

The answer is 0.101 kg/s.

We have:

$$\dot{m} = \frac{p\,V}{R\,T}$$

$$= \frac{100 \times 0.002\,088 \times 5000}{0.2871 \times 300 \times 60 \times 2}$$

$$= 0.101 \text{ [kg/s]}$$

With an air/fuel ratio of 16 to 1 by mass, the fuel flow will be:

$$0.101/16 = 0.0063 \text{ [kg/s]}$$

With a calorific value of 42 MJ/kg, the rate of energy supply to the engine is then:

$$0.0063 \text{ [kg/s]} \times 42 \text{ [MJ/kg]} \times 1000 \text{ [kJ/MJ]} = 264.6 \text{ kW}$$

The engine brake thermal efficiency is then:

$$\eta_{\text{brake}} = \frac{85}{264.6} \times 100\%$$

$$= 32.12\%$$

A similar example to this will be found in Frame 74.

72

It is rather beyond our scope at this level to go any further into the performance of reciprocating engines. We have seen how we can find the efficiency and work ratio of ideal air cycles appropriate to spark-ignition and compression-ignition engines, from which we can find approximately the efficiency of the corresponding real engine. The ideal air cycle consists of isentropic compression of pure air, heat transfer at constant volume and/or at constant pressure, isentropic expansion, and heat transfer at constant volume. By contrast, in the practical cycle we have irreversible compression with heat loss through the cylinder walls of a fuel–air mixture or just air in the case of the diesel, followed by combustion and expansion again with heat loss, and completed with an unresisted expansion of the combustion products from the cylinder. As we stated in Frame 28, the efficiency of a real engine could be predicted directly from a detailed analysis of these actual processes.

Thus the efficiency of a real engine is always less than that of any corresponding ideal air cycle. But how efficient can a real engine be?

We will seek to answer this question in the next frame.

73

In Frame 83 of Programme 9 we saw that in a chemically reacting system:

$$W_{max} = -[\Delta G']_0$$

and consequently, the maximum thermal efficiency would be:

$$\eta_{th,max} = \frac{[\Delta G']_0}{[\Delta H']_0}$$

This maximum thermal efficiency occurs when the reaction takes place isothermally at the datum temperature of 25°C. With hydrogen fuel this maximum efficiency, from Frame 86 of the Programme 9, is 94%, and for the sort of fuel used in engines is around 90%. As we know, the chemical reactions of combustion in real engines do not take place isothermally at 25°C and so their practical efficiencies are limited by the corresponding air standard cycle efficiencies. These are the Carnot efficiency for the Stirling engine, the Otto efficiency for the petrol engine and the Dual combustion efficiency for the diesel engine. At any maximum ideal cycle temperature, the Carnot efficiency is always the highest value.

With combustion at high temperatures, the main routes to higher real engine efficiencies are to increase compression ratios so that the ideal cycle efficiency target is raised and to reduce heat losses per cycle so that the difference between the real and ideal cycles is narrowed. The question of heat transfer per cycle was mentioned in Frame 24. The adiabatic diesel engine is a new development prompted by these findings. It is a high compression engine with no forced cooling which uses ceramic components to withstand the high temperatures of operation.

The route to approaching $\eta_{th,max}$ given above is only possible if the chemical oxidation reactions are isothermal at 25°C, and do not involve high temperature combustion. The fuel cell, considered briefly at the end of this programme, is a hopeful step in this direction.

74

Before moving on to consider gas turbine engines, satisfy yourself that you can do the following problems on reciprocating engine air cycles, and practical engine performance. For air, take $c_P = 1.005$ kJ/kg K, $c_V = 0.718$ kJ/kg K, $R = 0.2871$ kJ/kg K and $\gamma = 1.4$. You may wish to refer to earlier programmes when dealing with ideal gas and combustion calculations.

1. The Otto cycle consists of isentropic compression from state 1 to state 2 over a volume ratio of r_C, constant volume heating to state 3, isentropic expansion to the original volume at state 4, and constant volume cooling to the original state 1. Without looking back, show that the cycle efficiency is:

$$\eta_{\text{cycle}} = 1 - \frac{1}{r_C^{\gamma-1}}$$

Given that $p_1 = 100$ kPa, $T_1 = 300$ K and $T_3 = 2700$ k, find p_2, T_2, p_3, p_4, T_4, the net specific work of the cycle and the IMEP, for $r_C = 7$.
[$p_2 = 1524.5$ kPa, $T_2 = 653.4$ K, $p_3 = 6299.9$ kPa, $p_4 = 413.2$ kPa, $T_4 = 1239.7$ K, specific work = 1106.9 kJ/kg, IMEP = 1076.4 kPa]

2. The Lenoir cycle consists of constant pressure rejection of heat over a volume ratio r_C from state 1 to state 2, constant volume heating to state 3, and isentropic expansion to the original state. Show that the cycle efficiency is:

$$\eta_{\text{cycle}} = 1 - \frac{\gamma\,(r_C - 1)}{(r_C^{\gamma} - 1)}$$

Calculate η_{cycle} when r_C is 2, with ignition at half the piston stroke. [0.1458]

3. The Atkinson cycle has an expansion stroke over a volume ratio r_E which is greater than compression ratio r_C. There is isentropic compression over the volume ratio r_C from state 1 to state 2, constant volume heating to state 3, isentropic expansion over the volume ratio r_E to state 4, and constant pressure cooling to the original state 1. Draw the cycle on p–V and T–s axes. Show that:

$$\eta_{\text{cycle}} = 1 - \frac{\gamma}{r_E^{\gamma-1}}\,\frac{(1 - r_C/r_E)}{(1 - (r_C/r_E)^{\gamma})}$$

If $p_1 = 100$ kPa, $T_1 = 300$ K, $r_C = 5$ and $r_E = 8$, find states 2, 3 and 4, the heat supplied and heat rejected, and show that η_{cycle} is confirmed by the above formula.
[$p_2 = 951.8$ kPa, $T_2 = 571.1$ K, $p_3 = 1837.9$ kPa, $T_3 = 1102.7$ K, $p_4 = 100$ kPa, $T_4 = 480$ K, heat supplied is 381.67 kJ/kg, heat rejected is 180.9 kJ/kg, $\eta_{\text{cycle}} = 0.526$ by both methods]

4. Show that the work ratio of the Atkinson cycle is:

$$r_W = 1 - \frac{(r_C^{\gamma}/r_E - 1) + \gamma(1 - r_C/r_E)}{(r_E^{\gamma-1} - 1)}$$

Calculate the value for r_W when $r_C = 5$ and $r_E = 8$. [0.4491]

5. An air cycle consists of isothermal compression over a volume ratio r_C, constant volume heating, and isentropic expansion to the original state. Show the cycle

on p–V and T–s axes. In which process is there heat rejection? Show that the Cycle efficiency is:

$$\eta_{\text{cycle}} = 1 - \frac{(\gamma - 1) \ln r_C}{(r_C^{\gamma-1} - 1)}$$

If p_1 = 100 kPa, T_1 = 300 K and r_C = 8, find p_2, p_3, T_3 and η_{cycle}.
[Isothermal compression, p_2 = 800 kPa, p_3 = 1837.9 kPa, T_3 = 689.2 K, η_{cycle} = 0.359]

6. An engine is required to take a load of 50 kW at 4000 rev/m. It may be assumed to have a mechanical efficiency of 84% and a volumetric efficiency of 87%, and it will produce an IMEP of 1100 kPa on an air/fuel ratio of 15 to 1. The calorific value of the fuel is 42 MJ/kg. Calculate the displacement of the required engine and its brake thermal efficiency. [1.623 litres, 33.2%]
7. A real engine cycle is assumed to comprise the following events: polytropic compression with n = 1.3 over a volume ratio of 9.5 to 1 from an initial state of 100 kPa, and 300 K to a state 2; constant volume adiabatic combustion of C_8H_{18} at an air/fuel ratio of 16 to 1 by mass giving a state 3; polytropic expansion with n = 1.6 to the original volume at state 4. A constant volume pressure drop is assumed to complete the cycle. For C_8H_{18}, $[\Delta H]_0$ is −5 116 180 kJ/kmol, with octane in the vapour state. Calculate the pressure and temperature at states 2, 3 and 4, and find the net indicated work from the polytropic work of compression and expansion. Hence find the IMEP and the indicated thermal efficiency. For your closed system, take one mole of fuel and 16 times the mass of air. Use the following values of c_V: Reactants, C_8H_{18} = 2.330, O_2 = 0.696, N_2 = 0.752; Products, CO_2 = 1.128, H_2O = 2.090, O_2 = 0.874, N_2 = 0.835, all in kJ/kg K units.
[p_2 = 1867 kPa, T_2 = 589.4 K, p_3 = 10 133 kPa, T_3 = 3199 K, p_4 = 276.3 kPa, T_4 = 828.7 K, W_{comp} = 515 566.3 kJ, W_{exp} = 2 110 749.6 kJ, IMEP = 1436.3 kPa, η = 40.25%. Note that the values for T_3 and the IMEP are high because we have not considered dissociation]

75

Now we shall think about gas turbine engines. These were developed in the 1930s simultaneously in the UK by Sir Frank Whittle, and in Germany by Dr Hans von Ohain, and were first used in fighter aircraft in the latter part of the Second World War. A Heinkel jet plane first flew in August 1939, some 20 months ahead of the Whittle jet.

Apart from their main use in aviation, gas turbine engines are also used in ship propulsion and in electrical power generation. We have already seen that there can be both open and closed cycle machines, which are shown below.

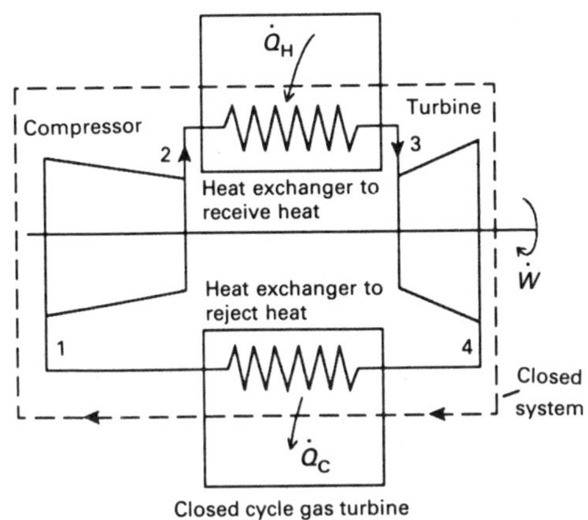

Closed cycle gas turbine

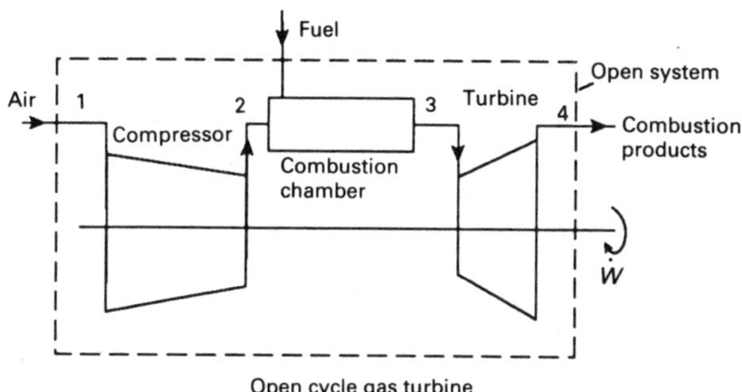

Open cycle gas turbine

The closed cycle machine is a true heat engine with gas circulating through the plant within the closed system. There is a compressor, a heat exchanger to receive heat, a turbine, and a heat exchanger to reject heat. Combustion of fuel is *external* to the cycle. In the open cycle machine, air enters the compressor, and fuel and air enter the combustion chamber so that combustion products enter the turbine which exhausts to the atmosphere. In either case, the steady flow energy equation may be applied in turn to each component of the plant. In Frame 54 of Programme 3 we introduced the idea of total head enthalpy in the steady flow energy equation; can you write this equation with total head enthalpy terms and with no potential energy terms?

76

$$\dot{Q} - \dot{W} = \dot{m}(h_{T2} - h_{T1})$$

In this equation, $\dot{m}$ is the mass flow rate, and:

$$\begin{aligned} h_T &= h + V^2/2 \\ &= c_P T + V^2/2 \\ &= c_P T_T \end{aligned}$$

for a gas, where T_T is the total head temperature. We shall use this form of the steady flow energy equation, involving total head enthalpy, since no other term is then involved except the heat and work transfers in the gas turbine cycle. The equation may be applied in turn to each component of the engine. Both the compressor and turbine are assumed to be adiabatic. For the compressor we can write:

$$-\dot{W}_{comp} = \dot{m}\, c_P(T_{T2} - T_{T1})$$

and for the turbine:

$$-\dot{W}_{exp} = \dot{m}\, c_P(T_{T4} - T_{T3})$$

The total temperature of the air increases from T_{T2} to T_{T3} between the compressor and the turbine and this is due to a heat transfer in the closed cycle machine, and due to combustion in the open cycle, which could be put in terms of an equivalent heat transfer in an ideal air cycle. What would then be the energy equation for both the heat exchanger in the closed cycle plant, and the combustion chamber in the open cycle plant? Remember that our equations here are *rate* equations in units of kW, or possibly MW.

77

$$\dot{Q} = \dot{m}\, c_P(T_{T3} - T_{T2})$$

The heat transfer rate in the closed cycle, and the equivalent heat transfer rate in the open cycle, are equal to the mass flow rate times the enthalpy increase. We can now find the cycle efficiency. The cycle shown below on T–s axes is the Brayton cycle, also called the Joule cycle, with heat transfers at constant pressure. The original Brayton engine of 1876 was a reciprocating engine with compressed air injected at the end of the compression stroke into the cylinder through a hot grating, with fuel burnt on the grating, so that combustion was at constant pressure.

In the ideal gas turbine cycle, the compression and expansion are both isentropic, and the pressure ratio across the compressor and turbine is r_P.

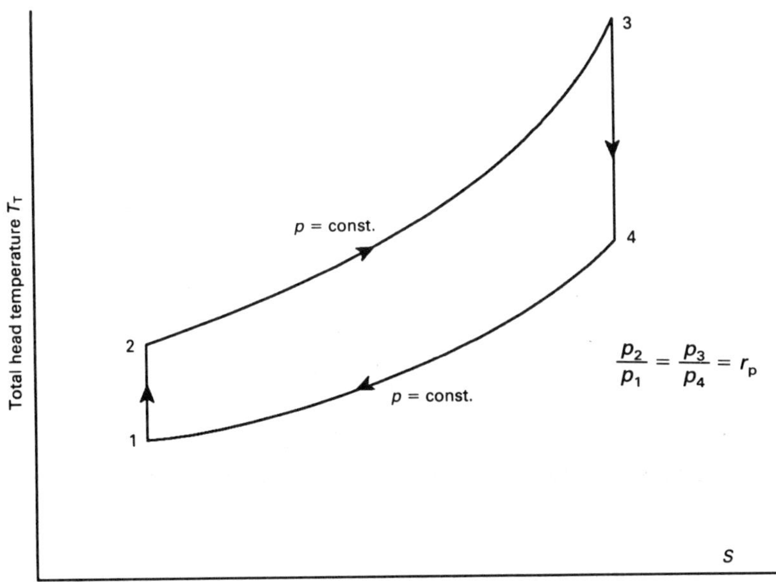

From the expressions for compressor and turbine work given in Frame 76 we can see that the positive net work available per unit mass flow is:

$$\dot{W}_{\text{net}} = c_P(T_{T3} - T_{T4}) - c_P(T_{T2} - T_{T1})$$

and the heat supplied is:

$$\dot{Q} = c_P(T_{T3} - T_{T2})$$

From these results, show that the cycle efficiency may be written as:

$$\eta_{\text{cycle}} = 1 - \frac{(T_{T4} - T_{T1})}{(T_{T3} - T_{T2})}$$

78

The efficiency of the cycle is:

$$\eta_{cycle} = \dot{W}_{net}/\dot{Q}$$

$$\therefore \qquad \eta_{cycle} = \frac{c_P(T_{T3} - T_{T4}) - c_P\,(T_{T2} - T_{T1})}{c_P(T_{T3} - T_{T2})}$$

This may be rearranged to give:

$$\eta_{cycle} = \frac{c_P(T_{T3} - T_{T2}) - c_P\,(T_{T4} - T_{T1})}{c_P(T_{T3} - T_{T2})}$$

$$= 1 - \frac{(T_{T4} - T_{T1})}{(T_{T3} - T_{T2})}$$

Next we can separate T_{T1} from the top, and T_{T2} from the bottom to give:

$$\eta_{cycle} = 1 - \frac{T_{T1}\,(T_{T4}/T_{T1} - 1)}{T_{T2}\,(T_{T3}/T_{T2} - 1)}$$

For the isentropic processes over the pressure ratio r_P we have:

$$\frac{T_{T2}}{T_{T1}} = \frac{T_{T3}}{T_{T4}} = r_P^{(\gamma-1)/\gamma}$$

In Frame 69 of Programme 5 we had a similar relationship between temperature ratio and pressure ratio, involving static properties, i.e. properties that were not of total head value. The same relationship may be used for total head properties in frictionless isentropic flow in an ideal gas turbine engine. Then by cross-multiplying we get:

$$\frac{T_{T4}}{T_{T1}} = \frac{T_{T3}}{T_{T2}}$$

Finally in the equation for cycle efficiency, given again below, we can then use these two results above to give the final result:

$$\eta_{cycle} = 1 - \frac{T_{T1}\,(T_{T4}/T_{T1} - 1)}{T_{T2}\,(T_{T3}/T_{T2} - 1)}$$

$$\therefore \qquad \boxed{\eta_{\text{cycle}} = 1 - \frac{1}{r_P^{(\gamma-1)/\gamma}}}$$

Using this result, what is the cycle efficiency for a pressure ratio of 10, with air for which γ is 1.4, and for helium for which γ is 1.67 ?

79

$$\boxed{0.482,\ 0.602}$$

For air with $\gamma = 1.4$:

$$\eta_{\text{cycle}} = 1 - 1/10^{(0.4/1.4)} = 0.482$$

With helium for which $\gamma = 1.667$:

$$\eta_{\text{cycle}} = 1 - 1/10^{(0.67/1.67)} = 0.602$$

This shows how the closed cycle machine, in which helium might be used, has its attractions. But this is only the ideal cycle efficiency; we must see what sort of values we get for the work ratio and the ratio of real to ideal cycle efficiency.

To find the work ratio for the Brayton, or Joule, cycle we remember that:

$$r_w = \frac{\text{Net work}}{\text{Positive work}}$$

The net work was given in Frame 77 and the positive work in Frame 76, from which we get:

$$r_w = \frac{c_P(T_{T3} - T_{T4}) - c_P\,(T_{T2} - T_{T1})}{c_P(T_{T3} - T_{T4})}$$

$$= 1 - \frac{(T_{T2} - T_{T1})}{(T_{T3} - T_{T4})}$$

$$= 1 - \frac{T_{T1}\,(T_{T2}/T_{T1} - 1)}{T_{T3}\,(1 - T_{T4}/T_{T3})}$$

This analysis to find the work ratio is concluded in the next frame.

80

From the relationships in Frame 78 we can then see that:

$$r_w = 1 - \frac{T_{T1}}{T_{T3}} \frac{(r_P{}^{(\gamma-1)/\gamma} - 1)}{(1 - 1/r_P^{(\gamma-1)/\gamma})}$$

$$\therefore \quad \boxed{r_w = 1 - \frac{T_{T1}}{T_{T3}} r_P{}^{(\gamma-1)/\gamma}}$$

Thus we may notice a similarity with the Otto cycle in that the cycle efficiency does not depend on the temperature limits while the work ratio does. What is the work ratio with $r_P = 10$, $T_{T1} = 300$ K and $T_{T3} = 1100$ K, for both air and helium for which γ was 1.4 and 1.67 respectively?

81

$$\boxed{0.473,\ 0.315}$$

With air: $$r_w = 1 - \frac{300}{1100} \times 10^{(0.4/1.4)}$$
$$= 0.473$$

and with helium: $$r_w = 1 - \frac{300}{1100} \times 10^{(0.67/1.67)}$$
$$= 0.315$$

Thus we see that the work ratio of the Joule or Brayton cycle is not particularly good and is poor with helium. From the figure in Frame 7, unless the compressor and turbine efficiencies are high, a real closed cycle engine with helium is not going to work at all, and the open cycle efficiency with air is going to be very low. But we can see that the work ratio will improve if we increase the turbine inlet temperature.

Because of the low work ratio, the practical gas turbine engine was relatively slow in being realised, and the earliest examples were very inefficient. Engines could only be made to run at all when compression and expansion efficiencies became good enough, and when turbines could be made to withstand sufficiently high values of entry temperature.

In spite of these apparent deficiencies, very great efforts were made to develop these engines during the Second World War so that they could be used to replace piston engines in fighter aircraft. They had one overwhelming advantage. Can you think what it was?

82

High power/weight ratio

Relative to a petrol engine, the gas turbine had a much smaller weight for the power it developed, so that engines of much greater power could be accommodated in aircraft. In jet aircraft applications, the turbine had only to power the compressor after which the gases expanded in a convergent nozzle to atmospheric pressure and forward thrust was obtained from the high exhaust jet velocity. It is important to see how the isentropic efficiencies of the turbine and compressor will affect the overall performance. We defined isentropic efficiency in flow systems when dealing with expansions in steam turbines and the compression of fluids in refrigeration plant. Now applied to gas turbine expansion, η_T, and compression, η_C, processes, we have:

$$\eta_T = \frac{\text{Actual turbine work}}{\text{Isentropic turbine work}}$$

$$= \frac{c_P\,(T_{T3} - T_{T4})}{c_P\,(T_{T3} - T'_{T4})} = \frac{T_{T3} - T_{T4}}{T_{T3} - T'_{T4}}$$

$$\eta_C = \frac{\text{Isentropic work}}{\text{Actual work to achieve same pressure}}$$

$$= \frac{c_P\,(T'_{T2} - T_{T1})}{c_P\,(T_{T2} - T_{T1})} = \frac{T'_{T2} - T_{T1}}{T_{T2} - T_{T1}}$$

The figure in the next frame shows the non-isentropic processes in the compressor and turbine on the T–s plane, and the isentropic temperatures T'_{T2} and T'_{T4} may be seen.

83

The figure below shows the simple gas turbine cycle with non-isentropic expansion and compression on the *T–s* plane.

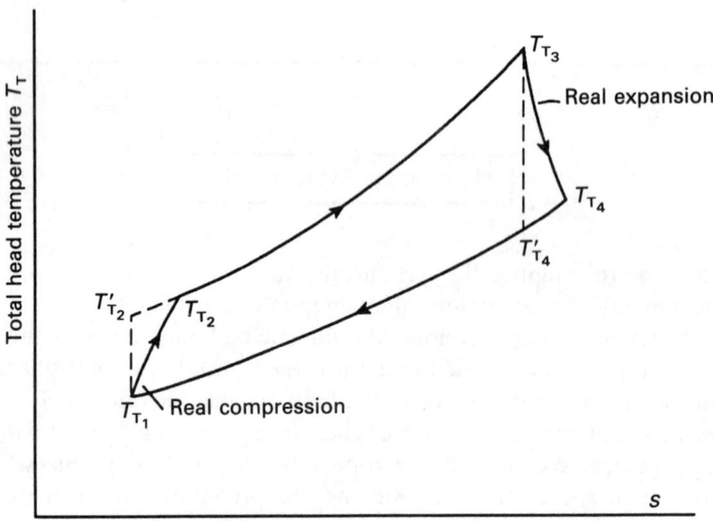

Dealing with isentropic values we have that:

$$\frac{T'_{T2}}{T_{T1}} = \frac{T_{T3}}{T'_{T4}} = r_P^{(\gamma-1)/\gamma}$$

so the actual temperature rise in the compressor is:

$$T_{T2} - T_{T1} = (T'_{T2} - T_{T1})/\eta_C$$

$$= T_{T1}(T'_{T2}/T_{T1} - 1)/\eta_C$$

$$\therefore \qquad T_{T2} - T_{T1} = \frac{T_{T1}}{\eta_C}\left(r_P^{(\gamma-1)/\eta} - 1\right)$$

In a similar way you can show that the temperature drop in the turbine is:

$$T_{T3} - T_{T4} = \eta_T T_{T3}\left(1 - 1/r_P^{(\gamma-1)/\gamma}\right)$$

These two results are important, and you should remember how to derive them. We can now use them to analyse a simple gas turbine cycle, and to find the specific work output and the cycle efficiency.

84

Typical values for the isentropic efficiencies are 0.88 for η_T and 0.85 for η_C. Let's now consider a gas turbine plant in which r_P is 12, T_{T1} is 300 K, T_{T3} is 1400 K, the power output is 10 MW, and the isentropic efficiencies are as given above. We have to calculate the specific work, the air mass flow rate and the thermal efficiency. We shall assume c_P for air and for air with combustion products is constant at 1.005 kJ/kg K. First, use the equations above to find the temperature rise in the compressor and the temperature drop in the turbine.

85

364.92 K, 626.26 K

Using:

$$T_{T2} - T_{T1} = \frac{T_{T1}}{\eta_C}\left(r_P^{(\gamma-1)/\gamma} - 1\right)$$

∴

$$T_{T2} - T_{T1} = \frac{300}{0.85} \times (12^{0.4/1.4} - 1) = 364.92 \text{ K}$$

and:

$$T_{T3} - T_{T4} = \eta_T \, T_{T3}\left(1 - 1/r_P^{(\gamma-1)/\gamma}\right)$$

$$= 0.88 \times 1400 \times (1 - 1/12^{0.4/1.4}) = 626.26 \text{ K}$$

The net specific work is then:

$$W_{sp} = c_P \times (626.26 - 364.92) = 1.005 \times 261.34$$

$$= 262.65 \text{ [kJ/kg]}$$

The air flow assumed in simple cycle analysis is the same through both compressor and turbine in a closed cycle, but remember that in the open cycle, although assumed the same, the mass flow of combustion products through the turbine is greater than the air flow through the compressor.

To give a power output of 10 MW, what is the required air mass flow rate?

86

38.07 kg/s

We have: $\dot{W} = 262.65 \times \dot{m} = 10\ 000$ kW

$\therefore$ $\dot{m} = 38.07$ [kg/s]

The temperature of the air leaving the compressor is:

$$T_{T1} + (T_{T2} - T_{T1}) = 300 + 364.92 = 664.92 \text{ K}$$

The temperature rise in the heat transfer or combustion process is then:

$$T_{T3} - T_{T2} = 1400 - 664.92 = 735.08 \text{ K}$$

From this, we can get the heat supplied per unit mass flow. The net specific work was 262.65 kJ/kg; what is then the thermal efficiency?

87

35.55%

The heat supplied per unit mass flow is:

$$Q_{\text{supplied}} = c_P \times (T_{T3} - T_{T2}) = 1.005 \times 735.08 = 738.76 \text{ [kJ/kg]}$$

The thermal efficiency is then simply:

$$\eta_{th} = (262.65/738.76) \times 100\%$$
$$= 35.55\%$$

With a pressure ratio of 12 to 1, the ideal cycle efficiency using:

$$\eta_{\text{cycle}} = 1 - \frac{1}{r_P^{(\gamma-1)/\gamma}}$$

gives a value of 50.83%. This is with isentropic processes in the compressor and turbine. Using the expression for work ratio from Frame 80:

$$r_w = 1 - \frac{T_{T1}}{T_{T3}} r_P^{(\gamma-1)/\gamma}$$

with $r_P = 12$, $T_{T1} = 300$ K and $T_{T3} = 1400$ K, what is the work ratio for the ideal cycle?

88

$$\boxed{r_w = 0.5642}$$

The work ratio is:

$$r_w = 1 - \frac{T_{T1}}{T_{T3}} r_P^{(\gamma-1)/\gamma}$$

$$= 1 - \left(\frac{300}{1400}\right) \times 12^{0.4/1.4}$$

$$= 0.5642$$

For this cycle we have obtained a real cycle efficiency of 35.55% and an ideal cycle efficiency of 50.83%. It would now be instructive to use the equation from Frame 6 to check this real cycle efficiency. Thus, using:

$$\eta_R = \eta_{Ideal} \times \left(\frac{\eta_{exp}}{r_w} \frac{1 - r_w}{\eta_{comp} r_w}\right)$$

with $\eta_{Ideal} = 50.83\%$, $r_w = 0.5642$, $\eta_{exp} = 0.88$ and $\eta_{comp} = 0.85$, what is η_R?

89

$$\boxed{\eta_R = 33.09\%}$$

The working follows in the next frame.

90

Using:

$$\eta_R = \eta_{Ideal} \times \left(\frac{\eta_{exp}}{r_w} - \frac{1 - r_w}{\eta_{comp} r_w} \right)$$

$\therefore \quad \eta_R = 0.5083 \times [(0.88/0.5642) - (1 - 0.5642)/(0.85 \times 0.5642)] \times 100\%$

$= 0.3309$

$= 33.09\%$

This compares with a figure of 35.55% obtained in Frame 87. Remember that the formula from Frame 6 given above was derived assuming the heat transfers to both ideal and real cycles were the same. In fact, as we have already seen, the heat transfer (or its equivalent) in the real cycle will be less to achieve the same turbine inlet temperature. This is because in the real cycle the compressor outlet temperature is higher than in the ideal cycle, so the real cycle efficiency predicted this way is less than it should be.

91

In the gas turbine engine example we have been considering, we found in Frame 85 that with the turbine inlet temperature of 1400 K the temperature drop in the turbine was 626.26 K, which means that the exhaust temperature will be 773.74 K, or 500.6°C. The energy this represents is much too valuable to go to waste, so an obvious improvement to the simple gas turbine cycle is to introduce a heat exchanger so that the hot exhaust may be used to heat the output from the compressor. Then less fuel will be needed to reach the top temperature of 1400 K, and the cycle efficiency will be improved.

The provision of such a heat exchanger is not possible with an aviation gas turbine, as the weight penalty would be too great. The diagram on the next page shows a modified plant layout and modified cycle on the *T–s* plane.

In the original cycle we had:

T_{T_1} = Compressor inlet temperature = 300 K

T_{T_2} = Compressor outlet temperature = 664.92 K

T_{T_3} = Turbine inlet temperature = 1400 K

T_{T_4} = Turbine outlet temperature = 773.74 K

As an ideal limit, the temperature of the output from the compressor could be raised to 773.74 K by the heat transfer from the exhaust.

The diagram of the modified plant layout and T–s diagram of the cycle are given below.

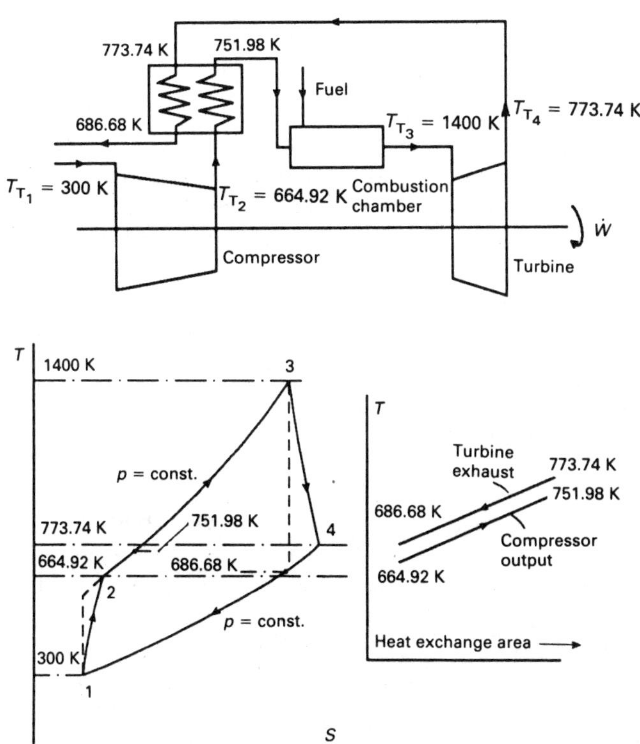

The maximum temperature difference across the heat exchanger, between the turbine exhaust and the compressor outlet, is 773.74 – 664.92 = 108.82 K. If we assume a heat exchanger efficiency of 80%, we are assuming that the compressor outlet is heated by 80% of this maximum difference, or by 87.06 K. Thus the compressor outlet temperature is 664.92 + 87.06 = 751.98 K, and the final exhaust temperature is 686.68 K. These temperatures are shown on the diagram above.

The heat supplied in the cycle is now to raise the temperature from 751.98 K to 1400 K, with c_p for air of 1.005 kJ/kg K. From Frame 85, the net specific work in the cycle was 262.65 kJ/kg. This will not change as a result of the modification. What is the new cycle efficiency?

93

$$\boxed{\eta = 40.33\%}$$

$$Q_{\text{supplied}} = 1.005 \times (1400 - 751.98) = 651.26 \text{ kJ/kg}$$

The cycle efficiency is then (262.65/651.26) × 100% = 40.33%.

The new exhaust temperature is still quite high at 686.68 K, or 413.53°C, so it has not been possible to make very good use of this source of energy.

Another possibility is to use the hot exhaust gases to generate steam for a steam power plant. When this is done it is called a **combined cycle plant**, since it is combining the Brayton or Joule cycle with the Rankine cycle.

Suppose the steam plant operates at 50 bar and 450°C with a condenser pressure of 0.1 bar. The figure below shows the open system of the Brayton or Joule cycle and the closed system of the Rankine cycle. The heat transfer between the two systems occurs across the common shared boundary as it passes through the heat exchanger/boiler, in which the enthalpy loss of the gas turbine exhaust is equal to the enthalpy gain of the steam.

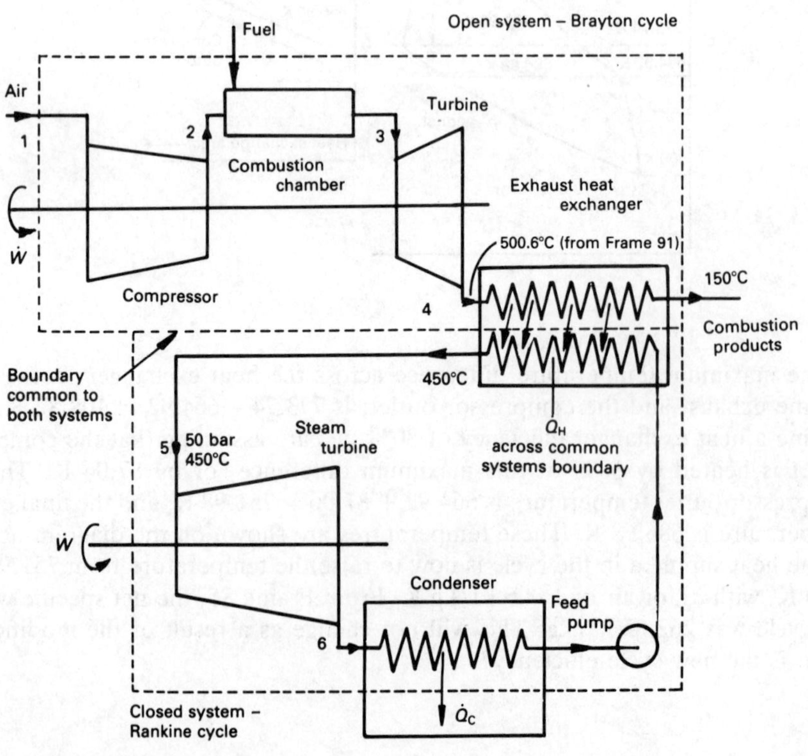

Steam is generated by counter flow heat transfer from the gas turbine exhaust, which we shall assume to be cooled to 150°C. Thus much better use can be made of this waste heat. The figure below shows the *T–s* diagrams for the gas and steam cycles, from which we can see the temperature profiles in the gas to steam heat exchanger. The heat transfer processes are fairly complex to calculate, so we are assuming this has been done and the figure of 150°C has been obtained for the final exhaust temperature.

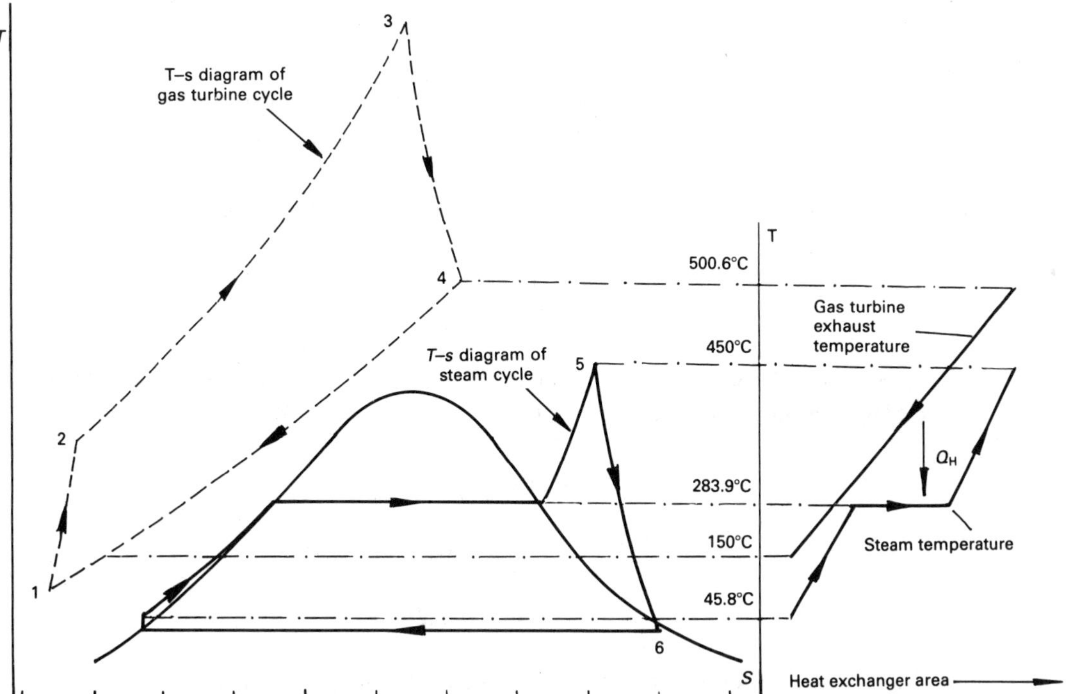

To simplify the calculations we shall neglect feed-pump work in the steam plant. The specific enthalpy of steam, h_5, at 50 bar and 450°C is 3316 kJ/kg (using the Rogers and Mayhew tables), and the enthalpy of water leaving the condenser at 0.1 bar is h_f at 0.1 bar, which is 192 kJ/kg. The enthalpy rise in the steam side of the heat exchanger is then:

$$3316 - 192 = 3124 \text{ [kJ/kg]}$$

The enthalpy drop in the gas turbine exhaust is:

$$c_P \times (500.6 - 150) = 1.005 \times 350.6 = 352.35 \text{ [kJ/kg]}$$

Now write down an equation in which the enthalpy changes in the two flows are numerically the same. The units of the equation are [kW].

95

$$\boxed{\dot{m}_{\text{steam}} \times 3124 = \dot{m}_{\text{gas}} \times 352.35 \text{ [kW]}}$$

The mass flow rate of steam × the specific enthalpy rise of the steam must equal the mass flow rate of exhaust × the specific enthalpy drop of the exhaust.

$$\therefore \quad \dot{m}_{\text{steam}}/\dot{m}_{\text{gas}} = 352.35/3124$$

$$= 0.1128 \text{ [kg steam/kg of gas]}$$

Now from your previous experience in Programme 7, calculate the enthalpy drop in the steam turbine, assuming an isentropic efficiency of 85%.

96

$$\boxed{983.02 \text{ kJ/kg}}$$

At 50 bar and 450°C, s = 6.818 kJ/kg K, from page 7 of the Rogers and Mayhew tables. For an isentropic expansion, the entropy at 0.1 bar is the same.

$$\therefore \quad 6.818 = s_f + x\, s_{fg}$$

$$= 0.649 + x\, 7.500$$

$$\therefore \quad x = 0.8225$$

The specific enthalpy after the isentropic expansion, h_6', is then:

$$h_6' = h_f + x\, h_{fg}$$

$$= 192 + 0.8225 \times 2392$$

$$= 2159.5 \text{ [kJ/kg]}$$

The isentropic enthalpy drop is then:

$$h_5 - h_6' = 3316 - 2159.5$$

$$= 1156.5 \text{ [kJ/kg]}$$

The actual enthalpy drop in the steam turbine is then:

$$1156.5 \times 0.85 = 983.02 \text{ [kJ/kg]}$$

Neglecting feed-pump work this is the net specific work for the steam turbine. We also have that the gas turbine specific work is 262.65 kJ/kg from Frame 85, the heat supplied is 738.76 kJ/kg of gas flow from Frame 87, and from the previous frame, in the combined cycle plant the ratio of steam to gas flow is 0.1128 kg/kg. So now calculate the total specific work per kg of gas flow, and hence find the thermal efficiency.

97

373.53 kJ/kg of gas flow, η = 50.56%

The total specific work is:

$$262.65 + 0.1128 \times 983.02 = 373.53 \text{ [kJ/kg of gas flow]}$$

The thermal efficiency is then:

$$373.53/738.76 \times 100\% = 50.56\%$$

The original requirement given in Frame 84 was for a 10 MW power plant, giving a gas flow of 38.07 kg/s in Frame 86. With a total specific work of 373.53 kJ/kg of gas flow for the combined cycle plant, the gas flow would now be:

$$\dot{m}_{gas} = 10\,000/373.53$$

$$= 26.77 \text{ [kg/s]}$$

The ratio of steam to gas flow was 0.1128 kg/kg, from Frame 95, so the steam flow is 3.02 kg/s. With the final gas turbine exhaust at 150°C there is still a valuable provision for process heat as well. With this included, we would have a **combined cycle (CHP) plant**.

Suppose the gas turbine exhaust is cooled to 80°C for process heat in another heat transfer; with an assumed c_p of 1.005 kJ/kg k, what is the amount of thermal power (defined in Frame 55, Programme 7), available for the gas flow of 26.77 kg/s?

98

1.883 MW

The thermal power is:

$$\begin{aligned}\dot{Q} &= \dot{m}\, c_P\,(150 - 80)\ [\text{kW}]\\ &= 26.77 \times 1.005 \times (150 - 80) \times 10^{-3}\ [\text{MW}]\\ &= 1.883\ [\text{MW}]\end{aligned}$$

The thermal efficiency of 35.55% for the gas turbine alone, obtained in Frame 87, and the figure of 50.56% for the combined cycle plant from Frame 97, are typical values for present practice. A turbine inlet temperature of 1400 K would mean blade cooling to about 1000 K would be necessary, using air bled off from the compressor. This entails a loss, but one which is offset by the benefits of the high turbine temperature.

Even though steel alloys melt at 1600 K and nickel alloys melt at 1700 K, it is confidently expected that developments in cooling will allow peak gas temperatures of 1800 K to be used. Cool air enters passages in each blade via the blade root and it escapes through small holes to flow over the blade surface, so that the blade is maintained at about 1000 K and is protected from the much hotter gas stream. Already demonstration units have run at 2000 K, but above 1800 K, nitrogen in the air reacts increasingly with oxygen to form oxides which are harmful to the environment, so 1800 K is at present regarded as a practical upper limit. With improvements in compressor and turbine blade design and the use of high pressure ratios with many stages of compression, it is predicted that combined cycle plant efficiencies will reach 60–62% within about 20 years. This is a very exciting challenge for today's student engineers.

We have considered only the barest outline of gas turbine theory, so that you can do some simple calculations. Examples are given in Frame 107. In practice, the steam cycle in combined cycle plant would involve high and low pressure turbines and reheat. Both topics were introduced in Programme 7. In the next few frames we shall conclude with a brief discussion of fuel cells.

99

Fuel cells were first mentioned in the previous programme and they are included here for the reasons given in Frame 73. They produce electrical work direct from the release of the chemical energy of fuel, and their efficiencies are not limited by the air cycle efficiencies of engines.

The fuel cell was invented by the British scientist, Sir William Grove, in 1839. It produces an electric current from a chemical reaction, usually between hydrogen and oxygen, and it demonstrates the reverse of the process of electrolysis in which an electric current produces hydrogen and oxygen from a weak solution of acid in water. Electrical work may also be produced from a chemical reaction inside an electric battery, but this is a **closed system** which needs a greater amount of electrical work to be done to 'recharge' it. The main use of the fuel cell to date has been in manned space flight as a source of both electric power and drinking water. But because of its special attractions of potentially high efficiency and clean exhaust, there is now considerable effort world wide to develop the fuel cell as a major source of power.

100

The essential features of a fuel cell are shown in the figure below.

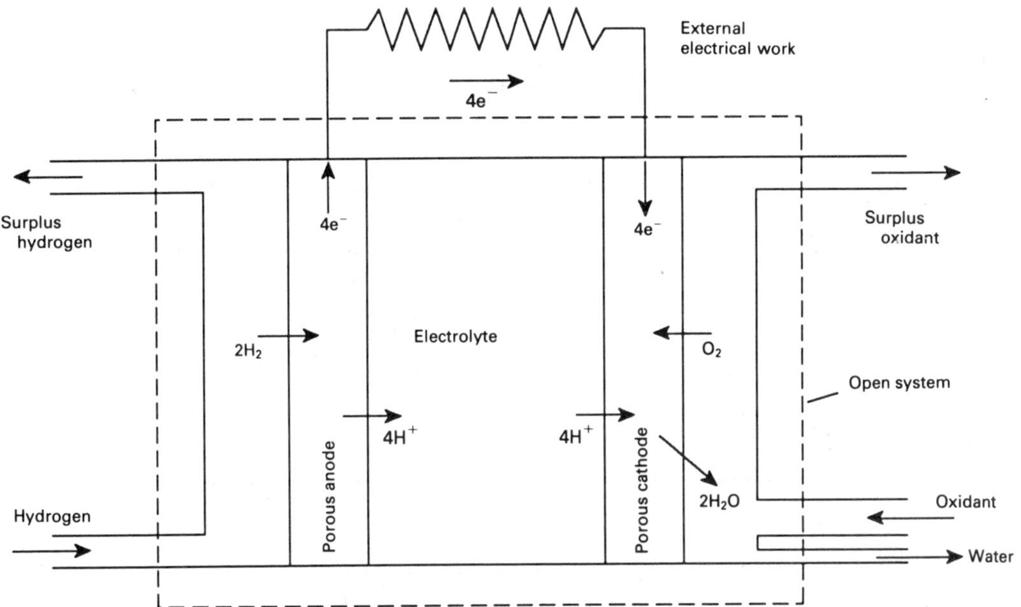

101

Hydrogen is introduced at the left to the surface of the porous anode, and oxygen (usually in air) is introduced at the right to the surface of the porous cathode. In between the anode and cathode is the electrolyte, which may be a solid. At the anode the hydrogen is ionised:

$$2\ H_2 = 4\ e^- + 4\ H^+$$

The term 4 e^- represents four electrons from the four hydrogen atoms which pass through the external electrical load to the cathode. The term 4 H^+ represents the four positive hydrogen ions (a hydrogen atom stripped of one electron), which pass through the electrolyte to the cathode.

At the cathode the hydrogen ions, having passed through the electrolyte, and the electrons having passed through the external load, all meet with the oxygen, and combine to form water. See if you can write the equation of this reaction.

102

$$\boxed{4\ e^- + 4\ H^+ + O_2 = 2\ H_2O}$$

The left-hand side is the right-hand side of the equation in the previous frame, plus the oxygen, which together give two molecules of water.

This and the equation from the previous frame may be added together:

$$2\ H_2 + O_2 = 2\ H_2O$$

The overall result is the chemical reaction between hydrogen and oxygen to form water which takes place at room temperature. In Frame 64 of Programme 9 we wrote an energy equation for this reaction at 25°C. Can you recall it?

103

$$\boxed{Q = H'_{P,0} - H_{R,0} = [\Delta H']_0}$$

This equation tells us that the change of enthalpy in the chemical reaction is $[\Delta H']_0$ which, in combustion with initial and end states at 25°C, was measured by the heat transfer Q. In the fuel cell, the reaction can take place at 25°C, and electrical work is produced. In Frame 83 of Programme 9 we saw that by the Second Law the maximum work obtainable in an isothermal chemical reaction at 25°C is:

$$W_{\text{max}} = [\Delta G']_0$$

where $[\Delta G]_0$ is the change in Gibbs function between reactants and products.
The efficiency of the work producing process is:

$$\eta_{\text{th,max}} = \frac{W_{\text{max}}}{[\Delta H']_0} = \frac{[\Delta G']_0}{[\Delta H']_0}$$

For the hydrogen–oxygen reaction we found that this was 94.11%, a figure that is the theoretical limit with hydrogen for any work-producing process in an open system. In the hydrogen fuel cell operating at room temperature, some 25% of the quantity $[\Delta H']_0$ is lost so that the efficiency is around 70%. Losses arise mainly owing to the electrical resistance of the cell itself, so a heating effect is also present.

104

The theoretical voltage for the hydrogen fuel cell without an external load is 1.184 volts. In working conditions the voltage is less, and the output of an individual cell is between 10 and 100W. Cells are stacked together in series and parallel to give higher voltages and powers. Thus theoretically it is possible for fuel cells using hydrogen to power an electric road vehicle.

105

The challenge is to find the right niche for the fuel cell in the whole field of power needs. The main fuel is now regarded as methane, CH_4, with associated reforming plant to produce hydrogen. As a fuel cell has its own exhaust streams of unused fuel and oxidant, ideas are being developed for composite plant in which the fuel cell is part of a scheme involving both gas turbine and steam power plant. It is believed that such plant could be giving large scale generation of electric power at efficiencies of 56–65% by 2010. Already demonstration plants have been built in the USA and Japan giving 11 MW, but at lower efficiencies. The prospect for fuel cells in this type of composite plant is again a very exciting challenge, and it would be very interesting and rewarding to graduate into this field of advanced thermodynamic systems.

106

The equations for Cycle efficiency and work ratio of the various cycles considered in this programme are listed below for quick reference:

Carnot cycle
Frame 11

$$\eta_{\text{Carnot}} = \frac{T_H - T_C}{T_H}$$

$$r_w = \frac{(\gamma - 1)\ln V_1/V_2 \times \eta_{\text{Carnot}}}{(\gamma - 1)\ln V_1/V_2 + \eta_{\text{Carnot}}}$$

Stirling cycle
Frames 17 and 19

$$\eta_{\text{Stirling}} = r_w = \eta_{\text{Carnot}} = \frac{T_H - T_C}{T_H}$$

Otto cycle
Frames 33 and 36

$$\eta_{\text{Otto}} = 1 - \frac{1}{r_C^{\gamma-1}}$$

$$r_w = 1 - \frac{T_1}{T_3}\, r_C^{\gamma-1}$$

Diesel cycle
Frames 53 and 56

$$\eta_{\text{Diesel}} = 1 - \frac{1}{r_C^{\gamma-1}}\,\frac{(r_P^{\gamma} - 1)}{\gamma\,(r_P - 1)}$$

$$r_w = 1 - \frac{(1 - 1/r_C^{\gamma-1})}{\gamma\,(r_P - 1) + (1 - r_P^{\gamma}/r_C^{\gamma-1})}$$

Dual cycle
Frames 61 and 67

$$\eta_{\text{dual}} = 1 - \frac{1}{r_C^{\gamma-1}}\,\frac{(r_P^{\gamma}\, r_H - 1)}{(r_H - 1) + \gamma\, r_H\,(r_P - 1)}$$

$$r_w = 1 - \frac{1}{r_H}\,\frac{(1 - 1/r_C^{\gamma-1})}{\gamma(r_P - 1) + (1 - r_P^{\gamma}/r_C^{\gamma-1})}$$

Brayton, or Joule cycle
Frames 78 and 80

$$\eta_{\text{cycle}} = 1 - \frac{1}{r_P^{(\gamma-1)/\gamma}}$$

$$r_w = 1 - \frac{T_{T_1}}{T_{T_3}}\, r_P^{(\gamma-1)/\gamma}$$

107

To conclude this last programme, here are some problems in gas turbine and combined cycle plant.

1. In an ideal gas turbine cycle with air as the working fluid, the pressure ratio is 8 to 1, compressor inlet temperature is 300 K, turbine inlet temperature is 1100 K, and the power output is 0.8 MW. Find the compressor outlet temperature, turbine outlet temperature, cycle efficiency, work ratio and mass flow of air, assuming for air $c_P = 1.005$ kJ/kg K, $\gamma = 1.4$.
[543.42 K, 607.27 K, 44.79%, 0.506, 3.19 kg/s]

2. A heat exchanger is introduced into the plant of problem **1.** The turbine exhaust is used to heat the compressor output, with a heat exchange efficiency of 80%. Recalculate the cycle efficiency and the final turbine exhaust temperature.
[49.32%, 556.2 K]

3. Repeat problem **1** with compressor and turbine isentropic efficiencies of 0.78 and 0.86 respectively. All other details remain the same.
[612.07 K, 676.25 K, 22.89%, 7.13 kg/s]

4. Reconsider problem **1**, with compressor and turbine isentropic efficiencies of 0.78 and 0.86 respectively, and with the introduction of a heat exchanger of 80% efficiency as in problem **2**. Recalculate the cycle efficiency and the final turbine exhaust temperature. [25.58%, 624.9 K]

5. A closed cycle gas turbine uses argon as the working fluid, and it operates at the same conditions as in the previous problems: pressure ratio is 8 to 1, compressor inlet temperature is 300 K, turbine inlet temperature is 1100 K. (a) Find the ideal cycle results as previously, and (b) find the corresponding results with the isentropic efficiencies of 0.78 and 0.86 for compressor and turbine. In light of the results for the practical cycle, would you recommend using a closed cycle engine with argon or an open cycle engine with air? For argon, $c_P = 0.5203$ kJ/kg K, $\gamma = 1.667$.
[(a) 689.22 K, 478.8 K, 56.47%, 6.628 kg/s, (b) 799.0 K, 565.77 K, 22.50%, 22.71 kg/s, use an open engine cycle with air]

6. (a) A state-of-the-art gas turbine has a pressure ratio of 20 to 1, turbine inlet temperature of 1500 K and compressor inlet temperature of 300 K. The compressor and turbine isentropic efficiencies are 0.812 and 0.893 respectively. Taking air as the working fluid, with $c_P = 1.005$ and $\gamma = 1.4$, calculate the compressor and turbine outlet temperatures, the cycle efficiency, the fuel and air flow rates and air/fuel ratio for a power output of 22 MW. The calorific value of the fuel is 42 MJ/kg. Assume combustion products passing through the turbine have the same properties as air.

(b) Repeat for a future machine, with a pressure ratio of 30 to 1, turbine inlet temperature of 1800 K, compressor inlet temperature of 300°C, and compressor and turbine isentropic efficiencies of 0.825 and 0.923 respectively.

(c) Can heat exchangers be introduced into these cycles?
[(a) 800.04 K, 729.67 K, 38.61%, 1.357 kg/s, 80.99 kg/s, 59.71 to 1. (b) 897.27 K, 767.32 K, 48.23%, 1.086 kg/s, 50.28 kg/s, 46.3 to 1. (c) No,

the compressor outlet temperature is higher than the turbine exhaust temperature. This problem could be overcome in a more advanced cycle by compressing in two or more stages with cooling between the stages]

7. The same two turbines in the previous problem are used in combined cycle plant with steam. In (a), the steam turbine entry condition is 60 bar, 400°C, the condenser pressure is 0.1 bar, and the turbine isentropic efficiency is 0.88. In (b), the turbine entry condition is 80 bar, 425°C, the isentropic efficiency is 0.90, and the condenser pressure is 0.1 bar. In both cases take into account feed-pump work. In each case the gas turbine exhaust is cooled to 110°C in raising the steam. Calculate in each case the combined cycle efficiency, and the steam, air and fuel flows and air/fuel ratio for the same power output of 22 MW. In practice, the steam cycle would be more complicated than the superheat cycle with a single turbine used in these examples; you will find the final dryness of the steam is too low. What sort of cycle would you recommend?
[(a) 54.69%, 6.686 kg/s, 57.19 kg/s, 0.958 kg/s, 59.71 to 1. (b) 62.9%, 4.95 kg/s, 38.55 kg/s, 0.833 kg/s, 46.36 to 1, reheat cycle]

INDEX

Adiabatic
  combustion temperature 493
  flow 151–4, 407–28
  process *see under* Process
  surface 300
  throttling 153
Air
  composition of 160, 267
  conditioning plant 281–5
  standard cycles 534, 538–62, 571–80
Air/fuel ratio
  by mass 473
  by volume 470
  rich 471
  stoichiometric 470
  weak 471
Avogadro's Hypothesis 200

Barometer 25
Beam engine 362–3
Bernoulli equation 349
Boiling point 29, 165
Boundary, of a system 10
Bourdon tube pressure gauge 21
Boyle's Law 81, 195
Brake power *see under* Power
Brayton cycle 570–80
British Thermal Unit 96

Calorific value 501, 512
Carnot 311
Carnot cycle
  with gases 304–6, 525–8
  with steam 310–14, 339, 354–62
Centigrade heat unit 96
Chemically reacting system 484
CHP cycle 385–6
Clausius 28, 298
  inequality 327, 328–32
Clausius–Clapeyron equation 312–13
Coefficient of performance 299, 391, 395
Combined cycle 387, 582–6
Combustion 462–516
  chemistry 464–83
  First Law, applied to 484–508
  Second Law, applied to 509–15
Compressible flow 405–34
Compression
  fully resisted 55
  irreversible 56–7
  ratio, in engines 528, 537, 542, 549
  reversible 55
Control
  surface and volume 13
Critical pressure ratio 416
Critical state 166
  of steam 168, 174, 217, 338
Cycle, of processes, definition 114

Dalton's Law of Partial Pressure 262
Density, of a system 14
Diesel cycle 549–55
Diesel engine 520
Differential
  exact 209–10
  partial 208
  total 208
Diffuser flow *see under* Flow
Displacement work 52–83
  definition of 52
  fully resisted 55–7
Dissociation 510
Dryness fraction, definition 170
Dual cycle 556–62

Efficiency
  air standard cycles 540, 553, 558, 573
  blade 442
  combustion 378
  compression 522
  expansion 521
  generator 377
  heat transfer 108
  isentropic 374, 400, 425, 575
  maximum 565–6
  mechanical 64, 543
  stage 443, 457
  steam plant 378–80
  thermal 9, 513–14, 585–6
    brake and indicated 542–3

Electrolysis 510
Emissivity 105
Energy 120–7
  equation
    closed system 120
    open system 138
    steady flow 139, 145, 154, 274–85, 357, 394, 407, 570
    steady flow, chemically reacting system 499–500
  internal
    definition 125
    of reaction 485
    of steam 169
  kinetic 125
  potential 125
Engine cycle, ideal 521–62
Engine indicator 61–5, 363–4
Enthalpy
  chemically reacting system 500
  definition 131
  stagnation 147, 152, 418
  of steam 168
Entropy 321–48
  property relationships 343–8
Euler equation 348
Expansion
  fully resisted 55
  irreversible 55–6
  reversible 54–5
  sudden 57
  unresisted 58–60

Feed-pump work 358–61
First Law of Thermodynamics 113–19
Flow
  in diffusers 151–2, 410
  in nozzles 151–2, 410–34
  in turbine wheel
    axial 435–58
    radial 435
Force, definition of 17
Four stroke cycle 38, 520, 534–6
Fourier 99
Fuel cells 586–89
Fuel/air equivalence ratio 476

Gas
  constant
    definition 195–6
    gas mixture 266
    universal 201
  ideal 193
  ideal rule 196, 208
  mixtures 159, 162, 261–85
    in steady flow 274–85
  perfect, definition 208
  'permanent' 194, 204
  real 193
  semi-perfect, definition 208
  thermometer, constant volume 203–5
  turbine engine 463, 506, 569–86
Gibbs function 512, 589
Gibbs–Dalton Law 264
Gravimetric analysis 477

Heat
  as an interaction 11, 13, 40, 94–110
  conduction 99–101
  convection 102–3
  definition 94
  engine
    definition 293
    real, efficiency 316–19
    reversed 296
    reversible 298
    reversible, efficiency 309, 314–16, 340
  exchangers 148–51
    in gas turbines 580–6
  flux 100
  pump 297, 317–19, 401–3
  radiation 104–7
  reservoir 98, 295
  sign convention 95
  sink 98, 118, 295
  source 98, 118, 295
  specific 129–34
    constant pressure, definition 132, 207
    constant volume, definition 130, 207
    gas mixture 265
Humidity 194
  relative 280
  specific 279

Impulse blading 438
Impulse turbine 438–51
Indicated mean effective pressure 62–3, 536, 563–4
Indicated power *see under* Power
Indicator diagram 61–3, 563–4
Insulation
  in a heat interaction 42

Interpolation procedure 182–5
Irreversible process 44
Isentropic *see under* Efficiency *and* Process
Isotherm 175, 259

Joule 28
  cycle 570–80
  experiments 97, 115, 205–7
Joule, definition of 46

Kelvin 28, *also see under* Temperature *and* Thermodynamics
Knock, in petrol engines 549

Lenoir engine 533–4
Liquid
  saturated 164, 166, 173
  subcooled 164, 187–8

Mach number 424
Manometer 22–4
Mass, of a system 14
Mechanical efficiency *see under* Efficiency
Mechanical equivalent of heat 97–8, 115
Mechanical work *see under* Work
Mixture strength 475
Mole, definition 199
Molecular mass 198–200
  of a gas mixture 268
Mollier diagram 340–1

Newcomen engine 363
Newton, definition 17
Newton's law of convection 102
Newton's Second Law of Motion 17, 437
Nozzle flow *see under* Flow

Otto cycle 534, 537–49

Paraffins 466
Pascal, definition 19
Path function 53, 109
Petrol engine 38, 520–1, 535–7
Phase diagram 176, 190–2
Photosynthesis 4
Polytropic *see under* Process
Power
  brake 64, 563–4
  definition of 49
  indicated 62, 64, 563–4
  shaft 86
  unit of 49
Power/weight ratio 575
Pressure
  absolute 20
  atmospheric 20
  critical 166
  definition 19
  difference, in work 41
  gauge 20–2
  partial 262
  of a system 14
  total 418–19
Probability 325–6
Problems
  combustion equations 483
  displacement work 83–4
  electrical work 94
  engine cycles 566–8
  entropy 349
  First Law 123–4, 134–5, 154–5
    applied to combustion 517
  frictional work 94
  gas turbines, combined cycles 591–2
  heat 111
  heat engines 320
  ideal gases 212, 285–7
  introductory 32–3
  nozzle flow, gases 428–9
  nozzle flow, steam 434–5
  pure substances 163
  Rankine cycles 390
  refrigerator and heat pump 404
  shear work 94
  Steam Tables, use of 188–9
  steam turbine stages 459–60
  vapours 243–4
Process 43
  adiabatic 117, 119, 337, 344
    with gases (isentropic) 251–5
  constant pressure 67–71
    with gases 248–51
    with steam 224–6
  constant volume 66, 117
    with gases 245–7, 345–7
    with steam 216–24
  flow
    with gases 274–85, 407–28
    with refrigerants 239–41
    with steam 233–8, 241–2, 429–34
  isentropic, definition 325, 337

isothermal
with gases 259–60
with steam 232–3, 324
polytropic 73–9
with gases 256–9
with steam 227–32
Properties, of a system 13–15
extensive 15, 126, 131
ice/water/steam 163–92
intensive 15, 38, 126
properties of 121
Property diagram, definition 43
Pure substance, definition 158

Rankine cycle 366–89
with ammonia 387–9
with reheat 371–3, 375–6
with superheat 369–73, 375–6, *also see under* Regenerative cycle
Reaction
blading 438, 451–2
degree of 452
turbine 451–7
condition line 457
reheat factor 457
Real process 44
Refrigerating effect 394–5
Refrigeration load 402
Refrigerator 297, 317–19
and heat pump cycles 391–403
Regenerative cycle 382–5
Regenerative feed heating 381–3
Reheat cycle *see under* Rankine cycle
Reheat factor *see under* Reaction turbine
Reversible
compression 55
expansion 54–5
heat transfer 107
process 44
Rotary machinery 145–8, *also see under* Gas turbine engine *and* Steam power cycles

Saturated
liquid 164
vapour 164
Second Law of Thermodynamics 9, 289–306
Solar energy 4
Sonic velocity 412

Specific
energy 126
enthalpy 131
heat, constant pressure 132
heat, constant volume 130
entropy 336
steam consumption 364
volume 14
work 69
State
disordered 291, 325
equilibrium 40, 43, 322
non-equilibrium 39, 43, 127, 322
ordered 291, 325
of a system 13
State path 43
Steam
critical state 168, 174, 217, 338
dryness fraction 170–2
engine 6, 76–80, 362–4
enthalpy of 168
internal energy of 169
power cycles 351–87, 582–5
saturated 166, 173, 180–3
supercritical 174
superheated 76, 164, 184–7
Tables 180–8, 341–2
wet 164, 170–2
Stefan–Boltzmann law of radiation 105
Stirling cycle 528–33
Subsonic flow 410, 414
Summaries
air standard cycles 590
combustion 515–16
First Law and energy 156
gas mixture formulae 273
gases 288
heat and work 112
introductory 34
pure substances 212–13
Second Law and entropy 350
Supersonic flow 410, 424
Surroundings, of a system 10
System
chemically reacting 484
closed, definition 10
open, definition 13

Temperature
absolute 31, 32, 307
absolute zero 205, 313
adiabatic combustion 493

Celsius, scale of 29
Centigrade, scale of 29
critical 166
dew point 280
difference, in heat flow 41
dynamic 147
–entropy diagram for gases 345–8
for steam 338
Fahrenheit, scale of 29, 35
fixed points 29, 30
ideal gas, scale of 29, 202–5
International, scale of 30, 314
Kelvin, thermodynamic, scale of 31, 307–9
stagnation 147, 152, 570
static 147
of a system 14, 26–8
Thermal conductivity, definition 99
Thermal power 385, 585–6
Thermodynamic relations 208–11
Thermodynamics
definition of 5
First Law 116, 119, 124, 215–88
closed system 116–23, 215–33, 244–60
open system 136–54, 233–42
reacting system 484–508
Second Law 289–306
Clausius statement 298
Kelvin–Planck statement 295
Zeroth Law 28
Thermometer
electrical resistance 26
equality in 27
ideal gas 32, 203–5, 306
mercury in glass 26
Throat 410
Throttling calorimeter 241–2
Torque 85
Triple point 159, 176
Turbine stage 438, 448, 452
Turbine work *see under* Work
Two property rule 169
Two stroke cycle 39, 536

Units, S.I. 16
Universal gas constant 201

Vacuum
definition 22
expansion into 58–60
Volume, of a system 14

Water
saturated 164, 166, 173
subcooled 164, 187–8
vapour 194
Watt, definition 49
Watt hour, definition 50
Weight, definition 18
Whittle, Sir Frank 568
Work
as an interaction 11, 13, 40
turbine wheel 357–60, 367, 370–6, 441–57, 570, 576
displacement 51, 52–82
electrical 51, 90–3
fluid shear 51, 88–9
maximum, in isothermal reaction 512–13, 589
ratio 361, 521–4, 527, 531, 541, 555, 562, 574
shaft 51, 84–8
sign convention 47, 95
solid friction 51, 90
specific, definition 69
stirring 87–8, 115
thermodynamic, definition 51
unit of 46

Zeroth Law of Thermodynamics 28